P9-CPY-029

WITHDRAWN

Clean Coal Engineering Technology

Bruce G. Miller

AMSTERDAM • BOSTON • HEIDELBERG • LONDON
NEW YORK • OXFORD • PARIS • SAN DIEGO
SAN FRANCISCO • SINGAPORE • SYDNEY • TOKYO

Butterworth-Heinemann is an imprint of Elsevier

CUYAHOGA COMMUNITY COLLEGE
EASTERN CAMPUS LIBRARY

Butterworth-Heinemann is an imprint of Elsevier
30 Corporate Drive, Suite 400
Burlington, MA 01803, USA

The Boulevard, Langford Lane
Kidlington, Oxford, OX5 1GB, UK

⃝ 2011 Elsevier Inc. All rights reserved.

No part of this publication may be reproduced or transmitted in any form or by any means, electronic
or mechanical, including photocopying, recording, or any information storage and retrieval system,
without permission in writing from the publisher. Details on how to seek permission, further
information about the Publisher's permissions policies and our arrangements with organizations such
as the Copyright Clearance Center and the Copyright Licensing Agency, can be found at our website:
www.elsevier.com/permissions.

This book and the individual contributions contained in it are protected under copyright by the
Publisher (other than as may be noted herein).

Notices
Knowledge and best practice in this field are constantly changing. As new research and experience
broaden our understanding, changes in research methods, professional practices, or medical treatment
may become necessary.

Practitioners and researchers must always rely on their own experience and knowledge in evaluating
and using any information, methods, compounds, or experiments described herein. In using such
information or methods they should be mindful of their own safety and the safety of others, including
parties for whom they have a professional responsibility.

To the fullest extent of the law, neither the Publisher nor the authors, contributors, or editors, assume
any liability for any injury and/or damage to persons or property as a matter of products liability,
negligence or otherwise, or from any use or operation of any methods, products, instructions, or ideas
contained in the material herein.

Library of Congress Cataloging-in-Publication Data
Miller, Bruce G., M.S.
    Clean coal engineering technology / Bruce G. Miller.
        p. cm.
Includes bibliographical references.
ISBN 978-1-85617-710-8
1. Clean coal technologies. 2. Coal—Environmental aspects. 3. Coal-fired
    power plants. 4. Coal-fired furnaces. I. Title.
TP325.M548 2011
662.6'20286—dc22   2010020592

British Library Cataloguing-in-Publication Data
A catalogue record for this book is available from the British Library.

For information on all Butterworth–Heinemann publications
visit our Web site at *www.elsevierdirect.com*

Printed in the United States
10   11   12   13   14   10 9 8 7 6 5 4 3 2 1

Working together to grow
libraries in developing countries

www.elsevier.com | www.bookaid.org | www.sabre.org

ELSEVIER   BOOK AID
             International   Sabre Foundat

Dedication

In memory of my father, Fred Miller, who worked for the Knife River Coal Company near Beulah, North Dakota, for 20 years until his disability retirement because of an on-the-job injury. In the photograph, he is standing beside the loading shovel he operated at the mine, one that he used to load coal-hauling trucks like the one in Figure 6.2. During summer vacation while I was attending college, I drove the same kind of truck when working with my father at the mine. Both my father's career in "coal" and the importance of coal to the community that I grew up in sparked my lifelong interest in coal. In addition, his role as husband and father and his strong work ethic are attributes that I continually strive to match.

Contents

Part IV – Clean Coal Technology Programs and Energy Security

11 U.S. and International Activities for Near-Zero Emissions during Electricity Generation

Preface

This book is the second edition of *Coal Energy Systems,* which was published in 2005 and is the culmination of a 15-month effort to update the first edition. When first approached by my editor, Kenneth McCombs, I was not sure if there would be enough new information to warrant undertaking this exercise. However, after carefully reviewing what had transpired since I wrote the first edition, especially with respect to environmental regulations (specifically mercury emissions and carbon management, technology advances in carbon dioxide capture and storage, and advanced coal utilization technologies), along with the growing demand for, and utilization of, coal in the developing world, it became apparent that a second edition was necessary.

As a result of the major worldwide activities in developing technologies for utilizing coal more efficiently and with less of an environmental impact, it was decided to modify the title of the book to *Clean Coal Engineering Technology.* Although this title may draw some skepticism, it is an accepted fact that coal is a major energy source today and will continue to be so for the next several generations, especially with the developing nations striving to improve their standard of living. Therefore, it is imperative that we continue to advance clean coal initiatives. Many governments around the world recognize that coal "is here to stay" and are working diligently with industry and academia to develop clean coal technologies.

In this preface, I want to repeat the first two paragraphs of the first edition preface, which are just as important now as they were when the first edition was published (with maybe the exception of the reduced emphasis on the hydrogen economy):

> *Coal is currently a major energy source in the United States as well as throughout the world, especially among many developing countries, and will continue to be so for many years. Fossil fuels will continue to be the dominant energy source for fueling the U.S. economy, with coal playing a major role for decades. Coal provides stability in price and availability, will continue to be a major source of electricity generation, will be the major source of hydrogen for the coming hydrogen economy, and has the potential to become an important source of liquid fuels. Conservation and renewable/sustainable energy are important in the overall energy picture, but they will play a lesser role in helping us satisfy our energy demands.*
>
> *It is recognized in the energy industry that the manner in which coal is used must, and will, change. Concerns over the environmental effects of coal utilization are resulting in better methods for controlling emissions during combustion, and more research and development into technologies to utilize coal more efficiently especially in nontraditional methods. While major advances have been made in reducing the environmental impact when using coal, we have other technologies*

in hand, either near commercialization or under development, that will allow coal to be used in an even more environmentally friendly manner. The roadblocks to implementing these technologies are the financial risks associated with new technologies and the resulting higher costs of energy to the consumers. Consumers in the United States, for example, have become accustomed to low energy prices and are reluctant to pay more for their energy, whether it is transportation fuels for their vehicles, natural gas or propane for domestic heating, or electricity for their homes. The implementation of these technologies that increase energy efficiency or reduce pollution will be driven by legislative mandate and, to a lesser extent, the willingness of the consumer to pay more for energy.'

These words are as true today as they were in 2004.

Consequently, the first edition has undergone some major rework, mainly with the addition of updated information and the inclusion of the many activities underway in the clean use of coal. The book has been expanded from 8 to 12 chapters and from 4 to 5 appendices. It has been "informally" divided into five sections. The first part, which consists of Chapters 1 through 4, provides detailed background information on coal. It discusses the historical use of coal, how it is currently being used, future trends of coal usage, and a comparison of coal's contribution and role in the overall energy picture. Coal distribution throughout the world, the chemical and physical characteristics of coal, and the effect of coal usage on human health and the environment are presented.

The second part focuses on coal utilization technologies. Chapter 5 introduces the four major technologies for utilizing coal: carbonization/pyrolysis, combustion, gasification, and liquefaction. Chapter 6 looks at the anatomy of a coal-fired power plant because this is the technology that developing countries are currently installing at a rapid pace. A discussion of clean coal technologies, such as fluidized-bed combustion, advanced pulverized coal power plants, and integrated gasification combined cycle systems, is provided in Chapter 7.

The third part of the book focuses on emissions and carbon management. Coal-fired emissions regulations are presented in Chapter 8. Emissions control strategies are presented in Chapter 9 with carbon capture, and storage is discussed in Chapter 10.

In the book's fourth part, a discussion of clean coal technology programs for near-zero emissions when generating electricity is provided in Chapter 11. This includes international activities as well as those in the United States. Energy security and sustainable development are also discussed in Part IV. Coal's role in providing energy security for the world, with an emphasis on the United States, is presented in Chapter 12. The book concludes with the five appendices in Part V that contain supporting information for the various chapters.

I will conclude by first stating that all errors or omissions are entirely my own. I want to thank my wife, Sharon, and children, Konrad and Anna, for supporting me these last 15 months. Thanks also go to Konrad for helping me with my literature searches, and Anna for providing some of her "powerplant" photographs. A very special thank you goes to Ruth Krebs and Elizabeth Wood for their work on the figures.

1 Coal as Fuel

Past, Present, and Future

1.1 Organization of this Book

Coal is the largest source of solid fuel in the world. It exists in almost every country, and approximately 70 countries mine it. Coal is the primary fuel source for power generation, and as developing countries become more industrialized, more coal is being used, and its use is projected to increase over at least the next couple of decades. Coal is also being considered as a way to produce liquid fuels. More countries are considering installing facilities to produce liquid fuels via indirect liquefaction (i.e., gasifying coal and producing liquid fuels from the gases), which South Africa has done since the 1950s. Countries are also seriously exploring direct coal liquefaction technologies for fuel production. China has taken the lead in constructing direct coal-to-liquid plants, and it is likely that more countries will do the same.

Along with increased coal usage, however, come concerns about the environment. System efficiencies must be improved; coal cannot be used as a "dirty" fuel. Emissions, including carbon dioxide, must be controlled. *Clean coal technology* is not an oxymoron, and using these technologies for efficient coal consumption in an environmentally acceptable manner must be ramped up immediately. This book provides significant information on all of the issues associated with clean coal technologies.

Chapters 1 through 3 discuss coal use in the past and present and the potential future trends of coal usage, coal distribution throughout the world, and coal chemistry. The effect of coal as a fuel on human health and the environment is presented in Chapter 4. Chapter 5 introduces the four major technologies for utilizing coal: carbonization/pyrolysis, combustion, gasification, and liquefaction. Chapter 6 looks at the anatomy of a coal-fired power plant, which developing countries are installing at a rapid pace.

A discussion of clean coal technologies, such as fluidized-bed combustion, advanced pulverized coal power plants, and integrated gasification combined cycle systems, is provided in Chapter 7. Coal-fired emission regulations are presented in Chapter 8. Emissions-control strategies are presented in Chapter 9, and the process of carbon capture and storage is examined in Chapter 10. A discussion of clean coal technology programs for near-zero emissions when generating electricity is provided in Chapter 11. Finally, Chapter 12 discusses coal's role in providing energy security for the world, especially in the United States.

Clean Coal Engineering Technology. DOI: 10.1016/B978-1-85617-710-8.00001-7
Copyright © 2011 by Elsevier Inc. All rights of reproduction in any form reserved.

1.2   The History of Coal Use

The next sections present a brief history of the use of coal, including a comparison to other energy sources. Although the concentration here is on coal use in the United States, a global perspective is also presented, especially with regard to comparing overall energy consumption. In addition, the sections look at the different types of technologies that were used in the past and were developed as part of the Industrial Revolution. A more in-depth discussion of major coal technologies, however, is provided in subsequent chapters.

1.3   Coal Use before the Industrial Revolution

Coal has been used as an energy source since ancient times, but it was considered a minor resource until the Industrial Revolution. The first mention of coal in European literature dates from the fourth century B.C. [1], but scholars are certain that coal was first used in China as early as 1000 B.C. [2]. By A.D. 1000, coal was the primary fuel source in China, and its use was reported by the Venetian traveler Marco Polo in the thirteenth century [3, 4].

The first documented use of coal in Western civilization was by the Greek philosophers Pliny, Aristotle, and Theophrastus—Aristotle's pupil [1]. The first definitive record of coal being used for fuel is found in Aristotle's *Meteorology,* where he writes of combustible bodies [1]. Theophrastus, in his fourth-century *Treatise on Stones,* describes a fossil substance used as a fuel [5]. Theophrastus and Pliney both mention blacksmiths using coal [1, 6]. This was most likely brown coal from Thrace in northern Greece and from Ligurai in northwestern Italy. Because it was high in impurities and did not create a very hot fire, it was not commonly used in iron smelting furnaces. Pliny, however, refers to its use in copper casting, which can be done at considerably lower temperatures [6].

Although the Greeks and Romans knew about coal by 400 B.C., they mostly used wood for fuel, since it was so plentiful. Coal was used as a domestic heating fuel in some parts of the Roman Empire, particularly in Britain, but it never made more than a marginal contribution as a fuel resource [6]. As the Romans invaded northward, they saw coal being mined and used in the vicinity of St. Etienne in Gaul (France) and in Britain, where coal cinders that have been discovered in Roman ruins show that coal was used during the Roman occupation from approximately A.D. 50 to 450 [5].

During the Middle Ages, coal had to be rediscovered in Europe, and for some time coal was of little local importance. Although coal was used on a small scale in Western Europe for thousands of years, the remains of a body that had been cremated with coal during the Bronze Age were discovered in South Wales. Evidence of Roman coal-fueled fires has also been discovered on the northern English frontier along Hadrian's Wall. It does not appear that European coal was used for hundreds of years after the fall of the Roman Empire [7]. Records from the Middle Ages show that coal was given to the monks in the Abbey of Peterborough in A.D. 852 as an offering or as a settlement of a claim [2]. Coal was mined in Germany as early as the 900s and was mentioned in the charter (dated 1025) of the French priory of St. Sauveuren-Rue [8].

Even though peasants probably continued to use surface coal for domestic heating fuel, no evidence of coal use for industry has been found until around 1200. Coal was discovered to be a very good fuel for iron forges and metalworking because it burned almost as slowly as charcoal, which is produced from wood and was the primary fuel of choice for village smiths. Shipping records from this period show that coal was marketed in Western Europe; smiths preferred coal over charcoal if they could get it at a reasonable price [7]. Liège in Belgium, Lyonnais in France, and Newcastle in England all became important mining centers [7, 8].

At first, only coal that was near the areas where it was mined was used because wood and charcoal were much lighter and thus less expensive than coal to transport by land. However, as water transportation on rivers and the sea increased, and as wood became scarcer, particularly in the cities, coal became the fuel of choice. By the mid-1200s, *sea-coal*—as it was called to distinguish it from charcoal—was being transported to London by sea from Newcastle. By the 1370s, 84 coal-boats were traveling down the east coast of England to ten different ports along the European coast between France and Denmark, returning with iron, salt, cloth, and tiles [7].

Although coal became the fuel of choice for blacksmiths during medieval times, it had limited use as a heat source because of its fumes. However, as wood became increasingly scarce and coal became less expensive in the cities, coal use increased significantly. With the increased use of coal (mainly in fireplaces designed to burn wood) came increased pollution problems, mainly black smoke and fumes. In 1257, Queen Eleanor was forced to flee Nottingham Castle as a result of smoke and fumes rising from the city below. In 1283 and in 1288, the air quality in London suffered because coal was being used in lime kilns. In 1307, a Royal Proclamation forbade the use of coal in lime-burners in parts of South London [7]. This proclamation did not work, however, and a later commission was ordered to punish offenders with fines and ransoms for a first offense and the demolition of their furnaces for a second offense.

Eventually, economics and a change in government policy won out over the populace's comfort, and London's air continued to be polluted by coal fumes for another 600 years [7]. The price of firewood increased, and it became more profitable to transport coal over longer distances. In addition, during late sixteenth- and seventeenth-century England faced the dilemma of conserving its remaining forests and using the only available substitute: coal. In 1615, the English government encouraged the substitution of coal for wood whenever possible. A fundamental change in English domestic building followed with more brick chimneys constructed to accommodate the fumes from coal [7].

1.3.1 *The Early History of U.S. Coal Mining and Use*

Coal was reportedly used by the American Indians of the Southwest before the arrival of the early explorers in America [9]. The first mention of coal in the United States is a map prepared around 1673–1674 by the Frenchman Louis Joliet that indicates *charbon de terra* along the Illinois River. In 1701, coal was discovered near Richmond, Virginia, and a map drawn in 1736 shows the location of several "cole mines" on the upper Potomac River, near what is now the border of Maryland and West Virginia. By the mid-1700s, coal was also reported in Pennsylvania, Ohio, and Kentucky, with the first commercial U.S. coal production beginning near Richmond, Virginia [9].

Blacksmiths in colonial days used small amounts of coal to supplement the charcoal they used in their forges. Farmers dug coal from beds exposed at the surface and sold it. Although most of the coal for the larger cities along the eastern seaboard was imported from England and Nova Scotia, some of it came from Virginia [9].

1.4   Coal Use during the Industrial Revolution

Several developments during the eighteenth century caused an expanded use of coal in England and culminated with the Industrial Revolution from 1750 to 1850. These developments included the transport revolution, the iron industry revolution, and the demise of the forests [10].

In eighteenth-century England, coal was the only available fuel because the supply of wood had become exhausted in the populated areas. Demand for coal increased as a source of fuel for domestic needs and the small businesses that developed in the towns, such as bakeries, smithies, tanneries, sugar refineries, and breweries [10]. Transporting the coal by sea was not sufficient to meet the demand for coal because sea transport was unreliable, and it could not accommodate the demand in inland areas. Canals, however, could meet this need. More than half of the Navigation Acts passed between 1758 and 1802 to establish a canal or river improvement company were for those whose primary business was to transport coal [10]. Establishing a transport system was crucial to the success of the Industrial Revolution, which would ultimately be driven by the coal and iron industries. Bulky raw materials and the finished products needed to be transported quickly and cheaply across England [10].

Both the iron industry and the increased need for coal to produce coke for iron smelting were important contributing factors to the development of the Industrial Revolution. Abraham Darby successfully smelted iron with coke as early as 1709, and this technological innovation became very important in the 1750s as the price of charcoal rose and the price of coal declined [10]. Once the switch from wood to coal was complete, an ironmaster's constraint on his output was not his fuel supply, but his power supply to provide an adequate blast in his furnace. In 1775, Boulton and Watt's invention, the steam engine, provided an unlimited source of power that up to that time had been water power and, to a lesser extent, wind power. The steam engine, which was fueled by coal, removed any restrictions on the ironmasters as to the size or location of iron works. The ironmasters could now move into areas that were rich in coal and iron resources and reap the economies of scale of a modern industry [10]. With the invention of the steam engine came the locomotive, another means for mass transportation of raw materials and products, which also consumed coal as a fuel source.

The uses for coal did not stop with coking and solid-fuel combustion for transportation. It was discovered that the gases released from coal during the coking process could be burned as well. This in turn led to the establishment of the manufactured gas industry to take advantage of the illuminating power of coal gas. In 1810, an Act of Parliament was obtained to form a company to supply coal gas to London [2].

The Industrial Revolution began in England, but it soon spread to continental Europe, mainly France and Germany, and eventually to the United States. These countries were able to benefit from the discoveries that were driving the Industrial Revolution because they had ample supplies of coal as well. During the 1800s in the United States, coal became the principal fuel used by locomotives, and as the railroads branched into the coal fields, they became a vital link between the mines and the markets. Coal also found growing markets as fuel for homes and steamboats and in the production of illuminating oil and gas. In fact, shortly after London began using coal gas, Baltimore became the first city in the United States to light streets with coal gas in 1816. As in England, coke soon replaced charcoal as the fuel for iron blast furnaces in the latter half of the 1800s.

1.5   The Post–Industrial Revolution Use of Coal

The technologies that used the most coal—combustion, gasification, liquefaction, and carbonization, which is the production of coke—either got their start during the 100-year-plus period of the Industrial Revolution/post–Industrial Revolution era or, in the case of coke production, they made major strides in technology development and usage. These technologies are discussed in detail in Chapter 5, but an introduction to their history is provided here. The demand for coal increased as additional fuel chemistry and engineering technologies were developed. In 1855, R.W. Bunsen invented the atmospheric gas burner, which led to a wide range of heating applications.

Electric lighting, which had its start in the late 1800s, led to widespread coal combustion in the process of producing steam for power generation. The generation of electricity through the use of coal has undergone various stages of technology development since the late 1800s, and examples include areas such as advanced combustion technologies; new materials of construction; innovative system designs; and new developments in steam production, electricity generation, and pollution control. These are discussed in detail in later chapters.

As the use of coal gas became more widespread, the production of gas for heating purposes was also developing. The first gas producer that created low-Btu gas was built in 1832, and at the turn of the century was an important method for heating furnaces [2]. Another development in the gas production field was the discovery of the carbon–steam reaction, where steam is reacted with carbon to produce carbon monoxide and hydrogen. This gasification technology had its major start in the mid- to late 1800s and increased until less expensive natural gas replaced manufactured gas [2].

Coal hydrogenation began in 1913 with the Bergius concept of direct hydrogenation of coal under hydrogen pressure at an elevated temperature [2]. The production of liquid hydrocarbons through indirect liquefaction—known as the Fischer-Tropsch synthesis—was also conceived at this time.

1.6   An Overview of Energy in the United States

The United States has always been a resource-rich nation, but in Colonial times, nearly all of the energy was supplied by muscle power—both human and animal; water; wind; and wood. The history of energy use in the United States began with wood being the dominant energy source from the founding of the earliest colonies until late last century, as shown in Figure 1.1 [11]. Consumption is illustrated in quadrillion (i.e., 10^{15}) Btu in Figure 1.1. Although wood use continued to expand along with the nation's economic growth, energy shortages led to the search for other energy sources. Hence, coal began to be used in blast furnaces for coke production and in the making of coal-gas for illumination in the early 1800s. Natural gas found limited application in lighting. It was still not until well after midcentury that the total work output from engines exceeded that of work animals.

Manifest Destiny, or the westward expansion from the seacoast to the heart of the nation, was a major factor in the increased use of coal. As railroads drove west to the plains and mountains, they left behind the plentiful wood resources along the East Coast. Coal became more attractive as deposits were found along the railroad right-of-way and it was found to have a higher energy content than wood. This meant more train-miles traveled per pound of fuel. Demand for coal in coke production also rose because the railroads were laying thousands of miles of new track, and iron and steel were needed for the rails and spikes. The rapid growth of the transportation and industrial sectors was fueled by coal.

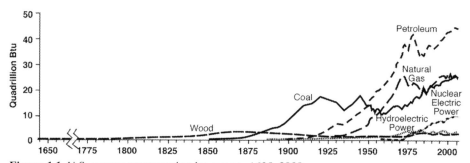

Figure 1.1 U.S. energy consumption by source, 1635–2008.
Source: From Energy Information Administration, Annual Energy Review 2008.

Around 1885, coal ended the long dominance of wood in the United States, only to be surpassed in 1951 by petroleum and then by natural gas a few years later. Hydroelectric power and nuclear electric power appeared about 1890 and 1957, respectively. Solar photovoltaic, advanced solar thermal, and geothermal technologies represent further recent developments in energy sources.

Petroleum was initially used as an illuminant and an ingredient in medicines, but was not used as a fuel for many years. At the end of World War I, coal still accounted for approximately 75 percent of the United States' total energy use. This changed, however, after World War II. Coal relinquished its place as the premier fuel in the United States as railroads lost business to trucks that operated on gasoline and diesel fuel. The railroads themselves began switching to diesel locomotives. Natural gas also started replacing coal in home stoves and furnaces. The coal industry survived, however, mainly because nationwide electrification created new demand for coal among electric utilities.

Most of the energy produced today in the United States comes from fossil fuels: coal, natural gas, crude oil, and natural gas plants liquids (Figure 1.2) [11]. Although U.S. energy production takes many forms, fossil fuels together far exceed all other forms of energy. In 2008, fossil fuels accounted for 79 percent of total energy production and were valued at an estimated $382 billion [11].

For most of its history, the United States was self-sufficient in energy, although small amounts of coal were imported from Britain and Nova Scotia during Colonial times. Through the late 1950s, production and consumption of energy were nearly in balance. However, beginning in the 1960s and continuing through today, consumption outpaced domestic production (Figure 1.3).

This is further shown in Figure 1.4, where, in 2008, the United States produced approximately 72 quadrillion Btu but consumed nearly 102 quadrillion Btu, with crude oil imports totaling nearly 35 quadrillion Btu [11]. Because of its insatiable

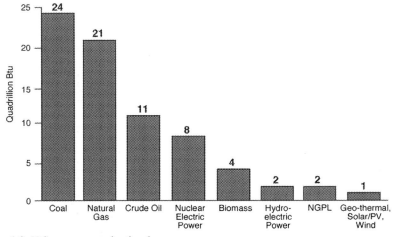

Figure 1.2 U.S. energy production by source.
Source: From Energy Information Administration, Annual Energy Review 2008.

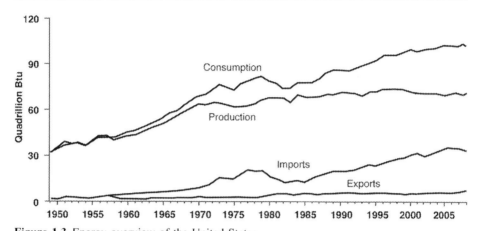

Figure 1.3 Energy overview of the United States.
Source: From Energy Information Administration, Annual Energy Review 2008.

demand for petroleum, the U.S. petroleum imports were nearly 13 million barrels per day in 2008. This is disturbing, especially when compared to U.S. petroleum imports in 1973, which totaled 6.3 million barrels per day. In October 1973, the Arab members of the Organization of Petroleum Exporting Countries (OPEC) embargoed the sale of oil to the United States, causing prices to rise sharply and sending the country into a recession. Although petroleum imports declined for two years, they increased again until prices rose dramatically from about 1979 through 1981, which suppressed imports. The increasing import trend resumed in 1986, and except for some slight dips, it has continued ever since. This dependency on foreign energy affects the security of the United States and must be addressed by its political leaders.

Energy is crucial in the operation of the industrialized U.S. economy, and energy spending is high. Since 2005, American consumers have spent more than $1 trillion a year on energy [11]. Energy is consumed in four major sectors: residential, commercial, industrial, and transportation.

Industry is historically the largest user of energy, and the most vulnerable to fluctuating prices, and consequently shows the greatest volatility (Figure 1.5). In particular, steep drops occurred in 1975, from 1980 to 1983, and again in 2001 in response to high oil prices and economic slowdown. Transportation was the next largest energy-consuming sector, followed by residential and commercial use.

Energy sources have changed over time for the various sectors. In the residential and commercial sectors, coal was the leading source as late as 1951, but then it decreased (Figures 1.6 and 1.7). Coal was then replaced by other forms of energy. Meanwhile, electricity's use and related losses during generation, transmission, and distribution increased dramatically. The expansion of electricity reflects the increased electrification of American households, which typically rely on a wide range of electrical appliances and systems. Home heating in the United States also

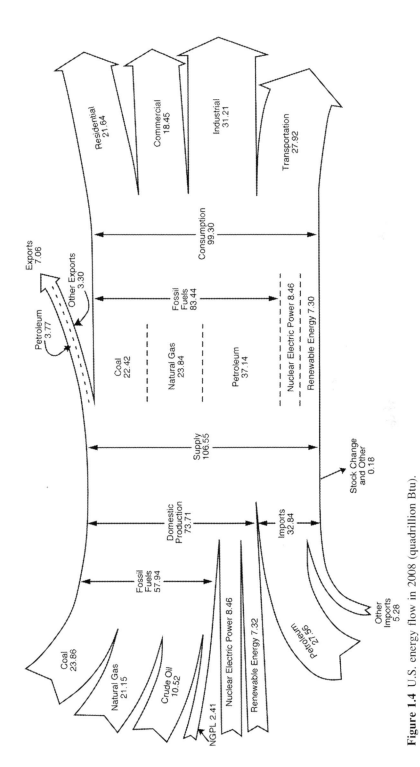

Figure 1.4 U.S. energy flow in 2008 (quadrillion Btu).

Source: From Energy Information Administration, Annual Energy Review 2008.

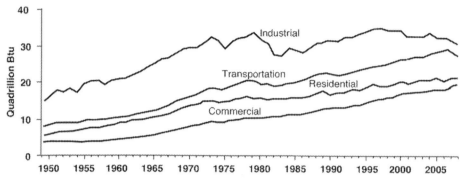

Figure 1.5 U.S. energy consumption by sector.
Source: From Energy Information Administration, Annual Energy Review 2008.

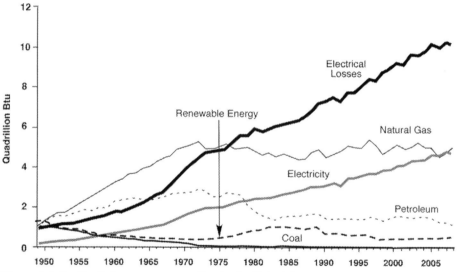

Figure 1.6 Residential energy consumption in the United States.
Source: From Energy Information Administration, Annual Energy Review 2008.

underwent a big change. More than one-third of all housing units were heated by coal in 1950, but less than 0.5 percent were coal heated in 2005. Similarly, distillate fuel oil lost a significant share of the home heating market, dropping from 22 to 9 percent over the same period. Home heating by natural gas and electricity, on the other hand, rose from 26 to 52 percent and from 1 to 31 percent, respectively [11].

In the industrial sector, the use of coal—once the leading source of energy—decreased as the consumption of both natural gas and petroleum increased (see

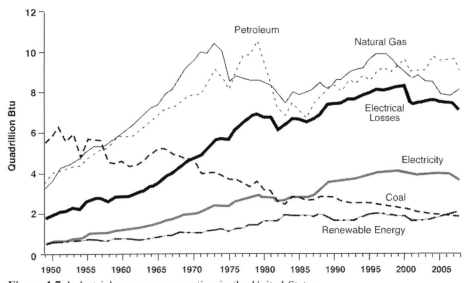

Figure 1.7 Industrial energy consumption in the United States.
Source: From Energy Information Administration, Annual Energy Review 2008.

Figure 1.7). Electricity and its associated losses also grew steadily. Approximately 71 percent of the energy consumed in the industrial sector is used for manufacturing. The remainder goes to mining, construction, agriculture, fisheries, and forestry. The large consumers of energy in the manufacturing industries, for which the fuel of choice is primarily natural gas, include petroleum and coal products, chemicals and associated products, paper and associated products, and metal industries. Less than 6 percent of all the energy consumed in the United States is used for nonfuel purposes such as asphalt and road oil for roofing products; road construction and road conditioning; liquefied petroleum gases for feedstocks at petrochemical plants; waxes for packaging, cosmetics, pharmaceuticals, inks, and adhesives; and gases for chemical and rubber manufacturing [11].

The transportation sector's use of energy, which is mainly petroleum—it accounted for 95 percent of the sector's energy in 2008—has more than tripled over the last 50 years, as shown in Figure 1.8. Motor gasoline accounts for about 67 percent of the petroleum consumed in this sector (in barrels), with distillate fuel oil and jet fuel the main other petroleum products used in this sector.

Figure 1.9 provides a summary of primary energy consumption in the United States by source and sector in 2008. Power generation is the largest consumer of energy (40 percent at 40.1 quadrillion Btu), and coal is the largest fuel source for this sector, providing 22.5 quadrillion Btu of energy or 51 percent of the total energy consumed for electrical power generation.

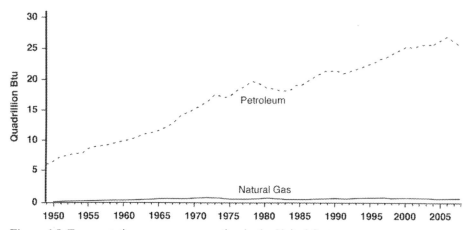

Figure 1.8 Transportation energy consumption in the United States.
Source: From Energy Information Administration, Annual Energy Review 2008.

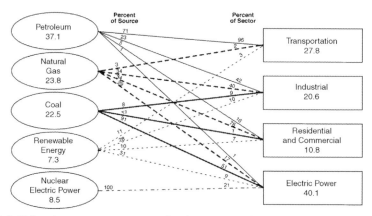

Figure 1.9 U.S. primary energy consumption by source and sector in 2008 (quadrillion Btu).
Source: From Energy Information Administration, Annual Energy Review 2008.

1.7 Coal Production in the United States

The total amount of coal consumed in the United States in all the years before 1800 was an estimated 108,000 short tons, with much of it imported [12]. However, production and consumption began to increase as a result of the Industrial Revolution and the development of the railroads. From 1881 through 1951, coal was the leading energy source produced in the United States [12]. Coal was surpassed by crude oil and natural gas until 1982 and 1984, respectively, at which time coal regained its position as the top energy resource (Figure 1.10).

Enormous quantities of coal are mined in the United States, and 1.17 billion short tons of coal were mined in 2008 [13]. Coal flow in the United States for 2007 is

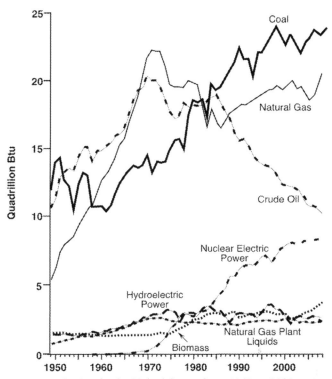

Figure 1.10 Energy production in the United States from 1949 to 2008.
Source: From Energy Information Administration, Annual Energy Review 2008.

summarized in Figure 1.11. The trend in U.S. coal production for the last six decades is illustrated in Figure 1.12. Coal production by region and state is listed in Table 1.1 for 2008 [11]. Coal production has shifted from mainly underground mines to surface mines, as illustrated in Figure 1.13. Also, coal resources west of the Mississippi River, especially those in Wyoming, underwent tremendous development (see Figure 1.14, page 17). Coal was produced in 25 states in 2008, with 14 of the states located west of the Mississippi River.

The shift toward surface-mined coal, especially west of the Mississippi River, came about because of the technological improvements in mining and the geological nature of the deposits—that is, thick coal seams located near the surface. Coal in the eastern United States generally occurs in seams that tend to be less than 15 feet thick. Thicker coalbeds are common in the western United States, particularly Wyoming, where coal seams average about 65 feet, as illustrated in Figure 1.15 (see page 17) [9]. Coal seams that are more than 200 feet under the surface are mined by underground methods. Most underground mines are less than 1,000 feet deep, although several reach depths of about 2,000 feet [9]. The largest coal-producing western mines are surface mines.

Individual coalbeds commonly cover large geographical regions. For example, the heavily mined Pittsburgh coalbed is found in parts of Pennsylvania, West

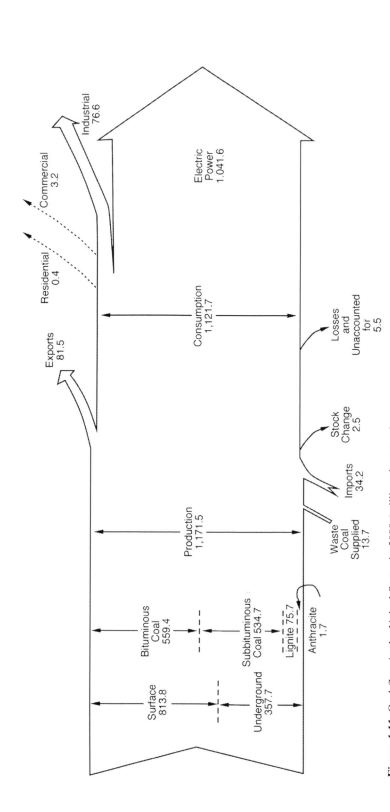

Figure 1.11 Coal flow in the United States in 2008 (million short tons).
Source: From Energy Information Administration, Annual Energy Review 2008.

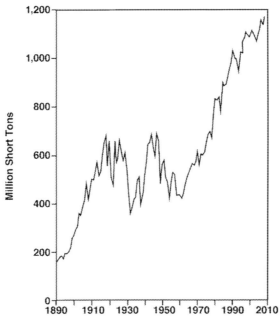

Figure 1.12 Coal production in the United States for the period 1890–2008.
Source: From Energy Information Administration, Coal Data: A Reference (1995).

Table 1.1 U.S. Coal Production during 2008 by Region and State

Coal-Producing Region and State	Millions of Short Tons
Appalachian Total	**389.8**
Alabama	20.6
Kentucky, Eastern	89.9
Maryland	2.8
Ohio	26.3
Pennsylvania	65.3
Anthracite	*1.7*
Bituminous	*63.6*
Tennessee	2.3
Virginia	24.6
West Virginia	158.0
Interior Total	**146.7**
Arkansas	0.1
Illinois	33.0
Indiana	36.2
Kansas	0.2
Kentucky, Western	30.0
Louisiana	3.8
Mississippi	2.8

Continued

Table 1.1 U.S. Coal Production during 2008 by Region and State—*Cont'd*

Coal-Producing Region and State	Millions of Short Tons
Missouri	0.2
Oklahoma	1.4
Texas	39.0
Western Total	**633.6**
Alaska	1.5
Arizona	8.0
Colorado	32.0
Montana	44.8
New Mexico	25.6
North Dakota	29.6
Utah	24.4
Wyoming	467.6
Refuse Recovery	**1.4**
U.S. Total	**1,171.5**

Source: From Freme (2009) and National Mining Association (2008) [14].

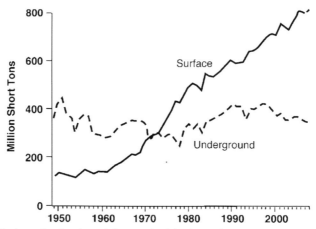

Figure 1.13 Coal production by mining method in the United States.
Source: From Energy Information Administration, Annual Energy Review 2008.

Virginia, Ohio, and Maryland. Similarly, the Wyodak coalbed, which is the leading source of coal in the United States, is estimated to cover at least 10,000 square miles in the Powder River Basin of Wyoming and Montana [9]. Consequently, although there were about 300 coalbeds mined in the United States in 1993, nearly half of the coal produced that year was from only the ten seams listed in Table 1.2. Note that the data in Table 1.2 are for 2007 because detailed production by seam was not available for 2008.

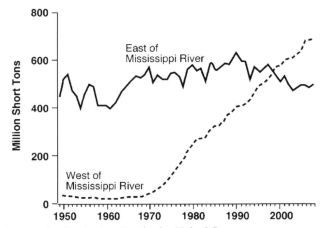

Figure 1.14 Coal production by location in the United States.
Source: From Energy Information Administration, Annual Energy Review 2008.

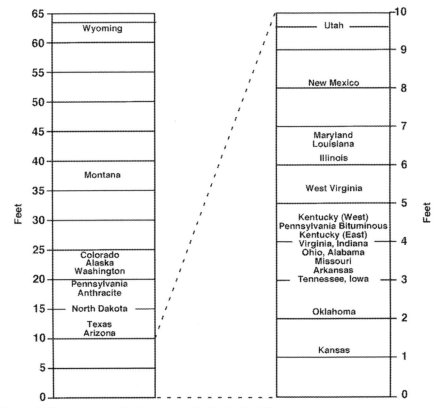

Figure 1.15 Average coalbed thickness in the United States.
Source: From Energy Information Administration, Coal Data: A Reference (1995).

Table 1.2 U.S. Coal Production during 2007 from the Ten Leading Coalbeds

Coalbed Name	Production (in thousands of short tons)	State with the Largest Production in the Coalbed
Wyodak	389.2	Wyoming
Pittsburgh	85.6	West Virginia
No. 9	48.3	Kentucky, Western
Coalburg	34.0	Kentucky, Eastern
Canyon	28.5	Wyoming
Beulah-Zap	27.6	North Dakota
Herrin (Illinois No. 6)	23.1	Illinois
Anderson-Dietz 1–Dietz 2	20.5	Wyoming
Roland	18.8	Wyoming
Upper Elkhorn No. 3	18.6	Kentucky, Eastern
Total	**694.2**	—
Percentage of U.S. Total	**60.5**	—

The most important coal deposits in the eastern United States are in the Appalachian Region, an area that encompasses more than 72,000 square miles and parts of nine states (Figures 1.16 and 1.17). This region contains large deposits of low- and medium-volatile bituminous coal and the principal deposits of anthracite. (A discussion of coal ranks is provided in Chapter 2.) Historically, this region has been the major source of U.S. coal, with approximately 75 percent of the total annual production as recently as 1970 [9]. Now the region produces about a third of the United States' total, with 390 million short tons mined in 2008. The reduction is due to the increased coal production in the western United States (Figures 1.17 and 1.18). This region, however, is still the principal source of bituminous coal and anthracite, and three of the top four coal-producing states (West Virginia, eastern Kentucky, and Pennsylvania) come from this region. (Note that eastern Kentucky is in the Appalachian Region, while western Kentucky is in the Interior Region.) These states, listed in Table 1.3, are among the top ten producing states (of the 25 producing coal in 2008), which as a group produced nearly 88 percent of all the coal in the United States.

The Interior Region consists of several separate basins located from Michigan to Texas, and it produced approximately 147 million short tons in 2008. Four states—Texas, Indiana, Illinois, and western Kentucky—produce the majority of the coal in this region. These coals range in rank from high-volatile bituminous coal in the northern part of the region to lignite in Texas.

The Western Region of the United States has several coal basins that contain all ranks of coal. More than half of the coal produced in the United States comes from this region, with four of the states—Wyoming, Montana, Colorado, and North Dakota—producing 49 percent of the total coal produced (Table 1.3). Of these, Wyoming is by far the largest coal mining state in the United States and produces approximately 40 percent of the country's total coal. Lignite is mined in North

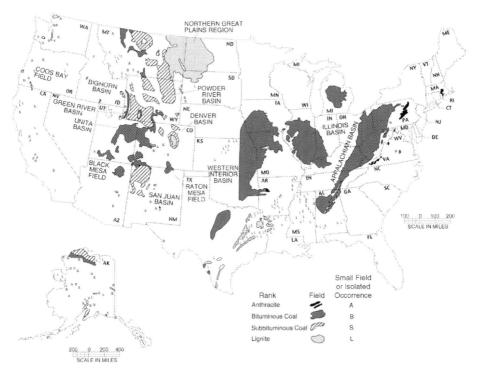

Figure 1.16 Major coal-bearing areas of the United States.
Source: From Averitt (1975) [15].

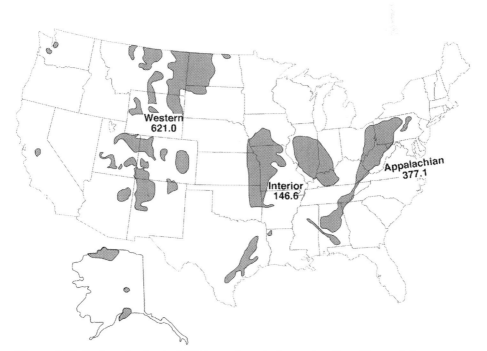

Figure 1.17 Coal production in 2007 by coal-producing region (in millions of short tons).
Source: From Freme (2009).

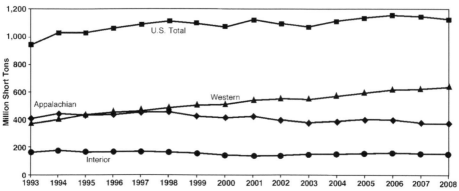

Figure 1.18 Coal production by region (1993–2007).
Source: Modified from Freme (2009).

Table 1.3 Ten Top Coal-Producing States in 2007

State	Production (in millions of short tons)
Wyoming	467.6
West Virginia	158.0
Kentucky—total	119.9
Eastern	*89.9*
Western	*30.0*
Pennsylvania	65.3
Montana	44.8
Texas	39.0
Colorado	32.0
Indiana	36.2
Illinois	33.0
North Dakota	29.6
Total Ten Top States	**1,025.4**
Percentage of U.S. Total	**87.5**

Sources: From Freme (2009) and Energy Information Administration, Annual Coal Report 2008.

Dakota and Montana, subbituminous coal is mined in southeastern Montana and northeastern Wyoming, and the principal bituminous coal mining production is in Utah, Colorado, and Arizona.

1.8   Coal Consumption in the United States

Coal is used in all 50 states and the District of Columbia. In 2007, ten states consumed approximately 50 percent of the total coal produced in the United States (Table 1.4) [16]. In 2007, most of the coal was consumed in the electric power

Table 1.4 Ten Top Coal-Consuming States in 2007

State	Production (in Short Tons)
Texas	104,784
Indiana	66,593
Ohio	61,468
Illinois	60,323
Pennsylvania	59,054
Missouri	45,376
Kentucky	42,547
Georgia	42,317
West Virginia	39,215
Alabama	38,939
Total	**560,616**
Percentage of U.S. Total	**48.3**

sector as shown in Figures 1.11 and 1.19. Nearly 93 percent of all coal consumed in 2008 was in the electric power sector, which includes both the electric utilities and independent power producers. The breakdown for coal consumption by sector for 2008 is shown in Table 1.5 [11].

The resurgence in coal consumption, after decreasing around 1950 (see Figures 1.1 and 1.10), was driven by the Arab oil embargo in the first half of the 1970s, which caused significant price increases for petroleum, and by a natural gas shortage in the second half of the 1970s. Virtually all of this growth has been due to the increasing amounts of coal used to generate electricity. Consequently, coal became the dominant energy source in the electrical power industry, and approximately 49 percent of the electricity generated in the United States in 2008 was from coal, as illustrated in Figure 1.20 [17]. In 2002, coal's share of electricity

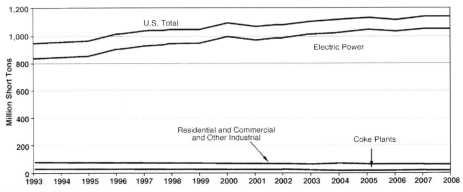

Figure 1.19 Coal consumption by sector.
Source: Modified from Freme (2009).

Table 1.5 Coal Consumption by Sector in 2008

Sector	Consumption (in short tons)
Electric utilities/power	1,041,600
Coking	22,100
Other Industrial	54,500
Residential/commercial	3,600
Total	**1,121,800**

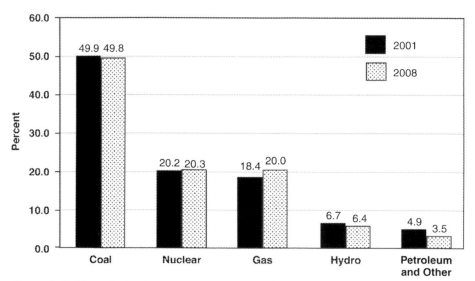

Figure 1.20 Share of electric power sector net generation by energy source, 2002 versus 2008. *Source:* From Freme (2009) and U.S. DOE (2002).

generation dropped below 50 percent for the first time since 1979 due to gains by natural gas and hydroelectric plants. While a balanced energy portfolio makes sense from an energy security stance, the increasing use of natural gas for baseloaded electricity generation is questionable.

When compared to the electric power sector, coal consumption in the nonelectric power sectors is low. Although consumption fluctuates slightly, total coal usage for coke production and the industrial/commercial/residential sectors has changed little over the last 15 years.

1.9 U.S. Coal Exports and Imports

The U.S. coal export and import markets are relatively small (see Figure 1.11), with the export market steadily declining for many years until 2002, when slight increases occurred (Figure 1.21 [17]). In 2008, the United States exported approximately 81.5

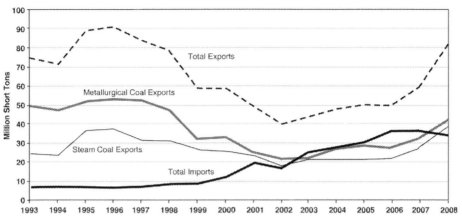

Figure 1.21 U.S. coal imports and exports.
Source: From Freme (2009) and U.S. DOE (2002).

million short tons, which consisted of approximately 42.5 million short tons of met-allurgical coal and approximately 39.0 millions short tons of steam coal.

Europe is the main destination of U.S. metallurgical coal, with 25.5 million short tons exported in 2008. The Netherlands was the primary European destination of U.S. metallurgical coal exports, with a total of 3.4 million short tons. Italy received 2.9 million short tons of U.S. metallurgical coal, making it the second largest European destination. Both France and the United Kingdom each received 2.1 million short tons of metallurgical coal. Other major European destinations of metallurgical coal were Belgium, Croatia, Germany, Poland, Romania, Spain, and Turkey, with each receiving more than 1 million short tons of coal.

The primary destinations for U.S. metallurgical coal exports in North and South America were Canada and Brazil, respectively. Canada received 3.6 million short tons of metallurgical coal, and Brazil received 6 million short tons. The Asian mar-ket for U.S. metallurgical coal totaled 4.2 million short tons exported, which accounted for almost 10 percent of U.S. exports. Australia, India, Japan, and South Korea all imported U.S. coal. Africa accounted for 2 million short tons of U.S. metallurgical coal exports, with most going to Egypt.

Canada is the single largest market for all U.S. coal exports, as well as the pri-mary North American destination of steam coal exports. In 2008, Canada received 19.4 million short tons of steam coal exports, which accounted for 49.7 percent of all 2008 steam coal exports.

Europe is the second largest market for U.S. steam coal exports due to the declin-ing coal production in many of the countries. A total of 14.8 millions short tons of U.S. steam coal were exported with the primary markets being the United Kingdom (3.7 million short tons), the Netherlands (3.6 million short tons), Germany (1.4 million short tons), France (1.4 million short tons), Belgium (1.3 million short tons), and Spain (1.1 million short tons). U.S. steam coal exports to the African continent decreased in 2008 to 1.9 million short tons, of which 1.7 million short tons was exported to Morocco. U.S. steam coal exports to South America and Asia were 1.3 and 1.1 million short tons, respectively.

The United States imports only a small amount of coal compared to total U.S. coal consumption, with 34.2 million short tons imported out of more than 1 billion short tons consumed. Colombia dominates the U.S. import market, with its 26.3 million short tons accounting for more than 75 percent of all coal imported in 2008, and it is followed by Indonesia (3.4 million short tons), Venezuela (2.3 million short tons), and Canada (2.0 million short tons). These four countries accounted for approximately 99 percent of total U.S. imports.

1.10   World Primary Energy Production and Consumption

The world's energy production and consumption is summarized in Tables 1.6 and 1.7, respectively. These tables contain information on the world's primary energy—petroleum, natural gas, coal, and electric power (hydro, nuclear, geothermal, solar, wind, wood, and waste)—production and consumption by world region and country for 2006. These data were tabulated from information published by the U.S. Department of Energy's Energy Information Administration (EIA) [18]. As of November 2009, data only up to 2006 were available.

1.10.1   World Primary Energy Production

The world's output of primary energy totaled more than 469 quadrillion Btu in 2006. Petroleum (crude oil and natural gas plants) continued to be the world's most important primary energy source, accounting for more than 157 quadrillion Btu, or 33.5 percent of world primary energy production. Petroleum production in 2006 was 81.3 million barrels per day. The Middle East, as a region, produced about one-third of this petroleum at 49 quadrillion Btu (with Saudi Arabia being the leading producer at 20 quadrillion Btu), followed by Eurasia at 25 quadrillion Btu (with Russia the leading producer at nearly 20 quadrillion Btu), North America at 23 quadrillion Btu (with the United States producing nearly 11 quadrillion Btu), and Africa at 20 quadrillion Btu, with the other regions of the world producing approximately 10 to 15 quadrillion Btu each.

Coal ranked second as a primary energy source in 2006, accounting for 27.8 percent of world primary energy production. More than 128 quadrillion Btu of coal (6.8 billion short tons) were produced. As a region, Asia/Oceania was the largest producer at more than 77 quadrillion Btu. In this region, China was the largest producer at nearly 58 quadrillion Btu (2.6 billion short tons), which also made it the largest coal producer in the world. Australia and India ranked a very distant second and third in coal production in this region at approximately 8.5 and 8.2 quadrillion Btu, respectively. This is equivalent to 420 and 497 million short tons, respectively. The second largest quantity of coal produced was on the North American continent accounting for more than 24 quadrillion Btu, of which more than nearly

Table 1.6 World Energy Production during 2006 (in quadrillion Btu)

Region/ Country	Total Primary Energy	Crude Oil[a]	Dry Natural Gas	Coal	Net Hydroelectric Power	Net Nuclear Power	Other[b]
North America							
Canada	19.25	5.36	6.73	1.52	3.49	1.05	0.11
Mexico	10.35	7.14	1.86	0.22	0.30	0.10	0.16
United States	71.02	10.80	18.99	23.79	2.87	8.21	1.25
Total	**100.62**	**23.30**	**25.58**	**25.53**	**6.66**	**9.36**	**1.52**
Central and South America							
Argentina	3.80	1.52	1.70	0.001	0.37	0.08	0.014
Bolivia	0.62	0.10	0.48	0	0.02	0	0.002
Brazil	8.03	3.72	0.36	0.08	3.43	0.14	0.17
Chile	0.39	0.01	0.07	0.01	0.28	0	0.01
Colombia	3.54	1.17	0.24	1.72	0.40	0	0.01
Ecuador	1.26	1.17	0.01	0	0.07	0	0
Paraguay	0.53	0[c]	0	0	0.53	0	0
Peru	0.48	0.16	0.06	0	0.19	0	0.002
Trinidad and Tobago	1.74	0.31	1.34	0	0	0	0
Venezuela	8.08	5.62	1.09	0.23	0.81	0	0
Other	0.53	0.20	0.03	0.01	0.24	0.00	0.07
Total	**29.00**	**13.98**	**5.38**	**2.05**	**6.34**	**0.22**	**0.28**
Europe							
Austria	0.50	0.04	0	0	0.34	0	0.05
Belgium	0.50	0	0.07	0.001	0.004	0.46	0.03
Bosnia and Herzegovina	0.19	0	0	0.13	0.06	0	0
Bulgaria	0.42	0.002	0	0.18	0.04	0.20	0
Croatia	0.17	0.04	0.06	0	0.06	0	0
Czech Republic	1.15	0.02	0.01	0.82	0.02	0.28	0.01
Denmark	1.21	0.71	0.41	0	0	0	0.10
Finland	0.44	0	0	0	0.11	0.22	0.10
France	5.13	0.05	0.04	0	0.55	4.42	0.07
Germany	5.25	0.15	0.62	2.18	0.20	1.59	0.51
Greece	0.41	0.004	0.001	0	0.06	0	0.02
Hungary	0.39	0.04	0.10	0.33	0.002	0.13	0.01
Italy	1.22	0.25	0.40	0.002	0.36	0	0.21
Netherlands	2.65	0.06	2.44	0	0.001	0.03	0.09
Norway	10.23	5.11	3.42	0.07	1.17	0	0.01
Poland	2.93	0.06	0.17	2.65	0.02	0	0.02
Romania	1.17	0.20	0.44	0.27	0.18	0.06	0

Continued

Table 1.6 World Energy Production during 2006 (in quadrillion Btu)—*Cont'd*

Region/ Country	Total Primary Energy	Crude Oil[a]	Dry Natural Gas	Coal	Net Hydroelectric Power	Net Nuclear Power	Other[b]
Slovakia	0.31	0.02	0.01	0.02	0.04	0.20	0.004
Spain	1.39	0.01	0.002	0.29	0.25	0.57	0.26
Sweden	1.34	0[c]	0	0	0.61	0.64	0.10
Switzerland	0.60	0	0	0	0.30	0.27	0.02
Turkey	1.17	0.09	0.03	0.61	0.43	0	0.005
United Kingdom	7.87	3.16	3.02	0.45	0.04	0.82	0.15
Other	1.10	0.02	0.04	0.52	0.43	0.07	0.14
Total	**47.74**	**10.04**	**11.28**	**8.52**	**5.28**	**9.96**	**1.91**
Eurasia							
Azerbaijan	1.66	1.37	0.25	0	0.02	0	0
Kazakhstan	5.71	2.82	0.95	1.76	0.08	0	0
Lithuania	0.10	0.01	0	0	0.004	0.09	0
Russia	53.34	19.85	23.38	6.14	1.72	1.59	0.03
Tajikistan	0.17	0.001	0.001	0.002	0.16	0	0
Turkmenistan	2.71	0.35	2.34	0	0	0	0
Ukraine	3.33	0.20	0.72	1.34	0.13	0.93	0
Uzbekistan	2.56	0.13	2.25	0.05	0.06	0	0
Other	0.48	0.06	0.01	0.13	0.25	0.03	0.01
Total	**70.06**	**24.80**	**29.90**	**9.42**	**2.42**	**2.64**	**0.04**
Middle East							
Bahrain	0.51	0.07	0.42	0	0	0	0
Iran	13.12	8.66	4.05	0.04	0.18	0	0.001
Iraq	4.34	4.24	0.07	0	0.005	0	0
Israel	0.03	0	0.03	0	0	0	0
Kuwait	6.14	5.48	0.46	0	0.001	0	0
Oman	2.47	1.58	0.88	0	0	0	0
Qatar	4.05	1.79	1.88	0	0	0	0
Saudi Arabia	24.68	19.74	2.72	0	0	0	0
Syria	1.24	0.91	0.21	0	0.04	0	0
United Arab Emirates	7.87	5.57	1.80	0	0	0	0
Yemen	0.78	0.78	0	0	0	0	0
Other	0.03	0.01	0.01	0	Minor	0	Minor
Total	**65.26**	**48.83**	**12.53**	**0.04**	**0.23**	**0.00**	**0.001**
Africa							
Algeria	7.73	3.68	3.47	0	0.002	0	0
Angola	3.09	3.01	0.02	0	0.03	0	0
Cameroon	0.22	0.19	0.001	0	0.04	0	0

Table 1.6 *Cont'd*

Region/ Country	Total Primary Energy	Crude Oil[a]	Dry Natural Gas	Coal	Net Hydroelectric Power	Net Nuclear Power	Other[b]
Congo (Brazzaville)	0.53	0.51	0.007	0	0.004	0	0
Congo (Kinshasa)	0.12	0.04	0	0.003	0.07	0	0
Egypt	3.25	1.38	1.67	0.001	0.13	0	0.006
Gabon	0.52	0.51	0.004	0	0.01	0	0
Libya	4.29	3.54	0.55	0	0	0	0
Nigeria	6.37	5.24	1.05	0	0.08	0	0
South Africa	6.01	0.04	0.11	5.74	0.01	0.10	0.003
Tunisia	0.27	0.16	0.11	0	0.001	0	0
Zambia	0.10	0	0	0.006	0.09	0	0
Other	2.76	1.95	0.17	0.13	0.43	0.00	0.02
Total	**35.26**	**20.25**	**7.16**	**5.88**	**0.90**	**0.10**	**0.03**
Asia and Oceania							
Australia	11.40	0.87	1.63	8.57	0.15	0	0.03
Bangladesh	0.56	0.01	0.53	0	0.01	0	0
Brunei	0.97	0.43	0.51	0.001	0	0	0
Burma	0.51	0.05	0.40	0.03	0.03	0	0
China	67.74	7.88	2.16	52.80	4.28	0.55	0.06
India	12.39	1.44	1.14	8.20	1.12	0.19	0.09
Indonesia	9.62	2.13	2.20	4.95	0.09	0	0.13
Japan	4.29	0.01	0.20	0	0.84	2.93	0.29
Korea, North	0.97	0	0	0.84	0.12	0	0
Korea, South	1.51	0	0.02	0.05	0.03	1.40	0.006
Malaysia	3.80	1.27	2.31	0.01	0.06	0	0
Mongolia	0.09	0	0	0.09	0	0	0
New Zealand	0.66	0.03	0.14	0.16	0.23	0	0.08
Pakistan	1.61	0.14	1.04	0.09	0.30	0.03	0
Papua New Guinea	0.10	0.09	0.005	0	0.01	0	0
Philippines	0.48	0.05	0.08	0.05	0.10	0	0.21
Taiwan	0.47	0.002	0.01	0	0.08	0.37	0.003
Thailand	1.82	0.47	0.84	0.23	0.08	0	0.03
Vietnam	2.15	0.76	0.21	0.95	0.23	0	0
Other	0.33	0.22	Minor	0.03	0.14	0.00	0.09
Total	**121.47**	**15.85**	**13.42**	**77.05**	**7.90**	**5.47**	**0.94**
World Total	*469.41*	*157.05*	*107.23*	*128.50*	*29.73*	*27.76*	*4.72*

[a]Including lease condensate production.
[b]Electric power generation from geothermal, solar, wind, wood, and wastes.
[c]Values less than 0.001 are reported as zero (0); totals may not add up due to rounding.

Table 1.7 World Energy Consumption in 2006 (in quadrillion Btu)

Region/ Country	Total Primary Energy	Petroleum	Dry Natural Gas	Coal	Net Hydroelectric Power	Net Nuclear Power	Other[a]
North America							
Canada	13.95	4.51	3.42	1.43	3.49	1.05	0.11
Mexico	7.36	4.04	2.35	0.40	0.30	0.10	0.16
United States	99.86	39.96	22.19	22.51	2.87	8.21	1.25
Other	0.01	0.03	0.00	0.00	0.00	0.00	0.00
Total	**121.18**	**48.54**	**27.96**	**24.34**	**6.66**	**9.37**	**1.52**
Central and South America							
Argentina	3.15	1.11	1.54	0.02	0.37	0.08	0.014
Bolivia	0.22	0.11	0.08	0	0.02	0	0.002
Brazil	9.63	4.61	0.71	0.44	3.42	0.14	0.17
Chile	1.25	0.54	0.28	0.13	0.28	0	0.01
Colombia	1.30	0.56	0.24	0.11	0.40	0	0.01
Ecuador	0.42	0.33	0.01	0	0.07	0	0
Paraguay	0.43	0.05	0	0	0.53	0	0
Peru	0.61	0.33	0.06	0.03	0.19	0	0.002
Trinidad and Tobago	0.82	0.07	0.74	0	0	0	0
Venezuela	3.19	1.29	1.09	0.002	0.81	0	0
Other	3.16	3.06	0.06	0.09	0.25	0.00	0.07
Total	**24.18**	**11.68**	**4.81**	**0.82**	**6.34**	**0.22**	**0.28**
Europe							
Austria	1.53	0.63	0.33	0.16	0.34	0	0.05
Belgium	2.75	1.37	0.66	0.20	0.004	0.46	0.03
Bosnia and Herzegovina	0.25	0.05	0.01	0.13	0.06	0	0
Bulgaria	0.90	0.24	0.20	0.24	0.04	0.20	0
Croatia	0.41	0.21	0.10	0.02	0.06	0	0
Czech Republic	1.81	0.43	0.33	0.77	0.02	0.28	0.01
Denmark	0.88	0.40	0.20	0.21	0	0	0.10
Finland	1.32	0.46	0.17	0.21	0.11	0.22	0.10
France	11.44	4.11	1.92	0.59	0.55	4.42	0.07
Germany	14.63	5.57	3.48	3.34	0.20	1.59	0.51
Greece	1.49	0.93	0.12	0.34	0.06	0	0.02
Hungary	1.14	0.33	0.51	0.14	0.002	0.13	0.01
Italy	8.07	3.62	3.05	0.67	0.36	0	0.21
Netherlands	4.14	2.11	1.51	0.31	0.001	0.03	0.09
Norway	1.89	0.44	0.23	0.03	1.17	0	0.01
Poland	3.86	1.00	0.54	2.30	0.02	0	0.02
Romania	1.68	0.44	0.66	0.35	0.18	0.06	0

Table 1.7 *Cont'd*

Region/ Country	Total Primary Energy	Petroleum	Dry Natural Gas	Coal	Net Hydroelectric Power	Net Nuclear Power	Other[a]
Slovakia	0.81	0.16	0.24	0.17	0.04	0.20	0.004
Spain	6.51	3.33	1.32	0.78	0.25	0.57	0.26
Sweden	2.22	0.72	0.04	0.10	0.61	0.64	0.10
Switzerland	1.28	0.55	0.12	0.005	0.30	0.27	0.02
Turkey	3.91	1.37	1.14	0.96	0.43	0[b]	0.005
United Kingdom	9.80	3.65	3.42	1.69	0.04	0.82	0.15
Other	2.70	1.88	0.53	0.71	0.43	0.07	0.14
Total	**85.42**	**34.00**	**20.83**	**14.42**	**5.28**	**9.96**	**1.91**
Eurasia							
Azerbaijan	0.70	0.25	0.42	0	0.02	0	0
Kazakhstan	2.97	0.48	1.15	1.27	0.08	0	0
Lithuania	0.35	0.14	0.10	0.01	0.004	0.09	0
Russia	30.39	5.78	16.75	4.55	1.72	1.59	0.03
Tajikistan	0.28	0.06	0.05	0.002	0.16	0	0
Turkmenistan	0.87	0.21	0.67	0	0	0	0
Ukraine	5.87	0.70	2.68	1.47	0.13	0.93	0
Uzbekistan	2.21	0.30	1.80	0.04	0.06	0	0
Other	2.24	0.69	1.12	0.16	0.25	0.03	0.01
Total	**45.88**	**8.61**	**24.74**	**7.50**	**2.42**	**2.64**	**0.04**
Middle East							
Bahrain	0.49	0.07	0.42	0	0	0	0
Iran	7.69	3.40	4.05	0.05	0.18	0	0.001
Iraq	1.25	1.17	0.07	0	0.005	0	0
Israel	0.85	0.50	0.03	0.32	0	0	0
Kuwait	1.14	0.67	0.46	0	0	0	0
Oman	0.55	0.15	0.40	0	0	0	0
Qatar	0.91	0.18	0.72	0	0	0	0
Saudi Arabia	6.89	4.17	2.72	0	0	0	0
Syria	0.81	0.56	0.21	0	0.04	0	0
United Arab Emirates	2.46	0.87	1.59	0	0	0	0
Yemen	0.27	0.27	0	0	0	0	0
Other	0.50	0.42	0.09	0.00	0.01	0.00	0.00
Total	**23.81**	**12.43**	**10.76**	**0.37**	**0.23**	**0.00**	**0.001**
Africa							
Algeria	1.54	0.50	1.02	0.01	0.002	0	0
Angola	0.16	0.11	0.02	0	0.03	0	0
Cameroon	0.09	0.05	0.001	0	0.04	0	0

Continued

Table 1.7 World Energy Consumption in 2006 (in quadrillion Btu)—*Cont'd*

Region/ Country	Total Primary Energy	Petroleum	Dry Natural Gas	Coal	Net Hydroelectric Power	Net Nuclear Power	Other[a]
Congo (Brazzaville)	0.03	0.02	0.007	0	0.004	0	0
Congo (Kinshasa)	0.10	0.02	0	0.01	0.07	0	0
Egypt	2.54	1.34	1.05	0.03	0.13	0	0.006
Gabon	0.04	0.03	0.004	0	0.01	0	0
Libya	0.78	0.54	0.24	0	0	0	0
Nigeria	1.02	0.54	0.40	0	0.08	0	0
South Africa	5.18	1.11	0.11	3.85	0.01	0.10	0.003
Tunisia	0.33	0.18	0.15	0	0.001	0	0
Zambia	0.13	0.03	0	0.004	0.09	0	0
Other	2.55	1.65	0.17	0.27	0.43	0.00	0.02
Total	**14.49**	**6.12**	**3.17**	**4.17**	**0.90**	**0.10**	**0.03**
Asia and Oceania							
Australia	5.61	1.89	1.07	2.47	0.15	0	0.03
Bangladesh	0.74	0.19	0.53	0.015	0.01	0	0
Brunei	0.18	0.03	0.15	0	0	0	0
Burma	0.24	0.08	0.11	0.004	0.03	0	0
China	73.81	14.82	2.08	52.03	4.28	0.55	0.06
India	17.68	5.41	1.43	9.42	1.12	0.19	0.09
Indonesia	4.15	2.49	0.87	0.56	0.09	0	0.13
Japan	22.79	10.51	3.59	4.63	0.84	2.93	0.29
Korea, North	0.95	0.03	0	0.79	0.12	0	0
Korea, South	9.45	4.50	1.29	2.21	0.03	1.40	0.006
Malaysia	2.56	1.04	1.20	0.27	0.06	0	0
Mongolia	0.10	0.03	0	0.07	0	0	0
New Zealand	0.86	0.32	0.15	0.09	0.23	0	0.08
Pakistan	2.30	0.77	1.04	0.16	0.30	0.03	0
Papua New Guinea	0.07	0.06	0.005	0	0.01	0	0
Philippines	1.27	0.65	0.08	0.24	0.10	0	0.21
Taiwan	4.57	1.96	0.41	1.73	0.08	0.37	0.003
Thailand	3.74	1.99	1.15	0.48	0.08	0	0.03
Vietnam	1.40	0.53	0.21	0.44	0.23	0	0
Other	3.84	3.04	0.34	0.32	0.23	0.00	0.01
Total	**156.31**	**50.34**	**15.71**	**75.93**	**7.90**	**5.47**	**0.94**
World Total	*472.27*	*171.72*	*108.00*	*127.55*	*29.73*	*27.76*	*4.72*

[a]Electric power generation from geothermal, solar, wind, wood, and wastes.
[b]Values less than 0.001 are reported as zero (0); totals may not add up due to rounding.

23 quadrillion Btu (or 1.2 billion short tons) were produced by the United States. The United States produced the second largest quantity of coal in 2006, second only to China.

Dry natural gas ranked third as a primary energy source, accounting for 22.8 percent of world primary energy production in 2006 with more than 107 quadrillion Btu (104 trillion cubic feet) produced. Eurasia was the largest dry natural gas-producing region at about 30 quadrillion Btu, followed closely by North America, with nearly 26 quadrillion Btu produced. All other regions of the world produced between about 5 and 13 quadrillion Btu in 2006. Russia, the United States, and Canada were the three largest dry natural gas producing countries at 23.8, 19.0, and 6.7 quadrillion Btu, respectively. All other countries produced, individually, from less than 5 trillion Btu up to about 4 quadrillion Btu.

The remaining primary energy sources listed in Table 1.6—hydroelectric, nuclear, and other (geothermal, solar, wind, wood, and waste) electric power generation—ranked fourth, fifth, and sixth, respectively. They accounted for 6.3, 5.9, and 1.0 percent, respectively, of total primary energy sources in 2006. Combined, they accounted for a total of 62 quadrillion Btu (6.1 trillion kilowatthours).

The United States produced 2.7 quadrillion Btu of renewable energy that was not used for electricity generation [18]. This included ethanol blended into motor gasoline and geothermal, solar, wood, and waste energy not used for electricity generation. This renewable energy accounted for 0.6 percent of world primary energy production and ranked seventh as a primary energy source.

Three countries—the United States, China, and Russia—were the leading energy producers in 2006, together producing 41 percent of the world's total energy. Individually they produced 71.0, 67.7, and 53.3 quadrillion Btu, respectively. When Saudi- Arabia and Canada are included, these five countries produced over half (i.e., 50.3 percent) of the world's total energy. The next five leading producers of primary energy were Iran, India, Australia, Mexico, and Norway, and together they supplied an additional 12.2 percent of the world's total energy [18].

1.10.2 World Primary Energy Consumption

As with primary energy production, the three largest consumers of world energy in 2006 were the United States, China, and Russia. In 2006 these countries consumed 99.9, 30.4, and 73.8 quadrillion Btu, respectively, which was about 43 percent of the world's total energy. When Japan and India are included, these five largest consumers of primary energy in 2006 accounted for 51.8 percent of the world's total. The leading five countries were followed by Germany, Canada, France, and the United Kingdom, which together accounted for an additional 12.6 percent of the world's total consumption [18].

The United States consumed nearly 100 quadrillion Btu, by far the most of any country. The United States was followed by China at nearly 74 quadrillion Btu and Russia at more than 30 quadrillion Btu. However, on a per capita basis, the United Arab Emirates was the largest consumer in the world, with 576 million Btu consumed per person followed by Kuwait (471 million Btu/person), Canada

(427 million Btu/person), the United States (334 million Btu/person), and Australia (277 million Btu/person). Table 1.8 lists per capita energy consumption for the regions of the world along with selected countries. By region, the per capita energy consumption ranges from about 276 million Btu per person for North America to 18 million Btu per person for Africa. As expected, the per capita consumption is highest for the industrialized nations. Table 1.9 lists the per capita energy consumption for the regions of the world for 2001 for comparison. Over the five-year period, some regions of the world were fairly constant in per capita energy consumption (e. g., North America), some exhibited slight increases (e.g., Europe and Africa), and others showed significant increases on a percentage basis (e.g., Central and South America, and Asia and Oceania). It is recognized that the regions of the world with the largest increases included the less developed countries that are becoming more industrialized and are consuming more energy.

China was the largest consumer of coal in 2006, using 2.6 billion short tons, which is an increase of nearly 100 percent from 1.4 billion short tons consumed in 2001. In 2006, the United States consumed 1.1 billion short tons (comparable to 2001 usage). China and the United States were followed by India (543 million short tons), Germany (272 million short tons), and Russia (264 million short tons). These five countries accounted for 71 percent of the world coal consumption in 2006, an increase from 64 percent in 2001.

Table 1.8 Per Capita Primary Energy Consumption in 2006

Region/Country	Energy Consumption (in quadrillions Btu)	Population (in millions)	Per Capita Energy Consumption (in millions Btu per person)
North America	*121.18*	*438.68*	*276.2*
United States	99.86	298.44	334.6
Canada	13.95	32.66	427.1
Mexico	7.36	107.45	68.5
Central and South America	*24.18*	*330.28*	*73.2*
Brazil	9.63	188.08	51.2
Venezuela	3.19	25.64	124.4
Bolivia	0.22	8.99	24.5
Europe	*85.42*	*590.97*	*144.5*
France	11.44	63.33	180.6
Germany	14.63	82.42	177.5
United Kingdom	9.80	60.61	161.7
Italy	8.07	58.16	138.8
Netherlands	4.14	16.49	251.0
Poland	3.54	38.64	91.6
Hungary	1.09	9.92	109.9

Table 1.8 Cont'd

Region/Country	Energy Consumption (in quadrillions Btu)	Population (in millions)	Per Capita Energy Consumption (in millions Btu per person)
Denmark	0.88	5.45	161.5
Croatia	0.41	4.49	91.3
Eurasia	*45.88*	*285.39*	*160.8*
Russia	30.39	142.07	213.9
Ukraine	5.87	46.62	125.9
Kazakhstan	2.97	15.23	195.0
Turkmenistan	0.87	5.01	173.6
Lithuania	0.35	3.59	97.5
Middle East	*23.81*	*187.21*	*127.2*
Iran	7.69	65.03	118.2
Saudi Arabia	6.89	27.02	255.0
United Arab Emirates	2.46	4.27	576.1
Iraq	1.25	26.78	46.7
Kuwait	1.14	2.42	471.1
Israel	0.85	6.87	123.7
Syria	0.81	18.88	42.9
Oman	0.55	3.10	177.4
Africa	*14.49*	*799.99*	*18.1*
South Africa	5.18	44.19	117.2
Egypt	2.54	78.95	32.2
Nigeria	1.02	131.86	7.7
Libya	0.78	5.90	132.2
Angola	0.16	11.99	13.3
Gabon	0.04	1.43	28.0
Asia and Oceania	*156.31*	*3,649.36*	*42.8*
China	73.81	1,313.97	56.2
Japan	22.79	127.52	178.7
India	17.68	1,111.71	15.9
South Korea	9.45	48.85	193.4
Australia	5.61	20.26	276.9
Indonesia	4.15	231.82	17.9
Pakistan	2.30	161.74	14.2
Vietnam	1.40	84.40	16.59
North Korea	0.95	23.11	41.1
New Zealand	0.86	4.09	210.3
Bangladesh	0.74	147.37	5.0

Table 1.9 Per Capita Primary Energy Consumption in 2001

Region/Country	Energy Consumption (in quadrillions Btu)	Population (in millions)	Per Capita Energy Consumption (in millions Btu per person)
North America	*115.58*	*416.93*	*277.2*
United States	97.05	283.97	341.8
Canada	12.51	31.08	402.5
Mexico	6.00	101.75	59.0
Central and South America	*20.92*	*426.20*	*49.1*
Brazil	8.78	172.39	50.9
Venezuela	2.95	24.63	119.8
Cuba	0.39	11.22	34.8
Western Europe	*72.76*	*482.42*	*150.8*
France	10.52	59.19	177.7
Germany	14.35	82.36	174.2
United Kingdom	9.81	59.54	164.8
Italy	8.11	57.95	139.9
Netherlands	4.23	16.04	263.7
Portugal	1.09	10.02	108.8
Denmark	0.90	5.33	168.9
Croatia	0.63	4.66	135.2
Eastern Europe and Former USSR	*51.54*	*386.25*	*133.4*
Russia	28.20	144.40	195.3
Ukraine	6.08	49.11	123.8
Poland	3.54	38.64	91.6
Hungary	1.09	9.92	109.9
Turkmenistan	0.48	4.88	98.4
Lithuania	0.33	3.49	94.6
Middle East	*17.92*	*171.21*	*104.7*
Iran	5.18	64.53	80.3
Saudi Arabia	4.91	21.03	233.5
Syria	2.06	16.72	123.2
Iraq	1.08	23.58	45.8
Kuwait	0.92	1.97	467.0
Israel	0.79	6.45	122.5
Oman	0.34	2.62	129.8
United Arab Emirates	0.15	2.65	56.6

Table 1.9 *Cont'd*

Region/Country	Energy Consumption (in quadrillions Btu)	Population (in millions)	Per Capita Energy Consumption (in millions Btu per person)
Africa	*12.45*	*811.69*	*15.3*
South Africa	4.60	44.33	103.8
Egypt	2.13	67.89	31.4
Nigeria	0.92	116.93	7.9
Zimbabwe	0.24	13.96	17.2
Asia and Oceania	*112.76*	*3,450.11*	*32.7*
China	39.67	1,285.00	30.9
Japan	21.92	127.34	172.1
India	12.80	1,017.54	12.6
South Korea	8.06	47.34	170.3
Australia	4.97	19.49	255.0
Indonesia	4.63	214.84	26.2
North Korea	2.84	22.30	127.4
Pakistan	1.87	144.97	12.9
New Zealand	0.84	3.85	218.2
Vietnam	0.79	79.18	10.0
Bangladesh	0.51	140.37	3.6

1.11   Projections of Energy Use and Coal's Contribution to the Energy Mix

World energy consumption is projected by the EIA to increase by 44 percent from 472 quadrillion Btu in 2005 to 678 quadrillion Btu in 2030 (Figure 1.22) [19]. Although high prices for oil and natural gas, which are expected to continue throughout the period, are likely to slow the growth of energy demand in the long term, world energy consumption is projected to continue increasing greatly as a result of the robust economic growth and expanding populations in the world's developing countries. OECD (Organization for Economic Cooperation and Development) member countries are, for the most part, more advanced energy consumers. Appendix A provides a complete list of OECD and non-OECD member countries, as well as other regional definitions, as defined by the EIA's Office of Integrated Analysis and Forecasting [18]. Energy demand in the OECD economies is expected to grow slowly over the projection period, at an average annual rate of 0.6 percent, whereas energy consumption in the emerging economies of non-EOCD countries is expected to expand by an average of 2.3 percent per year, as shown in Figure 1.23.

China and India—the fastest-growing non-OECD economies—will be key contributors to world energy consumption in the future. Over the past decades, their energy consumption as a share of total world energy use has increased significantly.

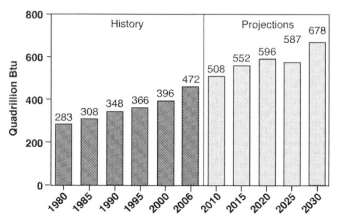

Figure 1.22 Projected total world energy consumption.
Source: From Energy Information Administration, Annual Energy Review 2009.

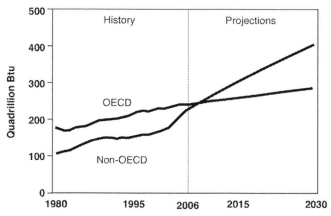

Figure 1.23 Projected world energy consumption by region.
Source: From Energy Information Administration, Annual Energy Review 2009.

In 1980, China and India together accounted for less than 8 percent of the world's total energy consumption; in 2006 their share had grown to 19 percent. Even stronger growth is projected over the next 20 to 25 years, with their combined energy use more than doubling and their share increasing to 28 percent of world energy consumption in 2030. In contrast, the U.S. share of total world energy consumption is projected to contract from 21 percent in 2006 to about 17 percent in 2030.

Energy consumption in other non-OECD regions also is expected to grow extensively from 2006 to 2030, with increases of around 60 percent projected for the Middle East and Central and South America and of around 104 percent for Africa (Figure 1.24). A smaller increase—about 25 percent—is expected for non-OECD Europe and Eurasia, including Russia and the other former Soviet Republics.

The use of all energy sources is projected to increase over time, as shown in Figure 1.25 [19]. Given the expectations that that world oil prices will remain

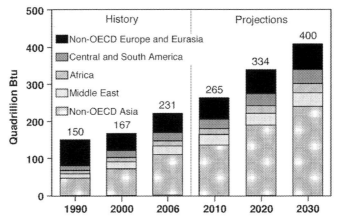

Figure 1.24 Projected energy use in non-OECD nations by region.
Source: From Energy Information Administration, Annual Energy Review 2009.

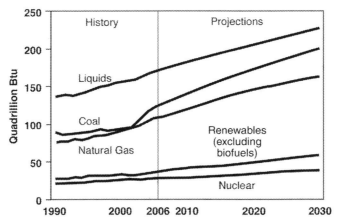

Figure 1.25 Projected world energy consumption by fuel type in quadrillion Btu.
Source: From Energy Information Administration, Annual Energy Review 2009.

relatively high throughout the projection, liquid fuels are the world's slowest growing source of energy; liquids consumption increases at an average annual rate of 0.9 percent from 2006 to 2030. Renewable energy and coal are the fastest growing energy sources, with consumption increasing by 3.0 and 1.7 percent, respectively. Projected high prices for oil and natural gas, as well as rising concerns about the environmental impacts of fossil fuel use, improve prospects for renewable energy sources. Coal's costs are comparatively low in relation to the costs of liquids and natural gas, and abundant resources in large energy-consuming countries, such as China, India, and the United States, make coal an economical fuel choice.

1.11.1   World Consumption of Liquid Fuels

Liquid fuels, as defined by IEA, include both conventional and nonconventional liquid products. Conventional liquids include crude oil and lease condensate, natural gas plant liquids, and refinery gain. Unconventional liquids include biofuels, gas-to-liquids, coal-to-liquids, and unconventional petroleum products (e.g., extra-heavy oils, oil shale, and bitumen). Also included is petroleum coke, which is a solid.

The demand for liquid fuels will increase from 85 million barrels of oil equivalent per day in 2006 to 91 million in 2015 and 107 million in 2030. Much of the increase in total liquids consumption is projected for nations of non-OECD Asia and the Middle East, where strong economic growth is expected, and approximately 50 percent is projected for use in the transportation sector.

Unconventional sources from both OPEC and non-OPEC (see Appendix A for a list of OPEC countries) sources are expected to become increasingly competitive, as illustrated in Figure 1.26. World production of unconventional resources, which totaled only 2.7 million barrels per day in 2006, are projected to increase to 13.0 million barrels per day in 2030, accounting for approximately 12 percent of total world liquids.

Liquids remain the most important fuels for transportation because there are few alternatives that can compete widely with liquid fuels. With world oil prices remaining relatively high through 2030, the increasing cost-competitiveness of non-liquid fuels causes many stationary uses of liquids—that is, electric power generation and use in the industrial sectors—to be replaced by alternative energy sources, which increases the transportation share of liquids consumption. On a global basis, the transportation sector accounts for 79 percent of the total projected increase in liquids use from 2005 to 2030, with the industrial sector accounting for nearly all of the remainder, as illustrated in Figure 1.27 [19].

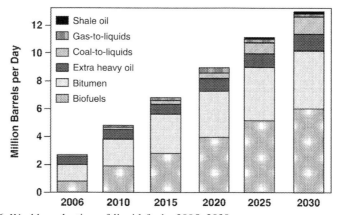

Figure 1.26 World production of liquid fuels, 2005–2030.
Source: From Energy Information Administration, Annual Energy Review 2009.

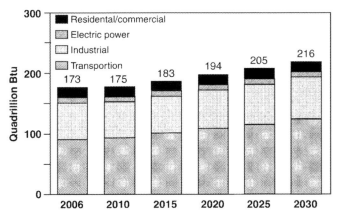

Figure 1.27 World liquids' consumption by sector, 2006–2030.
Source: From Energy Information Administration, Annual Energy Review 2009.

1.11.2 World Consumption of Natural Gas

Worldwide, total natural gas consumption is projected to increase from 104 trillion cubic feet in 2006 to 153 trillion cubic feet in 2030 [19]. World oil prices are expected to remain high, and this causes natural gas to be used instead of oil whenever possible. In addition, because natural gas produces less carbon dioxide when it is burned than coal or petroleum, governments implementing national or regional plans to reduce greenhouse gas emissions may encourage its use to displace fossil fuels.

Natural gas remains a major energy source for industrial sector uses and electricity generation throughout IEA's projection. The industrial sector, which is the world's largest consumer of natural gas, accounts for 40 percent of projected natural gas use in 2030. Electricity generation accounts for 35 percent of the world's total natural gas consumption in 2030 due to its relative fuel efficiency and low carbon dioxide intensity. In 2006, OECD member countries and non-OECD countries each consumed 52 trillion cubic feet of natural gas. By 2030, it is projected that non-OECD member countries will be using nearly 80 trillion cubic feet of natural gas compared to 50 trillion cubic feet for OECD countries.

1.11.3 World Consumption of Coal

World coal consumption is projected to increase from 127.5 quadrillion Btu in 2006 to 190.2 quadrillion Btu in 2030. Regionally, increased use of coal in non-OECD countries accounts for 94 percent of the total growth in world coal consumption over the forecast period [19]. In 2006, coal accounted for 27 percent of world energy consumption, as shown in Figure 1.28 [19]. Of the coal produced worldwide in 2005, 62 percent was shipped to electricity producers, 34 percent to industrial consumers, and most of the remaining 4 percent to the residential and commercial sectors. Coal's share of total world energy consumption is projected to increase to 28 percent in 2030, and its share in the electric power sector is projected to remain relatively constant at 42 percent from 2006 to 46 percent in 2030.

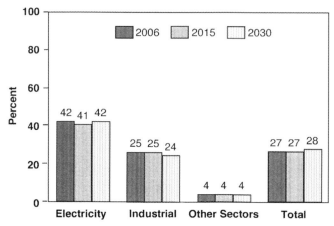

Figure 1.28 Coal share of world energy consumption by sector, 2005, 2015, and 2030.
Source: From Energy Information Administration, Annual Energy Review 2009.

Projected Coal Consumption in North America

Coal consumption in OECD countries is projected to increase from 47.3 quadrillion Btu in 2005 to 49.9 quadrillion Btu in 2015 and 55.0 quadrillion Btu in 2030 [19]. Of the OECD regions, North America is the major consumer, as illustrated in Figure 1.29.

Coal use in the United States totaled 22.5 quadrillion Btu in 2006, accounting for 92 percent of total coal use in North America and 48 percent of the OECD total. Coal demand in the United States is projected to rise to 26.6 quadrillion Btu in 2030. This projection is due to substantial coal reserves and the heavy reliance on it for electricity generation. Coal's share of electricity generation is expected to decline from 49 percent in 2006 to 47 percent in 2030.

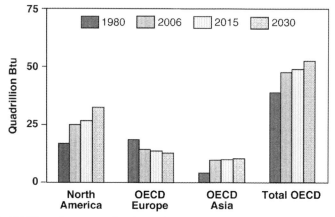

Figure 1.29 OECD coal consumption by region.
Source: From Energy Information Administration, Annual Energy Review 2009.

Smaller quantities of coal are consumed in other sectors such as Independent Power Producers, metallurgical coal consumption for coke production, and residential/commercial and other industrial users, but each of these sectors uses only about 2 to 7 percent per category [20]. Although environmental issues and legislation will always impact the use of coal (which will be discussed in detail in subsequent chapters), the overall scenario for coal production/consumption is that it will continue to increase through 2030. Key features of the U.S. coal market over the next 20 years include the following [20]:

- Increased production of low-sulfur coal from the western coalfields
- A continuing decline in coal prices
- Steady improvements in productivity
- Decreasing transportation costs

U.S. coal consumption is projected to increase modestly at an average of 0.7 percent/year from 2006 to 2030 as a result of increasing coal use for electricity generation at new and existing plants, combined with the start-up of several coal-to-liquids (CTL) plants. Although an assumed risk premium for carbon-intensive technologies dampens investment in new coal-fired power plants, increased generation from coal-fired power plants accounts for 39 percent of the growth in total U.S. electricity generation from 2006 to 2030 [19].

In Canada and Mexico, small increases in coal consumption are projected over the period—that is, 0.2 quadrillion Btu for each country. As a result, the two countries have a combined 8 percent share of North America's total coal consumption through 2030.

Projected Coal Consumption in OECD Europe

Total coal consumption in the countries of OECD Europe is projected to remain fairly flat, with a slight decline from 13.2 quadrillion Btu in 2006 to 12.0 quadrillion Btu in 2030 [19]. In 2006, the major coal-consuming countries of OECD Europe included the Czech Republic, Germany, Italy, Poland, the United Kingdom, Spain, and Turkey [19]. Low-Btu coal (called brown coal or lignite) is an important domestic source of energy in this region, with lignite accounting for 47 percent of their total coal consumption on a tonnage basis in 2006. Plans to replace or refurbish existing coal-fired generating capacity in a number of countries are an indication that coal will continue to play an important role in the overall energy mix, although governments are enacting policies to discourage the use of the fuel, primarily in response to environmental concerns.

Other factors that influence coal usage are the relatively slow growth in overall energy consumption, penetration of natural gas in the electric power and industrial sectors, increasing use of renewable fuels, and pressure on member countries of the European Union to reduce subsidies that support domestic production of hard coal (i.e., anthracite, bituminous coal, and, in some definitions, subbituminous coals).

Germany is one of the world's largest energy consumers and one of the largest coal-consuming countries. However, the country has only limited energy resources, and security of supply remains an issue. Germany is highly dependent on imported coal (27 percent of coal used is imported), oil (40 percent of oil used is imported), and natural gas (78 percent of natural gas used is imported); and the country's decision to phase out nuclear power will increase its reliance on imported energy even further [21].

Germany is the third largest oil importer in the world and the second largest European natural gas market after the United Kingdom. Increasing costs in these fuels are causing concern. Consequently, coal provides energy supply security as well as important employment within the power and mining industries. These are important to Germany, and the federal government and governments of Germany's two major coal producing states, Saarland and North Rhine-Westphalia, continue to provide large subsidies for mining to preserve jobs in the domestic coal industry.

Poland is the second largest consumer of coal in OECD Europe, and coal has a dominant role in the Polish energy supply, providing approximately 63 percent of the country's primary energy demand and most of its electricity demand. This role is expected to be maintained as the country possesses large hard coal and lignite reserves and an increasingly efficient infrastructure for its recovery [21].

Projected Coal Consumption in OECD Asia

OECD Asia countries are prominent consumers of coal. In 2006, they used 9.4 quadrillion Btu of coal, which represented 20 percent of total OECD coal consumption. OECD Asia's coal demand is projected to increase to 9.8 quadrillion Btu in 2030, as was shown in Figure 1.29. In this region, Australia is the world's leading coal exporter, while Japan and South Korea are the world's leading coal importers. Over the projection period, Japan's coal consumption is projected to decrease, while Australia, New Zealand, and South Korea account for nearly all of the projected growth in OECD Asia's demand for coal [19].

Coal consumption normally increases in Australia and New Zealand due to the substantial coal reserves in Australia and the region's heavy reliance on coal's use for electricity generation. However, natural gas consumption for electricity generation is expected to increase with coal-fired power plants supplying 70 percent of the region's total electricity generation in 2006, as compared to a projected 58 percent share in 2030. South Korea's total coal consumption is projected to increase by 0.7 quadrillion Btu from 2006 to 2030, primarily to fuel existing and planned electric power plants.

Projected Coal Consumption in Non-OECD Asia

Led by strong economic growth and rising demand for energy in China and India, non-OECD coal consumption is projected to rise to 139.6 quadrillion Btu in 2030, an increase of 73 percent over that consumed in 2006 (Figure 1.30) [19]. The increase of 59 quadrillion Btu, which represents 94 percent of the projected increase in total world coal consumption, shows the importance of coal in meeting overall demand in the non-OECD countries. China and India together account for 61 percent of the projected increase in world coal consumption from 2006 to 2030. Much

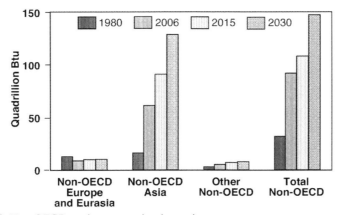

Figure 1.30 Non-OECD coal consumption by region.
Source: From Energy Information Administration, Annual Energy Review 2009.

of the increase in their demand for energy, particularly in the electric power and industrial sectors, is expected to be met by coal.

Coal usage in China's electricity sector is projected to increase from 24.9 quadrillion Btu in 2006 to 57.3 quadrillion But in 2030. At the beginning of 2006, China had an estimated 350 gigawatts of coal-fired electric generation. An additional 600 gigawatts of coal-fired capacity are projected to be brought on line by 2030 to meet the demand for electricity that is expected to accompany its economic growth.

More than 52 percent of China's coal use in 2006 was in the nonelectricity sectors, primarily in the industrial sector, since China was the world's leading producer of steel and pig iron. Coal remains the primary source of energy in the industrial sector because of China's limited reserves of oil and natural gas.

Rapid development of a CTL industry in China, which was a potential source of future coal demand, was anticipated several years ago. However, the government's concerns about several issues, including potential strains on water resources and shortages of coal supply, have led to increased oversight. As a result, many proposed CTL projects do not have official government approval and are now on hold [19].

In India, nearly 71 percent of the projected coal growth is expected to be in the electric power sector, with most of the remaining in the industrial sector [19]. In 2006, India's coal-fired power plants consumed 5.9 quadrillion Btu of coal, which represents 63 percent of the total coal demand. Coal consumption for electric power generation is projected to increase to 9.3 quadrillion Btu in 2030 as an additional 64 gigawatts of coal-fired capacity is brought on line. This will increase India's coal-fired generating capacity from 78 gigawatts in 2006 to 142 gigawatts in 2030.

Coal consumption in the other countries of non-OECD Asia is projected to increase from 5.1 quadrillion Btu in 2006 to 10.4 quadrillion Btu in 2030. Similar to China and India, these increases will be in the electric power generation and industrial sectors. Significant growth in coal consumption in the electric power section is expected in Taiwan, Vietnam, Indonesia, and Malaysia.

Projected Coal Consumption in Non-OECD Europe and Eurasia

Coal consumption in non-OECD Europe and Eurasia is projected to increase from 8.7 quadrillion Btu in 2006 to 9.4 quadrillion Btu in 2030, as shown in Figure 1.30 [19]. Russia is the largest consumer in this region. In 2006, Russia consumed 4.6 quadrillion Btu, or 52 percent of the region's total coal consumed. In 2030, Russia's coal consumption is projected to total 5.2 quadrillion Btu, or 55 percent of the total coal consumed in the region.

In other countries in this region, coal consumption is projected to remain near the 2006 level of 4 quadrillion Btu in 2006 through 2030. Plans are in place for both new coal-fired generating capacity and the refurbishment of the existing capacity in Albania, Bosnia and Herzegovina, Bulgaria, Montenegro, Romania, Serbia, and Ukraine.

Projected Coal Consumption in Africa

Coal consumption in Africa is projected to increase from the 2006 level of about 6 quadrillion Btu by 1.1 quadrillion Btu by 2030, with South Africa accounting for about 92 percent of the total for the projection period. In South Africa, the increasing demand for electricity in recent years has led to the decision to restart three large coal-fired power plants with 3.8 gigawatts of capacity that have been closed for more than a decade [19]. In addition, new power plants with 9.6 gigawatts capacity will be constructed. Increased coal consumption is also expected in the industrial sector for the production of steam and process heat, production of coke for the steel industry, and production of coal-based synthetic liquids. Currently, two commercial CTL plans in South Africa supply 150,000 barrels of synthetic liquids per day or approximately 20 percent of the country's total liquid fuel requirements.

Projected Coal Consumption in Central and South America

Coal consumption in Central and South America was 0.8 quadrillion Btu in 2006, with Brazil, the world's tenth largest steel industry, accounting for 54 percent of the region's demand. Most of the remaining demand was accounted for by Argentina, Chile, Colombia, and Puerto Rico. Coal consumption in this region is projected to increase to 1.9 quadrillion Btu in 2030, with Brazil accounting for 75 percent of the region's demand. Brazil's demand will be primarily for coke manufacture and electricity generation.

Projected Coal Consumption in the Middle East

Projected coal consumption in the Middle East remains relatively flat, increasing from 0.4 quadrillion Btu in 2005 to 0.5 quadrillion Btu in 2030. Israel accounts for about 85 percent of the total, and Iran accounts for the remainder.

1.11.4   World Consumption of Nuclear Energy

Electricity generation from nuclear power is projected to increase from about 2.7 trillion kilowatthours in 2006 to 3.8 trillion kilowatthours in 2030, as concerns about rising fossil fuel prices, energy security, and greenhouse gas emissions support the

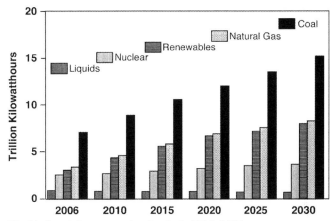

Figure 1.31 World electricity generation by fuel, 2006–2030.
Source: From Energy Information Administration, Annual Energy Review 2009.

development of new nuclear generation capacity (Figure 1.31) [19]. Each of the three largest coal-consuming nations—China, United States, and India—is projected to expand nuclear capacity significantly by 2030. Declines in nuclear capacity are projected only for OECD Europe, where several countries, including Germany and Belgium, have either plans or mandates to phase out nuclear power and where some older reactors are expected to be retired and not replaced. On a regional basis, the strongest growth is projected in non-OECD Asia countries. Outside of Asia, the largest increase in installed nuclear capacity among the non-OECD nations is projected for Russia.

1.11.5 World Consumption of Renewable Energy

High prices for oil and natural gas, which are expected to persist over the projection period, should encourage expanded use of renewable fuels [18]. Renewable energy sources are attractive for environmental reasons, especially in countries where reducing greenhouse gas emissions is of concern. Government policies and incentives to increase the use of renewable energy sources for electricity generation are expected to encourage the development of renewable fuels. Worldwide, the consumption of hydroelectricity and other renewable energy sources are projected to increase from 35 quadrillion Btu in 2005 to 59 quadrillion Btu in 2030.

In the non-OECD countries, much of the growth in renewable energy consumption is projected to come from hydroelectric facilities in Asia and in Central and South America. Among the OECD countries, where hydroelectricity is fairly well established, plans to undertake major hydroelectric power projects are in place only in Canada and Turkey. Instead, increases in renewable energy consumption are expected to be in biomass and wind.

1.11.6 Energy Outlook for the United States

Projections of U.S. energy consumption and production up to 2030 have been made by the EIA [22] and are shown in Figures 1.32 and 1.33, respectively. These projections have been developed to reflect new trends that are expected to persist in the economy and in energy markets, including higher price projections for crude oil and natural gas, higher price projections for delivered energy prices, slower projected growth in energy demand, faster projected growth in the use on non-hydroelectric renewable energy (resulting from state renewable portfolio standards), higher projections for domestic oil production in the near term, and slower projected growth in energy imports of natural gas and oil. Less emphasis has been placed on near-term influences such as supply disruptions or political actions because of the difficulty in predicting what new regulations will be or how they will be implemented.

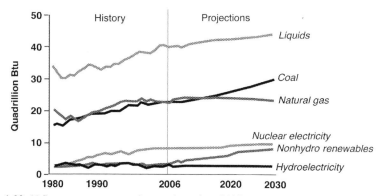

Figure 1.32 U.S. energy consumption projected to 2030.
Source: From Energy Information Administration, Annual Energy Outlook 2008.

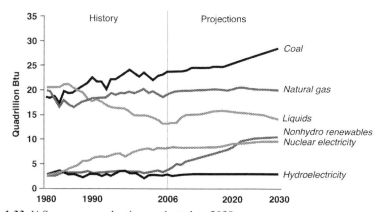

Figure 1.33 U.S. energy production projected to 2030.
Source: From Energy Information Administration, Annual Energy Outlook 2008.

Projected Energy Consumption in the United States

Energy consumption in the United States is projected to increase in all of the end-use sectors, with the demand for transportation energy expected to increase the most [22]. Residential energy demand is expected to increase to 12.9 quadrillion Btu in 2030 from 2006 levels of about 10.8 quadrillion Btu (i.e., an increase of 0.7 percent/year). Commercial energy demand is expected to grow from about 8.3 quadrillion Btu in 2006 to more than 11.3 quadrillion Btu in 2030. Industrial energy demand is projected to increase by 0.7 percent/year and will increase from about 25 quadrillion Btu to 27.7 quadrillion Btu in 2030. Transportation energy demand is projected to grow at an average annual rate of 2.0 percent/year, increasing from about 28 quadrillion Btu in 2006 to 33 quadrillion Btu by 2030. Demand for electricity is projected to increase by 1.1 percent/year from 3,814 billion kilowatthours in 2006 to 4,972 billion kilowatthours in 2030.

Projected energy consumption, by fuel, is shown in Figure 1.32 [22]. Total consumption of liquid fuels, including both fossil liquids and biofuels, is expected to grow from 20.7 million barrels per day in 2006 to 22.8 million barrels per day in 2030. Natural gas consumption is projected to increase from 21.7 trillion cubic feet in 2006 to 23.8 trillion cubic feet in 2016 and then decline to 22.7 trillion cubic feet in 2030 due to higher prices and slower growth in electric demand.

Coal consumption is projected to increase from about 22.5 quadrillion Btu (1,114 million short tons) in 2006 to 29.9 quadrillion Btu (1,545 million short tons) in 2030, mainly due to projected growth in the electric power sector (Figure 1.34). Total demand for industrial steam coal is projected to rise slightly, but coal demand for the residential and commercial sectors is expected to remain relatively constant, and demand for coking coal is projected to decline slightly. The projected increase in coal use for CTL plants is 0.6 quadrillion Btu in 2020 to 1.0 quadrillion Btu in 2030.

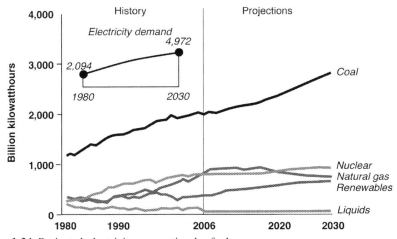

Figure 1.34 Projected electricity generation by fuel.
Source: From Energy Information Administration, Annual Energy Outlook 2008.

Coal is projected to maintain its fuel cost advantage over both oil and natural gas during the projection period. Coal prices should remain relatively unchanged, while natural gas and petroleum prices are expected to increase from the 2006 baseline.

Projected Energy Production in the United States

Total energy consumption in the United States is expected to increase more rapidly than domestic energy production through 2030, increasing to approximately 120 and 80 quadrillion Btu, respectively [23]. The shortfall of about 40 quadrillion Btu in 2030 is projected to be met primarily through imports of petroleum and natural gas. Crude oil production is projected to increase from 5.1 million barrels per day in 2006 to a peak of 6.3 million barrels per day in 2018 (as a result of production increases from the deep waters of the Gulf of Mexico and from onshore enhanced oil recovery projects) declining to 5.6 million barrels per day by 2030.

Total domestic natural gas production is projected to increase from 18.6 trillion cubic feet in 2006 to 20.0 trillion cubic feet in 2022 before declining to 19.5 trillion cubic feet in 2030, primarily because of higher costs associated with the exploration and development of natural gas and lower net pipeline imports from Canada and Mexico, despite the expectation that the Alaskan natural gas pipeline will be completed in 2020 and the fact that there will a slight increase in liquefied natural gas (LNG) imports. Nuclear power generation is expected to increase through the project period from 787 billion kilowatthours in 2006 to 917 billion kilowatthours in 2030. Renewable energy production is projected to increase to approximately 10 quadrillion Btu in 2030 from current levels of about 3.5 quadrillion Btu. The increase will be primarily a result of the use of renewable energy in power generation. U.S. coal production is projected to increase an average of 0.3 percent per year from 2007 to 2015, and then increase at 1 percent per year to 2030, with a total of 1,455 million short tons produced [23]. Higher electricity demand and lower prices will be the primary reasons for the coal demand.

Another emerging market for coal is CTL. Western coal production, which has grown steadily since 1970, will continue to increase through 2030. Much of the production will come from the Powder River Basin, with its thick seams and low overburden ratios, resulting in higher labor productivity than other coalfields. This advantage is expected to be maintained throughout the forecast period. Production of low-Btu lignite in the Western supply regions should also increase substantially, primarily to meet the energy and feedstock requirements of new coal-fired power plants and CTL plants in Texas, Montana, and North Dakota. Appalachian coal production should decline slightly because this basin has been mined extensively and production costs have been increasing more rapidly than in other regions. The eastern portion of the Interior coal basin (Illinois, Indiana, and western Kentucky), with extensive reserves of mid- and high-sulfur bituminous coals, benefits from the new coal-fired generating capacity in the Southeast.

1.12   Coal's Role in the U.S.'s 2001 Energy Policy

In May 2001, the National Energy Policy Development (NEPD) Group, chaired by then Vice President Richard Cheney, unveiled a National Energy Policy for President George W. Bush. This report, titled *Reliable, Affordable, and Environmentally Sound Energy for America's Future*, was the first attempt at an energy policy in the United States since President Jimmy Carter's administration. President Bush directed the NEPD Group to "develop a national energy policy designed to help bring together business, government, local communities, and citizens to promote dependable, affordable, and environmentally sound energy for the future" [23]. In the report's cover letter, Cheney states that the report "envisions a comprehensive long-term strategy that uses leading-edge technology to produce an integrated energy, environmental, and economic policy. To achieve a twenty-first-century quality of life—enhanced by reliable energy and a clean environment—we must modernize conservation, modernize our infrastructure, increase our energy supplies, including renewables, accelerate the protection and improvement of our environment, and increase our energy security" [23]. While the outcome of the exercise is debated on its usefulness as a true energy policy, the NEPD Group recognized the importance of coal in the U.S.'s energy policy and security. Coal's role in a U.S. energy policy is introduced in this section, and its contribution to the energy security of the United States is discussed in detail in Chapter 12.

The energy strength in the United States lies in the abundance and diversity of its energy resources and in its technological leadership in developing and efficiently using these resources. The United States has significant domestic energy resources—coal, oil, and natural gas—and it remains a major producer. However, as outlined earlier in this chapter, the United States will need more energy supply than it produces. This shortfall can be made up in only three ways: import more energy; improve energy efficiency even more than expected; and increase domestic energy supply. Coal, an abundant resource in the United States, can be used to increase the domestic energy supply even more than projected through the use of advanced utilization technologies (which are discussed in more detail in later chapters). The NEPD Group recognized this, so coal has a significant role in the U.S. energy picture.

The demand for electricity in the United States is projected to increase substantially from 3,659 to 4,705 billion kilowatt hours by 2030 to meet the country's demand [22]. Coal, nuclear energy, natural gas, and hydropower account for about 95 percent of total electricity generation, with oil and renewable energy contributing the remainder. Coal is used almost exclusively to generate electricity, with 100 gigawatts of new coal-fired generating capacity projected to be added by 2030 out of a total of 185 gigawatts projected to be added during this period. Coal-fired power plants account for about 51 percent of all U.S. electricity generation (refer to Figure 1.9). Coal is attractive because electricity generation costs are low and coal prices have proved stable.

Although coal is the most abundant fossil energy source in the United States, production and market issues affect the adequacy of supply. As highlighted in the energy policy, issues such as protection of public health, safety, property, and the environment can limit production and utilization of some coal resources and has lead, over the last several years, to using natural gas for power generation due to higher efficiencies, lower capital and operating costs, and fewer emissions. However, technological advances in cleaner coal technology have allowed for significant progress toward reducing these barriers, and coal, with its more stable prices than natural gas, must play a significant role in meeting the rising electricity demand in the United States.

Technology has been and will continue to be a key for the United States to achieve its energy, economic, and environmental goals. In recent years, technological advancements have led to substantial reductions in the cost of controlling sulfur dioxide and nitrogen oxide emissions [23]. The U.S. Department of Energy (DOE), through its Clean Coal Technology Program, has worked to provide effective control technologies. Clean Coal Technology is a category of technologies that allow for the use of coal to generate electricity while meeting environmental regulations at low cost. In the short term, the goal of the program is to meet existing and emerging environmental regulations, which will dramatically reduce compliance costs for controlling sulfur dioxide, nitrogen oxides, fine particulate matter, and mercury at new and existing coal-fired power plants. In the midterm, the goal of the program is to develop low-cost, superclean coal-fired power plants with efficiencies 50 percent higher than today's average [23]. The higher efficiencies will reduce emissions at minimal costs. In the long term, the goal of the program is to develop low-cost, zero-emission power plants with efficiencies close to double that of today's fleet (see Chapters 7 and 11). The long-term goals also include coproduction of fuels and chemicals along with electricity, which is leading to the development of technologies that may result in coal being a major source of hydrogen for the much publicized "hydrogen economy" that the United States is moving toward.

The NEPD Group recognized the importance of looking toward technology to help meet the goals of increasing electricity generation while protecting the environment. As a result, the NEPD Group had recommended that President Bush direct DOE to continue to develop advanced clean coal technology by doing the following [23]:

- Investing $2 billion over ten years to fund research in clean coal technologies
- Supporting a permanent extension of the existing R&D tax credit
- Directing agencies to explore regulatory approaches that will encourage advancements in environmental technology

The NEPD Group also recommended that George W. Bush direct federal agencies to provide greater regulatory certainty relating to coal electricity generation through clearer policies that could be more easily applied to business decisions. Some of the recommendations, which will be discussed in later chapters, were implemented, including discussions on the several rounds of the clean coal technologies being demonstrated and regulatory/environmental legislation that was proposed and, in some cases, enacted.

References

[1] E.S. Moore, Coal: Its Properties, Analysis, Classification, Geology, Extraction, Uses, and Distribution, John Wiley & Sons, 1922.

[2] M.A. Elliot (Ed.), Chemistry of Coal Utilization Second Supplementary Volume, John Wiley & Sons, 1981.

[3] *World Book Encyclopedia*, Coal, 4 World Book, Inc, vol. 4, 2001a, pp. 716–733.

[4] *World Book Encyclopedia*, Marco Polo, World Book, Inc, vol. 15, 2001b, pp. 648–649.

[5] Environmental Literacy Council, *www.enviroliteracy.org/article.php/18.html*, printed May 2003.

[6] J.G. Landels, Engineering in the Ancient World, University of California Press, 1978.

[7] U.C. Davis (University of California Davis), *www-geology.ucdavis.edu/-GEL115/115CH11coal.html*, printed May 2003.

[8] H.H. Schobert, Coal: The Energy Source of the Past and Future, American Chemical Society, 1987.

[9] EIA (Energy Information Administration), Coal Data: A Reference, U.S. Department of Energy, February 1995.

[10] P. Deane, The First Industrial Revolution, second ed., Cambridge University Press, 1979, pp. 78, 103–105.

[11] EIA, Annual Energy Review 2008, U.S. Department of Energy, June 2009.

[12] EIA, Annual Energy Review 2001, U.S. Department of Energy, November 2002.

[13] F. Freme, U.S. Coal Supply and Demand: 2008 Review, U.S. Department of Energy, 2009.

[14] National Mining Association, Most Requested Statistics—U.S. Coal Industry, National Mining Association, November 2008.

[15] P. Averitt, Coal Resources of the U.S., January 1, 1974, U.S. Geological Survey Bulletin No. 1412 (1975) (reprinted 1976), pp. 131.

[16] EIA, Annual Coal Report 2008, U.S. Department of Energy, September 2009.

[17] EIA, International Energy Outlook 2002, DOE (U.S. Department of Energy), March 2002.

[18] EIA, International Energy Annual 2006, U.S. Department of Energy, December 2008.

[19] EIA, International Energy Outlook 2009, U.S. Department of Energy, May 2008.

[20] M.C. Grimston, Coal as an Energy Source, IEA Coal Research, 1999.

[21] S.J. Mills, Coal in an Enlarged European Union, IEA Coal Research, 2004.

[22] EIA, Annual Energy Outlook 2008, U.S. Department of Energy, June 2008.

[23] NEPD (National Energy Policy Development) Group, National Energy Policy, U.S. Government Printing Office, May 2001.

2 The Chemical and Physical Characteristics of Coal

Chapter 1 provided an introduction on the use of coal, specifically its past use, current importance, and future role in the world economy. This chapter presents an introductory overview of the chemical and physical characteristics of coal, including a description of coal and how it is formed, coal resources and recoverable reserves in the world, U.S. coals and coalfields, the types and characteristics of coal, and coal classification systems relevant to commercial coal use. The purpose of this chapter is to provide the reader with basic coal information as a prelude to the subsequent chapters.

2.1 The Definition of Coal

Van Krevelen [1] offers an encompassing definition of coal:

> *Coal is a rock, a sediment, a conglomerate, a biological fossil, a complex colloidal system, an enigma in solid-state physics, and an intriguing object for chemical and physical analyses.*

In short, coal is a chemically and physically heterogeneous, "combustible" sedimentary rock that consists of both organic and inorganic material. Organically, coal consists primarily of carbon, hydrogen, and oxygen, with lesser amounts of sulfur and nitrogen. Inorganically, coal consists of a diverse range of ash-forming compounds distributed throughout the coal. The inorganic constituents can vary in concentration from several percentage points down to parts per billion of the coal.

2.2 Origin of Coal

Coal is found in deposits called seams that originated through the accumulation of vegetation that has undergone physical and chemical changes. These changes include decaying of the vegetation, deposition and burying by sedimentation, compaction, and transformation of the plant remains into the organic rock found today. Coals differ throughout the world in the kinds of plant materials deposited (type of coal), in the degree of metamorphism or coalification (rank of coal), and in the range of impurities included (grade of coal).

Clean Coal Engineering Technology. DOI: 10.1016/B978-1-85617-710-8.00002-9
Copyright © 2011 by Elsevier Inc. All rights of reproduction in any form reserved.

There are two main theories for the accumulation of the vegetal matter that creates coal seams [2]. The first theory, and the one that is most accepted because it explains the origin of most coals, is that the coal formed in situ—that is, where the vegetation grew and fell—and such a deposit is said to be autochthonous in origin. Most coal deposits began with thick peat bogs where the water was nearly stagnant and plant debris accumulated. Vegetation tended to grow for many generations, with plant material settling on the swamp bottom and being converted into peat by microbiological action. After some time, the swamps became submerged and were covered by sedimentary deposits, and a future coal seam was formed. When this cycle was repeated, over hundreds of thousands of years, additional coal seams were formed. These cycles of accumulation and deposition were followed by diagenetic (i.e., biological) and tectonic (i.e., geological) actions, and depending on the extent of temperature, time, and forces exerted, they formed the different ranks of coal observed today.

While the formation of most coals can be explained by the autochthonous process, some deposits are not easily explained by this model. Some coals appear to have been formed through the accumulation of vegetal matter that was transported by water. According to this theory (i.e., allochthonous origin) the fragments of plants were carried by streams and deposited on the bottom of the sea or lakes and built up strata, which later become compressed into solid rock.

Major coal deposits formed in every geological period since the Upper Carboniferous Period (350–270 million years ago), with the main coal-forming periods shown in Figure 2.1 [3]. The figure shows the relative ages of the world's major coal deposits. The considerable diversity of various coals is due to the differing climatic and botanical conditions that existed during the main coal-forming periods, along with subsequent geophysical actions.

2.3   Coalification

The geochemical process that transforms plant material into coal is called coalification and is often expressed as:

peat → lignite → subbituminous coal → bituminous coal → anthracite

This is a simplistic classification, and more elaborate systems have evolved and are discussed later in this chapter. Coalification can be described geochemically as consisting of three processes: the microbiological degradation of the cellulose of the initial plant material; the conversion of the lignin of the plants into humic substances; and the condensation of these humic substances into larger coal molecules [4]. The kind of decaying vegetation, the conditions of decay, the depositional environment, and the movements of the earth's crust are all important factors in determining the nature, quality, and relative position of the coal seams [1]. Of these, the physical forces exerted upon the deposits play the largest role in the coalification

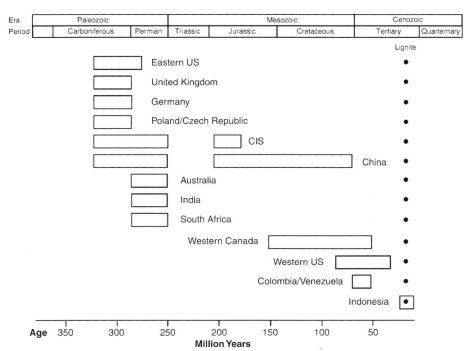

Figure 2.1 Comparison of the geological ages of the world's hard coal and lignite deposits. *Source:* From Walker (2000).

process. Variations in the chemical composition of the original plant material contributed to the variability in coal composition [1, 5]. The vegetation of various geologic periods differed both biologically and chemically. The conditions under which the vegetation decayed are important. The depth, temperature, degree of acidity, and the natural movement of water in the original swamp are important factors in the formation of the coal [1, 6].

The geochemical phase of the coalification process is the application of temperature and pressure over millions of years, and it is the most important factor of the coalification process. While there is some disagreement as to which is more important in promoting the chemical and physical changes—the high pressures exerted by massive overburdening strata or time–temperature factors—the changes are characterized physically by decreasing porosity and increasing gelification and vitrification [7].

The process of coalification involves several sequential reactions from the base material, the plant life, that form the basis of coal. Highly reactive matter (e.g., some extractives, hemicelluloses) can undergo low-grade oxidation or volatilization processes in the presence of heat and pressure. This material can "leave" the fuel matrix. The lignins, which are the "glues" that hold the biomass material together, remain in the fuel matrix, and they contain aromatic structures as single units. Under pressure and temperature, lignins and other remaining biomass matter undergo condensation to form peat and then the various coals. The condensation processes produce coals with fused aromatic rings ranging in number from two to six, depending on coal rank.

Coalification also involves changes in functional groups attached to the fused aromatic structures, or coal backbones. As rank increases, the number and type of functional groups decrease. For example, methoxy ($-OCH_3$) functional groups are commonly found in lignites and some subbituminous coals in significant concentrations. As rank increases, the concentration of such functionalities decreases; bituminous coals do not contain such functional groups. The characteristics of oxygen, which Van Krevelen and Schuyer [8] identified, are important considerations in understanding the chemical changes in coal as a function of rank caused by increasing coalification processes. In Table 2.1, note the higher total concentration of oxygen in the lower-rank coals and also the presence of oxygen in the highly reactive functional groups (O_{COOH}, O_{OCH3}, and O_{OH}). As rank (which is indicated by carbon content) increases, the concentration of oxygen in the coal decreases, and the presence of oxygen in the most reactive functionalities decreases to nearly zero.

Chemically, moisture and volatile matter (i.e., methane, carbon dioxide) content decrease, the percentage of carbon increases, the percentage of oxygen gradually decreases, and, ultimately, as the anthracitic stage is approached, the amount of hydrogen decreases [5, 7]. For example, carbon content (on a dry, mineral-matter free basis) increases from approximately 50 percent in herbaceous plants and wood to 60 percent in peat, 70 percent in lignite, 75 percent in subbituminous coal, 80 to 90 percent in bituminous coal, and over 90 percent in anthracite [5, 9, 10, 11]. This change in carbon content is known as carbonification. The coalification/carbonization process is shown in Figure 2.2, where some of the main chemical reactions that occur during coalification are listed [1].

Table 2.1 Oxygen Content by Functional Group

Carbon Content (%)	Oxygen Content by Functional Group (Wt %)				
	O_{COOH}	O_{OCH3}	O_{OH}	$O_{C=O}$	O_{NR}
65.5	8.0	1.1	7.2	1.9	9.6
70.5	5.1	0.4	7.8	1.1	8.2
75.5	0.6	0.3	7.5	1.4	6.4
81.5	0.3	0.0	6.1	0.5	4.2
85.5	0.05	0.0	5.6	0.5	1.75
87.0	0.0	0.0	3.2	0.6	1.3
88.6	0.0	0.0	1.9	0.25	0.85
90.3	0.0	0.0	0.5	0.2	2.2

Source: From Van Krevelen and Schuyer (1957).

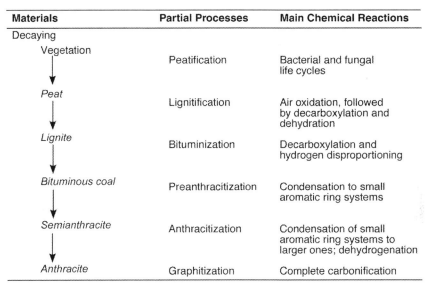

Materials	Partial Processes	Main Chemical Reactions
Decaying		
Vegetation	Peatification	Bacterial and fungal life cycles
↓		
Peat	Lignitification	Air oxidation, followed by decarboxylation and dehydration
↓		
Lignite	Bituminization	Decarboxylation and hydrogen disproportioning
↓		
Bituminous coal	Preanthracitization	Condensation to small aromatic ring systems
↓		
Semianthracite	Anthracitization	Condensation of small aromatic ring systems to larger ones; dehydrogenation
↓		
Anthracite	Graphitization	Complete carbonification

Figure 2.2 The coalification process.
Source: From Van Krevelen (1993).

2.4   The Classification of Coal

Efforts to classify coals began more than 180 years ago and were prompted by the need to establish some order in the confusion in identifying different coals. Two types of classification systems arose, one to aid scientific studies and one to assist coal producers and users. The scientific systems of classification are concerned with origin, composition, and fundamental properties of coals, while the commercial systems address trade and market issues, utilization, technological properties, and suitability for certain end uses. It is the latter classification systems that will be discussed here. Excellent discussions on scientific classifications are available elsewhere [1, 9].

2.4.1   Basic Coal Analysis

Before discussing the rank, type, grade, and classification systems of coal, there is a discussion of basic coal analyses, on which the classification schemes are based. These analyses do not yield any information on coal structure but do provide important information on coal behavior and are used in the marketing of coals.

Three analyses are used in classifying coal: two chemical analyses and one calorific determination. The chemical analyses include proximate and ultimate analysis. The proximate analysis gives the relative amounts of moisture, volatile matter, and ash (i.e., the inorganic material) that remain after all of the combustible matter has been burned off and, indirectly, the fixed carbon content of the coal. The ultimate analysis gives the amounts of carbon, hydrogen, nitrogen, sulfur, and oxygen comprising the coal. Oxygen is typically determined by subtracting the total percentages of carbon, hydrogen, nitrogen, and sulfur from 100 because of the complexity in determining oxygen directly. However, this technique does accumulate all of the errors in determining the other elements into the calculated value for oxygen. The third important analysis, the calorific value (also known as the heating value), is a measure of the amount of energy that a given quantity of coal will produce when burned.

Since moisture and mineral matter (or ash) are extraneous to the coal substance, analytical data can be expressed on several different bases to reflect the composition of as-received, air-dried, or fully water-saturated coal, or of dry, ash-free (daf) or dry, mineral matter–free (dmmf) coal. The most commonly used bases in the various classification schemes are shown in Figure 2.3 [12]. The most commonly used bases are defined as follows [1]:

- *as-received*—data are expressed as percentages of the coal with the moisture. This is also sometimes referred to as *as-fired* and is commonly used by the combustion engineer to monitor operations and for performing calculations, since it is the whole coal that is being utilized
- *dry basis* (db)—data are expressed as percentages of the coal after the moisture has been removed

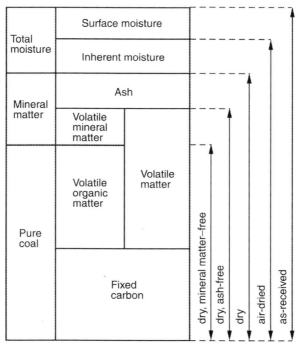

Figure 2.3 Relationship of different analytical bases to coal components.
Source: From Ward (1982).

- *dry, ash-free* (daf) basis—data are expressed as percentages of the coal with the moisture and ash removed
- *dry, mineral matter–free* (dmmf) basis—the coal is assumed to be free of both moisture and mineral matter, and the data are a measure of only the organic portion of the coal
- *moist, ash-free* (maf) basis—the coal is assumed to be free of ash but still contains moisture
- *moist, mineral matter–free* (mmmf) basis—the coal is assumed to be free of mineral matter but still contains moisture

2.4.2 The Ranks of Coal

The degree of coal maturation is known as the rank of coal, and it is an indication of the extent of metamorphism the coal has undergone. Because vitrinite precursors (see the next section for a discussion on maceral types) such as humates and humic acids were the major constituents in peat, the extent of metamorphism is often determined by noting the changes in the properties of vitrinite [7]. Some of the more important properties of vitrinite that did change with metamorphism are listed in Table 2.2 [7]. Metamorphism did affect the other macerals, but the relationship between the severity of metamorphism and the magnitude of change is different from those of vitrinite.

Table 2.2 Properties of Vitrinite Affected Progressively by Metamorphism

Increase with Increasing Rank
Reflectance (microscopic) and optical anisotropy
Carbon content
Aromaticity, $f_a = C_{aromatic}/C_{total}$
Condensed-ring fusion
Parallelization of molecular moieties
Heating value (a slight decrease at very high rank)

Decrease with Increasing Rank
Volatile matter (especially oxygenated compounds)
Oxygen content (especially as functional groups)
Oxidizability
Solubility (especially in aqueous alkalis and polar hydrocarbons)

Increase Initially to a Maximum, then Decrease
Hardness (minor increase at very high rank)
Plastic properties
Hydrogen content

Decrease Initially to a Minimum, then Increase
Surface area
Porosity (and moisture holding capacity)
Density (in helium)

Source: Adapted from Elliott (1981).

Several useful rank-defining properties are elemental carbon content, volatile matter content, moisture-holding capacity, heating value, and microscopic reflectance of vitrinite [7]. Figure 2.4 illustrates the relationship between rank and fixed carbon content [13], the one chemical property used most commonly to express coal rank. In the United States, lignites and subbituminous coals are referred to as being low in rank, while bituminous coals and anthracites are classified as high-rank coals. In Europe, low-rank coals are referred to as brown coal or lignite, and hard coals are anthracite and bituminous coals, with subbituminous coals included in certain cases as well. Figure 2.4 illustrates this relationship between rank and fixed carbon content [13]. The fixed carbon content shown in the figure is calculated on a dry, mineral matter–free basis. Figure 2.4 also shows the comparison between heating value and rank, with the heating value calculated on a moist, mineral matter–free basis. Note that the heating value increases with the increase in rank but begins to decrease with semianthractic and higher-rank coals. This decrease in heating

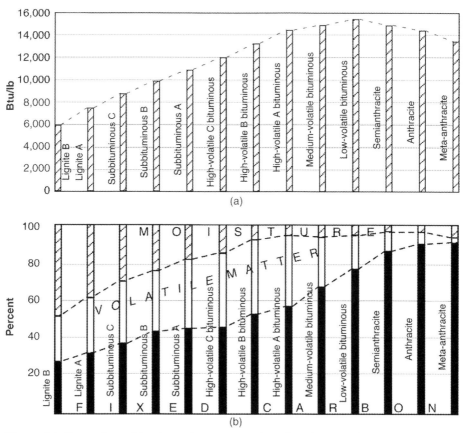

(a)

(b)

Figure 2.4 Comparison of heating values (on a moist, mineral matter–free basis) and proximate analyses of coals of different ranks.
Source: From Averitt (1975).

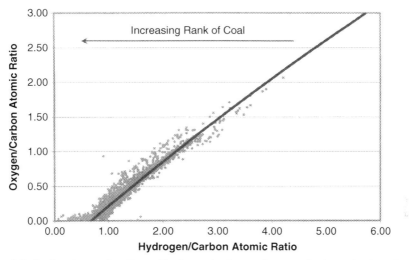

Figure 2.5 Coal rank as a function of hydrogen/carbon and oxygen/carbon atomic ratios. *Source:* From U.S. Geological Survey, U.S. Coal Resource Databases (2006).

value is a result of the significant decrease in volatile matter, which is also shown in Figure 2.4 [13].

Another example of the coal composition/rank relationship is shown in Figure 2.5, which is a type of coalification diagram based on an analysis of the U.S. Geological Survey coal database [14]. In general, as the coal rank increases, the O/C and H/C atomic ratios decrease.

Porosity decreases with an increasing level of metamorphism, thereby reducing the moisture-holding capacity of the coal. Moisture-holding capacity is also affected by the functional group characteristics, and in coals where cationic elements replace protons on acid functional groups, moisture-holding capacity is reduced [7].

Vitrinite reflectance is another import rank-measuring parameter. The advantage of this technique is that it measures a rank-sensitive property on only one petrographic constituent; therefore, it is applicable even when the coal type is atypical [7].

2.4.3 The Types of Coal

As mentioned previously, coal is composed of macerals, discrete minerals, inorganic elements held molecularly by the organic matter, and water and gases contained in submicroscopic pores. Macerals are organic substances derived from plant tissues that have been incorporated into the sedimentary strata, subjected to decay, compacted, and chemically altered by geological processes. This organic matter is extremely heterogeneous, and a classification system has been developed to characterize it [7, 15, 16]. Classifying the coal, known as petrography, was primarily used to characterize and correlate coal seams and resolve questions about coal diagenesis and metamorphism, but today it is an important tool for assessing coals for industrial applications [17, 18].

All macerals are classified into three maceral groups—vitrinite, liptinite (sometimes also referred to as exinite), and intertinite—and they are characterized by their appearance, chemical composition, and optical properties. Each maceral group includes a number of macerals and other subcategories, but only the preceding three maceral groups are introduced here; more extensive discussions of petrography can be found elsewhere [7, 9, 15, 16].

In most cases, the constituents in the coal can be traced back to specific components of the plant debris from which the coal formed [7, 9, 15, 16]. Figure 2.6 is a simplified overview of these various substances that accumulated as peat deposited and the components they represent in coal [7]. Adding to the complexity of the source materials were inorganic substances (also shown in Figure 2.6) that entered the environment as mineral grains and dissolved ions. Many of the dissolved ions either combined with the organic fraction or were precipitated in place to form discrete mineral grains [7].

Vitrinite group macerals are derived from the humification of woody tissues and can either possess remnant cell structures or be structureless [6]. Vitrinite contains more oxygen than the other macerals at any given rank level, and they are characterized by a higher aromatic fraction.

Liptinite group macerals are not derived from humifiable materials but rather from relatively hydrogen-rich plant remains such as resins, spores, cuticles, waxes, fats, and algal remains, which are fairly resistant to bacterial and fungal decay [9, 17]. Liptinites are distinguishable by a higher aliphatic (i.e., paraffin) fraction and a correspondingly higher hydrogen content, especially at lower ranks [15].

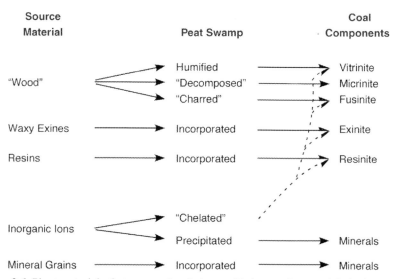

Figure 2.6 Plant materials that accumulated along with inorganic materials in the peat swamp retain their identity as distinctive macerals in the coal.
Source: Modified from Elliott (1981).

The inertinite group macerals were derived mostly from woody tissues, plant degradation products, or fungal remains. While they were derived from the same original plant substances as vitrinite and liptinite, they have experienced a different primary transformation [15]. Inertinite group macerals are characterized by a high carbon content that resulted from thermal or biological oxidation, as well as a low hydrogen content and an increased level of aromatization [8, 15].

Although petrographic analysis has many uses, it was initially used primarily to characterize and correlate seams and resolve questions about coal diagenesis and metamorphism. Later, it influenced developments in coal preparation (i.e., crushing, grinding, and removal of mineral constituents) and conversion technologies [9]. Industrially, petrographic analysis can provide insight into the hardness of a coal (i.e., its mechanical strength) as well as a coal's thermoplastic properties, which is of significant importance in the coking industry.

2.4.4   The Grades of Coal

The grades of coal refer to the amount of mineral matter that is present in the coal and are a measure of coal quality. Sulfur content, ash fusion temperatures (i.e., the measure of the behavior of ash at high temperatures), and the quantity of trace elements in coal are also used to grade coal. Formal classification systems have not been developed for the grades of coal, but grade is still important to the coal user.

Mineral matter may occur as finely dispersed throughout the coal or in discrete partings in the coal. Some of the inorganic matter and trace elements are derived from the original vegetation, but the majority is introduced during coalification by wind or water to the peat swamp or through movement of solutions in cracks, fissures, and cavities [19]. Coal mineralogy can affect the ability to remove minerals during coal preparation/cleaning, coal combustion and conversion (i.e., the production of liquid fuels or chemicals) characteristics, and metallurgical coke properties.

2.4.5   Classification Systems

An excellent discussion of the many classification systems, scientific as well as commercial, is provided by van Krevelen [1]. The commercial systems that are discussed here consist of two primary systems: the ASTM (American Society of Testing Materials) system used in the United States and North America and an international ECE (Economic Commission for Europe) codification system developed in Europe. It is interesting to note that in all countries the classification systems used commercially are primarily based on the content of volatile matter [1]. In some countries, a second parameter is also used, and in the United States, for example, this is the heating value (see Figure 2.4). In many European countries this parameter is either the caking or the coking property. Caking coals are coals that pass through a plastic state upon heating in which they soften, swell, and resolidify into a coherent carbonaceous matrix. Noncaking coals, on the other hand, do not become plastic when heated and produce a weakly coherent char residue. Coking coals are strongly caking coals that exhibit characteristics making them suitable for conversion into metallurgical and other industrial cokes [9].

The ASTM Classification System

The ASTM classification system (ASTM D388) distinguishes among four coal classes, each of which is subdivided into several groups, as shown in Table 2.3. As previously mentioned, high-rank coals (i.e., medium volatile bituminous coals or those of higher rank) are classified based on their fixed carbon and volatile matter contents (expressed on a dmmf basis), while low-rank coals are classified in terms of their heating value (expressed on a mmmf basis).

This classification system was developed for commercial applications, but it has proven to be satisfactory for certain scientific uses as well [7]. For example, if a given coal is described as being a certain rank, then an estimate of some

Table 2.3 ASTM Coal Classification by Rank

Class/Group	Fixed Carbon[a] (%)	Volatile Matter[b] (%)	Heating Value[b] (Btu/lb)
Anthracitic			
Metaanthracite	>98	<2	
Anthracite	92–98	2–8	
Semianthracite	86–92	8–14	
Bituminous			
Low volatile	78–86	14–22	
Medium volatile	69–78	22–31	
High volatile A	<69	>31	>14,000
High volatile B			13,000–14,000
High volatile C			10,500–13,000[c]
Subbituminous			
Subbituminous A			10,500–11,500[c]
Subbituminous B			9,500–10,500
Subbituminous C			8,300–9,500
Lignitic			
Lignite A			6,300–8,300
Lignite B			<6,300

[a]Calculated on dry, mineral matter-free coal; correction from ash to mineral matter is made by means of the Parr formula: mineral matter = 1.08[percent ash + 0.55(percent sulfur)]. Ash and sulfur are on a dry basis.
[b]Calculated on mineral matter-free coal with bed moisture content.
[c]Coals with heating values between 10,500 and 11,500 Btu/lb are classified as high volatile C bituminous if they possess caking properties or as subbituminous A if they do not.
Source: From Berkowitz (1979).

properties can be made, and if the coal is classified as subbituminous/lignitic or anthracitic, then it would not be considered for certain applications such as for coke production.

International Classification/Codification Systems

Because of the increasing amount of coal trade in the world, the ECE Coal Committee developed a new classification system in 1988 for higher-rank coals [1]. The original international system had deficiencies in that it was primarily developed for trading Northern Hemisphere coals, which have had distinctly different characteristics than those from the Southern Hemisphere (e.g., Australia and South Africa). As trade of Southern Hemisphere coals increased, it became apparent that a new classification system was needed. This new system, which in reality is a system of codes, is better known as a codification system. The codification system for hard coals, combined with the ISO (International Organization for Standardization) codification of brown coals and lignites (which was established in 1974), provides a complete codification for coals in the international trade.

The ISO codification of brown coals and lignites is given in Table 2.4 [1]. Total moisture content of run-of-mine coal and tar yield (i.e., determining the yields of tar, water, gas, and coke residue by low-temperature distillation) are the two parameters coded.

The ECE international codification of higher-rank coals is much more complicated and is listed in Table 2.5. Eight basic parameters define the main properties of the coal, which are represented by a 14-digit code number. The codification, which is for medium- and high-rank coals; blends and single coals; raw and washed coals; and all end-use applications, is commercial, and includes petrographic, rank, grade, and environmental information [1]. The major drawback of this system is that it is very complicated.

Table 2.4 Codification of Brown Coals and Lignites

Parameter	Total Moisture Content (run-of-mine coal)		Tar Yield (dry, ash-free)	
Digit	*1*		*2*	
Coding	Code	Weight %	Code	Weight %
	1	≤20	0	≤10
	2	>20–30	1	>10–15
	3	>30–40		
	4	>40–50	2	>15–20
	5	>50–60	3	>20–25
	6	>60	4	>25

Source: From Van Krevelen (1993).

Petrographic Tests

Table 2.5 International Codification of Higher-Rank Coals[a]

Parameter	Vitrinite Reflectance (mean random)		Characteristics of Reflectogram[b]			Maceral Group Composition (mmf)			
						Inertinite[c]		Liptinite	
Digit	1, 2		3			4		5	
Coding	Code	R$_{random}$ %	Code	Standard Deviation	Type	Code	Volume %	Code	Volume %
	02	0.2–0.29	0	≤1	No gap — Seam coal	0	0–<10	1	0–<5
	03	0.3–0.39	1	>0.1≤0.2	No gap — Simple blend	1	10–<20	2	5–<10
	04	0.4–0.49	2	>0.2	No gap — Complex blend	2	20–<30	3	10–<15
	—	—	3		1 gap — Blend with 1 gap	—	—	—	—
	48	4.8–4.89	4		2 gaps — Blend with 2 gaps	7	70–<80	7	30–<35
	49	4.9–4.99	5		>2 gaps — Blend with >2 gaps	8	80–<90	8	35–<40
	50	≥ 5.0				9	≥90	9	≥40

Technological Tests

Parameter	Crucible Swelling No.		Volatile Matter[d], daf		Ash, dry		Total Sulfur, dry		Gross Calorific Value, daf	
Digit	6		7, 8		9, 10		11, 12		13, 14	
Coding	Code	Number	Code	wt.%	Code	wt.%	Code	wt.%	Code	MJ/kg
	0	0–0.5	48	≥48	00	0–<1	00	0–<0.1	21	<22
	1	1–1.5	46	46–<48	01	1–<2	01	0.1–<0.2	22	22–<23
	2	2–2.5	44	44–<46	02	2–<3	02	0.2–<0.3	23	23–<24
	—	—	—	—	—	—	—	—	—	—
	7	7–7.5	12	12–<14	20	20–<21	29	2.9–<3.0	37	37–<38
	8	8–8.5	10	10–<12	—	—	30	3.0–<3.1	38	38–<39
	9	9–9.5	09	9–<10			—	—	39	≥39
			—	—						
			02	2–<3						
			01	1–<2						

[a]Higher-rank coals are coals with gross calorific value (maf) ≥24 MJ/kg and those with gross calorific value (maf) <24 MJ/kg provided mean random vitrinitic reflectance ≥0.6%. To convert from MJ/kg to Btu/lb, multiply by 429.23.

[b]A reflectogram as characterized by code number 2 can also result from a high-rank seam coal.

[c]It should be noted that some of the intertinite may be reactive.

[d]Where the ash content of the coal is more than 10 percent, it must be reduced, before analysis, to below 10 percent by dense medium separation. In these cases, the cutting density and resulting ash content should be noted.

Source: From Van Krevelen (1993).

REFERENCES

[1] D.W. Van Krevelen, Coal: Typology–Physics–Chemistry–Constitution, third edition, Elsevier Science, 1993.

[2] E.S. Moore, Coal: Its Properties, Analysis, Classification, Geology, Extraction, Uses, and Distribution, John Wiley & Sons, 1922.

[3] S. Walker, Major Coalfields of the World, IEA Coal Research, 2000.

[4] J.H. Tatsch, Coal Deposits: Origin, Evolution, and Present Characteristics, Tatsch Associates, 1980.

[5] H.H. Schobert, Coal: The Energy Source of the Past and Future, American Chemical Society, 1987.

[6] G. Mitchell, Basics of Coal and Coal Characteristics, Iron & Steel Society, Selecting Coals for Quality Coke Short Course, 1997.

[7] M.A. Elliott (Ed.), Chemistry of Coal Utilization, Second Supplementary Volume, John Wiley & Sons, 1981.

[8] D.W. Van Krevelen, J. Schuyer, Coal Science: Aspects of Coal Constitution, Elsevier Science, 1957.

[9] N. Berkowitz, An Introduction to Coal Technology, Academic Press, 1979.

[10] J.G. Singer (Ed.), Combustion: Fossil Power Systems, Combustion Engineering, Inc., 1981.

[11] B.G. Miller, S. Falcone Miller, R. Cooper, J. Gaudlip, M. Lapinsky, R. McLaren, et al., Feasibility Analysis for Installing a Circulating Fluidized Bed Boiler for Cofiring Multiple Biofuels and Other Wastes with Coal at Penn State University, U.S. Department of Energy, National Energy Technology Laboratory, DE-FG26-00NT40809, 2003, Appendix J.

[12] C.R. Ward (Ed.), Coal Geology and Coal Technology, Blackwell, 1984, p. 66.

[13] P. Averitt, Coal Resources of the U.S., January 1, 1974, U.S. Geological Survey Bulletin No. 1412, 1975 (reprinted 1976), p. 131.

[14] U.S. Geological Survey, U.S. Coal Resource Databases (USCOAL), (2006).

[15] G.H. Taylor, M. Teichmüller, A. Davis, C.F.K. Diessel, R. Littke, P. Robert, Organic Petrology, Bebrüder Borntraeger, 1998.

[16] R.M. Bustin, A.R. Cameron, D.A. Grieve, W.D. Kalkreuth, Coal Petrology Its Principles, Methods, and Applications, Geological Association of Canada, 1983.

[17] I. Suárez-Ruiz, J.C. Crelling (Eds.), Applied Coal Petrology: The Role of Petrology in Coal Utilization, Elsevier, 2008.

[18] B.G. Miller, D.A. Tillman (Eds.), Combustion Engineering Issues for Solid Fuel Systems, Academic Press, 2005.

[19] M.T. Mackowsky, in: D. Murchson, T.S. Westoll (Eds.), Mineral Matter in Coal: in Coal and Coal-Bearing Strata, Oliver & Boyd, Ltd., 1968, pp. 309–321.

3 The Worldwide Distribution of Coal

Coal is the dominant fuel source in the entire world. Current production and consumption, along with projected usage, was discussed in Chapter 1. This chapter presents information on coal resources and recoverable reserves in the world, with an emphasis on coals and coalfields in the United States, to illustrate the vast resources of this fuel source. By the end of 2007, recoverable coal reserves in the United States, which contains the world's largest coal reserves, totaled 264 billion short tons [1] compared to a total world reserve of 930 billion short tons (as of 2006) [2]. On an oil-equivalent basis, there is more than twice as much recoverable coal in the world as there is oil and natural gas combined [3]. Consequently, coal has been and will continue to be a major economic/energy resource.

3.1 Coal Distribution and Resources

Coal deposits are broadly categorized into resources and reserves. Resources refer to the quantity of coal that may be present in a deposit or coalfield but may not take into account the feasibility of mining the coal economically. Reserves generally tend to be classified as proven or measured and probable or indicated, depending on the level of exploration of the coalfield. The basis for computing resources and reserves varies among countries, which makes it difficult to do direct comparisons. In addition, the techniques are constantly being refined, thereby resulting in variability from year to year. Walker [4] discusses some of the various measurement criteria used by the major coal-producing countries in the world in detail. Similarly, EIA discusses measurement criteria as well [2].

Figure 3.1 illustrates the relationship between coal resources (as of 1997) and reserves (as of 2007) in the United States [1, 5]. The United States has a total of nearly 4,000 billion short tons of coal resources, with approximately 19 billion short tons classified as recoverable reserves at active mines out of 263 billion short tons that are economically recoverable.

The definitions used in Figure 3.1 are as follows:

- *Total resources*—coal that can currently or potentially in the future be extracted economically
- *Measured resources*—the quantity of coal that has been determined to a high degree of geologic assurance
- *Indicated resources*—the quantity of coal that has been determined to a moderate degree of geological assurance

Clean Coal Engineering Technology. DOI: 10.1016/B978-1-85617-710-8.00003-0
Copyright © 2011 by Elsevier Inc. All rights of reproduction in any form reserved.

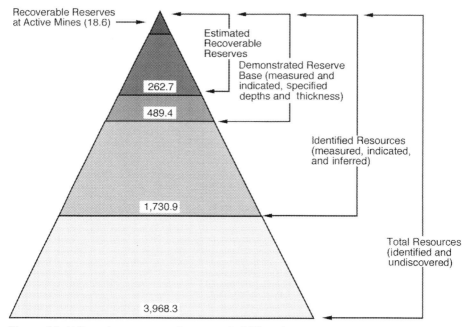

Figure 3.1 U.S. coal resources and reserves in billion short tons.
Source: From EIA, Annual Energy Review 2007 (2008), and EIA, U.S. Coal Reserves: 1997 Update (1999).

- *Inferred resources*—the quantity of coal that has been determined with a low degree of geologic assurance
- *Recoverable reserves*—coal that can be recovered economically with technology that is currently available or in the foreseeable future

Terminology also varies among countries and can contribute to confusion when comparing coal resources and reserves. For purposes of discussion in this section, recoverable coal reserves will primarily be used when comparing world coal deposits to lessen confusion.

3.1.1   Coal Reserves throughout the World

Coal is the most abundant fossil fuel in the world. BP [2008] reports that as of 2008, oil reserves were 186 gigatons (Gt) (converted to short tons from metric tons), representing a reserve-to-production (R/P) ratio of 42 years, while natural gas reserves were 175 gigatons of oil equivalent (Gtoe), with a R/P ratio of 60 years. Coal was calculated to have reserves of approximately 444 Gtoe (based on a hard coal:brown coal/lignite ratio of 0.6:0.4) and a R/P ratio of 147 years—roughly 45 percent more than oil and natural gas combined. Coal reserves are also more widely distributed throughout the world, as shown in Figure 3.2. All major regions of the world contain appreciable quantities of coal except for the Middle East, which is

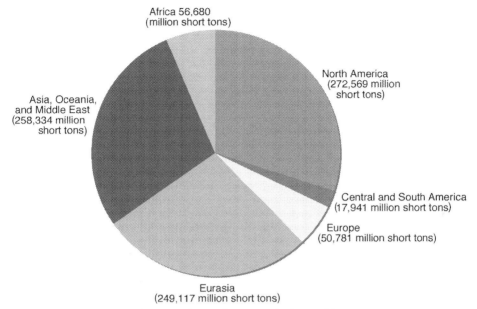

Figure 3.2 Distribution of recoverable coal reserves in the world.
Source: From Energy Information Administration, Annual Energy Review 2007.

not shown separately on Figure 3.2 because only Iran has coal reserves, totaling 462 million short tons, which is only 0.04 percent of the world total. The Middle East, on the other hand, contains almost two-thirds of the world oil reserves and over 41 percent of the natural gas reserves [3].

Coal is available in almost every country worldwide, with recoverable reserves in around 70 countries. Total recoverable reserves of coal around the world are estimated at more than 930 billion short tons [1]. According to the Energy Information Administration [1], this is enough coal to last approximately 150 years at current consumption levels. However, this could be extended still further by the discovery of new reserves through ongoing and improved exploration activities and by advances in mining techniques that will allow previously inaccessible reserves to be reached.

Historically, estimates of world recoverable coal reserves have declined from 1,174 billion short tons in 1990 to 1,001 billion short tons in 2003 to 930 billion short tons in 2007 [2, 6]. Recent assessments of world coal reserves include a significant downward adjustment for Germany, from about 73 billion short tons of recoverable coal reserves to 7 billion short tons. The reassessment reflects more restrictive criteria for various parameters. This downward trend is also observed in other countries as well.

A detailed breakdown of EIA's estimated recoverable world coal reserves of 930 billion short tons is provided in Table 3.1 [1]. This table classifies the recoverable coal reserves in two major categories: recoverable anthracite and bituminous coal (i.e., hard coal) and recoverable lignite and subbituminous coal—for the major regions and countries of the world.

Table 3.1 World Estimated Recoverable Coal Reserves (million short tons)

Region/Country	Anthracite and Bituminous Coal	Lignite and Subbituminous Coal	Total
North America			
Canada	3,826	3,425	7,251
Greenland	0	202	202
Mexico	948	387	1,335
United States	122,001	141,780	263,781
Total	**126,776**	**141,780**	**272,569**
Central and South America			
Brazil	0	7,791	7,791
Chile	34	1,268	1,302
Colombia	7,251	420	7,671
Peru	154	0	154
Other	529	494	1,023
Total	**7,969**	**9,973**	**17,941**
Europe[a]			
Bulgaria	6	2,195	2,200
Czech Republic	1,844	3,117	4,962
Former Serbia and Montenegro	7	15,299	15,306
Germany	168	7,227	7,394
Greece	0	4,299	4,299
Hungary	219	3,420	3,640
Poland	6,627	1,642	8,270
Romania	13	452	465
Turkey	0	2,000	2,000
United Kingdom	171	0	171
Other	241	1,834	2,076
Total	**9,296**	**41,485**	**50,781**
Eurasia[b]			
Kazakhstan	31,052	3,450	34,502
Russia	54,110	118,964	173,074
Ukraine	16,922	20,417	37,339
Uzbekistan	1,102	2,205	3,307
Other	0	895	895
Total	**103,186**	**145,931**	**249,117**
Africa			
Botswana	44	0	44
South Africa	52,911	0	52,911

Table 3.1 *Cont'd*

Region/Country	Anthracite and Bituminous Coal	Lignite and Subbituminous Coal	Total
Zimbabwe	553	0	553
Other	980	192	1,172
Total	**54,488**	**192**	**54,680**
Middle East, Asia, and Oceania[a]			
Australia	40,896	43,541	84,437
China	68,564	57,651	126,215
India	57,585	4,694	62,278
Indonesia	1,897	2,874	4,771
Korea, North	331	331	661
Pakistan	1	2,184	2,185
Thailand	0	1,493	1,493
Other	2,249	1,046	3,295
Total	**171,522**	**113,813**	**285,334**
World Total	**473,236**	**457,186**	**930,423**

[a]Excludes countries that were part of the former U.S.S.R.
[b]Includes only countries that were part of the former U.S.S.R.

Source: From Energy Information Administration, Annual Energy Review 2007.

Although coal deposits are widely distributed, 82 percent of the world's recoverable reserves are located in six countries: the United States (~28 percent; ~264 billion short tons), Russia (~19 percent; ~172 billion short tons), China (13 percent; 126 billion short tons), Australia (~9 percent; 84 billion short tons), India (~7 percent; ~62 billion short tons), and South Africa (~6 percent; ~53 billion short tons). Figure 3.3 shows the ten countries that have the largest recoverable coal reserves. Approximately 70 countries contain recoverable coal, but those in the figure contain more than 856 billion short tons, or more than 92 percent of the world's total.

Table 3.2 lists the major coal-producing countries/regions and their reserves-to-production (R/P) ratio in years. These data are based on recoverable reserves and production from IEA [2] as of 2009, and they differ slightly from BP data [3] (e.g., BP reports a world R/P ratio of 147 years, and IEA reports a world R/P ratio of 143 years). The IEA data are reported here because they provide a clearer breakdown of coal rank (i.e., subdividing subbituminous coal and lignite), whereas BP combines the two lower-rank coals into one category. The data are comparable between the two data sets, but what is especially worth noting is that the R/P ratio varies from 30 to 50 years for some countries/regions with heavy usage and/or low reserves to more than 1,000 years for countries with little or no production. Most major coal-producing countries are in the almost 100- to 250-year range, except for China. Note that China, which is a country with the third largest recoverable reserves in the world, has an R/P ratio of 52 years. This is low compared to

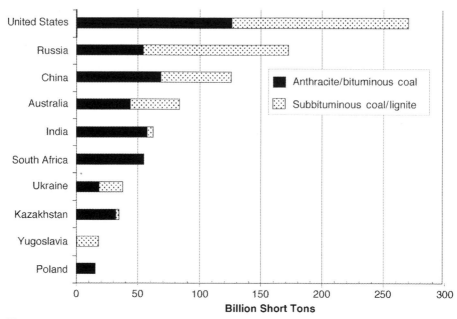

Figure 3.3 Countries with the largest recoverable coal reserves.

countries such as the United States (233 years) or Russia (540 years), which are the two leading countries with recoverable reserves.

3.2   Major Coal-Producing Regions in the World

Coal is found on all inhabited continents of the world. It is very likely that coal is also on Antarctica, particularly when one looks at the coal-forming periods in history and the corresponding locations of the present-day continents. A review of the major coal-producing countries in the world, summarized by coal-producing region, follows.

3.2.1   North America

The recoverable coal reserves of North America are the second largest in the world, with more than 272,000 million short tons identified (see Table 3.1). Coal is found in the United States, Canada, Mexico, and, to a much lesser extent, Greenland.

United States

The coal reserves of the United States are the largest of any country in the world, with about 263,000 to 264,000 million short tons [1, 7]. Recoverable coal reserves are found in 32 of the states, with the major coalfields shown in Figure 1.16. The 11 states with the largest recoverable coal reserves are listed in Table 3.3, and they contain approximately 91 percent of the total coal in the United States [7].

Table 3.2 World Recoverable Coal Reserves (billion short tons) and Reserves-to-Production Ratio (years)

	Bituminous Coal and Anthracite	Subbituminous Coal	Lignite	Total	2005 Production	R/P Ratio (years)
2006–2008 Recoverable Reserves by Coal Rank						
World Total	**471.8**	**293.6**	**165.0**	**930.4**	**6.5**	**143**
United States	120.6	109.8	33.4	263.8	1.1	233
Russia	54.1	107.4	11.5	173.1	0.3	540
China	68.6	37.1	20.5	126.2	2.4	52
Other Non-OECD Europe and Eurasia	**49.1**	**19.0**	**27.3**	**95.3**	**0.3**	**307**
Australia and New Zealand	**40.9**	**2.5**	**41.6**	**85.1**	**0.4**	**203**
India	57.6	0.0	4.7	62.3	0.5	132
Africa	54.5	0.2	0.0	54.7	0.3	196
OECD Europe	9.3	3.4	19.0	31.7	0.7	47
Other, Central and South America	8.0	2.2	0.0	10.2	0.1	138
Other Non-OECD Asia	**3.8**	**2.7**	**4.5**	**9.7**	**0.3**	**34**
Brazil	0.0	7.8	0.0	7.8	<0.1	1,131
Canada	3.8	1.0	2.5	7.3	0.1	101
Other	2.9	0.5	0.1	3.4	<0.1	207

Source: From Energy Information Administration, Annual Energy Review 2007.

The top five states contain more than 70 percent of the total recoverable coal reserves in the United States. After New Mexico and North Dakota, which each have recoverable reserves of about 6,900 million short tons, the state with the next largest reserve total is Missouri, with almost 3,800 million short tons. After that, recoverable reserve totals for the other states are less than 3,000 million short tons per state.

Of the four ranks of coal in the United States—that is, anthracite, bituminous, subbituminous, and lignite—bituminous coal accounts for 53 percent of demonstrated reserve base (see Figure 3.1). Bituminous coal is concentrated primarily east of the Mississippi River, with the greatest amounts in Illinois, West Virginia, Kentucky, Pennsylvania, and Ohio (see Table 3.3). All subbituminous coal, which accounts for 37 percent of the demonstrated reserve base, is located west of the Mississippi River and is concentrated in Montana and Wyoming. Lignite accounts for about 9 percent of the demonstrated reserve base and is found mostly in

Table 3.3 Top 11 States with the Largest Recoverable Coal Reserves
(million short tons), 2007

State	Underground Mineable Coal	Surface Mineable Coal	Total
Montana	35,922	38,934	74,856
Wyoming	22,946	16,728	39,674
Illinois	27,893	10,064	37,957
West Virginia	15,395	2,274	17,669
Kentucky	7,265	7,417	14,682
Pennsylvania	10,595	1,026	11,621
Ohio	7,692	3,755	11,447
Colorado	5,946	3,746	9,692
Texas	—	9,449	9,449
New Mexico	2,788	4,156	6,944
North Dakota	—	6,849	6,849
Total	**136,442**	**104,398**	**240,840**
Percentage of U.S. Total	**91.3**	**92.5**	**91.2**

Source: From Energy Information Administration, Coal Reserves Current and Back Issues (2009).

Montana, Texas, and North Dakota. Anthracite only accounts for about 1.5 percent of the demonstrated reserve base, and is found only in Pennsylvania.

Estimated low-sulfur coal comprises the largest portion of the total recoverable coal reserves at 36 percent [5]. Low-sulfur coal is defined as less than 0.8 and 0.5 percent by weight (as received) sulfur for high-grade bituminous coal and high-grade lignite, respectively. These sulfur contents are a quantitative rating and have been correlated with U.S. sulfur emissions regulations from coal-fired power plants and the various stages of control that are required [5]. Estimated medium (0.8–2.2 percent for bituminous coal and 0.5–1.3 percent for lignite) and high (>2.2 and >1.3 percent for bituminous coal and lignite, respectively) sulfur recoverable reserves account for 31 and 33 percent of the total, respectively.

The U.S. Geological Survey has divided the reserves into seven provinces: Eastern Province, Interior Province, Gulf Province, Northern Great Plains Province, Rocky Mountain Province, Pacific Coast Province, and Alaskan Province. The provinces are further subdivided into regions, fields, and districts. Carboniferous coal deposits in the eastern United States occur in a band of coal-bearing sediments that include the Appalachian and Illinois basins. Coal deposits in the western United States range from Upper Jurassic to Tertiary in age.

The Eastern Province includes the anthracite regions of Pennsylvania and Rhode Island, the Atlantic Coast region of middle Virginia and North Carolina, the vast Appalachian basin that extends from Pennsylvania to eastern Ohio, western Kentucky, West Virginia, western Virginia, Tennessee, and Alabama. The Eastern Province is about 900 miles long and 200 miles wide at is broadest point [8]. This province also contains the greatest reserves of anthracite in the United States, with more than 760 million short tons in eastern Pennsylvania.

The Appalachian basin contains the largest deposits of bituminous coal in the United States. In the northern region of the Appalachian basin the coal rank ranges from high volatile bituminous in the west to low-volatile bituminous coal in the east. In the central region of the basin, the coal includes low- to high-volatile bituminous ranks. In the southern region, the coals are mainly of high-volatile bituminous rank, with some medium- and low-volatile bituminous coals [4]. Coals are used for steam production, electricity generation, and metallurgical coke production. These coals have high heating values, low to medium ash content (up to 20 percent), and variable sulfur content, with much of the coal that contains sulfur in the 2 to 4 percent range.

The Interior Province is subdivided into three regions: the Northern region consists of Michigan; the Eastern region or Illinois basin consists of Illinois, southern Indiana, and western Kentucky; and the Western region consists of Iowa, Missouri, Nebraska, Kansas, Oklahoma, Arkansas, and western Texas. The Eastern region is the most important region of this province, with vast reserves in Illinois (i.e., nearly 38,000 million short tons) and western Kentucky (nearly 9,000 million short tons of the approximate 15,000 million short tons listed in Table 3.3). The coal in the Interior Province is mainly bituminous in rank and tends to be lower in rank and higher in sulfur than the Eastern Province bituminous coals. Coals are used for steam production, electricity generation, and metallurgical coke production. Coal composition in this province is quite variable, with coals from the Illinois basin noted for having high-sulfur content (3 to 7 percent). The ash content is variable.

The Gulf Province consists of the Mississippi region in the east and the Texas region in the west. The coals in this province, which extends from Alabama through Mississippi, Louisiana, into Texas, are lignitic in rank and are the lowest-rank coals in the United States, with moisture contents of up to 40 percent.

The Northern Great Plains Province contains the large lignite deposits of North and South Dakota and eastern Montana, along with the subbituminous fields of northern and eastern Montana and northern Wyoming. These lignite deposits are contained in the Fort Union Region and are the largest lignite deposits in the world [8]. The coals are used primarily as power station fuels. The lignite contains high moisture (38 percent), low ash (6 percent), and medium sulfur (<1 percent) contents and a heating value of approximately 6,800 Btu/lb.

The Northern Great Plains Province also contains extensive subbituminous coal reserves from the Powder River basin. Wyoming and Montana have the largest recoverable coal reserves in the United States. Wyoming's coal reserves are split between the Northern Great Plains Province and the Rocky Mountain Province. The Powder River basin coals are used primarily as power station fuels and average about 1 percent sulfur with generally low ash content (3–10 percent).

The Rocky Mountain Province includes the coalfields of the mountainous districts of Montana, Wyoming, Utah, Colorado, and New Mexico. The coals range in rank from lignite through anthracite in this province. The most important Rocky Mountain Province coals are the coals from Wyoming, primarily those from the Green River, Hanna, and Hanna Fork coalfields. These coals are subbituminous in rank, typically contain low sulfur, and are used in power generation stations.

The Pacific Coast Province is limited to small deposits in Washington, Oregon, and California. The coals range in rank from lignite to anthracite. The fields are small and scattered and are not being utilized to any great extent.

The Alaskan Province contains coal in several regions [9]. These coals vary in rank from lignite to bituminous, with a small amount of anthracite. The total reserves are estimated to be 15 percent bituminous coal and 85 percent subbituminous coal and lignite. However, extensive mining is not performed due to the low population density and pristine wilderness environment. Only fields close to main lines of transportation have been developed. The coals are used primarily for steam generation and as power station fuels.

Canada

Canada has about 7,300 million short tons of recoverable coal ranging in rank from anthracite to lignite. The coal deposits formed in late Jurassic, Cretaceous, and early Tertiary times. Most of the recoverable reserves are in British Columbia, Alberta, and Saskatchewan, which is an extension of the Great Plains Province coals from the United States. Coals from western Canada tend to be low in sulfur, with those from Alberta and Saskatchewan used as power station fuels, while British Columbia metallurgical coal is exported to the Far East.

Coals from eastern Canada, primarily the Cape Breton Island coalfield in Nova Scotia, are the most important in the Atlantic region. The coals are of high-volatile bituminous rank and vary from medium to high sulfur. Coal production in Nova Scotia is a small percentage of the national output and is expected to decline further [4].

3.2.2 Eurasia

Eastern Europe and the FSU contain extensive recoverable coal reserves totaling over 279,000 million short tons, or 27 percent of the world's total. Russia, Ukraine, and Kazakhstan contain over 98 percent of the recoverable reserves for this region.

Russia

Russia has extensive coal reserves, more than 173,000 million short tons ($\sim$19 percent of the world total), of which 119,000 million short tons are subbituminous and lignitic in rank. The coal resources in eastern Siberia and the Russian Far East remain largely unused because of their remoteness and lack of infrastructure [4].

Russia's main coal basins contain coals ranging from Carboniferous to Jurassic in age. Most hard coal reserves are in numerous coalfields in European and central Asian Russia, particularly in the Kuznetsk and Pechora basins and the Russian sector of the Dontesk basin. The Kansk-Achinsk basin in eastern Siberia is the country's main source of subbituminous coal. The Moscow basin contains significant lignite reserves, but production there has virtually stopped [4].

The Kuznetsk basin, which is located to the east of Novosibirsk, contains coals exhibiting a wide range in quality and rank from brown coal to semianthracite. The ash content of the coal is variable, and the sulfur content is generally low.

High-quality coals with low moisture, ash, and sulfur contents are used for coking and steam coal production. This basin is now the largest single producer in Russia of coking and steam coal.

The Pechora basin is located in the extreme northeast of European Russia. The coal rank in the basin increases from brown coal in the west to bituminous coal and anthracite in the east. Ash content varies considerably from 9 to 43 percent, while sulfur content, for the most part, does not exceed 1.5 percent. This basin is the principal supplier of coking coal.

The Dontesk basin is located in eastern Russia and western Ukraine and contains the whole range of coal rank from brown coal to anthracite, which increases toward its central and eastern sections. These coals tend to have ash contents of 15 to 20 percent and sulfur contents of 2 to 4 percent and are used as coking and steam coals.

The Kansk-Achinsk basin, located adjacent to the east side of the Kuznetsk basin, contains brown coals that are described as lignites or subbituminous coals; however, their heating value is higher than that of most lignites. These coals have low to medium ash contents (6–20 percent) and low sulfur contents (<1 percent), which make them attractive for power station fuel.

Ukraine

The Ukraine has significant coal reserves totaling more than 37,000 million short tons, which is nearly evenly split between hard coal (bituminous and anthracite) and brown coal, as shown earlier in Table 3.2 [1]. Most of the coal resources are found in two coal basins: the Donetsk and Dneiper basins.

The Donetsk basin, which is Carboniferous in age, is located in the east (and crosses over into Russia) and contains most of the country's hard coal resources. These coals contain medium ash (15–20 percent) and medium-to-high sulfur (2–4 percent) content, and they are used for steam production, power station fuels, and metallurgical applications. The Dneiper basin is adjacent to the eastern edge of the Donetsk basin and it stretches across much of central Ukraine. This basin contains Ukraine's brown coal reserves and currently is of relatively minor importance [4].

Kazakhstan

Kazakhstan contains total recoverable coal reserves similar to Ukraine, with approximately 34,500 million short tons. Unlike Ukraine, however, most of Kazakhstan's reserves are hard coals that total more than 31,000 million short tons. The coal deposits are late Carboniferous and Jurassic in age and are located mainly in the Karaganda and Ekibastuz basins, which produce hard coal. The coal deposits of these basins lie along the southern edge of the Siberian platform [4]. In the Karaganda basin, coking and steam coals are produced that have sulfur contents ranging from 1.5 to 2.5 percent and high ash content (20–35 percent). Coals from the Ekibastuz basin typically have high ash (39 percent on average) and low sulfur (<1 percent) contents and are predominately used for thermal power generation.

3.2.3   Middle East, Asia, and Oceania

This region contains significant recoverable coal reserves totaling over 285,000 million short tons, or approximately 31 percent of the world total. China, Australia, and India comprise most of this total, with more than 126,000, 84,000, and 62,000 million short tons, respectively.

China

China contains more than 126,000 million short tons of recoverable coal reserves in the world, third behind only the United States and Russia [1]. These recoverable reserves are nearly equally divided between hard coal and lignite deposits (i.e., 68,600 and 57,700 million short tons, respectively), with the hard coals being of Carboniferous, Permian, and Jurassic age and the lignite Tertiary in age. Coalfields are scattered throughout China, with the largest deposits in western China stretching from north to south, with most of the reserves in the northern part, specifically in the Inner Mongolia, Shanxi, and Shaanxi Provinces. Significant anthracite deposits are found in the Shanxi and Guizhou Provinces. Bituminous coal deposits occur in the Heilongjiang, Shanxi, Jiangxi, Shandong, Henan, Anhui, and Guizhou Provinces [4].

China is the world's largest coal producer, with most of the coal being used internally for industry and electricity generation. The rank of hard coal appears to increase slightly northward from the Yangtze River, while locally seam quality is very variable [4].

Australia

Australian recoverable coal reserves total more than 84,000 million short tons, which is nearly equally divided between hard coal and lignite deposits (i.e., 41,000 and 43,500 million short tons, respectively), with the hard coals being of Carboniferous and Permian age and the lignite Tertiary in age.

Coal is mined in all of the states except for the Northern Territory. New South Wales and Queensland produce both steam and metallurgical coal for export, while production in Victoria, South Australia, and Western Australia is for thermal electricity generation [4]. Hard coal is mined in New South Wales, Queensland, and Western Australia, while subbituminous and brown coal is mined in South Australia and Victoria. The major coal reserves are found in eastern Australia, with the Bowen, Sydney, and Gippsland basins being the most important.

The Bowen basin is located in Queensland and developed during early Permian times. The rank varies in this basin, increasing from west to east, with the higher-rank coals ranging from low-volatile bituminous coal in the west to semianthracites and anthracites in the east. The coals have low sulfur content (typically 0.3–0.8 percent) and ash contents of 8 to 10 percent and 8 to 16 percent for coking and thermal coals, respectively.

The Sydney basin is located in New South Wales, is of Permian age, and consists of several coalfields. In general, the Sydney basin coals are medium- to high-volatile bituminous coal, with the highest rank contained in the northern portion of the basin. The coals in this basin have low sulfur content (<1 percent), and ash contents typically ranging from 6 to 24 percent, with one coalfield exceeding 40 percent ash.

The brown coal resources found in the Gippsland basin lie within the Latrobe Valley in Victoria and are of Tertiary age. This area is noted for its thick coal seams ranging from 330 to 460 feet in thickness. The brown coals have low heating values (3,400–5,200 Btu/lb) due to high and very variable moisture contents, which range from 49 to 70 percent. Ash contents, on the other hand, are low and range from 0.5 to 2 percent.

India

India's recoverable coal reserves rank fifth in the world, with more than 62,000 million short tons. These reserves vary in rank from lignites to bituminous coal, with most of it being hard coal (i.e., nearly 56,000 million short tons), although the coal quality is generally poor. India's coalfields are located mainly in the east in the states of Assam, Bihar, Uttar, Pradesh, Madhya, Pradesh, Andhra Pradesh, Orissa, and West Bengal [4]. India's coals are principally of Permian age, with some of Tertiary age.

The most significant deposits are in the Raniganj and Jharia basins of northeast India. In the Raniganj basin, the rank increases from noncaking bituminous coal in the east to medium coking coal in the west. Ash content is variable though, varying from 15 to 35 percent. Sulfur content is low (<1 percent). The Jharia coalfield is India's major source of prime coking coal, although it also contains significant noncoking coal as well. As with Raniganj basin coals, ash content varies from 15 to 35 percent, with low sulfur contents in the Jharia basin.

Most of India's lignite mining occurs in southern India in the Neyveli coalfield, although other areas contain larger resources. The lignite contains low ash (2–12 percent) and low sulfur (<1 percent) contents, but the moisture content is high, varying between 45 and 55 percent. India's coal is used primarily for power production. Although India has substantial recoverable resources, coal imports are steadily rising to meet demands for coking coal as well as for steam coal as new power plants begin operation [4].

3.2.4 Europe

Europe contains approximately 51,000 million short tons of recoverable coal reserves, with the majority of the reserves spread fairly evenly among eight countries except for Former Serbia and Montenegro (at ≈15,000 million short tons) and Poland (at 8,300 million short tons). The recoverable reserves in the other six primary countries vary from 2,000 to 7,000 million short tons.

Poland

Poland contains recoverable coal reserves of more than 8,000 million short tons, of which than 6,600 million short tons are hard coal. The hard coal deposits are found in three main basins located in the southern half of the country: the Upper Silesian, the Lower Silesian, and the Lublin basins. These basins are of Carboniferous age. Poland uses its hard coal in world export markets.

Poland's lignite deposits are found in a number of Tertiary basins across the central and southwestern parts of the country. Poland ranks fourth in world lignite production and is the second largest European producer after Germany. The lignite is used as a fuel for electricity generation. Polish lignite has variable ash contents (4–25 percent) and low to medium sulfur contents (0.2–1.7 percent).

Germany

For decades, Germany has been a major European coal producer and consumer. In the last several years, Germany has revised its recoverable coal reserves downward due to more restrictive criteria, although some question whether the new figures are accurate. Germany revised its estimates from nearly 73,000 million short tons of recoverable coal reserves in 2001 to approximately 7,000 million short tons in 2007. Germany's three main areas of lignite resources are the Rhineland, Lusatian, and Central German basins, which are of Tertiary age. In addition, Germany has hard coal capacity, which is of Carboniferous age, located in the Ruhr and Saar basins.

Of the three main lignite basins, the Rhineland deposits are now the most important and are located between the River Rhine and the German/Dutch/Belgian border. The Central German and Lusatian basins are located in eastern Germany. The lignites have heating values of 3,350 to 5,400 Btu/lb and moisture contents that vary from 40 to 60 percent. Ash and sulfur contents vary from 1.5 to 8.5 percent and 0.2 to 2.1 percent, respectively, with the Rhineland basin lignite containing sulfur contents of less than 0.5 percent. These coals are used for producing electricity in generating stations.

Because of restructuring of the hard coal mining sector, which began in 1999, the Ruhr coalfield has greater economic significance than the Saar coalfield, as mines continue to close and overall production declines [4]. The Ruhr coalfield primarily consists of bituminous coal, much of which is coking coal. The basin contains two small areas of anthracite, where the ash and sulfur contents of the coals are 4 to 9 percent and less than 1 percent, respectively. The coals are used primarily for electricity generation along with some industrial applications.

3.2.5 Africa

Africa contains 61,000 million short tons of recoverable coal, with approximately 55,000 million short tons of those reserves in South Africa. Thirteen other countries contain the balance [10], but only Zimbabwe contains almost 550 million short tons, with the rest containing a total of almost 1,200 million short tons (refer to Table 3.2).

South Africa

South Africa's recoverable coal reserves of 53,000 million short tons consist entirely of hard coal. These coals are of Carboniferous and Permian age, with significant deposits in the Great Karoo basin. This basin extends about 300 miles from west to east across northern Free State Province and south and east Mpumalanga, and about 700 miles from southern Mpumalanga in the north to the center of Kwazulu-Natal in the south [4]. Although the Great Karoo basin is the largest, several other basins and 19 coalfields exist throughout South Africa.

The hard coal consists of bituminous coals, anthracite, and semianthracite. The ash content ranges from 7 percent for some anthracites to more than 30 percent for bituminous coals. Sulfur contents range from less than 1 percent to nearly 3 percent. Domestically, the coal is used for electricity generation and conversion into synthetic liquid fuels and chemical feedstocks. South Africa exports significant quantities of steam coal, with minor amounts of coking coal and anthracite.

3.2.6 Central and South America

Central and South America contain approximately 18,000 million short tons of recoverable coal reserves, or 2.2 percent of the world's total. Coal is found in several countries, including Argentina, Bolivia, Brazil, Chile, Colombia, Ecuador, Peru, and Venezuela. Two of these countries, however, contain the majority of these reserves: Brazil with 7,800 million short tons and Colombia with 7,700 million short tons. Brazil's coals are subbituminous and lignitic in rank, while Colombia's coals are primarily high-volatile bituminous with a small amount of subbituminous coals. These coals formed during late Cretaceous to Tertiary times.

REFERENCES

[1] EIA (Energy Information Administration), Annual Energy Review 2007, U.S. Department of Energy, June 2008.

[2] EIA, International Energy Outlook 2008, U.S. Department of Energy, September 2008.

[3] British Petroleum, BP Statistical Review of World Energy, British Petroleum, June 2008.

[4] S. Walker, Major Coalfields of the World, IEA Coal Research, 2000.

[5] EIA, U.S. Coal Reserves: 1997 Update, U.S. Department of Energy, February 1999, Appendix A.

[6] EIA, International Energy Outlook 2006, U.S. Department of Energy, June 2006.

[7] EIA, Coal Reserves Current and Back Issues, U.S. Department of Energy, February 2009.

[8] H.H. Schobert, Coal: The Energy Source of the Past and Future, American Chemical Society, 1987.

[9] J.G. Singer (Ed.), Combustion: Fossil Power Systems, Combustion Engineering, Inc., 1981.

[10] EIA, International Energy Annual 2001, U.S. Department of Energy, March 2003.

4 The Effect of Coal Usage on Human Health and the Environment

Coal has played a significant role in the advancement of civilization, and, as discussed in Chapter 1, it will continue to be a major fuel source for at least the next quarter century. The value of coal, however, is partially offset by the environmental issues it raises. Some of these environmental issues also impact human health.

Coal mining has a direct impact on the environment, disturbing large areas of land and, in the case of surface mining, potentially contaminating surface and groundwater. In some surface mines, the generation of acid mine drainage (AMD) is a major problem. Other significant impacts include fugitive dust emissions and the disposal of overburden and waste rock.

In underground mining, the surface disturbance is less obvious, but the effect of subsidence can be large. The generation and release of methane and other gases can be a problem. As with surface mining, groundwater can also be disturbed, and AMD can become an issue. In addition, underground miners often suffer from respiratory ailments and, until the last few decades, have faced the high risk of injuries and death from accidents.

Coal beneficiation is primarily based on wet physical processes that produce waste streams that must be controlled. These include fine materials that are discharged as a slurry to a tailings impoundment and a coarse material that is hauled away as a solid waste. The storage, handling, and transportation of coal all produce fugitive dust.

Coal utilization, specifically combustion, on which this chapter focuses, produces several types of emissions that adversely affect the environment, particularly ground-level air quality. In addition, the generation of coal combustion by-products must be addressed, whether they are disposed or reused/recycled. Concern for the environment has always contributed to policies that affect the consumption of coal, and it will continue to do so in the future. The main emissions from coal combustion are sulfur dioxide (SO_2), nitrogen oxides (NO_x), particulate matter (PM), and carbon dioxide (CO_2). Recent studies on the health effects of mercury have raised concerns about mercury emissions from coal-fired power plants. The environmental and health effects of these pollutants, along with other pollutants, such as carbon monoxide (CO), lead, ozone, and organic emissions, are discussed in this chapter.

This chapter also examines the effect of coal use on human health and the environment. The impacts of mining, storage, handling, transportation, beneficiation, combustion by-products, and emissions from coal-fired power plants are discussed, with a focus on activities in the United States.

Clean Coal Engineering Technology. DOI: 10.1016/B978-1-85617-710-8.00004-2
Copyright © 2011 by Elsevier Inc. All rights of reproduction in any form reserved.

4.1 Coal Mining

Many believe that mining operations can cause major problems, and this has led to confrontations among citizen groups, governmental agencies, and the mining industry. These conflicts tend to be centered on the following issues [1]:

*   Destruction of the landscape
*   Degradation of the visual environment
*   Disturbance of surface water and groundwater
*   Destruction of agricultural and forest lands
*   Damage to recreational lands
*   Noise pollution
*   Dust
*   Truck traffic
*   Sedimentation and erosion
*   Land subsidence
*   Vibration from blasting

These issues, along with past operating procedures that often involved unsafe working conditions, deaths and injuries in mines, high incidences of respiratory diseases, use of child labor, scarred landscapes, poor miner living conditions, contentious and sometimes extremely violent labor relations gave coal mining a bad reputation. Consequently, major effort has gone into addressing these issues so that today coal mining is a highly regulated industry that has seen a number of significant changes in the approach to mining and resource development, along with improved miner safety:

*   Environmental impact assessment and public inquiries
*   Conditions for the approval of mining permits
*   Resource management and land-use planning
*   Land reclamation and rehabilitation
*   Regulations specifically addressing miner safety and training

Coal mining falls into two categories: underground mining and surface mining. Surface mining is used when the coal seam to be mined lies close to the surface—typically less than 200 feet. Each mining technique has its own set of technical and economic advantages and disadvantages, and each mining technique has its own set of health and environmental impacts. In 2007, nearly 352 million short tons of coal were mined by underground methods, while more than 793 million short tons of coal were removed by surface mining [2].

4.1.1 Underground Mining

Underground mining is used for deep seams, and mining methods vary according to the site conditions. Underground mines can be classified by the types of access used to reach the coal, such as shaft mines, slope mines, and drift mines, but they are primarily classified by the coal removal system involved: room and pillar mining, pitch mining, or longwall mining [2]. A shaft mine uses a vertical hole dug straight down from the surface to the coal seam. A slope mine provides an access at a slant and is used to follow a seam along its pitch or to cut through a mountain to reach the coal. A drift mine accesses coal outcrops on a mountainside.

Room and pillar mining is used when the coal seam lies relatively level, and is accomplished by either conventional mining or continuous mining [2]. In 2007, more than 2 million short tons of the coal produced in underground mines were obtained by conventional mining, and it is being replaced by other mining methods [2]. In conventional mining, explosives are used to shatter the face of the seam and the broken pieces are manually loaded onto tram cars or conveyors and hauled out of the mine. In continuous mining, where nearly 174 million short tons of coal were produced in 2007 [2], a machine moving along caterpillar tracks cuts the coal from the face of the seam and automatically loads the coal onto tram cars or a conveyor.

Pitch mining is a technique used when the coal seams are inclined, and is frequently used in anthracite mining [4]. In pitch mining, the bottom of the seam is accessed, and the coal is dropped into chutes that gravity-feed tram cars. Longwall mining is the removal of coal from a long, continuous face rather than removing it from several short faces as occurs in room and pillar, or pitch mining. Yearly longwall mining production is similar to that of continuous mining, and it accounted for nearly 176 million short tons of coal produced in the United States in 2007 [2].

Regardless of the mining system used, the health and environmental impacts are common to every underground mine. These include land subsidence, generation of methane and other gases, liquid effluents, dust, solid waste, and miner safety.

Subsidence

Subsidence can have a major effect on the topography of the land surface. Following the removal of coal from an underground mine, the roof materials may cave, causing collapse of the overlying rock strata, which leads to subsidence of the surface. The degree of collapse of the overlying rock strata can vary from practically no collapse with no resulting surface impacts to total collapse with more pronounced changes at the surface [5]. In general, mine subsidence problems develop where postmining pillar support systems and coal barriers ultimately fail. Many interrelated factors control when, where, and how failure will occur, including the following [1]:

- Thickness of coal removed
- Size, shape, and distribution of pillars and rooms
- Depth of mining
- Percent extraction of coal
- Thickness and physical characteristics (i.e., strength) of the overburden
- Method of mining—for example, longwall, shortwall (which is a slight modification of longwall mining), room and pillar, or room and pillar with full or partial retreat
- Dry or flooded conditions in the mine
- Actual or potential level and degree of fracturing in the overburden
- Mineralogy of the overburden (e.g., clay minerals that swell when water is added, sulfide minerals that chemically and physically change in the presence of oxygen and moisture, and minerals that react with water to form new minerals)

Over areas that have been longwall mined, the subsidence is often a shallow trough. In flat terrain, this trough is usually quite visible and can cause local changes in drainage. Subsidence from active coal mining results in the largest effect to the land surface in terms of area undermined, although the effects are often small in terms of overall topography [5]. Subsidence impacts on surface structures and sub-surface hydrologic resources are generally of more importance than the impacts on topography or surface features.

Subsidence from shallow, abandoned coal mines often results in abrupt but localized changes in topography that reflects the collapse of individual rooms or voids. This type of subsidence can be an isolated, single collapse, or it can involve a larger area with many individual subsidence pits.

Generation of Gases

Methane (CH_4) is produced during coalification, and only a fraction of this remains trapped under pressure in the coal seam and surrounding rock strata. This trapped methane is released during mining when the coal seam is fractured. The amount of methane released during coal mining depends on a number of factors, including coal rank, coal seam depth, and method of mining [6]. As coal rank increases, the amount of methane produced also increases. The adsorption capacity of coal increases with pressure, and pressure increases proportionately with the depth of the coal seam; consequently, deeper coal seams generally contain more methane than shallow seams of the same rank.

Underground coal mining releases more methane than surface mining because of the higher gas content of deeper seams. The methane that is contained in the coal seams is called coalbed methane (CBM). The CBM that is released from the coal during coal mining is referred to as coal mine methane (CMM) and is a subset of CBM. CMM is emitted from several sources:

1. Degasification systems at underground coal mines, which may employ vertical and/or horizontal wells to recover methane in advance of mining or after mining.
2. Ventilation air from underground mines, which contains dilute concentrations of methane.
3. Abandoned or closed mines, from which methane may sweep out through vent hole or through fissures or cracks in the ground.
4. Surface mines, from which methane in the coal seams is directly exposed to the atmosphere.
5. Fugitive emissions from postmining operations, from both underground and surface mines, in which coal continues to emit methane as it is stored, processed, or transported.

Figure 4.1 shows an example of this distribution for CMM in the United States in 2006 [7].

Methane is highly explosive in air in concentrations between 5 and 15 percent, and operators have developed two types of systems for removing methane from underground mines: ventilation and degasification [6]. At present, almost all venti-lation air is emitted into the atmosphere. Although the methane concentrations are low, the amount of methane released into the atmosphere each year is significant. Emissions factors for underground mining range from 10 to 25 cubic meters emitted

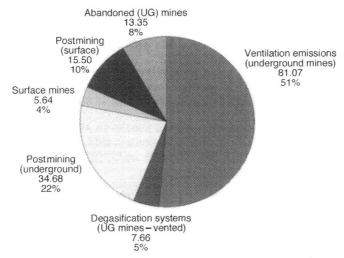

Figure 4.1 Sources of CMM in the United States in 2006 (billion cubic feet).

per ton of coal mined, compared to 0.3 to 2 cubic meters per ton of coal that is surface mined. Table 4.1 lists the methane emissions factors for underground mining for selected countries [8].

As a greenhouse gas, methane is more than 21 times more potent than carbon dioxide (CO_2). (A discussion of greenhouse gases and their global warming potential relative to CO_2 is provided later in the chapter.) Table 4.2 lists global estimates of methane emissions from coal mining in 2005, along with projections for 2010 [8]. This is further illustrated in Figure 4.2 for 2005 methane emissions [8]. Globally, CMM accounts for 6 percent of total methane emissions resulting from human activities [9]. In 2005, estimated worldwide CMM emissions totaled more than 388 million metric tons of CO_2 equivalent ($MtCO_2eq$), or about 30 billion cubic meters. (CO_2 equivalents are also discussed later in the chapter.)

Table 4.1 Methane Emission Factors for Underground Mining in Selected Countries

Country	Emissions Factor (m³/ton coal)	Emissions Factor[a] (tCO₂eq/ton coal)
Australia	15.6	0.22
Czech Republic	23.9	0.34
Former Soviet Union	17.8–22.2	0.25–0.32
Germany	22.4	0.32
Poland	6.8–12.0	0.10–0.17
United Kingdom	15.3	0.22
United States	11.0–15.3	0.16–0.22

[a]Conversion factor of 1 m³ = 0.0143 tCO₂eq = 35.31 ft³ × 0.00404 tCO₂eq

Source: From U.S. Environmental Protection Agency, Global Migration of Non-CO₂ Greenhouse Gases (2006).

Table 4.2 Global Estimates for CMM Emissions (MtCO$_2$eq) for 2005 and 2010

Country	2005	2010
China	135.7	153.8
United States	55.3	51.1
India	19.5	23.2
Australia	21.8	26.4
Russia	26.3	27.5
Ukraine	26.3	24.5
North Korea	25.6	24.3
Poland	11.3	10.8
South Africa	7.4	7.2
United Kingdom	6.7	6.6
Germany	8.4	7.7
Kazakhstan	6.7	6.4
Colombia	3.4	4.0
Mexico	2.5	2.8
Czech Republic	4.8	3.9
Rest of the World	26.5	27.5
World Total	**388.1**	**407.6**

Source: From U.S. EPA, Global Migration of Non-CO$_2$ Greenhouse Gases (2006).

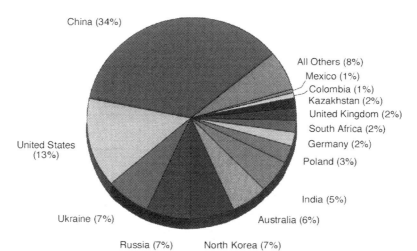

Figure 4.2 Global CMM emissions for 2005 (MtCO$_2$eq).

Some of the mines in the United States that produce very high levels of gas (as well as mines in Russia, Australia, and other countries) have installed degasification systems to extract the methane from the coal seams in advance of mining, during coal recovery, and during postmining operations [6]. Depending on the quality of this gas, mine operators can use the recovered methane for on-site electricity generation or sell it to a local pipeline.

The total volume of CMM liberated in the United States in 2000 was about 196 billion cubic feet (Bcf), with underground mining activities liberating 142 Bcf [10]. By 2006, CMM emissions in the United States dropped to 158 billion cubic feet. CMM emissions account for approximately 10 percent of total U.S. methane emissions. Globally, coal mines account for 6 percent of all methane emissions. Underground mines are the largest source of CMM, accounting for approximately 78 percent of the total CMM liberated.

Since 1994, the U.S. Environmental Protection Agency's (EPA) Coalbed Methane Outreach Program has worked cooperatively with the coal mining industry both in the United States and internationally to reduce CMM emissions [7]. Technology is readily available to recover methane. Worldwide, CMM is most often used for power generation, district heating, boiler fuel, or town gas, or it is sold to natural gas pipeline systems. CMM emitted through ventilation systems each year represents a considerable source of greenhouse gas emissions and is a major focus area for recovery and use. The U.S. coal industry has made substantial progress in recovering and using CMM through drainage systems. Coal mines in the United States recovered nearly 90 percent of the gas liberated through the drainage systems.

Some coal seams, particularly in Australia as well as in France and Poland, contain high carbon dioxide concentrations, which can comprise as much as 100 percent of the gases in the coal seam [6]. It is thought to have a magmatic origin. Although carbon dioxide is not toxic, it can cause asphyxiation by displacing breathable oxygen.

Liquid Effluents/Acid Mine Drainage

The generation of liquid effluents from major mining techniques tends to be higher for underground mining than for surface mining. For underground mining, liquid effluent rates of 1.0 and 1.6 tons per 1,000 short tons of coal produced have been documented for conventional and longwall mining, respectively [11]. If groundwater systems are disturbed, the possibility then exists for serious pollution from highly saline or highly acidic water.

The highly acidic water, commonly known as acid mine drainage (AMD), is produced by the exposure of sulfide minerals, most commonly pyrite, to air and water, resulting in the oxidation of sulfur and the production of acidity and elevated concentrations of iron, sulfate, and other metals [1]. Pyrite and other sulfide minerals are generally contained in the coal, overburden, and coal processing wastes.

Historically, coal extraction in the northern Appalachian coalfields has resulted in serious problems related to contaminated mine drainage [12]. Acid drainage from closed and abandoned mines (both underground and surface) has far-ranging effects on water quality and, therefore, on fish and wildlife. Drainage from closed mines is particularly acidic in Pennsylvania, Ohio, northern West Virginia, and Maryland.

The formation of AMD is primarily a function of the geology, hydrology, and mining technology employed for the mine site [13, 14]. AMD is formed by a series of complex geochemical and microbial reactions that occur when water comes into contact with pyrite (iron disulfide minerals) in coal, refuse, or the overburden of a mine operation. The resulting water is usually high in acidity and dissolved metals.

The metals stay dissolved in solution until the pH raises to a level where precipitation occurs.

The chemistry of pyrite weathering to form AMD is commonly represented by the chemical reactions in the following four equations.

$$4FeS_2 + 15O_2 + 14H_2O \rightarrow 4Fe(OH)_3^- + 8H_2SO_4 \tag{4.1}$$

Pyrite + Oxygen + Water → Yellowboy + Sulfuric Acid

The first reaction in the weathering of pyrite includes the oxidation of pyrite by oxygen. Sulfur is oxidized to sulfate, and ferrous iron is released. This reaction generates two moles of acidity for each mole of pyrite oxidized:

$$2FeS_2 + 7O_2 + 2H_2O \rightarrow 2Fe^{2+} + 4SO_4^{2-} + 4H^+ \tag{4.2}$$

Pyrite + Oxygen + Water → Ferrous Iron + Sulfate + Acidity

The second reaction involves the conversion of ferrous iron to ferric iron, which consumes one mole of acidity. Certain bacteria increase the rate of oxidation from ferrous to ferric iron. This reaction rate is pH dependent, with the reaction proceeding slowly under acidic conditions (pH of 2 to 3) with no bacteria present and several orders of magnitude faster than pH values near 5. This reaction is often referred to as the "rate determining step" in the overall acid-generating sequence:

$$4Fe^{2+} + O_2 + 4H^+ \rightarrow 4Fe^{3+} + 2H_2O \tag{4.3}$$

Ferrous Iron + Oxygen + Acidity → Ferric Iron + Water

The third reaction that can occur is the hydrolysis of iron. Hydrolysis is a reaction that splits the water molecule. Three moles of acidity are generated as a by-product. Many metals are capable of undergoing hydrolysis. The formation of ferric hydroxide precipitate (i.e., a solid product) is pH dependent. If the pH is above about 3.5, solids may form, but below pH 2.5, little or no solids will precipitate. The third reaction is:

$$4Fe^{3+} + 12H_2O \rightarrow 4Fe(OH)_3^- + 12H^+ \tag{4.4}$$

Ferric Iron + Water → Ferric Hydroxide (Yellowboy) + Acidity

The fourth reaction is the oxidation of the additional pyrite by ferric iron. The ferric iron is generated in reactions (4.3) and (4.4). This is the cyclic and self-propagating part of the overall reaction, and it takes place very rapidly and continues until either ferric iron or pyrite is depleted. Note that in this reaction ferric iron, not oxygen, is the oxidizing agent:

$$FeS_2 + 14Fe^{3+} + 8H_2O \rightarrow 15Fe^{2+} + 2SO_4^{2-} + 16H^+ \tag{4.5}$$

Pyrite + Ferric Iron + Water → Ferrous Iron + Sulfate + Acidity

Treatment of AMD includes both chemical and passive techniques. In Pennsylvania, for example, strict effluent discharge limitations were placed on mine operations in 1968 [13]. Many companies used chemical treatment methods to meet these new effluent limits. In these systems, the acidity is buffered by the addition of alkaline chemicals such as calcium carbonate, sodium hydroxide, sodium bicarbonate, or anhydrous ammonia. These chemicals raise the pH to acceptable levels and decrease the solubility of dissolved metals. Precipitates form from the solution. These chemicals are expensive, however, and the treatment system requires additional costs associated with operation and maintenance, as well as the disposal of metal-laden sludges.

Many variations of AMD passive treatment systems were studied as early as 1978 [14]. During the last 15 years, passive treatment systems have been implemented on full-scale sites throughout the United States, with promising results. The concept behind passive treatment is to allow the naturally occurring chemical and biological reactions that aid in AMD treatment to take place in the controlled environment of the treatment system and not in the receiving water body.

Passive systems do not require the expensive chemicals of the chemical treatment systems, and operation and maintenance requirements are considerably less. Passive AMD treatment technologies being implemented include the following:

- Aerobic wetland
- Compost or anaerobic wetland
- Open limestone channels
- Diversion wells
- Anoxic limestone drains
- Vertical flow reactors
- The Pyrolusite process

Hydrologic Impact

With underground mining, subsidence and fracturing of overlying strata may cause surface runoff to be diverted underground and may disrupt aquifers, causing local water level declines, as well as a possible change in the direction of groundwater flow near the mine [1]. Dewatering required by mining operations affects groundwater quantity by depleting aquifers when mine features extend below the water table and become a drain [5]. Although groundwater depletion is mining's most obvious and immediate effect on groundwater, other, longer-term effects on the environment are equally important. Redistribution and/or change in groundwater recharge rates may affect the time and degree to which aquifers will recover to a static condition.

Conscientious management practices minimize water-related environmental impacts [15]. Coal mining activities are highly regulated, requiring extensive surface and groundwater sampling and monitoring to ensure compliance with federal, state, and local statutes. Also, hydrological impacts must be taken into consideration as part of the permitting process. Therefore, coal company hydrologists study and monitor the quality of surface and underground water resources before, during, and after mining activity to ensure minimal hydrological impacts.

Health Effects and Miner Safety

It is no argument that mining is a dangerous occupation and historically there has always been a tremendous social cost associated with coal mining. Miners frequently suffered from accidents and diseases. In the United States, more than 100,000 miners died over the period of about 1885 to 1985 [4]. In 1900, the annual death rate among underground miners was 3.5 deaths per 1,000 miners [4]. However, a commitment to enhanced safety training and the development of new technologies has yielded remarkable improvements in the mine as a workplace. In addition, coal mines are subject to regular, comprehensive inspections by the federal Mine Safety & Health Administration (MSHA), as well as to safety and health reporting requirements that are much more stringent than those required by the Occupational Safety and Health Administration (OSHA), which regulates most of the other U.S. industries.

According to MSHA, the twentieth century saw remarkable improvements in safety and health for U.S. miners, and the rate of fatal injuries for underground miners declined by 92 percent over the period 1960 to 1999 [15]. Data from the U.S. Department of Labor, MSHA, show that in 2008, 13, 9, and 5 fatalities occurred in underground mining, surface mining, and preparation plants, respectively [16]. This is an annual death rate of 0.2 per 1,000 mining personnel, nearly a 20-fold decrease since 1900. Figure 4.3 further illustrates this improvement by comparing fatalities and coal production trends from 1970 to 2008 [17]. Figure 4.4 shows the ten-year trend in U.S. mining injury rates from 1997 to 2007 [16].

According to the U.S. Department of Labor, mining is not even among the top ten most dangerous occupations in America. Pilots, truck and taxi drivers, loggers, fishermen, roofers, and other occupations face greater on-the-job risks than coal miners do. Table 4.3 lists the Bureau of Labor Statistics occupational and illness rates per 100 full-time workers by industry division for 2007 [16].

Figure 4.3 U.S. coal mine safety and production (millions of short tons) trends from 1970 to 2008.
Source: From Mine Safety and Health Administration (2008).

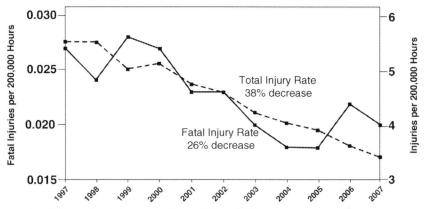

Figure 4.4 U.S. mining injury rates from 1997 to 2007.
Source: From Mine Safety and Health Administration (2008).

Table 4.3 Bureau of Labor Statistics Nonfatal Occupational Injury
and Illness Rates by Industry Division, 2007

Industry Division	Rates per 100 Full-Time Workers
Trade, transportation, and public utilities	4.9
Manufacturing	5.6
Construction	5.4
Agriculture, forestry, and fishing	5.4
Education and health services	5.2
Retail trade	4.8
Leisure and hospitality	4.5
Service providing	3.8
Mining	3.7
Financial activities	1.4
Private sector average	4.2

Source: From Mine Safety and Health Administration (2008).

Dust issues in mines have had a negative impact on the coal industry in general, and this is particularly true of underground mines. Many underground miners contract respiratory diseases. Diseases like pneumoconiosis (black lung) and silicosis have taken years to manifest themselves, and the industry has been working diligently to minimize the dust in mines. In an underground mine, the walls of the tunnels or shafts are covered with pulverized limestone to help settle the coal dust [15]. Water sprayers on the mechanical equipment, such as continuous and longwall miners, help reduce dust concentration in the mine. The ventilation fans that remove methane also remove the lingering dust and bring a continuous supply of fresh air into the mine. Miners also wear air-purifying systems. Dust production in underground mines is 0.0006 and 0.01 short tons per 1,000 short tons of coal produced for conventional and longwall mining, respectively [11].

4.1.2 Surface Mining

Surface mining of coal is an alternative to underground mining, and it is practiced as strip mining, the most common form, and auger mining. Some coal is also produced through dredging and culm bank (i.e., anthracite waste piles) recovery, but these quantities tend to be relatively small compared to underground, strip, and auger mining. Table 4.4 lists the quantities of coal produced by the various methods, as reported by MSHA, in 2007 [16].

Strip mining is favored when the overburden—the overlying rock strata—is typically 200 feet or less in thickness, but it can still be economical with an overburden thickness of up to 600 feet [4, 18]. Another factor that is considered when determining the economics of strip mining is the stripping ratio, which is the ratio of overburden thickness to coal seam thickness. Typically the maximum stripping ratio that is considered economical is 20 to 1 [4].

The two general methods of strip mining are area mining and contour mining. In area mining, a trench is dug, the overburden is piled to one side, and the coal is removed. As mining progresses, the overburden from the new trench is dumped into the first trench. In contour mining, which is used when the coal lies beneath hilly terrain or outcrops on a hillside, the overburden is removed and dumped on the downhill side of the mining operation [3].

Auger mining can be used for coal outcrops on a hillside or in strip mines where the stripping ratio has become too high to be economical for strip mining [4]. In auger mining, an auger drills into the face of the coal seam, and the coal is removed via conveyor and loaded directly onto a truck.

Potentially, many adverse environmental impacts result from area surface mining of coal if no reclamation practices are used. Such measures are used with varying degrees of effectiveness in the United States [1]. Air quality can be affected in several ways. Fugitive dust from coal haul roads, unvegetated spoil surfaces, topsoil

Table 4.4 U.S. Coal Production by Type during 2007

Mining Type	Production (short tons)
Total underground	351,996,028
Total surface	795,748,272
Strip	785,000,352
Auger	9,592,411
Culm bank recovery	847,163
Dredge	308,346
Total Coal	**1,147,744,300**

Source: From Mine Safety and Health Administration (2008).

stockpiles, and coal stockpiles can be a problem. Overburden blasting can produce troublesome noise, air shocks, and ground vibrations. The quality and quantity of surface water and groundwater can be affected if efficient reclamation practices are not used. Sedimentation of surface waters may occur. Erosion of reclaimed slopes often occurs due to the unconsolidated nature of the reclaimed materials. Aesthetically, the disturbed land is very unsightly prior to reclamation.

Surface Disturbance

Surface mining causes large-scale disturbances on the earth's surface. The natural land surface is drastically changed by the mining activities through the removal of soil, rock, and coal. Depending on the mining conditions and equipment, widespread changes in the locations of materials will occur [5]. For example, an inadequate amount of material may be available to fill the final pit of a surface coal mine. As a result, the areas will usually be graded to a topography that includes a lake or basin where at least a portion of the area has relatively steep slopes. Reclamation of contour mines and sometimes mountaintop removal mines often results in some very steep slopes [5]. These areas are prone to erosion and mass failure due to the steepness of the slopes and the loose, nonhomogeneous nature of the materials present. Unfortunately, surface mining often causes erosion and mass wasting because these reclaimed land forms are often in states of disequilibrium compared to the natural environment in which they were created [5]. The primary role of reclamation, therefore, is to achieve a landscape that approximates premining conditions that are considered to be near equilibrium with the local environmental factors.

Generation of Gases

Although a surface mine releases less methane than an underground mine, the amount emitted into the atmosphere is still significant due to the large amount of surface-mined coal. Measurements of actual emissions of methane from surface mining are technically difficult, costly, and generally not available for inventory preparation, but emission factors have been developed from a number of country-specific studies [6, 8]. Irving and Tailakov estimate that surface mining releases 0.3 to 2 cubic meters of methane per metric ton of coal mined [6].

Liquid Effluents and Acid Mine Drainage

Surface mining operations experience some of the same issues as underground mining. Contour and area mining generate 0.24 and 1.2 tons of liquid effluents per 1,000 tons of coal produced, respectively, compared to 1 and 1.6 tons, respectively, for conventional and longwall mining operations [11].

The presence of soluble salts such as sodium in discarded overburden can cause saline and caustic conditions in topsoils if conditions allow the upward migration of these salts. Also, oxidation processes result in significant changes in chemistry. When sulfides such as pyrite are present, acid is produced, and the solubility of elements tends to increase, leading to acid mine drainage.

Hydrologic Impact

Surface mining affects surface stream runoff. The runoff may be increased and sub-sequently channel erosion experienced as a result of reduced infiltration rates [1]. Conversely, streams may also be affected by decreased surface runoff, where more permeable rock strata become exposed by the surface mining.

Modifications of the local or regional recharge zones involve changing the infiltra-tion rates by removal of the vegetative cover, alteration of soil profiles, and com-paction. Reduced infiltration rates decrease groundwater storage and reduce water availability. These disruptions are of particular concern in the semiarid western region of the United States. Shallow and coal seam aquifers can be drained by mining activ-ity, causing temporary or permanent loss of existing wells near mined areas [1].

The disturbance of the overburden during surface mining also causes significant changes in the chemical nature of the system [5]. Such changes are due to the influ-ence of water on the now available soluble salts and to the changing redox (reduc-tion of oxidation) conditions resulting from the influx of oxygen into the system that was previously oxygen depleted. The movement of high concentrations of salts and/or elements into existing or reestablished groundwater aquifer systems can occur as a result of the disruption of the consolidated overburden and increased water pene-tration into reclaimed land.

Solid Waste and Dust

Waste rock is one product of the mining process that influences the postmining land surface. In the case of mountaintop removal and contour mining methods, waste materials are often used to fill adjacent canyons or hollow areas. When associated with canyon fill, steep slopes that can be very erosive are formed. Surface mining produces more solid waste than underground mining techniques, with 10 tons of solid waste produced per 1,000 tons of coal removed for both contour and area mining [11]. Conventional and longwall underground mining, on the other hand, produce 3 and 5 tons of solid waste, respectively, per 1,000 tons of coal removed. Similarly, dustiness associated with surface mining is significantly greater than that of undergound mining. The World Bank Group [11] reports dust generation of 0.1 and 0.06 tons per 1,000 tons of coal produced for contour and area mining, respec-tively, while only 0.0006 and 0.01 tons of dust are generated during conventional and longwall underground mining, respectively.

Health Effects and Miner Safety

Surface miners experience different safety issues than underground miners, and his-torically the annual death rate for surface miners was significantly less than that of underground miners [16]. Underground miners typically experienced more respira-tory diseases than surface miners until regulations improved dust control practices and mandated the use of personal air-purifying systems. Now the fatality rates are similar for surface and underground miners [16].

4.1.3 Legislation and Reclamation

Because coal mining can have a significant impact on both the environment and the health of miners, coal producers are required to go through a complicated process for obtaining local, state, and federal permits to mine. Coal mining is one of the most extensively regulated industries in the United States [19, 20]. A company must comply with many laws and regulations, and meeting all of those requirements can be a long and difficult process [19, 21]. As many as 10 years can pass between the start of planning a mine and mining the first ton of coal.

The coal company must provide detailed information on how the coal will be mined, how the land will be reclaimed, the quality and quantity of surface and underground water sources and how the mining operations will affect them, and the method used to transport the coal from the mine and how the area will be affected by the transportation. The coal company has to return the land to approximately the same physical contour and to a state of productivity equal to or better than premining conditions. Wildlife habitats cannot be permanently disrupted, and archeological sites must be protected. Companies must post bonds to ensure that the sites will be reclaimed.

Unfortunately, in the past, concern for the environment was not always a high priority, and as a result, many abandoned mines exist today. Mined coal is taxed, and the funds go into the Federal Abandoned Mines Land Fund to finance reclamation projects of these orphaned mines.

Some states had reclamation laws on their books since the 1930s, but in 1977 Congress enacted the Surface Mining Control and Reclamation Act (SMCRA), which mandated strict regulation of surface mining [19]. This act, as well as the Clean Air Act, the Clean Water Act, and the National Environmental Policy Act, has had a significant impact on surface mining. In addition, many other legislative acts affect surface mining in the United States, and it is obvious from their names that they cover a wide range of subject areas [19]:

- American Indian Religious Freedom Act of 1978
- Antiquities Act of 1906
- Archeological and Historical Preservation Act of 1974
- Archeological Salvage Act
- Bald Eagle Protection Act of 1969
- Endangered Species Act of 1963
- Fish and Wildlife Coordination Act of 1934
- Forest and Rangeland Resources Planning Act of 1974
- Historic Preservation Act of 1966
- Migratory Bird Treaty Act of 1918
- Mining and Minerals Policy Act of 1970
- Multiple Use–Sustained Yield Act of 1960
- National Forests Management Act of 1976
- National Trails System Act
- Noise Control Act of 1976
- Resource Conservation and Recovery Act
- Safe Drinking Water Act of 1974

- Soil and Water Resources Conservation Act of 1977
- Wild and Scenic Rivers Act
- Wilderness Act of 1964

Although the mining industry must conform to many laws and regulations, and even though past mining operations were not environmentally conscientious, mine reclamation has become a success story in the United States. Mine operators are addressing the issues discussed earlier in this chapter, and as coal continues to be a major energy source for the United States, the mining industry; federal, state, and local governments; and the general public must continue to work together to minimize the environmental and health impacts of coal mining.

Surface mining of coal should be regarded as only a temporary land use. The enactment of stringent laws requiring that the land be returned to its original condition or better has meant that restoration of the land surface and rehabilitation of the soil materials have now become normal parts of the planning, approval, and operation of most surface mines. Many examples of high-quality restored land can now be found, and the Office of Surface Mining annually recognizes companies and individuals whose efforts are exemplary. These companies and individuals are honored for not only doing the reclamation required of them but also for going beyond the requirements to achieve outstanding landscape restoration [22].

Internationally, many countries have their own legislation governing the restoration and rehabilitation of the land [19]. Major coal-producing countries like Canada, Germany, the United Kingdom, Australia, and South Africa practice reclamation. Unfortunately, due to economic reasons, less-developed countries like China, India, and Indonesia practice little reclamation.

4.2 Coal Preparation

The purpose of coal preparation is to improve the quality of the coal to make it suitable for a specific purpose by doing the following [23, 24]:

- Crushing
- Screening
- Conventional cleaning
- Deep cleaning
- Blending
- Dedusting

Run-of-mine (ROM) coal generally falls into two major groups: coal from underground mining and coal from strip mining. Underground mining tends to produce a finer product than strip mining, but both products are crushed further to produce the desired size for a coal-cleaning process or directly for utilization (e.g., combustion or gasification) [23].

In addition, increased mechanization in the underground mining industry has deceased selectivity and increased the volume of refuse [24]. Equipment like continuous miners and longwall shearers often gather roof and floor rock in addition to the coal. Equipment currently used to mine and transport coal produces more fine coal particles than did earlier equipment [24].

End-use facilities such as power plants are designed for optimal combustion to burn a coal with a specific composition of ash, sulfur, energy, and, sometimes, volatile matter content. These requirements are becoming more difficult to meet with ROM coal or coal from one specific source because of several factors: coal variability (e.g., ash and sulfur contents) within the coal seam, considerable variation in underground mined coal quality due to the inclusion of roof and floor layers, the need for reduced sulfur content due to limitations on sulfur dioxide (SO_2) emissions from power plants, and the declining quality of coal being mined in the eastern United States as higher-quality reserves have been depleted. Consequently, techniques have been implemented to upgrade the coal quality, and the need for advanced coal cleaning processes for both coarse coal and fine coal has increased [24].

In the past, coal was predominately cleaned by dry methods, but in recent years these methods have been abandoned in favor of wet cleaning processes. There are several factors for the abandonment of the methods: particle size requirements (i.e., grinding the coal to finer sizes liberates more ash from the coal); dust emissions (i.e., the use of water to control dust in underground mines); transportation issues; health, safety, and noise impacts; and the better performance of wet processes for coal cleaning [24]. A coal preparation plant separates the material it receives into a product stream and a reject stream, which may be further divided into coarse and fine refuse streams. Depending on the source, 20 to 50 percent of the material delivered to a coal preparation plant may be rejected [23]. One of the reject streams is a slurry, a blend of water, coal fines, silt, sand, and clay particles, which is most commonly disposed of in an impoundment [24].

The coal-cleaning processes that are predominately used today include dense medium separation, hydraulic separation, froth flotation, and agglomeration [23]. Dense medium separations include coal preparation processes that clean raw coal, coarse or fine, by immersing it in a fluid that has a density intermediate between clean coal and the rejects. Since there is a general correlation between ash content and specific gravity, it is possible to achieve the required degree of removal of impurities from raw coal by regulating the specific gravity of the separating fluid, which can be organic liquids, dissolved salts in water, aerated solids, and suspensions of fine solids suspended in water [23].

Hydraulic separation jigging is a process of coarse particle stratification in which the particle rearrangement results from an alternative expansion and compaction of a bed of particles by a pulsating flow. The rearrangement results in layers of particles that are arranged by increasing density from top to bottom of the bed, with the coal near the top. Hydraulic concentration of fine coal is performed using wet concentrating tables, cyclones, launders, feldspar jigs, and hydrorotators. These processes depend on the physical characteristics—size, shape, and density—of particles suspended in a liquid medium to effect a concentration of desired quality [23].

Froth flotation is a chemical process that depends on the selective adhesion of some solids (i.e., fine coal) to air and the simultaneous adhesion of other particles (i.e., refuse) to water. A separation of coal from coal waste then occurs as finely disseminated air bubbles are passed through a feed coal slurry. Agglomeration works on a principle similar to froth flotation (i.e., differences in the surface properties of coal and inorganic matter), and fine coal particles in a suspension can readily be agglomerated by the addition of a bridging liquid (many different oils), under agitation, and then recovered while the inorganic constituents remain in the aqueous suspension and are rejected.

The preceding processes impact the environment in several ways. This includes dealing with the contamination aspects of fine coal cleaning and "blackwater" disposal, air contamination, refuse disposal and control, and operator health and safety.

4.2.1   Water Contamination from Preparation Plants

The effluents from coal preparation plants and water draining from plant site surfaces contain fine coal and coal refuse materials in suspension. At older plants, the disposal of effluent continues to present a serious problem, since it is becoming increasingly difficult to comply with the standards required by many water authorities and pollution control agencies [23]. New plants, however, are installing more complex clarification facilities, and are designed to operate on a closed circuit to satisfy pollution requirements.

4.2.2   Air Contamination from Preparation Plants

Preparation of fine coal can cause air pollution if the proper dust and gas removal equipment is not installed. The air effluent from a fine coal preparation plant consists of entrained dust, both coal and ash, and various gases, primarily consisting of products of coal combustion from thermal dryers [23]. Sources of particulates include thermal dryers, pneumatic coal-cleaning equipment, coal processing and conveying equipment, screening equipment, coal storage, coal transfer points, and coal handling facilities. The coal dust and ash can be controlled by a combination of mechanical separators, wet scrubbers, electrical precipitators, and filters.

4.2.3   Refuse Contaminants from Preparation Plants

Coal refuse, which varies in composition and size, is a function of the seam mined, the mining system, and the preparation system. Coal refuse consists mainly of unsalable coal, shale, bone, calcite, gypsum, clay, pyrite, or marcasite [23].

Disposal methods for coarse and fine coal refuse developed differently. Prior to modern coal preparation plants, coarse refuse was handpicked from the coal and either discarded back into the mine or deposited on the surface. When fine coal cleaning came into widespread use, disposal became a bigger issue. Early practices were to discharge the "blackwater," or the liquid effluent containing solids, into the nation's stream systems [23]. This is no longer acceptable, and the effluent must now be filtered to remove the fine solids. Embankments are constructed using compacted coarse refuse material to impound fine coal slurry, and the impoundments serve as a filter to clarify this effluent as it flows through the permeable structure. These impoundments are also used as settlement basins and water supply reservoirs.

The environmental impacts for coarse coal refuse sites include the nonproductive use of land, the loss of aesthetic value, water pollution, and air pollution. Many burning coal refuse sites exist in the United States, and they can create local air pollution and safety hazards.

Coal refuse disposal impoundments are constructed for the permanent disposal of any coal, rock, and related material removed from a coal mine in the process of mining. The environmental impacts of impoundments include the nonproductive use of land, the loss of aesthetic value, the danger of slides, dam failures, and water pollution [25].

Most coal waste impoundments in the United States, of which there are approximately 700, are found in the eastern part of the country, mostly in West Virginia, Pennsylvania, Kentucky, and Virginia [24]. Most of the coal mined in the western United States is shipped without extensive processing, so coal waste impoundments are rarely used there. The majority of coal from underground mines is processed before sale. Of the more than 1 billion short tons of coal mined each year in the United States, about 600 to 650 million short tons are processed to some degree, 350 to 400 million short tons are handled in wet-processing systems, and 70 to 90 million short tons of fine refuse are produced [24]. It is estimated that 2 billion short tons of refuse are contained in impoundments in the United States [25].

4.2.4 Health and Safety Issues

The incidence of injuries and fatalities at coal preparation facilities is similar to coal mining, but the largest impact of coal-cleaning facilities is on the environment. Coal waste facilities have been involved in several accidents or incidents over the last 30 years. The dramatic failure on Buffalo Creek in West Virginia (in 1972) brought notice of the hazards associated with embankments. Prior to this accident, little consideration was given to the control of water entering an impoundment from a preparation plant or as runoff or on the discharge of contaminated effluent to the stream system. This accident, which occurred when a coal waste impounding structure collapsed on a Buffalo Creek tributary, resulted in a flood that killed 125 people; injured 1,100 people; left more than 4,000 people homeless; demolished 1,000 cars and trucks, 502 houses, and 44 mobile homes; damaged 943 houses and mobile homes; and caused $50 million in property damages [24].

At the time of the Buffalo Creek accident, there were no federal standards requiring that either impoundments or hazardous refuse piles be constructed and maintained in an approved manner. This changed, however, as a result of this accident, and today numerous federal and state statutes and regulations apply to disposal of coal waste impoundments. The Mining Enforcement and Safety Administration developed standards for impoundments and refuse piles. Nearly every coal waste facility is subject to regulatory requirements imposed by MSHA, OSHA, or a state with a regulatory program approved under the Surface Mining and Reclamation Act of 1977 [24].

Although the industry has become more regulated, embankment failures have still occurred since 1972 [24]. These failures have been smaller than the Buffalo Creek incident in that the quantity of slurry released, the damages incurred, and the number of injuries and deaths (only one person died) have been less. A notable exception is the October 2000 impoundment failure near Inez, Kentucky, in which a seven-acre surface impoundment failed, and approximately 250 million gallons of slurry were released into a nearby underground coal mine. The slurry flowed through the mine and into nearby creeks and rivers, flooding stream banks to a depth of five feet [24]. No loss of life was experienced, but the environmental impact was significant and local water supplies from the rivers were disrupted for days. This incident resulted in Congress requesting the National Research Council to examine ways to reduce the potential for similar accidents in the future [24].

The National Research Council appointed the Committee on Coal Waste Impoundments to do the following:

- Examine the engineering practices and standards currently being applied to coal waste impoundments
- Evaluate the accuracy of mine maps and explore ways to improve surveying and mapping of underground mines to determine how they relate to current or planned slurry impoundments
- Evaluate alternative technologies that could reduce the amount of coal waste generated or allow productive use of the waste
- Examine alternative disposal options for coal slurry

The committee offered many conclusions and recommendations and the implementation of them should substantially reduce the potential for uncontrolled release of coal slurry from impoundments [24].

Despite these efforts, impoundment failures still occur. The most recent incident occurred at the Tennessee Valley Authority's (TVA) Kingston Fossil Plant in East Tennessee. On December 24, 2008, approximately 3.1 million cubic feet of fly ash and water were released onto land adjacent to the plant and into the nearby Clinch and Emory Rivers [26]. Although no loss of life was experienced and no constituents of concern were detected in soil and water samples (water samples from private wells and finished water from water treatment facilities), the impact from this dam failure was, and still is, significant. Concerns about erosion and dust emissions are ongoing, and steps have been taken to minimize them. Fish kill in the river has not been quantified but is considered significant. Also, the coal industry in general suffered a lot of negative publicity as a result of this accident.

4.3 Coal Transportation

Environmental impacts result from the transport of coal. Coal transportation is accomplished through rail, truck, water, slurry pipeline, or conveyor, but most of it is done by rail. Environmental impacts occur during loading, en route, or during unloading, and they can affect natural systems, man-made buildings and installations, and people (i.e., injuries or deaths) [27].

All forms of coal transportation have certain common environmental impacts, including the use of land, structural damage to facilities such as buildings or highways, air pollution from engines that power the transportation systems, and injuries and deaths related to accidents involving workers and the general public (e.g., railway crossing accidents). In addition, fugitive dust emissions are experienced with all forms of coal transport, although more precautionary measures are being practiced [27]. It is estimated that 0.02 percent of the coal loaded is lost as fugitive dust, and an equal amount is lost when it is unloaded. It is estimated that 0.05 to 1 percent of the coal is lost during transit. The actual amount depends on the mode of transit and the length of the trip, but it can be sizeable, especially with unit train coal transit across the country.

4.4   Coal Combustion By-Products

More than 120 million tons of coal-related residue is generated annually by coal-burning plants [28]. These materials have many names and are referred to as *fossil fuel combustion wastes* (FFCW) by the EPA; *coal combustion products* (CCP) by the utility industry, ash marketers, and ash users; and *coal combustion by-products* (CCB) by the U.S. Department of Energy and other federal agencies; they become products when they are used and wastes when they are discarded [29].

The residue includes fly ash, bottom ash, boiler slag, and flue gas desulfurization (FGD) material (i.e., synthetic gypsum). The fly ash is the fine fraction of the CCBs that is entrained in the flue gas that exits a boiler and is captured by particulate control devices. Bottom ash is the large ash particles that accumulate at the bottom of a boiler. Boiler slag is the molten inorganic material that is collected at the bottom of some boilers and discharged into a water-filled pit, where it is quenched and removed as glassy particles. FGD units, which remove sulfur dioxide (SO_2) using calcium-based reagents, generate large quantities of synthetic gypsum, which is a mixture of mainly gypsum ($CaSO_4$) and calcium sulfite ($CaSO_3$) but can also contain fly ash and unreacted lime or limestone [29, 30]. In 2007, CCB production in the United States was approximately 126 million short tons, as shown in Table 4.5 [28]. In the United States, approximately 40 percent of the CCBs are used in a variety of applications, with the remainder discarded [28]. The components of the CCBs have different uses because they have distinct chemical and physical properties that make them suitable for a specific application. CCBs are used in cement and concrete; mine backfill, agriculture, blasting grit, and roofing applications; waste stabilization; wallboard production; acid mine drainage control; and as road base/subbase, antiskid material, fillers, and extenders [29, 30].

Globally, CCB use varies significantly. In Europe, more CCB is used than in the United States, as shown in Table 4.5 [31]. For example, in 2005 89 percent of the CCBs were profitably used in Europe compared to about 40 percent in the United States [28, 31]. The CCBs are used in a number of applications, primarily in concrete, Portland cement manufacture, and road construction. Raw material shortages and favorable state regulations account for higher usage in Europe than in the United States. Countries such as Canada, India, and Japan utilize 33, 13, and 84 percent of their CCBs, respectively [30, 32]. CCB usage in India is low (i.e., about 13 percent)

Table 4.5 Coal Combustion By-Product Production

Category	United States—2007		European Union—2005	
	Million Short Tons	Percent Utilized	Kilo Tonnes[a]	Percent Utilized
Fly ash	71.70	44.1	42.75	91
Bottom ash	18.10	40.4	6.13	84
Boiler slag	2.07	80.3	1.98	100
FGD gypsum	12.30	75.3	11.54	85
FGD material—wet scrubbers	16.60	4.9	—[b]	—
FGD material—dry scrubbers	1.81	8.3	0.05	81
FGD other	2.45	4.6	—	—
FBC ash	1.27	25.4	0.95	81
Other[c]	—	—	0.05	100
Totals	**126.31**	**40.6**	**63.91**	**89**

[a]1 short ton = 0.907 metric tonnes
[b]Not reported
[c]Cenospheres and fly ash and slag from coal gasification
Source: From American Coal Ash Association, 2007 Coal Combustion Product (2008), and European Coal Combustion Products Association, Product and Utilisation of CCPs in 2995 in Europe (2009).

due to the relatively large amount of CCBs produced because of the coal's high ash content [33]. Japan, meanwhile, uses up most of its CCBs due to the high cost of disposal in Japan.

CCBs primarily contain elements such as iron, aluminum, magnesium, manganese, calcium, potassium, sodium, and silica, which for the most part are innocuous. CCBs also contain small amounts of trace elements such as arsenic, barium, beryllium, cadmium, cobalt, chromium, copper, nickel, lead, selenium, zinc, and mercury. These elements can be classified as essential nutrients, toxic elements, or priority pollutants and are considered to have some environmental or public health impacts [29]. The risks include potential groundwater contamination of trace elements and above-ground human health impacts through inhalation and ingestion of contaminants that are released through wind erosion, surface water erosion, and runoff [34].

The Resource Conservation and Recovery Act (RCRA) has been the primary statute governing the management and use of CCBs. The EPA was considering some form of Subtitle C regulation (i.e., classify CCBs as hazardous wastes) under the RCRA for CCBs used in mine backfill or for agricultural applications [34]. During an investigation of the dangers of CCBs to human health and the environment, the EPA concluded that CCBs do not pose sufficient danger to the environment to warrant regulations under RCRA Subtitle C. However, the EPA did determine that CCBs are disposed in landfills and that surface impoundments should be regulated under RCRA Subtitle D (nonhazardous solid waste) [35].

4.5 Emissions from Coal Combustion

Coal combustion produces large quantities of products that may be released into the atmosphere. The emissions are largely steam (i.e., water vapor), which is what we see coming out of a smokestack on a power plant, carbon dioxide, and nitrogen from the air, and they do not present any direct health hazards. However, the emissions do contain small concentrations of atmospheric pollutants, which translate into large quantities emitted due to the vast amount of coal consumed. The principal pollutants that can cause health problems are sulfur and nitrogen oxides; particulate matter; trace elements, including arsenic, lead, mercury, fluorine, selenium, and radionuclides; and organic compounds. The environmental impacts and health effects of these pollutants, along with carbon monoxide and carbon dioxide, are discussed below.

The definition of *air pollution* is the addition to the atmosphere of any material that may have a deleterious effect to life [36]. It results from natural processes or from anthropogenic, or man-made, sources. The legal definition of an *air pollutant* is any air pollution agent or combination of such agents, including any physical, chemical, biological, or radioactive substance, or matter that is emitted into or otherwise enters the ambient air [36]. This includes primary and secondary pollutants, which are classifications that indicate how the various pollutants are formed. Primary pollutants are those that have the same state and chemical composition in the ambient atmosphere as when emitted from sources. Secondary pollutants are those that changed form after leaving the source due to oxidation, decay, or reaction with other primary pollutants.

4.5.1 Sulfur Oxides

Gaseous emissions of sulfur oxides from coal combustion are mainly sulfur dioxide (SO_2) and, to a much lesser extent, sulfur trioxide (SO_3) and gaseous sulfates. The sulfur in the coal reacts with oxygen to form the sulfur oxides:

$$S + O_2 \rightarrow SO_2 \tag{4.6}$$

$$S + 1.5O_2 \rightarrow SO_3 \tag{4.7}$$

Sulfur dioxide is a nonflammable, nonexplosive, colorless gas that causes a taste sensation at concentrations from 0.1 to 1 ppmv (part per million by volume) in air [36]. At concentrations greater than 3.0 ppm, the gas has a pungent, irritating odor. Sulfur dioxide is partly converted to sulfur trioxide or to sulfuric acid (H_2SO_4), and its salts are converted by photochemical or catalytic processes in the atmosphere. When sulfur trioxide combines with water, sulfuric acid is formed.

Environmental Effects

The environmental effects of sulfur compounds include impaired visibility, damage to materials, damage to vegetation, and deposition as acid rain. Fine particles in the atmosphere reduce the visual range by scattering and absorbing light [37].

Aerosols of sulfuric acid and other sulfates make up from 5 to 20 percent of all suspended particulate matter in urban air and thus contribute to the reduction in visibility.

Sulfur compounds are responsible for major damage to materials. Sulfur oxides generally accelerate metal corrosion by first forming sulfuric acid either in the atmosphere or on the metal surface. Sulfur dioxide is the most damaging pollutant when it comes to metal corrosion [37]. Temperature and relative humidity also significantly influence the rate of corrosion. Sulfurous and sulfuric acids are capable of damaging a wide variety of building materials, including limestone, marble, roofing slate, and mortar. Textiles made of nylon are also susceptible to pollutants in the atmosphere.

In general, damage to plants from air pollution usually manifests itself in the leaf structure, since the leaf contains the building blocks for the entire plant [37]. Sulfur dioxide enters the leaf, and the plant cells convert it to sulfite and then to sulfate. Apparently, when excessive sulfur dioxide is present, the cells are unable to oxidize sulfite to sulfate fast enough, and disruption of the cell structure begins. Spinach, lettuce, and other leafy vegetables are most sensitive, as are cotton and alfalfa. Pine needles are also affected, with either the needle tip or the whole needle becoming brown and brittle.

Acidic deposition, or acid rain, occurs when emissions of sulfur dioxide and oxides of nitrogen in the atmosphere react with water, oxygen, and oxidants to form acidic compounds. These compounds fall to the earth in either dry form (i.e., gas and particles) or wet form (i.e., rain, snow, and fog).

The acidity of water is reported in terms of the pH, where pH is the logarithm (base 10) of the molar concentration of hydrogen ions:

$$pH = -\log_{10}[H^+] \tag{4.8}$$

Pure water contains a hydrogen ion concentration that is approximately 10^{-7} molar or pH = 7, which is referred to as neutral pH. Water droplets formed in the atmosphere, however, normally have a pH of about 5.6 because atmospheric carbon dioxide is dissolved in the rain and forms carbonic acid (H_2CO_3). When sulfur dioxide or nitrogen oxides are also dissolved in the water, the pH becomes lower, and average yearly pH values of 4 to 4.5 are reported in the eastern United States [37]. Sulfur dioxide can be absorbed from the gas phase into an aqueous droplet, creating acidic conditions as follows [37]:

$$SO_2(g) \Leftrightarrow SO_2(aq) \tag{4.9}$$

$$SO_2(aq) + H_2O \Leftrightarrow HSO_3^- + H^+ \tag{4.10}$$

$$HSO_3^- \Leftrightarrow SO_3^{2-} + H^+ \tag{4.11}$$

$$SO_3^{2+} + H_2O \Leftrightarrow SO_3^{2-} + 2H^+ \tag{4.12}$$

Acid rain can have many deleterious effects. One is the acidification of natural water sources, which can have a devastating effect on fish. Trout and salmon are particularly sensitive to a low pH [37]. Reproduction in many fish fails to occur at a pH less than 5.5. A decrease in plankton and bottom fauna is also observed as the pH is lowered, which reduces the food supply for the fish. Nutrients are leached from the soil, which can lead to loss in productivity of crops and forests or changes in natural vegetation. The vegetation itself can be directly damaged. An increase in corrosion to materials is also observed.

Health Effects

Sulfur dioxide and other oxides of sulfur have been studied extensively, but many questions remain about the effects of sulfur dioxide on people's health [37]. Few epidemiological studies have been able to differentiate adequately the effects of individual pollutants because sulfur oxides tend to occur in the same kinds of atmospheres as particulate matter and high humidity.

High concentrations of sulfur dioxide can cause temporary breathing impairments in asthmatic children and adults who are active outdoors [38]. Short-term exposures of asthmatic individuals to elevated sulfur dioxide levels, while at moderate exertion, may result in reduced lung function that may be accompanied by such symptoms as wheezing, chest tightness, or shortness of breath. Other effects that have been associated with longer-term exposures to high concentrations of sulfur dioxide, in conjunction with high levels of particulate matter, include respiratory illnesses, alterations in the lungs' defenses, and aggravation of existing cardiovascular diseases. Those that may be affected under these conditions include individuals with cardiovascular disease or chronic lung disease, as well as children and the elderly.

The relationship between human health and the concentration of sulfur dioxide in the atmosphere is very difficult to decipher, especially with the compounding factor that other pollutants, such as particulate matter, and high humidity often occur simultaneously. Consequently, the Electric Power Research Institute (EPRI) has designed a program called the Aerosol Research and Inhalation Epidemiology Study (ARIES) to address the issue of air pollution components by coupling an extensive air quality monitoring effort with five health studies: daily mortality, emergency room visits, heart rate variability, arrhythmic events, and unscheduled physician visits [39].

ARIES started in 1998 with Atlanta, Georgia, being selected by the EPA as the first site. In Atlanta, the health effects of suspended particulate matter (PM)—fine particulate matter less than 2.5 μm ($PM_{2.5}$)—were studied, and preliminary results suggested that adverse health effects appeared to be associated with carbon-containing PM and not the sulfate and nitrate components primarily derived from coal combustion. In Atlanta, the study is to continue through the end of 2010, and "replicate" studies have been started in Dallas (Texas ARIES), Detroit (Detroit Cardiovascular Health Study), Pittsburgh (Pittsburgh ARIES), New York City (Children's Air Pollution Asthma Study), and St. Louis (St. Louis ARIES) [40, 41].

4.5.2 Nitrogen Oxides

Seven oxides of nitrogen are present in ambient air: nitric oxide (NO); nitrogen dioxide (NO_2); and nitrous oxide (N_2O), NO_3, N_2O_3, N_2O_4, and N_2O_5 [37]. Nitric oxide and nitrogen dioxide are collectively referred to as nitrogen oxides (NO_x) because of their interconvertibility in photochemical smog reactions. The term NO_y is often used to represent the sum of the reactive oxides of nitrogen and all other compounds that are atmospheric products of NO_x. NO_y includes compounds such as nitric acid (HNO_3), nitrous acid (HNO_2), nitrate radical (NO_3), dinitrogen pentoxide (N_2O_5), and peroxyacetyl nitrate (PAN). It excludes N_2O and ammonia (NH_3) because they are not normally the products of NO_x reactions [37].

Nitrogen oxide emissions from coal combustion are produced from three sources: thermal NO_x, fuel NO_x, and prompt NO_x. Nitrogen oxides are primarily produced as a result of the fixation of atmospheric nitrogen at high temperatures (thermal NO_x) and the oxidation of coal nitrogen compounds (fuel NO_x). Prompt NO_x is formed when hydrocarbon radical fragments in the flame zone react with nitrogen to form nitrogen atoms, which then form NO. The majority of the oxide species produced is NO, with NO_2 accounting for less than 5 percent of the total, which is usually referred to as NO_x [42].

The production of thermal NO is a function of the combustion temperature and fuel-to-air ratio, and it increases exponentially with temperatures above 2,650°F. Thermal NO can be predicted by the following equation [43]:

$$[NO] = K_l e^{-k_2/T} [N_2][O_2]^{1/2} t \tag{4.13}$$

where T is temperature, t is time, K_1 and K_2 are constants, and $[N_2]$ and $[O_2]$ are concentrations in moles. Accordingly, thermal NO can be decreased by reducing the time, temperature, and concentration of N_2 and O_2. The principal reactions in the formation of thermal NO, which are referred to as the extended Zeldovich mechanism, are:

$$N_2 + O_2 \Leftrightarrow NO + N \tag{4.14}$$

$$N + O_2 \Leftrightarrow NO + O \tag{4.15}$$

$$N + OH \Leftrightarrow NO + H \tag{4.16}$$

Reaction (4.14) is assumed to be the rate-determining step due to the high activation energy required to break the triple bond in the nitrogen molecule. Reaction (4.16) has been found to contribute under fuel-rich conditions. The general conclusion is that very little thermal NO is formed in the combustion zone and that the majority is formed in the post-flame region where the residence time is longer.

Prompt NO is produced by the reaction of hydrocarbon fragments and molecular nitrogen in the flame front. Prompt NO is most significant in fuel-rich flames where the concentration of radicals such as O and OH can exceed equilibrium values, thereby enhancing the rate of NO formation. Prompt NO occurs by the collision of hydrocarbons with molecular nitrogen in the fuel-rich flames to form HCN (hydrogen cyanide) and N. The HCN is then converted to NO by a series of reactions between NCO, H, O, OH, NH, and N. The amount of prompt NO generated is proportional to the concentration of N_2 and the number of carbon atoms present in the gas phase, but the total amount produced is low compared to the total thermal and fuel NO in coal combustion. The two reactions believed to be the most significant to the mechanism for formation of prompt NO are [44]:

$$CH + N_2 \Leftrightarrow HCN + N \tag{4.17}$$

$$C + N_2 \Leftrightarrow CN + N \tag{4.18}$$

Reaction (4.17) was originally proposed, while reaction (4.18) is considered to be only a minor but nonnegligible contributor to prompt NO, with its importance increasing along with increasing temperatures.

Fuel NO is the primary source of NO_x in the flue gas from coal combustion and is formed from the gas-phase oxidation of devolatilized nitrogen-containing species and the heterogeneous combustion of nitrogen-containing char in the tail of the flame [42]. At temperatures below 2,650°F, fuel NO can account for more than 75 percent of the measured NO in coal flames and can be as high as 95 percent. This fuel NO dominance in coal systems is due to the moderate temperatures (2,240–3,140°F) and the locally fuel-rich nature of most coal flames. Fuel NO is produced more readily than thermal NO because the N-H and N-C bonds common in fuel-bound nitrogen are weaker than the triple bond in molecular nitrogen in the air, which must be dissociated to produce thermal NO. Combustion conditions and the nitrogen content of a coal affect the quantity of NO emissions. During devolatilization, a portion of the coal nitrogen is released as HCN and, to a lesser extent, as NH_3. HCN readily reacts with oxygen to form NO, but some of this NO can be converted to N_2 by a reaction with hydrocarbon radicals in fuel-rich zones:

$$Fuel \ N \begin{pmatrix} NH_3 \\ HCN \end{pmatrix} \xrightarrow[NO/O_x]{} \begin{pmatrix} N_2 \\ NO \end{pmatrix} \tag{4.19}$$

Nitrogen retained in the char is also oxidized to NO, which may react with the char surface or hydrocarbon radicals to form N_2. Much of the coal nitrogen is initially converted to NO. The final NO emissions, however, are determined largely by the extent of conversion of NO to N_2 in the various regions of the combustion unit. Unlike thermal NO, the production of fuel NO is relatively insensitive to temperatures over the range found in pulverized coal flames and more sensitive to the air-to-fuel ratio [43, 45].

Environmental Effects

Both NO_x and NO_y (i.e., HNO_3) have been shown to accelerate damage to materials in the ambient air. NO_x affects dyes and fabrics, causing fading, discoloration of archival and artistic materials and textile fibers, and loss of textile fabric strength [37]. NO$_2$ absorbs visible light and at a concentration of 0.25 ppmv will cause appreciable reduction in visibility. Studies have shown that NO$_2$ can suppress the growth of pinto beans and tomatoes and reduce orange yields. In combination with unburned hydrocarbon—that is, volatile organic compounds, (VOCs), which are emitted primarily from motor vehicles but also from chemical plants, refineries, factories, consumer and commercial products, and other industrial sources—nitrogen oxides react in the presence of sunlight to form photochemical smog.

Nitrogen oxides also contribute to the formation of acid rain. NO and NO$_2$ in the ambient air can react with moisture to form NO_3^- and H^+ in the aqueous phase (i.e., nitric acid), which can cause considerable corrosion of metal surfaces. The kinetics of nitric acid formation are not as well understood as those for the formation of sulfuric acid discussed earlier. Nitrogen oxides contribute to changes in the composition and competition of some species of vegetation in wetlands and terrestrial systems, acidification of freshwater bodies, eutrophication (i.e., explosive algae growth, leading to a depletion of oxygen in the water) of esturarine and coastal waters, and increases in levels of toxins that are harmful to fish and other aquatic life [38].

Health Effects

Nitrogen dioxide acts as an acute irritant, and in equal concentrations it is more injurious than NO. However, at concentration found in the atmosphere, NO$_2$ is only potentially irritating and potentially related to chronic obstructive pulmonary disease [37]. EPRI has shown from its early results from the ARIES study that the nitrate components of coal combustion were not the cause for adverse health effects [39]. The EPA reports that short-term exposures (e.g., less than three hours) to current NO$_2$ concentrations may lead to changes in airway responsiveness and lung function in individuals with preexisting respiratory illnesses and increases in respiratory illnesses in children from 5 to 12 years of age [38]. The EPA also reports that long-term exposures to NO$_2$ may lead to increased susceptibility to respiratory infections and may cause alterations in the lungs. Atmospheric transformation of NO$_x$ can lead to the formation of ozone and nitrogen-bearing particles (most notably in some western U.S. urban areas), which are associated with adverse health effects [38].

4.5.3 Particulate Matter

Particulate matter (PM) is the general term used for a mixture of solid particles and liquid droplets found in the air. Some particles are large or dark enough to be seen as soot or smoke, while others are so small that they cannot be seen with the naked eye. These small particles, which come in a wide range of sizes, originate from many different stationary and mobile sources as well as natural sources [38]. Fine particles—those less than 2.5 μm (i.e., PM$_{2.5}$)—are a result of fuel combustion from motor vehicles, power generation, industrial facilities, and residential fireplaces and woodstoves. Coarse particles—those larger than 2.5 μm but classified as less than 10 μm (i.e., PM_{10})—are generally emitted from sources such as vehicles traveling on unpaved roads, materials handling, crushing and grinding operations, and windblown dust [38]. Some particles are emitted directly from their sources, such as smokestacks and cars. In other cases, gases such as SO_2, NO_x, and VOCs react with other compounds in the air to form fine particles.

Coal generally contains from 5- to 20-weight percent of mineral matter (i.e., ash content per a proximate analysis). During combustion, most of the minerals are transformed into dust-sized glassy particles and, along with some unaltered mineral grains and unburned carbon, are emitted from smokestacks.

Particle composition and emission levels are a complex function of firing configuration, boiler operation, and coal properties [36]. In dry bottom, pulverized coal-fired systems (various types of combustion systems are described in Chapter 5), combustion is very good, and the particles are largely composed of inorganic ash residue. In wet bottom, pulverized coal-fired units and cyclone-fired boilers, the amount of fly ash is less than in dry bottom units, since some of the ash melts and is removed from the system as slag. Spreader stokers, which fire a mixture of fine and coarse coal, tend to have a significant quantity of unburned carbon in the fly ash. Overfeed and underfeed stokers emit considerably less particulate than pulverized coal-fired units or spreader stokers, since combustion takes place on a relatively undisturbed bed [36]. Fly ash reinjection for increased consumption of unburned carbon or load changes can also affect particulate emissions.

Environmental Effects

Particulate matter is responsible for the reduction in visibility. Visibility is principally affected by fine particles that are formed in the atmosphere from gas-phase reactions. Although these particles are not directly visible, carbon dioxide, water vapor, and ozone in increased concentrations change the absorption and transmission characteristics of the atmosphere [37].

Particulate matter can cause damage to materials, depending on its chemical composition and physical state [37]. Particles will soil painted surfaces, clothing, and curtains merely by settling on them. Particulate matter can cause corrosive damage to metals either by intrinsic corrosiveness or by the action of corrosive chemicals absorbed or adsorbed by inert particles.

Little is known of the effects of particulate matter in general on vegetation [37]. The combination of particulate matter and other pollutants such as sulfur dioxide may affect plant growth. Coarse particles, such as dust, may be deposited directly onto leaf surfaces and reduce gas exchange, increase leaf surface temperature, and decrease photosynthesis. Toxic particles that consist of elements such as arsenic or fluorine fall on agricultural soils or plants and are ingested by animals, which can be harmful to the animal's health.

Health Effects

Particulate matter alone or in combination with other pollutants constitutes a very serious health hazard. The pollutants enter the human body mainly via the respiratory system. Inhalable particulate matter includes both fine and coarse particles. These particles can accumulate in the respiratory system and are associated with numerous health effects [38]. Exposure to coarse particles is primarily associated with the aggravation of respiratory conditions such as asthma. Fine particles are most closely associated with health effects, including increased hospital admissions and emergency room visits for heart and lung diseases, increased respiratory symptoms and diseases, decreased lung function, and even premature death. Sensitive groups that appear to be at greatest risk include the elderly, individuals with cardiopulmonary diseases like asthma, and children [38].

As previously mentioned, because particulate matter has been linked with adverse health effects at levels currently observed in the United States, EPRI has initiated an epidemiological study (i.e., ARIES) coupled with comprehensive air quality monitoring to allow for identification of many of the components that may be associated with a particular health endpoint [39, 40]. This is critical because knowledge of the true causative agents enable for better protection of public health by regulating those sources that produce harmful pollutants, especially since health effects drive the regulatory agenda for proposed multipollutant legislation and air quality standards (see Chapter 8 for a discussion of legislation).

ARIES was initiated in 1998 in Atlanta, Georgia, and is designed to address the issue of air pollution components by coupling an extensive air quality monitoring effort with five health studies that focus on distinct endpoints to cover the following possible health effects [39]:

- A daily mortality study conducted by Klemm Analysis Group
- An emergency room visits study conducted by Emory University
- A heart rate variability study conducted by Harvard University
- A cardiac arrhythmic events study conducted by Emory University
- An unscheduled physician visits study conducted by Kaiser Permanente

The major findings from the Atlanta study are as follows [47]:

- Detailed characterization of $PM_{2.5}$ demonstrates the importance of carbonaceous matter.
- $PM_{2.5}$ composition varies from hour to hour, day to day, and season to season. Sulfate comprises the largest fraction of $PM_{2.5}$ during the summer, while carbonaceous matter comprises the largest fraction during the spring, fall, and winter.
- Particulate matter components are important, but gases should not be ignored.
- Daily mortality results show that the best model fit for all-cause mortality in those aged 65 and older are observed for CO, NO_2, $PM_{2.5}$, coarse PM, SO_2, and ozone, followed by elemental carbon (EC) and organic carbon (OC).
- Emergency department (ED) visits for cardiovascular disease are associated with NO_2, CO, EC, and OC.
- ED visits for respiratory diseases are associated with PM_{10}, $PM_{2.5}$, $PM_{2.5}$ water-soluble metals, ozone, NO_2, CO, and SO_2.
- Adverse cardiac events in defibrillator patients are associated with coarse PM.

- Ambulatory care visits for child asthma are associated with EC and water-soluble metals; for upper respiratory infections with SO_2; and for lower respiratory infections and oxygenated hydrocarbons.

It is generally recognized that coal-fired power plants can be important contributors to ambient fine particulate matter (i.e., $PM_{2.5}$) mass concentrations and regional haze. In 1999, coal-fired power plants emitted 1.5 percent of the total primary $PM_{2.5}$ in the United States [48]. In response to growing concerns over $PM_{2.5}$ emitted into the atmosphere from coal-fired power plants, the DOE's National Energy Technology Laboratory (DOE-NETL) initiated a research effort to provide the applied science needed to quantitatively relate the emissions from energy production to ambient $PM_{2.5}$ concentrations and composition at downwind receptors and to inform decision makers about management options applicable to coal-fired power generation to achieve the national standards for $PM_{2.5}$ and regional haze. Special emphasis is given to Pittsburgh and the surrounding upper Ohio River Valley region because, on many occasions during the year, this region is downwind of major coal-fired power plants and other industrial sources of air emissions and upwind of the Boston–Washington corridor, the largest regional complex of urban areas in the United States. In addition, DOE teamed with EPRI to perform the TERESA (Toxicological Evaluation of Realistic Emissions of Source Aerosols) project to evaluate the potential toxicity of coal combustion emissions [49, 50, 51]. The project involves exposing laboratory rats via inhalation to realistic coal-fired power plant emissions to help determine the relative toxicity of these PM sources. The TERESA study includes three coal-fired power plants with the following characteristics:

Plant 1: Upper Midwest, burning subbituminous coal, electrostatic precipitator (ESP) for PM control, no scrubber for SO_2 control, no selective catalytic reduction (SCR) for NO_x control

Plant 2: Southeast, burning low- to medium-sulfur eastern bituminous coal, ESP, SCR, no scrubber

Plant 3: Midwest, burning medium- to high-sulfur eastern bituminous coal, ESP, SCR, and scrubber

The emissions from the power plants were then introduced into reaction chambers to simulate several atmospheric conditions, including primary stack emissions; aged plume, oxidized stack emissions, sulfate aerosol formation; aged plume, unneutralized acidity, secondary organic aerosol (SOA) derived from biogenic emissions; aged plume, mixture of neutralized sulfate and SOA; primary emissions without particulate matter, oxidized; only SOA; and primary emissions without particulate matter, oxidized and SOA [50, 51]. The following are some of the results from the study:

- No effects observed at Plant 1.
- Some biological effects observed at Plant 2 under some conditions/scenarios.
- Very few effects observed.
- Results are subtle and lesser in magnitude than responses to concentrated ambient particles (such as ARIES).
- Effects do not appear to be correlated with mass.

4.5.4 Organic Compounds

Organic compounds are unburned gaseous combustibles that are emitted from coal-fired boilers but generally in very small amounts [52]. However, there are brief periods where unburned combustible emissions may increase significantly, such as during system start-ups or upsets. The organic emissions from pulverized coal-fired or cyclone-fired units are lower than from smaller, stoker-fired boilers where operating conditions are not as well controlled [37].

Organic emissions are either a result of constituents present in the coal or are formed as products of incomplete combustion. Polycyclic organic matter (POM) has also been referred to as polynuclear or polycyclic aromatic compounds (PACs). Nine major categories of POM have been identified by the EPA [53]. The most common organic compounds in the flue gas of coal-fired boilers are polycyclic aromatic hydrocarbons (PAHs). Hydrocarbons as a class are not listed as criteria pollutants, although a large number of specific hydrocarbon compounds are listed among the 188 hazardous air pollutants under Title III of the Clean Air Act Amendments of 1990 (see Chapter 8 for a discussion of regulations). Furthermore, the Environmental Protection Agency has specified 16 PAH compounds as priority pollutants: naphthalene, acenaphthylene, acenaphthene, fluorene, phenanthrene, fluoranthene, pyrene, chrysene, benz[a]anthracene, benzo[b]fluoranthene, benzo[a]pyrene, benzo[ghi]perylene, indeno[1,2,3-cd]pyrene, and dibenz[ah]anthracene.

Environmental Effects

Gaseous hydrocarbons as a broad class do not appear to cause any appreciable corrosive damage to materials [37]. Of all the hydrocarbons, only ethylene has adverse effects on plants at known ambient concentrations, which includes inhibiting plant growth and injury to orchids and cotton.

Health Effects

Studies of the effects of ambient air concentrations of many gaseous hydrocarbons have not demonstrated direct adverse effects upon human health [37]. Certain airborne PAHs, however, are known carcinogens. Also, studies of the carcinogenicity of certain classes of hydrocarbons indicate that some cancers appear to be caused by exposure to aromatic hydrocarbons found in soot and tars. An extreme example is the occurrence of highly elevated incidences of lung cancer in China from PAH exposure [54]. PAHs are released during unvented coal combustion of "smokey" coal in homes, resulting in lung cancer mortality five times the national average in China.

4.5.5 Carbon Monoxide

Carbon monoxide (CO) is a colorless, odorless gas that is very stable and has a life of two to four months in the atmosphere [37]. Like organic compounds, it is formed when fuel is not burned completely. It is a component of motor vehicle exhaust, which contributes about 56 percent of all CO emissions nationwide, with nonroad engines and vehicles accounting for another 22 percent [55]. High concentrations of CO occur in areas with heavy traffic congestion, where as much as 95 percent of all CO emissions may come from automobile exhaust. Other sources include industrial processes, nontransportation fuel combustion, and natural sources such as wildfires.

Carbon monoxide emissions from coal-fired boilers are generally low. Like organic hydrocarbon emissions, CO can be formed during system start-ups or upsets. Also, systems with good combustion control, which is typical of power generation plants, produce little CO.

Environmental Effects

Carbon monoxide appears to have no detrimental effects on material surfaces [37]. At ambient concentrations, experiments have not shown CO to produce any harmful effects on plant life. Carbon monoxide has been found to be a minor participant in photochemical reactions, leading to ozone formation.

Health Effects

High concentrations of CO can cause physiological and pathological changes and ultimately death. Carbon monoxide enters the bloodstream through the lungs and reduces oxygen delivery to the body's organs and tissues [38]. The health threat from lower levels of CO is most serious for those who suffer from cardiovascular diseases, such as angina pectoris. At much higher levels of exposure, CO can be poisonous, and even healthy individuals can be affected. Visual impairment, reduced work capacity, reduced manual dexterity, poor learning ability, and difficulty in performing complex tasks are all associated with exposure to elevated CO levels.

4.5.6   Trace Elements

All coals contain small concentrations of trace elements. Trace elements enter the atmosphere through natural processes, and sources of trace elements include soil, seawater, and volcanic eruptions. Human activities, such as power generation and industrial and commercial sectors, also lead to emissions of some elements. Although these elements are present in small concentrations in the coal—parts per million (ppm) by weight—the large amount of coal burned annually mobilizes tons of these pollutants as particles or gases.

Title III of the U.S. Clean Air Act Amendments (CAAA) of 1990 designates 188 hazardous air pollutants (HAPs). Included in the list are 11 trace elements: antimony (Sb), arsenic (As), beryllium (Be), cadmium (Cd), chromium (Cr), cobalt (Co), lead (Pb), manganese (Mn), mercury (Hg), nickel (Ni), and selenium (Se). In addition, barium (Ba) is regulated by the Resources Conservation and Recovery Act, and boron (B) and molybdenum (Mo) are regulated by Irrigation Water Standards [52]. Vanadium (V) is regulated based on its oxidation state, and vanadium pentoxide (V_2O_5) is a highly toxic regulated compound. Other elements, such as fluorince (F) and chlorine (Cl), which produce acid gases (i.e., HF and HCl) upon combustion, and radionuclides such as radon (Rn), thorium (Th), and uranium (U) are also of interest.

Partitioning of the trace elements in the bottom ash, ash collected in the air pollution control device, and fly ash and gaseous constituents emitted into the atmosphere depends on many factors, including the volatility of the elements, temperature profiles across the system, pollution control devices, and operating conditions [36, 56]. Numerous studies have shown that trace elements can be classified into three broad categories based on their partitioning during coal combustion. A summary of these studies is presented by Clarke and Sloss [56]. Figure 4.5 illustrates the classification scheme for selected elements.

Class I elements are the least volatile and are concentrated in the coarse residues (i.e., bottom ash) or equally partitioned between coarse residues and finer particles (i.e., fly ash). Class II elements will volatilize in the boiler but condense downstream and are concentrated in the finer-sized particles. Class III elements are the most volatile and exist entirely in the vapor phase. Overlap between classifications exits, which can be a function of fuel, combustion system design, and operating conditions, especially temperature [56].

Environmental Effects

The environmental effects of trace elements are a function of the chemical and physical form in which they are found [56]. Environmental effects may occur due to the element itself or as a result of a combination of the element and other compounds. Linking a specific environmental effect to an individual element can be difficult, as is determining the contribution from human activities, since the trace elements also occur naturally.

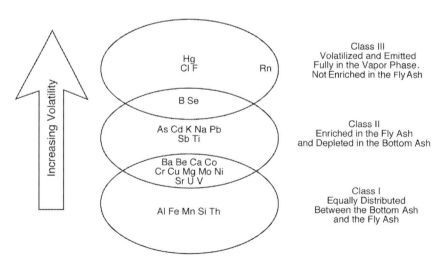

Figure 4.5 Classification scheme for selected trace elements relative to their volatility and partitioning in power plants.
Source: Modified from Miller et al. (1996) and Clarke and Sloss (1992).

Some trace elements may have an immediate effect on the atmosphere. Trace element metals such as Mn (II) and Fe (III) may contribute to acid rain by promoting oxidation of sulfur dioxide to sulfate in water droplets [56]. Trace elements may also be involved in the complex atmospheric chemistry that forms photochemical smog and may affect cloud formation.

Soils may contain high concentrations of certain trace elements due to natural minerals and ores. In addition, deposition of trace elements downwind from power plants can lead to high concentrations in the soils and uptake by plants. Some elements found in coal are major plant nutrients, specifically calcium, magnesium, and potassium [56]. These elements are not considered trace elements, since they occur in quantities greater than 1,000 ppm (i.e., >0.1 percent by weight). Other elements (both major and trace) are considered minor plant nutrients, such as iron, manganese, copper, zinc, molybdenum, cobalt, and selenium. Elements such as aluminum, sodium, and vanadium are considered essential for some species, while others are potentially toxic, including chromium, nickel, lead, arsenic, and cadmium.

Cadmium

Cadmium (Cd) is a silvery metal and is both toxic and carcinogenic [57]. The correlation between soil concentration of cadmium and plant uptake is not clear. Consequently, there is concern about cadmium concentrations in the environment and ingestion of plant-based foods by the general population.

Mercury

Mercury (Hg) is a liquid silvery metal that is considered toxic; when it is in the form of methylmercury (CH_3Hg), it is extremely toxic [46, 57]. The mercury directly emitted from power plants is measured as three forms: elemental ($Hg°$), oxidized (Hg^{+2}), and condensed on ash particles (Hg_p). In the natural environment, mercury can go through a series of chemical transformations to convert mercury to a highly toxic form, methylmercury, that is concentrated in fish and birds [12]. Methylation rates in the ecosystems are a function of mercury availability, bacterial population, nutrient load, acidity and oxidizing conditions, sediment load, and sedimentation rates. Methylmercury enters the food chain, particularly in aquatic organisms, and bioaccumulates.

Volatile elements that are emitted from power plants, such as mercury, are mostly found either in gaseous form or enriched on the surface of fine particles and physically should be available for uptake by plants [56]. There is evidence, however, that almost no mercury from the soil is taken into the shoots of the plants, so this appears to be an important barrier against entry of mercury into the aboveground ecosystem, even if accumulation in the soil has occurred.

Lead

Lead (Pb) is the only metal currently listed as a criteria pollutant. Lead is a gray metal with a low melting point that is soft, malleable, ductile, resistant to corrosion, and a relatively poor electrical conductor [37]. For these reasons, lead has been used for more than 4,000 years for plates and cups, food storage vessels, paints, piping, roofing, storage containers for corrosive materials, radiation shields, lead-acid batteries, and as an organolead additive in gasoline. As result, lead can be found throughout the world, including trace amounts in Antarctica and the Arctic [37]. In the past, automotive sources were the major contributors of lead emissions into the atmosphere [38]. With the EPA's regulatory efforts to remove lead from gasoline, along with banning lead from paint pigments and solder, a decline in lead emissions has been observed. The highest concentrations of lead are found in the vicinity of nonferrous and ferrous smelters and battery manufacturers. Lead can be deposited on the leaves of plants, presenting a hazard to grazing animals.

Selenium
Crops, including animal forages, are sensitive to the addition of small amounts of selenium (Se) in the soil [56]. Selenium is a silvery metallic allotrope or red amorphous powder [57]. Selenium is toxic to plants at low concentrations and has been shown to cause stunting and brown spots in some varieties of beans and to reduce germination and cause stunting in cereals and cotton.

Other Trace Elements
Clarke and Sloss also listed examples where trace elements from coal combustion may have beneficial effects in some areas [56]. Boron is a micronutrient and is required in trace amounts by many plants and animals, and the amount of boron release from coal-fired power plants is likely to be beneficial to local agriculture. Copper, iron, manganese, and zinc are necessary for normal growth of plants. It is recognized, however, that excessive concentrations of copper and zinc lead to damage to root formation and growth, but the quantities being deposited around coal-fired power stations are likely to be beneficial to local soils. Also, manganese can be harmful in large doses, especially in acidic soils.

Health Effects

Trace element emissions can potentially cause a number of harmful effects on human health. While there is no evidence that most trace elements from coal-fired power plants are causing health effects due to their low ambient air concentration, there is concern that pollutants may become accumulated through the food chain. This is especially true of mercury, which is discussed in more detail later in this section. The health impacts of various trace elements are also discussed.

Arsenic
The combustion of most coals is unlikely to contribute toxic amounts of arsenic (As) into the air [58]. There is some concern, however, that arsenic in fly ash disposal and coal cleaning wastes may be leached into the groundwater. Arsenic, which is gray, metallic, soft, and brittle, may be considered essential, but it is toxic in small doses [57]. Arsenic can cause anemia, gastric disturbances, renal symptoms, ulceration, and skin and lung cancers [56]. In addition, arsenic can damage peripheral nerves and blood vessels, and it is a suspected teratogen (i.e., causes damage to

embryos and fetuses). The chemical form of arsenic can affect its toxicity, with organic forms of arsenic being more toxic than elemental arsenic.

An extreme example of chronic arsenic poisoning is currently occurring in the Gizhou Province of China, [54, 59]. In this province, the villagers bring their chili pepper harvest indoors in the autumn to dry. They hang their peppers over open-burning stoves, where arsenic rich-coal—up to 35,000 ppm—is used to heat and cook. Chili peppers, which normally contain less than 1 ppm of arsenic, can contain as much as 500 ppm of arsenic after drying. About 3,000 people are exhibiting typical symptoms of arsenic poisoning, including hyperpigmentation (flushed appearance, freckles), hyperkeratosis (scaly lesions on the skin, generally concentrated on the hands and feet), Bowen's disease (dark, horny, precancerous lesions of the skin), and squamous cell carcinoma.

Boron
Boron (B) is similar to arsenic in that the concentration emitted into the atmosphere is small and is unlikely to cause problems as an airborne pollutant [58]. However, boron in the fly ash can become soluble in ash disposal sites. Boron, a dark powder, is essential for plants but can be toxic in excess [57].

Beryllium
Berryllium (Be), a silvery, lustrous, relatively soft metal, is toxic and carcinogenic [57]. It can cause respiratory disease and lymphatic, liver, spleen, and kidney problems [56].

Cadmium
Cadmium (Cd) has no known biological function and is therefore not a nutritional requirement [56]. Cadmium is toxic, carcinogenic, and teratogenic [57]. It can cause emphysema, fibrosis of the lung, renal injury, and possibly cardiovascular disease.

Chromium
Chromium (Cr), a hard, blue-white metal, is an essential trace element, but its chromates are toxic and carcinogenic [57]. Chromium, which is ingested by humans through food and drink, can be toxic when it accumulates in the liver and spleen [56]. The oxidation state of chromium affects its mobility and toxicity. Chromium (III) is nontoxic and has a tendency to absorb into clays, sediments, and organic matter and therefore is not very mobile. Chromium (IV), however, is more mobile and toxic and may make up approximately 5 percent of the total chromium particles emitted from power plants.

Fluorine
Fluorine (F) is a pale yellow gas and is the most reactive of all elements [57]. It is an essential element and is commonly used for protecting the enamel of teeth, but excess fluoride is toxic. Fluorosis includes mottling of tooth enamel (dental

fluorosis) and various forms of skeletal damage, including osteosclerosis, limited movement of the joints, and outward manifestations such as knock-knees, bowlegs, and spinal curvature [59]. Fluorosis combined with nutritional deficiencies in children can result in severe bone deformation. An extreme example of this is exhibited in China, where the health problems caused by fluorine volatilized during domestic coal use are far more extensive than those caused by arsenic [54, 59]. More than 10 million people in Guizhou Province and surrounding areas suffer from various forms of flurosis, where they dry corn over unvented ovens that burn coal that is very high (>200 ppm) in fluorine.

Mercury

Mercury (Hg) exists in trace amounts in fossil fuels, including coal, vegetation, crustal material, and waste products. Through combustion or natural processes, mercury vapor can be released into the atmosphere, where it can drift for a year or more, spreading with air currents over vast regions of the world [60]. Research indicates that mercury poses adverse human health effects, with fish consumption the primary pathway for human and wildlife exposure. Mercury bioaccumulates in fish as methylmercury (CH_3Hg) and poses a serious health hazard for humans.

Other research suggests that other forms of mercury may be harmful as well [61, 62]. Ingested mercury in elemental, organic, and inorganic form is converted to mercuric mercury, which is slowly eliminated from the kidneys but remains fixed in the brain indefinitely [62]. Exposure to high levels of metallic, inorganic, or organic mercury can permanently damage the brain, kidneys, and developing fetus [61]. Associations between low-dose prenatal exposure to methylmercury and neurodevelopmental effects on attention, motor function, language, visual-spatial, and verbal abilities have been documented [63].

Loss of sight has been associated with cases of extreme mercury ingestion [59]. Chronic thallium poisoning in the Guizhou Province, China, has been reported, where vegetables are grown on mercury/thallium-rich mining slag. Most symptoms that have been reported, such as hair loss, are typical of thallium poisoning. However, many patients from this region have lost their vision, which is being attributed to mercury poisoning, since the mercury concentration of this coal is 55 ppm, or about 200 times the average mercury concentration in U.S. coals.

Manganese

Manganese (Mn) is unlikely to cause health problems as an airborne pollutant from the combustion of most coals, but leaching of ash may be a concern [58]. Manganese, a hard, brittle, silvery metal, is considered an essential nutrient, is nontoxic and is a suspected carcinogen [57]. It is also reported to cause respiratory problems [56].

Molybdenum

Molybdenum (Mo), a lustrous, silvery, and fairly soft metal, is an essential nutrient, is moderately toxic, and a teratogenic [57]. Under some conditions, molybdenosis may occur in animals, notably ruminants, through eating vegetation with relatively high concentrations of molybdenum [58].

Nickel
Nickel (Ni), a silvery, lustrous, malleable, and ductile metal, has no known biological role, and nickel and nickel oxide are carcinogenic [57]. Nickel can cause dermatitis and intestinal disorders [56].

Lead
Exposure to lead (Pb) occurs mainly through inhalation of air and ingestion in food, water, soil, or dust. It accumulates in the blood, bones, and soft tissues [38]. Lead can adversely affect the kidneys, liver, nervous system, and other organs. Excessive exposure to lead may cause neurological impairments, such as seizures, mental retardation, and behavioral disorders. Even at low doses, lead exposure is associated with damage to the nervous systems of fetuses and young children. Lead may be a factor in high blood pressure and heart disease.

Selenium
Selenium (Se) is considered an essential element, but it is toxic in excess of dietary requirements and is also a carcinogen [57]. Livestock that consume plants with excessive amounts of selenium can suffer two diseases called alkali disease and blind staggers, as well as infertility and cirrhosis of the liver. In extreme cases, selenium can cause death [56]. In humans, selenium can cause gastrointestinal disturbance, liver and spleen damage, and anemia, and it is a suspected teratogen. Symptoms of selenium poisoning include hair and nail loss. Selenosis has been reported in southwest China where selenium-rich carbonaceous shales, locally known as "stone coals," are used for home heating and cooking [59]. The ash from his selenium-rich coal (as much as 8,400 ppm Se) is then used as a soil amendment, thereby introducing high concentrations of selenium into the soil, which is subsequently absorbed by crops.

Vanadium
Vanadium (V), a shiny, silvery, soft metal, is an essential trace element, although some compounds, specifically vanadium pentoxide (V_2O_5), are quite toxic [56]. Health effects associated with vanadium include acute and chronic respiratory dysfunction [56].

Radionuclides
Radionuclides are listed generically as a 1990 CAAA HAP. Radioactivity arises mainly from isotopes of lead, radium, radon, thorium, and uranium [52, 56, 58, 64]. Health effects from radiation are well documented including various forms of cancer; however, radionuclide emissions from power plants are quite low [52, 56]. During coal combustion, most of the uranium, thorium, and their decay products are released from the original coal matrix and are distributed between the gas phase and solid combustion products [64]. Virtually 100 percent of the radon gas present in the coal feed is transferred to the gas phase and is lost in stack emissions. In contrast, less volatile elements such as thorium and uranium, and the majority of their

decay products, are retained in the solid combustion wastes [52, 64]. Fly ash is commonly used as an additive to concrete building products, but the radioactivity of typical fly ash is not significantly different from that of more conventional concrete additives or other building materials such as granite or red brick [64].

4.5.7 Greenhouse Gases—Carbon Dioxide

The planet Earth naturally absorbs and reflects incoming solar radiation and emits longer-wavelength terrestrial (thermal) radiation back into space [65]. On average, the absorbed solar radiation is balanced by the outgoing terrestrial radiation emitted to space (Figure 4.6) [66, 67]. A portion of this terrestrial radiation, though, is absorbed by gases in the atmosphere. These gases, known as greenhouse gases, have molecules that have the right size and shape to absorb and retain heat. These gases include water vapor (H_2O), carbon dioxide (CO_2), methane (CH_4), nitrous oxide (N_2O), and to a lesser extent halocarbons consisting of hydrochlorofluorocarbons, perfluorocarbons, and sulfur hexafluoride (SF_6). The energy from this absorbed terrestrial radiation warms Earth's surface and atmosphere, creating what is known as the natural greenhouse effect, which is what makes Earth inhabitable. Without the natural heat-trapping properties of these atmospheric gases, the average surface temperature of Earth would be about 60°F lower.

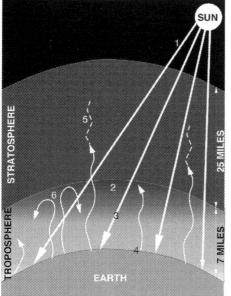

Figure 4.6 The greenhouse effect.
Source: Modified from U.S. DOE (2000) and Peabody Holding Company, Clearing the Air about the Greenhouse Effects (1997).

Although the Earth's atmosphere consists of mainly oxygen and nitrogen—more than 99 percent of the dry atmosphere—neither plays a significant role in enhancing the greenhouse effect because both are essentially transparent to terrestrial radiation [65]. The greenhouse gases comprise the remaining slightly less than 1 percent of the atmosphere, of which more than 97 percent is water vapor. Methane, carbon dioxide, nitrous oxide, and other gases comprise the balance of the greenhouse gases. The composition of global greenhouse gas emissions in 2000 includes the following [68]:

- CO_2 fuel and cement: 55 percent
- CO_2 land use change and forestry: 19 percent
- Methane (CH_4): 16 percent
- Nitrous oxide (N_2O): 9 percent
- High global warming potential gases: 1 percent

Carbon dioxide, methane, and nitrous oxide are continuously emitted to and removed from the atmosphere by the Earth's natural processes. Anthropogenic activities, however, can cause additional quantities of these and other greenhouse gases to be emitted or sequestered, thereby changing their global average atmospheric concentrations. Natural activities such as respiration by plants or animals and seasonal cycles of plant growth and decay generally do not alter average atmospheric greenhouse gas concentrations over decadal time frames [65]. Climatic changes, which are long-term fluctuations in temperature, precipitation, wind, and other elements of Earth's climate system, resulting from anthropogenic activities can have positive or negative feedback effects on these natural systems.

Overall, the most abundant and dominant greenhouse gas in the atmosphere is water vapor. Water vapor varies in concentration in the atmosphere both spatially and temporally, and it can range from 0 to 4 percent of the total atmosphere. The highest concentrations of water vapor are found near the equator over the oceans and tropical rain forests. Water vapor concentrations can approach 0 percent near cold polar regions and subtropical continental deserts. Human activities, however, are not believed to directly affect the average global concentration of water vapor, although this currently is being debated [65, 69].

The fifth most abundant gas in the atmosphere is carbon dioxide at 0.036 percent (note that Argon is the fourth at 0.93 percent, but Argon is not a greenhouse gas). In nature, carbon dioxide is cycled between various atmospheric, oceanic, land biotic, marine biotic, and mineral reservoirs. Of all the greenhouse gases, human activity has the largest influence on carbon dioxide, which is a product of the combustion of fossil fuels, the cement industry, and land-use change and forestry. Carbon dioxide concentrations in the atmosphere increased from approximately 280 ppmv in preindustrial times to approximately 385 ppmv by February 2009 [65, 70]. Figure 4.7 illustrates the rise in atmospheric CO_2 concentration for the past nearly 50 years [70].

Methane is primarily produced through anaerobic decomposition of organic matter in biological systems. Agricultural processes such as wetland rice cultivation,

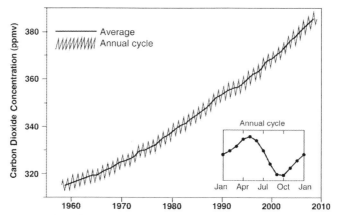

Figure 4.7 Atmospheric CO_2 concentration at Mauna Loa Observatory.
Source: From U.S. Department of Commerce, Trends in Atmospheric CO_2 (2009).

enteric fermentation in animals, and the decomposition of animal wastes emit methane, as does the decomposition of municipal solid wastes [65]. Methane is also emitted during the production and distribution of natural gas and petroleum. Methane is released as a by-product of coal mining and, to a lesser extent, incomplete fossil fuel combustion.

Anthropogenic sources of nitrous oxide include agricultural soils, especially through the use of fertilizers, fossil fuel combustion; especially from mobile sources; nylon and nitric acid production; wastewater treatment; waste combustion; and biomass burning [65]. Halocarbons that contain chlorine (e.g., chlorofluorocarbons, hydrofluorocarbons, methyl chloroform, and carbon tetrachloride) and bromine (e.g., halons, methyl bromide, and hydrobromofluorocarbons), perfluorocarbons, and sulfur hexafluoride (SF_6) are human-made chemicals and not products of combustion. They are, however, powerful greenhouse gases.

One concept, global warming potentials (GWPs), has been devised to evaluate the relative effects of emissions over a given time period in the future [71]. GWPs take account of the differing times that gases remain in the atmosphere, their greenhouse effect while in the atmosphere, and the time period over which climatic changes are of concern. GWPs are intended as a quantified measure of the globally averaged relative radiative forcing impacts of a particular greenhouse gas [65]. It is defined as the cumulative radiative forcing—both direct and indirect effects—integrated over a period of time from the emission of a unit mass of gas relative to some reference gas. Table 4.6 summarizes the greenhouse gases, their major anthropogenic sources, and their GWPs [65]. Carbon dioxide has been chosen as the reference gas, and, as an example, methane's GWP is 21, which means that methane is 21 times better at trapping heat in the atmosphere than carbon dioxide. GWPs are typically reported on a 100-year time horizon. The EPA uses a time period of 100 years for policy making and reporting purposes.

Table 4.6 Greenhouse Gases, Major Anthropogenic Sources, and Global Warming Potentials

Greenhouse Gas	Major Anthropogenic Sources	Global Warming Potential
Carbon dioxide (CO_2)	Fossil fuel combustion, iron and steel production, cement manufacture	1
Methane (CH_4)	Landfills, enteric fermentation, natural gas systems, coal mining, manure management	21
Nitrous oxide (N_2O)	Agriculture soil management, mobile sources, nitric and adipic acid production, manure management, stationary sources, human sewage	310
HFC-23	Substitution of ozone-depleting substances, semiconductor manufacture, mobile air conditioners, HCFC (hydrochlorofluorocarbons)-22 production	11,700
HFC-125		2,800
HFC-134a		1,300
HFC-143a		3,800
HFC-152a		140
HFC-227		2,900
HFC-236fa		6,300
HFC-431mee		1,300
CF_4	Substitution of ozone-depleting substances, semiconductor manufacture, aluminum production	6,500
C_2F_6		9,200
C_4F_{10}		7,000
C_6F_{14}		7,400
SF_6	Electrical transmission and distribution systems, magnesium casting	23,900

The greenhouse gas emissions of direct greenhouse gases in the U.S. inventory are reported in terms of equivalent emissions of carbon dioxide, using teragrams of carbon dioxide equivalents (Tg CO_2eq). The relationship between gigagrams (Gg) of a gas and Tg CO_2eq can be expressed as follows:

$$Tg\ CO_2eq = (Gg\ of\ gas)x(GWP)x\left(\frac{Tg}{1,000Gg}\right) \tag{4.20}$$

Quantities of emissions from coal-fired activities are provided in Chapter 8.

Environmental Effects

There is some uncertainty surrounding global climate change. The global climate is a massive and highly complex system with many interrelated subsystems. Evidence shows that in the past, the concentration of greenhouse gases in Earth's atmosphere has been both higher and lower than it is today, but it is difficult to separate causes and effects of those situations, and the data from earlier time periods are limited and imprecise [69].

One important cause for uncertainty in the area of global science lies in feedback loops. Complex climate change models have been developed, and different modelers judge different feedback to be more important, which leads to markedly different predictions about Earth's climate [69].

Many of the undisputed facts, however, have not changed over the last decade. For example, based on samples of air trapped in arctic ice, scientists have determined that before the Industrial Revolution, the concentration of carbon dioxide in the atmosphere had been stable at a level of around 280 to 290 ppmv. When people started to burn fossil fuels, the concentration of carbon dioxide began to increase and is now at approximately 385 ppmv [65, 70]. This correlation indicates that increased concentrations of greenhouse gases in the atmosphere have likely increased the amount of heat from the sun that stays within Earth's ecosystem, thereby contributing to increased global temperatures. The Goddard Institute for Space Studies has reported that the average temperature on Earth's surface has risen approximately 2°F from 1870 to 2007 [72]. Studies project that globally averaged surface temperatures will increase by 2.5 to 10.4°F between 1990 and 2100 at current rates of increase under a business as usual scenario [65].

Global warming has many potential environmental impacts, including effects on agriculture production, forests, water resources, coastal areas, and species and natural areas [73, 74]. Warmer temperatures can lead to more intense rainfall and flooding in some areas (e.g., the Pacific Northwest and the Midwest), with more frequent droughtlike conditions in other areas, such as the western United States.

Predictions of the agricultural effects of climate change remain uncertain, but models indicate potential changes in cereal grain production and irrigation demands in the United States. Models also predict that rising temperatures could affect current land use, such as the reduction of coffee-growing areas in countries such as Uganda.

Predictions indicate that forest areas could be seriously affected. As temperatures increase, the composition of a forest could change, with warmer-climate varieties moving into traditionally colder-climate areas. Forest health and productivity could be impacted.

The effect of global warming on freshwater resources is uncertain. Some studies indicate that global water conditions will worsen, while others suggest that climate change could have a net positive impact on global water resources. Impacts on water supply, water quality, and competition for water are all predicted.

Coastal areas are predicted to undergo erosion of beaches, inundation of coastal lands, and increased costs to protect coastal communities. Current rates of sea-level rise are expected to increase two to five times due to both thermal expansion of the oceans and the partial melting of mountain glaciers and polar ice caps. Low-lying areas along the Gulf of Mexico and estuaries like Chesapeake Bay are especially vulnerable. Bangladesh is the country most vulnerable to sea-level rise, while the Nile Delta, Egypt's only suitable agriculture area, would be flooded.

Changes in temperature and water availability could decimate nonintensively managed ecosystems, such as forests, rivers, and wetlands. For example, global warming could dry out the wetlands that support more than 50 percent of North American waterfowl.

Arctic and northern latitudes are likely to experience above-average warming and are especially vulnerable to its effects. This includes thinning of the arctic sea ice, and disruption may affect fisheries, human structures built on permafrost, and northern ecosystems.

Health Effects

Human health is also predicted to be impacted by global climate change [73, 74]. This includes increases in weather-related mortality, infectious diseases, and air-quality respiratory illnesses. Small increases in average temperatures can increase the spread of diseases like malaria and dengue fever and lead to a significant rise in the number of extreme heat waves. Elderly people are particularly vulnerable to heat stress. Heat waves could also aggravate local air quality problems, which pose threats to young children and asthmatics.

REFERENCES

[1] M. Sengupta, Environmental Impacts of Mining: Monitoring, Restoration, and Control, Lewis Publishers, 1993.

[2] National Mining Association, Most Requested Statistics—U.S. Coal Industry, National Mining Association, April 2008.

[3] World Book Encyclopedia, Coal, vol. 4, World Book, Inc., 2001, pp. 716–733.

[4] H.H. Schobert, Coal: The Energy Source of the Past and Future, American Chemical Society, 1987.

[5] J.J. Marcus (Ed.), Mining Environmental Handbook: Effects of Mining on the Environment and American Environmental Controls on Mining, Imperial College Press, 1997.

[6] W. Irving, O. Tailakov, CH_4 Emissions: Coal Mining and Handling, Good Practice Guidance and Uncertainty Management in National Greenhouse Gas Inventories, 1999.

[7] EPA (U.S. Environmental Protection Agency), Coalbed Methane Outreach Program, www.epa.gov/cmop/basic.html, last updated September 2, 2008.

[8] EPA, Global Mitigation of Non-CO_2 Greenhouse Gases, U.S. Government Printing Office, June 2006.

[9] EPA, Coalbed Methane Outreach Program, www.epa.gov/coalbed/about.htm#section2, March 3, 2003.

[10] EPA, Global Overview of CMM Opportunities, U.S. Government Printing Office, January 2009.

[11] World Bank Group, Coal Mining and Production, Pollution Prevention and Abatement Handbook, www.natural-resources.org.minerals/CD/docs/twb/PPAH/52.coal.pdf, July 1998.

[12] USGS (U.S. Geological Survey), Mercury in U.S. Coal—Abundance, Distribution, and Modes of Occurrence, USGS Fact Sheet FS-095-01, U.S. Government Printing Office, September 2001.

[13] PADEP (Pennsylvania Department of Environmental Protection), Science of Acid Mine Drainage and Passive Treatment, www.dep.state.pa.s/dep/deputate/minres/bamr/amd/science_of_amd.htm, April 5, 2002.

[14] OSM (U.S. Office of Surface Mining), Factors Controlling Acid Mine Drainage, Formation, *www.osmre.gov/amdform.htm*, March 5, 2002.

[15] CARE (Coalition for Affordable and Reliable Energy), *www.careenergy.com*, 2003.

[16] MSHA (Mine Safety & Health Administration), *www.msha.gov*.

[17] MSHA, 2008 Comparison of Year-to-Date and Total Fatalities for M/NM & Coal, *www.msha.gov*, last updated December 31, 2008.

[18] EIA (Energy Information Administration), Coal Data: A Reference, U.S. Department of Energy, February 1995.

[19] ACF (American Coal Foundation), Strict Regulations Govern Coal Mining, *www.ket.org/trip/Coal/AGSMM/agsnmregs.html*, 2001.

[20] L.J. Jackson, Surface Coal Mines—Restoration and Rehabilitation, IEA Coal Research, 1991.

[21] OSM, Chronology of the Office of Surface Mining and the Surface Mining Law Implementation, *www.osmre.gov*, February 13, 2003.

[22] OSM, Regulating Surface Coal Mining, Restoring Mining Landscapes, *www.doe.gov/energy_security/usCoalResources.html*, downloaded June 2003.

[23] J.W. Leonard (Ed.), Coal Preparation, fourth ed., The American Institute of Mining, Metallurgical, and Petroleum Engineers, Inc., 1979.

[24] National Research Council, Coal Waste Impoundments: Risks, Responses, and Alternatives, National Academy Press, 2002.

[25] S. Falcone Miller, J.L. Morrison, A.W. Scaroni, The Utilization of Coal Pond Fines as Feedstock for Coal-Water Slurry Fuels, in: Proceedings of the 20th International Technical Conference on Coal Utilization & Fuel Systems, Coal & Slurry Technology Association, 1995, pp. 535–558.

[26] Coal Age, TVA Works to Clear Kingston Ash Spill, vol. 114, No.1, January 2009, p. 4.

[27] M.J. Chadwick, N.H. Highton, N. Lindman, Environmental Impacts of Coal Mining and Utilization, Pergamon Press, The Beijer Institute, 1987.

[28] ACAA (American Coal Ash Association), 2007 Coal Combustion Product (CCP) Production & Use Survey Results, *www.ACAA-USA.org*, September 15, 2008.

[29] A.G. Kim, W. Aljoe, S. Renninger, Wastes from the Combustion of Fossils Fuels: Research Perspective on the Regulatory Determination, in: Proceedings of the 16th International Conference on Fluidized Bed Combustion, Council of Industrial Boiler Owners, 2001.

[30] R. Kalyoncu, Coal Combustion Products, Nineteenth Annual International Pittsburgh Coal Conference, Coal—Energy and the Environment, University of Pittsburgh, 2002.

[31] ECOBA (European Coal Combustion Products Association), Product and Utilisation of CCPs in 2005 in Europe, *www.ecoba.org* (accessed 17.02.09).

[32] CIRCA (Association of Canadian Industries Recycling Coal Ash), *www.circainfo.ca* (accessed 20.02.09).

[33] P. Asokan, M. Saxena, S.R. Asolekar, Coal combustion residues—environmental implications and recycling potentials, Resources Conservation & Recycling, vol. 43, 2005, pp. 239–262.

[34] EPA, Report to Congress: Wastes from the Combustion of Fossil Fuels, Volume 2—Methods, Findings, and Recommendations, U.S. Government Printing Office, March 1999 (Chapter 3).

[35] EPA, Fossil Fuel Combustion Waste, *www.epa.gov/osw/nonhaz/industrial/special/fossil/*, last updated January 12, 2009.

[36] W.T. Davis (Ed.), Air Pollution Engineering Manual, second ed., John Wiley & Sons, Inc., 2000, p. 9.

[37] K. Wark, C.F. Warner, W.T. Davis, Air Pollution Its Origin and Control, Second Edition, Addison Wesley Longman, 1998.
[38] EPA, Latest Findings on National Air Quality: 1997 Status and Trends, Office of Air Quality Planning and Standards, 1998.
[39] EPRI (Electric Power Research Institute), Air Pollution and Health Effects Research at EPRI: The ARIES Program, Strategic Overview Fact Sheet, EPRI, December 2002.
[40] EPRI, Atmospheric Research & Analysis, Fact Sheets, www.epri.com, August 2008.
[41] A. Rohr, Electric Power Research Institute, personal communiqué, (February 18, 2009).
[42] T.F. Wall, Principles of Combustion Engineering for Boilers, Harcourt Brace Javanovich, 1987.
[43] J.G. Singer (Ed.), Combustion: Fossil Power Systems, Combustion Engineering, Inc., 1981.
[44] J.A. Miller, C.T. Bowman, Mechanism and Modeling of Nitrogen Chemistry in Combustion, Progress Energy Combustion Science 15 (1989) 287–338.
[45] S.C. Wood, Select the Right NO_x Control Technology, Chemical Engineering Progress, vol. 90, No,1, 1994, pp. 32–38.
[46] EPA, Mercury Study Report to Congress, Office of Air Quality Planning & Standards, U.S. Government Printing Office, December 1997.
[47] EPRI, ARIES: Aerosol Research and Inhalation Epidemiology Study, Fact Sheet, EPRI, December 2004.
[48] W.W. Aljoe, T.J. Grahame, The DOE-NETL Air Quality Research Program: Airborne Fine Particulate Matter ($PM_{2.5}$), Conference on Air Quality III: Mercury, Trace Elements, and Particulate Matter, University of North Dakota, 2002.
[49] EPRI, TERESA: Toxicological Evaluation of Realistic Emissions of Source Aerosols, Fact Study, EPRI, January 2006.
[50] A.C. Rohr, Health Effects of Coal-Fired Power Plant Emissions, EUEC Energy & Environment Conference, Tucson, January 23, 2007.
[51] A.C. Rohr, Health Effects of Coal-Fired Power Plant Emissions: The TERESA Study, EUEC Energy & Environment Conference, Tucson, February 2, 2009.
[52] S.J. Miller, S.R. Ness, G.F. Weber, T.A. Erickson, D.J. Hassett, S.B. Hawthorne, et al., A Comprehensive Assessment of Toxic Emissions from Coal-Fired Power Plants: Phase I Results from the U. S. Department of Energy Study Final Report, 1996.
[53] A.W. Brooks, Estimating Air Toxic Emissions from Coal and Oil Combustion Sources, Report No. EPA-450/2-89-001, U.S. EPA, North Carolina, 1989.
[54] R.B. Finkelman, H.C.W. Skinner, G.S. Plumlee, J.E. Bunnell, Medical Geology, Geosciences and Human Health, American Geological Institute, 2001, pp. 20–23.
[55] EPA, Carbon Monixide, www.epa.gov.air, last updated May 9, 2008.
[56] L.E. Clarke, L.L. Sloss, Trace elements—emissions from coal combustion and gasification, IEA Coal Research, 1992.
[57] J. Emsley, The Elements, Clarendon Press, 1989.
[58] D.J. Swaine, F. Goodarzi (Eds.), Environmental Aspects of Trace Elements in Coal, Kluwer Academic, 1995.
[59] R.B. Finkelman, W. Orem, V. Castranova, C.A. Tatu, H.E. Belkin, B. Zheng, et al., Health impacts of coal and coal use: possible solution, International Journal of Coal Geology, Elsevier Science, vol. 50, 2002, pp. 425–443.
[60] T.J. Feeley, J. Murphy, J. Hoffmann, S.A. Renninger, A Review of DOE/NETL's Mercury Control Technology R&D Program for Coal-Fired Power Plants, U.S. Department of Energy, Pittsburgh, 2003.

[61] CERHR (Center for the Evaluation of Risks to Human Reproduction), CERHR: Mercury (5/16/02), *http://cerhr.niehs.nih.gov/genpub/topics/mercury2-ccae.html*, 2002.

[62] H.V. Aposhian, M.M. Aposhian, Elemental, mercuric, and methylmercury: biological interactions and dilemmas, in: Proceedings of the Air Quality II: Mercury, Trace Elements, and Particulate Matter Conference, University of North Dakota, 2000.

[63] National Research Council, Toxicological Effects of Methyl Mercury, National Academy Press, 2000.

[64] USGS, Radioactive Elements in Coal and Fly Ash: Abundance, Forms, and Environmental Significance, USGS Fact Sheet FS-163-97; U.S. Government Printing Office, 1997.

[65] EPA, Greenhouse Gases and Global Warming Potential Values, U.S. Greenhouse Gas Inventory Program Office of Atmospheric Programs, U.S. Government Printing Office, April 2002.

[66] DOE (U.S. Department of Energy), Flare for Fuel, Environmental Solutions, National Energy Technology Laboratory, Fall 2000.

[67] Peabody Holding Company, Clearing the Air about the Greenhouse Effects, *www.peabodygroup.com/menupage.html*, downloaded November 1997.

[68] EPA, Global Greenhouse Gas Data, *www.epa.gov/climatechange/emissions/*, last updated February 18, 2009.

[69] DOE, Greenhouse Gases and Global Climate Change Frequently Asked Questions, *www.netl.doe.gov/coalpower/sequestration/indedx.html* (accessed 27.06.03).

[70] U.S. Department of Commerce, National Oceanic & Atmospheric Administration, Earth System Research Laboratory, Trends in Atmospheric CO_2—Mauna Loa, *www.esrl.noaa.gov*, last updated February 2009.

[71] I.M. Smith, C. Nilsson, D.M.B. Adams, Greenhouse gases—perspectives on coal, IEA Coal Research, 1994.

[72] Goddard Institute for Space Studies, Datasets and Images, *http://data.giss.nasa.gov/*, 2007.

[73] NRDC (National Resources Defense Council), Global Warming: In Depth Testimony, *www.nrdc.org/globalwrming/tdl0300.asp*, March 20, 2000.

[74] IPPC (Intergovernmental Panel on Climate Change), Climate Change 2007: Synthesis Report, November 17, 2007.

5 Introduction to Coal Utilization Technologies

In the United States from the 1800s to the mid-1900s, coal was used primarily for iron and steel production, locomotives for transportation, and household heat. In addition, many chemicals, including medicines, dyes, flavorings, ammonia, and explosives, were produced from coal. Once electrification was introduced around 1950 and the feedstocks for chemical production shifted from coal to oil, the primary applications for coal use became electricity generation and the production of iron and steel. Coal is used in the industrial sector for both producing steam and, to a lesser extent electricity, and some chemicals are produced from coal. The technologies used for generating power, heat, coke, and chemicals are introduced in this chapter, including combustion, carbonization, gasification, and liquefaction, which have been referred to as the four "grand processes" of coal utilization [1]. This chapter emphasizes coal combustion, since this technology is the single largest user of coal. This chapter also sets the stage for later chapters in which clean coal technologies are discussed in more detail: advanced pulverized coal-fired systems (Chapter 7), fluidized-bed combustion technology (Chapter 7), and integrated gasification combined cycle technology (Chapter 7).

5.1 Coal Combustion

Burning coal for heat is the most straightforward way to use coal. The heat that is generated from burning coal is used for warmth, cooking, and industrial processes. The use of coal for warmth, cooking, and metalworks has been around for thousands of years. While the Chinese are credited with using coal as early as 1000 B.C. [2], and the first documented use of coal in Western civilization was by the Greek philosophers Pliny, Aristotle, and Theophrastus in the fourth century [3], coal was probably used by prehistoric man, since coal can be found at outcrops, and it is easy to ignite.

As discussed in Chapter 1, the use of coal increased substantially during the Industrial Revolution, but then it decreased with the discovery of oil, which became the dominant fuel for home heating and transportation. Today, the use of coal for direct residential heating and industrial processes is a small percentage of total coal consumption. The primary use of coal is for burning in boilers to generate electricity. The electrification of U.S. household in the 1950s, as well as the electrification of U.S. industries—electrometallurgical processes such as electric furnaces for steel

Clean Coal Engineering Technology. DOI: 10.1016/B978-1-85617-710-8.00005-4
Copyright © 2011 by Elsevier Inc. All rights of reproduction in any form reserved.

and aluminum manufacturing, electric motors, computerized control of processes, and the widespread conversion from shaft power to electrical power—created a large and ever-increasing demand for electricity.

A brief history of the key technological advances of boilers and combustion systems is presented next, followed by discussions of steam fundamentals and how they apply to boiler development, the chemistry of coal combustion, the types of combustion systems, and the influence of coal properties on utility boiler design, with an emphasis on coal ash properties.

5.1.1   Brief History of Boilers and Coal Combustion Systems

Harnessing steam power has been credited as being probably the most important technological advance that contributed to the rise of industrial nations. Steam power was key to the Industrial Revolution, and even today, after more than 100 years of development, boilers continue to dominate as a power source [4]. The first recorded use of a steam boiler dates back to 200 B.C., when Hero of Alexandria, a Greek mathematician and scientist, invented a steam machine (Figure 5.1) [5]. Hero's invention was simply a cauldron with a lid and a pipe for passing steam from the cauldron to a ball on a pivot. As steam exited the ball from the outlet pipes, the ball would spin [5].

This concept for using steam power is not very practical, since its overall efficiency (i.e., the conversion of heat in the fuel to power output, taking into account heat losses, friction, and steam leaking from joints) was as low as 1 percent, as demonstrated from working reconstructions [5]. It does, however, demonstrate a very early recognition of steam as a power source. Unfortunately, Hero considered his invention a toy, and the failure of the Greeks and Romans to harness steam as

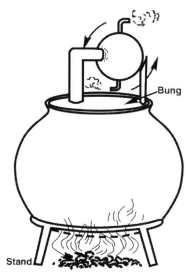

Figure 5.1 Hero's steam machine.
Source: From Landels (1978).

a power source was without a doubt one of the main factors that prevented the industrialization of their societies.

No other record of practical steam application exists until the seventeenth century in England [6]. At this time, England needed considerable fuel for space heating and cooking, and industrial and military growth. The forests were being rapidly depleted. It became necessary to find another source of energy, and coal mining expanded greatly. As large-scale coal mining developed, mines became deeper and often were flooded with water. England desperately needed a means to remove the water from the mines, and, consequently, the first large-scale use of steam was by mining engineers for steam-driven pumps.

In 1698, Thomas Savery invented the first commercially successful steam engine for the direct displacement of water [6]. This engine, however, was only moderately successful because the height of the water (i.e., the head pressure) that could be pumped was limited by the pressure that the boiler and vessels could withstand.

In 1711, Thomas Newcomen overcame Savery's limitations by inventing a steam-driven pump using a piston and cylinder system and incorporating Denis Papin's 1690 invention of the safety valve. Papin invented a steam digester for culinary purposes using a boiler under pressure and invented the valve to avoid an explosion. Boiler-related accidents began to take a terrible toll on lives and equipment, since boiler capacity was not keeping up with the demand for power. Papin is also credited with inventing a boiler with an internal firebox, the earliest record of such construction.

In the 1700s, developers of the steam engine noted that nearly half of the heat from the fire was lost because of short contact time between the hot gases and the boiler-heating surface. This concern for fuel efficiency led to the development of a separate boiler from the steam engine. In an effort to overcome the poor efficiency, John Allen developed, in 1730, an internal furnace with a flue winding through the water like a coil in a still and introduced the concept of forced draft by using bellows to force the gases through the flue [6].

During the latter half of the eighteenth century, James Watt made many significant improvements to the early steam engine and, along with Matthew Boulton, introduced the first boiler. In 1785, Watt took out a number of patents for variations in furnace construction. Richard Trevithick realized that one of the major problems of early steam systems was the manufacture of the boiler. From 1800 to 1804, Trevithick made a 650 psig (pounds per square inch gauge) engine with a high-pressure boiler [6]. The boiler was constructed of cast iron instead of copper, which was the material used up to this time, and the cast iron made the high-pressure steam engine possible.

Early boiler designs consisted of a simple shell for the boiler, with a feed pipe and steam outlet mounted on a brick setting. Fuel was burned on a grate within the setting, and the heat released was directed over the lower surface before most of it went out of the flue. Heating a single large vessel of water is inefficient, and it was necessary to bring more water into close contact with the heat. One way of doing this was to direct the hot combustion products through tubes in the boiler shell, which increased the heating surface and helped to distribute steam formation more uniformly through the water. This design, with multiple flue pipes submersed in water (i.e., firetube boiler; Figure 5.2), was in widespread use until about 1870.

Firetube boilers, however, were limited in capacity and pressure, and they could not fulfill the requirements that later developed for higher pressures and larger unit sizes. The development of watertube boiler designs for steam generation (Figure 5.3) allowed for larger units and higher pressures. Watertube designs feature one or more relatively small drums with multiple tubes in which a steam/water mixture circulates. Heat flows from outside the tubes to the mixture. This subdivision of pressure parts makes large capacities and high pressures possible.

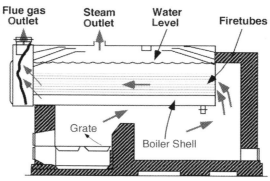

Figure 5.2 A typical firetube boiler design.
Source: From Power, Special Edition (June 1988).

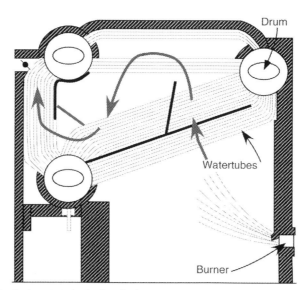

Figure 5.3 A typical watertube boiler design.
Source: From Power, Special Edition (June 1988).

The development of the watertube boiler started in 1766, when a patent was granted to William Blakely that included a form of watertube design for the steam generator [6]. The first successful user of the watertube boiler was James Rumsey, who patented several forms of watertube designs in 1788. A watertube boiler invention by John Stevens (1805), another form of watertube design by John Cox Stevens (1805), and a watertube boiler design by Jacob Perkins (1822) followed; it was the predecessor of the once-through steam generator. A significant breakthrough in watertube boilers occurred in 1856, when Stephen Wilcox proposed a design that allowed better water circulation and heat transfer. In 1866, George Babcock and Stephen Wilcox formed a partnership, and the first B&W (Babcock and Wilcox) boiler was patented in 1877.

B&W boilers were considered to be the best engineered during the late 1800s, and they powered two of the earliest electric power-generating stations [4]. In 1881, the Brush Electric Light Company in Philadelphia began operations, and the Pearl Street Station opened in New York City in 1882.

Once electricity was recognized as a safe and reliable power source, the demand for boilers increased and many generating stations were built to satisfy the needs of industrial and residential customers. With increasing demand for power came increasing demand on boiler manufacturers to improve output. From 1900 to 1920, major strides were made as utilities and other boiler customers tried larger boilers and multiple boiler arrangements to achieve greater output [4]. Steam pressures and temperatures nearly doubled as manufacturers began to optimize the boiler for electric power generation.

Key developments included using pulverized coal firing in place of stoker-fired boilers (i.e., small coal particle size versus large coal particle size) to take advantage of the pulverized coal's higher volumetric heat release rates; increasing system efficiencies by using superheaters (heat exchange surface to increase the steam temperature; see the following), economizers (heat exchange surface to preheat the boiler feedwater), and combustion air preheaters (heat exchange surface to preheat the combustion air); and improving construction materials, making it possible for steam generators to reach steam pressures in excess of 1,200 psig [4]. The relationship among system energy flows, superheated steam, and higher steam temperatures and pressures is discussed later in the chapter.

The incorporation of superheaters into boiler design resulted in great strides in increasing steam pressure and temperature in boiler systems in the United States. Much credit for incorporating superheaters is due to Earnest Foster, who founded the Power Specialty Company along with Pell Foster; Foster convinced boiler owners to install superheaters after he visited Europe and found that the United States was several years behind Europe in adopting superheated steam for power generation.

The development of superheaters, reheaters, economizers, and air preheaters played a significant role in improving overall system efficiency by utilizing as much of the heat generated from burning the coal as possible. The separation of the steam from the water and the use of superheaters and reheaters allowed for higher boiler pressures and larger capacities. The implementation of pulverized coal firing over stoker firing, which became widespread by the mid-1920s, resulted in increased boiler capacity and improved combustion and boiler efficiencies over stoker-fired boilers, which were commonplace up until that time.

Two well-known companies also got their start in the early 1900s: Combustion Engineering, now known as Alstom Power, and Riley Stoker Inc., now known as DB Riley Inc. [4]. These companies lead the technical development of fuel handling and the use of pulverized coal in the United States. Combustion Engineering was initially known for its stoker designs, including one for burning anthracite screenings and one for bituminous coal, but it expanded into a complete line of stokers. Similarly, Riley Stoker Inc., known for producing large multiple-retort underfeed stokers, expanded into a complete line of stokers for boilers of all sizes.

Because of limitations placed on boiler capacity by the size restriction of stokers, Combustion Engineering developed coal pulverizing. This was a major technological improvement in steam generation. Traveling grate stokers had met their technical limits of steam generation at around 200,000 pounds per hour, but by 1929, Combustion Engineering erected the first steam generator unit to produce steam at 1 million pounds per hour (using pulverized coal-firing) at New York Edison's East River Station [4].

In 1946, B&W introduced the cyclone furnace for use with slagging coals (i.e., coals that contain inorganic constituents that will form a liquid ash at temperatures of around 2,600°F or lower), which was the most significant advance in coal firing since the introduction of pulverized coal firing [4]. Cyclone furnaces provide the same benefits as pulverized coal firing, but they have the advantages of utilizing slagging coals, reducing costs due to less fuel preparation (i.e., the coal can be coarser and does not have to be pulverized), and reducing the furnace size.

Fluidized-bed boilers for utilizing coal were originally developed during the 1960s and 1970s, and they offer several inherent advantages to conventional combustion systems, such as the ability to burn coal cleanly by reducing sulfur dioxide emissions during combustion (i.e., in situ sulfur capture) and generating lower emissions of nitrogen oxides. In addition, fluidized-bed boilers provide fuel flexibility, since a range of low-grade fuels can be burned efficiently. Today, all of the major boiler manufacturers offer fluidized-bed boilers. Small, industrial-sized fluidized-bed boilers, however, have limited options, and they are not being produced at the very large steam capacities of pulverized coal-fired units.

Advances in materials of construction, system designs, and fuel firing have led to increasing capacity and higher steam-operating temperatures and pressures. In the United States, utilities typically choose between two basic pulverized coal-fired watertube steam generators: (1) subcritical drum-type boilers with nominal operating pressures of either 1,900 or 2,600 psig, or (2) once-through supercritical units operating at 3,800 psig [7]. These units typically range in capacity from 300 to 800 MW (i.e., producing steam in the range of 2 to 7 million pounds per hour). However, ultra-supercritical units were introduced in 1988 and operate at steam pressures of 4,500 psig and steam temperatures of 1,050°F, with capacities as high as 1,300 MW.

Comparison of Industrial and Utility Boilers

Most coal that is consumed in the United States is used for generating electricity, but a significant number of small, non-electricity-generating boilers that burn coal are also used. Therefore, a brief discussion on industrial boilers and a comparison to utility boilers follows.

Utility boilers and industrial boilers are quite different. The major differences between a utility boiler and an industrial boiler are:

- The size of the boiler
- The application of the steam the boiler generates
- The design of the boiler
- The diversity of the fuels, including the use of by-product fuels
- The global competition for the products created

Comparatively, the typical utility boiler is much larger than the average industrial boiler. As a result, industrial boilers do not enjoy the economies of scale that utility boilers do and, in the case of emissions reduction, must pay more to remove a given amount of emissions.

Size and Number of Units

The average new industrial boiler is considerably smaller than a utility boiler. A typical utility boiler produces about 3.5 million pounds of steam per hour (approximately 400 MW), while a typical industrial boiler produces about 100,000 pounds of steam per hour. Many industrial boilers are designed for less than 250,000 lb steam/h but can be designed for greater than 1 million lb steam/h [6, 8]. There are considerably more small-industrial boilers than large-utility boilers, and the industrial boilers are tailored to meet the needs and the constraints of widely varying industrial processes. The CIBO (Council of Industrial Boiler Owners) reports that the industrial boiler and process heater population (total and not just coal-fired units) consists of 70,000 and 15,000 units, respectively, ranging in size from 10,000 to 1,400,000 lb steam/h, with an average unit size of 100,000 lb steam/h [9, 10].

In comparison, about 4,000 utility units are in existence, of which about 1,250 boilers use conventional coal combustion technology (Table 5.1), with an additional 67 utility-scale fluidized-bed boilers [11]. This is further illustrated in Figure 5.4, which shows the distribution of coal-fired boilers by capacity for conventional utility, conventional nonutility, FBC utility, and FBC nonutility boilers [11]. EPA defines a *nonutility boiler* as a boiler whose primary product is not electricity but

Table 5.1 Distribution of Utility Boilers by Conventional Coal Combustion Technologies

Combustion Technology	Number of Boilers	Capacity (megawatts equivalent)
Pulverized coal boilers	1,068	294,035
Stokers	94	1,077
Cyclones	89	25,727
Total	**1,251**	**320,839**

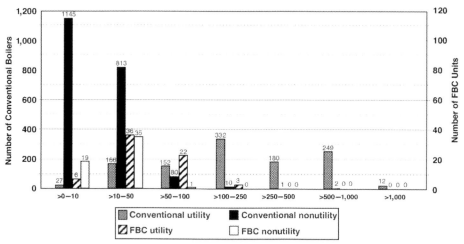

Figure 5.4 Distribution of coal-fired boilers by capacity (in MW).
Source: Modified from EPA (1999).

steam. Some of the nonutility boilers (both conventional and FBC) are cogeneration units in that they produce both steam and electricity.

Application of Steam

Industrial boilers are utilized in many different industries for a wide variety of purposes, and the main product is process steam. Industrial boiler operation can vary significantly among seasons, daily, and even hourly, depending on the steam demand. A utility boiler, however, generates steam for the sole purpose of powering turbines to produce electricity. A typical utility boiler—for example, a base-loaded unit—operates at a steady rate close to maximum capacity because of a constant demand for steam. Load swings from utility boilers that operate to meet utility load swings during the day or for seasonal peak demands—that is, peaking units—are more controlled than industrial boilers because they can balance their load over the complete electric production and distribution grid. Consequently, utility boilers tend to have lower operating costs than industrial boilers that are similarly equipped.

Utility units generally have a variety of backup alternatives for unscheduled outages. Industry, however, rarely has a backup system for steam generation because of the need to keep costs for steam production as low as possible. Hence, industrial boilers routinely operate with reliability factors of 98 percent.

Boiler Design

Utility boilers are primarily field-erected units designed for high-pressure and high-temperature steam. Boiler designs, capacities, steam pressures, and temperatures, among other parameters, vary with the fuel and service conditions. An expanded discussion of the effect of coal type and characteristics on utility boiler design is provided later in this chapter.

The two basic watertube boiler designs that are selected by utilities in the United States are the subcritical drum-type boilers designed for a nominal operating pressure of 1,900 or 2,600 psig steam and the once-through supercritical units designed for 3,800 psig steam [7]. Steam generators and their auxiliary components have many design criteria, but the important issues are efficiency, reliability/availability, and cost. While there are some stoker and cyclone boilers in operation, new designs are primarily pulverized coal and fluidized-bed units.

Industrial boilers and their incorporated combustion systems vary widely in their designs and their construction, including low- and high-pressure steam production, variability in sizes, shop-assembled packaged boilers or field-erected units, and in their capability to burn a wide variety of fuels. Industrial boilers consist of packaged and field-erected units of various boiler types: watertube, firetube, stokers, fluidized bed, pulverized coal, and cyclone units. Packaged units are available in capacities of up to about 600,000 lb steam/h, but boilers larger than 250,000 lb steam/h typically cannot be shipped by rail, although they can be shipped by barge or ocean vessels [6]. The industrial boiler industry is influenced by several factors [7]:

- The user's desire for fuel flexibility over the life of the unit
- The demands for ever-increasing emissions restrictions
- Significant interest in burning low-quality fuels
- Wider application of cogeneration
- The desire to optimize existing equipment in terms of efficiency, performance, and service life
- The recognition that turndown (i.e., the ability to operate efficiently at reduced steam output/fuel-firing rate) is as important over the boiler's lifetime as is maximum continuous rating

Fuel Diversity and Global Product Competition

Fuel diversity and global product competition are primarily of interest to the industrial sector, so they are not discussed in detail here. While electricity is being sold throughout the United States as a result of deregulation in many states, electricity is not a global product (excluding any power sales to Mexico or Canada), whereas many industrial products must compete with international markets. Fuel diversity does affect utilities but not to the extent it does industrial boilers. Coal is the cheapest energy feedstock available and is used extensively in the power-generation industry. Some fuels are cofired with coal, such as petroleum coke, tires, and biomass materials, but utility boilers that fire coal tend to use only coal.

This may change in the future if legislation is passed that requires power generators to produce a percentage of their electricity from renewable energy. Industrial boiler users, on the other hand, are interested in using a wider variety of fuels, since they experience more volatility in fuel availability and prices. Examples of industrial boiler fuels (not inclusive) include waste coals such as bituminous gob and anthracite culm, wood refuse, bagasse, digester (black) liquor, blast-furnace gas, petroleum coke, refining gas, carbon monoxide waste gas, peanut shells, palm fronds, rice husks, animal fats and proteins, and animal manure and litter [6, 9, 12–15].

5.1.2   Basic Steam Fundamentals and their Application
to Boiler Development

Figures 5.5 through 5.7 graphically explain the concept of superheated steam and
how it creates higher temperatures and pressures. Figure 5.5 is a general arrangement of a present-day watertube boiler that shows the location of the superheaters,
reheater, economizer, and air preheater. Figure 5.6 is a steam generator energy flow

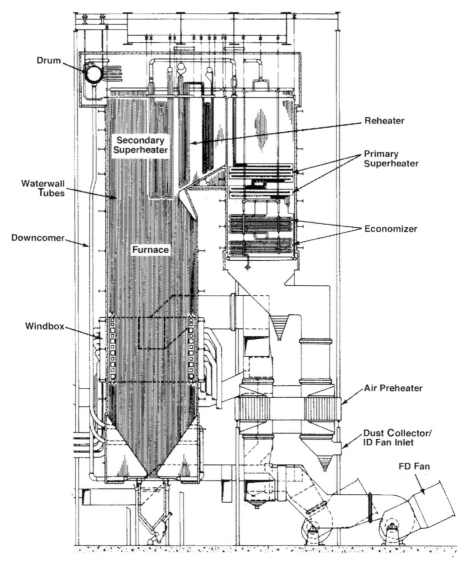

Figure 5.5 General arrangement of a watertube steam generator.
Source: From Elliot (1989).

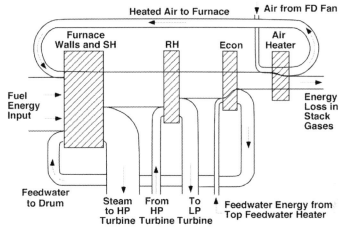

Figure 5.6 Steam generator energy flow.
Source: From Elliot (1989).

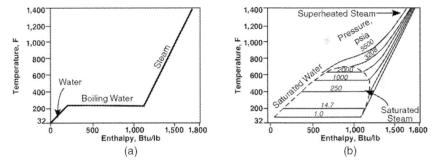

Figure 5.7 Water/steam enthalpy.
Source: From Power, Special Edition (June 1988).

that shows how the various heat exchange surfaces are integrated with one another and the steam turbine. Figure 5.7 illustrates the relationship among temperature, pressure, and enthalpy (i.e., heat content) of saturated and superheated steam.

Heating water at any given pressure eventually will cause it to boil and steam will be released. The heat required to bring the water from 32°F to the boiling point is the enthalpy, or heat content, of the liquid, measured in Btu/lb. When water boils, both it and the steam are at the same temperature, which is called the saturation temperature. For each boiling pressure, there is only one saturation temperature and vice versa. During the boiling process, temperature remains constant as more heat is added, which is being used to change the water from the liquid to the vapor state. This heat is the enthalpy of evaporation and when added to the enthalpy of the saturated liquid gives the enthalpy of the saturated steam, which is the total amount of heat added to bring 32°F water up to 100 percent steam [7]. The temperature of the steam and water will remain the same as long as the two are in contact. To raise the temperature of the steam, it must be heated out of contact with water—that is, it must be superheated. The enthalpy of the steam will increase by the amount of heat added and the temperature will rise.

The temperature–pressure relationship is shown in Figure 5.7. Figure 5.7a shows the energy (in Btu/lb of water) required to heat water from 32°F to its boiling point of 212°F at atmospheric pressure (or 14.7 psia—pounds per square inch absolute), the energy input to continue boiling the water until all the water is converted to steam, and the temperature rise of the steam (superheating) as more energy is put into the system. Figure 5.7b illustrates that as the pressure increases, the amount of heat required to raise the temperature of the water to its boiling point increases, while the amount of heat necessary to vaporize it decreases.

5.1.3   The Chemistry of Coal Combustion

Coal is burned in three ways: as large pieces in a fixed bed or on a grate, as smaller or crushed pieces in a fluidized bed, or as very fine particles in suspension. Theoretically, any particle size can be burned by any of these three methods, but engineering limitations establish preferred particle sizes for the three methods. Particle size has also been found to be the most important parameter with respect to the dominant reaction mechanism and other thermal behaviors (i.e., rate of heating, which can control volatiles yield and composition) [2]. The main characteristics of the three techniques are summarized in Table 5.2.

The combustion process consists of several steps. As the coal particles are heated, moisture is driven off the coal particles. Next, the coal particles undergo devolatilization, releasing volatile organic constituents. The volatile matter is combusted in the gas phase (homogenous reaction). This can occur prior to and simultaneously with combustion of the char particles, which is the last step. Combustion of the char is a surface (heterogeneous) reaction. These reactions are for the most part sequential, and the slowest of these will determine the rate of the overall process.

Devolatilization of Pulverized Coal and Volatiles Combustion

The design of coal burners and furnaces is very dependent on the volatile matter released by the coal as it heats [17]. In flames, pulverized coal heats primarily by convective heat transfer with hot gases, which are entrained and recirculated, with only the coarsest particles having its heating dominated by radiation from the hot regions of the flame. For large flames, in which coal remains for several hundred milliseconds, the extent of devolatilization is strongly influenced by temperature rather than by limitations due to heating times or devolatilization kinetics. Studies have indicated that changes in heating rate (in the range of $1–50 \times 10^{3}$°C/s have little effect on volatiles yield and that the yield is more strongly influenced by the final temperature, with an increase in final temperature producing an increased yield [2].

Table 5.2 Comparison of Characteristics of Combustion Methods

	Combustion Method		
Variables	**Fixed Bed (stoker)**	**Fluidized Bed**	**Suspension**
Particle size			
Approximate top size	<2 inches	<0.2 inches	180 μm
Average size	0.25 inches	0.04 inches	45 μm
System/bed temperature	<1,500°F	1,500–1,800°F	>2,200°F
Particle heating rate	≈1°/s	10^3–10^4 °/s	10^3–10^6 °/s
Reaction time			
Volatiles	≈100 seconds	10–50 seconds	<0.1 seconds
Char	≈1,000 seconds	100–500 seconds	<1 second
Reactive element description[a]	Diffusion-controlled combustion	Diffusion-controlled combustion	Chemically controlled combustion

[a]Described in the Char Combustion section and illustrated in Figure 5.8.
Source: Modified from Van Krevelen (1993) and Elliot (1987).

Volatiles yield is also found to depend on particle size, with smaller particles tending to yield more volatiles. Also, volatile yield can vary significantly within a given rank for coals that are similar in composition and are mined adjacent to each other in the same coal basin (e.g., subbituminous coals from neighboring Powder River Basin coal mines) [18].

The combustion of the volatiles is generally assumed to be a homogenous reaction, although the possibility of volatile matter burning heterogeneously has been suggested by Howard and Essenhigh [19]. The burning of the volatiles is a very fast process that is measured in milliseconds [20].

Char Combustion

Char combustion is a much slower process than devolatilization, and it therefore determines the time for complete combustion in a furnace, which is on the order of several seconds for pulverized coal at furnace temperatures. Studies have shown that the combustion of the char begins with chemisorption of oxygen at active sites on char surfaces and that the decomposition of the resultant surface oxides mainly generates CO [2, 21]. (Some researchers believe that some CO_2 may be released during this step also.) The CO is then oxidized to CO_2 in a gaseous boundary zone around the char particle. Fresh reaction sites are continuously exposed as the surface oxides are decomposed. CO_2 then either moves off into the gas stream or is reduced to CO if it impinges on the char. The overall reaction mechanism is complex [2, 22], but the combustion of char involves at least four carbon-oxygen reactions [21]:

$$C + \tfrac{1}{2}O_2 \rightarrow CO \tag{5.1}$$

$$CO + \tfrac{1}{2}O_2 \rightarrow CO_2 \tag{5.2}$$

$$CO_2 + C \rightarrow 2CO \tag{5.3}$$

$$C + O_2 \rightarrow CO_2 \tag{5.4}$$

as well as the oxidation of noncarbon atoms, mainly

$$S + O_2 \rightarrow SO_2 \tag{5.5}$$

$$H_2 + \tfrac{1}{2}O_2 \rightarrow H_2O \tag{5.6}$$

which may be followed by

$$H_2O + C \rightarrow CO + H_2 \tag{5.7}$$

$$CO + H_2O \rightarrow CO_2 + H_2 \tag{5.8}$$

Some species of the mineral matter can be volatilized during combustion, while others are left behind as ash. In both cases, the mineral matter is usually altered in composition and mineralogy.

The char combustion rate is a complicated process, since it is influenced by mass transfer by diffusion through pores and surface reactions. The diffusion coefficients are highly dependent on pore diameter and pressure, and surface reaction is influenced by the formation of activated adsorption complexes and their decomposition [1].

The rate of char combustion is controlled by two processes: the chemical reaction rate of carbon and oxygen on the char surface, and the rate of mass transfer of oxygen from the bulk gas stream through the boundary layer surrounding the particle to the particle surface. This is illustrated in Figure 5.8, where the general relationship between temperature and reaction rate for a heterogeneous gas/solid system is shown [22]. At low temperatures, region I, the chemical reaction rate is slow compared with the diffusion rate through pores; therefore, oxygen completely penetrates the char matrix. Combustion then takes place within the porous char—the char's density rather than diameter changes. In this case, the oxygen concentration at the particle surface would be the same as that in the bulk gas stream, and the overall reaction rate would be limited by the chemical reaction's inherent rate. In region I, the rate of surface reaction is rate determining, and oxygen molecules diffuse fast enough to reach the whole internal surface. The reaction rate is given by

$$q = -\frac{d}{6}\frac{\rho_p}{dt} \tag{5.9}$$

so $q \propto d$, where q is the char combustion reaction rate (kg/m^{2}s), d is the diameter of the particle (m), ρ_p is particle density (kg/m^3), and t is burning time (s).

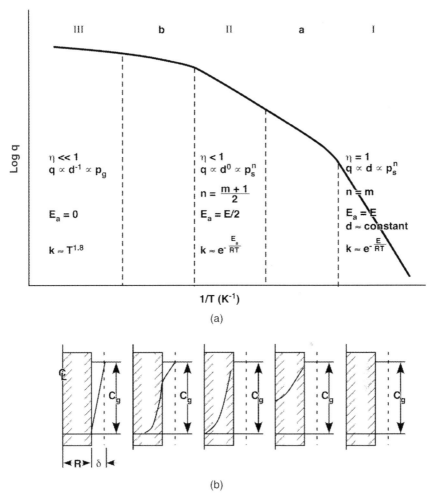

Figure 5.8 Relationship between temperature and reaction rate (C_g = concentration of oxygen in the bulk gas, δ = boundary layer thickness, η = effectiveness factor, E_a = apparent activation energy, E = true activation energy).
Source: From Walker, Rusinko, and Austin (1959).

The rate of chemical reaction may be expressed by a generalized expression of the type

$$q = k_c = A\exp(-E/RT_p)P_s^m \tag{5.10}$$

where A is the true preexponential constant (kg C/m^2 s (atm O$_2$)$^{-m}$); T is the particle temperature (K); R is the universal gas constant; and, since chemical reaction controls the rate, P is the partial pressure of oxygen at the surface (atm); E is the true activation energy (J/mol); and m is the true reaction order.

As the temperature is increased, the chemical reaction becomes sufficiently rapid for the diffusion of oxygen through the pores to exert a notable rate-limiting effect. Under these conditions, regime II, the diameter and the density of the particle, will both change. The apparent activation energy and apparent order of reaction (n) are approximated by

$$Ea \approx 2, n \approx (m + 1)/2 \qquad (5.11)$$

so the apparent reaction rate does not change as rapidly with temperature.

Any further increase in temperature eventually causes the chemical reactions to become so rapid that the oxygen is consumed as it reaches the outer surface of the particle. In this case, regime III, the reaction, is entirely controlled by the diffusion from the free stream to the particle, and only the diameter of the particle changes.

Field and colleagues [20] give an expression for the overall reaction rate coefficient k as

$$k = \frac{1}{1/k_d + 1/k_c} \qquad (5.12)$$

where k_d is the diffusional rate coefficient and k_c is the chemical rate coefficient defined in Eq. 5.10, and the diffusional rate coefficient can be defined as

$$k_d = \frac{24\phi D}{dRT_m} \qquad (5.13)$$

where ϕ is a mechanism factor that takes the value of 1 for the reaction to CO_2 and takes 2 for the reaction to CO, and D is the diffusion coefficient (cm^2/s) of oxygen through the boundary layer at temperature T_m given by

$$D = 3.49x\left(\frac{T_m}{1,600}\right)^{1.75} \qquad (5.14)$$

where x is the particle diameter (cm), R is the universal gas constant, and T_m is the mean temperature (K) for the boundary layer taken as the average of the surface temperature of the particle and the bulk gas temperature.

For the char sizes, porosities, internal surface areas, and temperatures typical to pulverized coal-fired furnaces, the char combustion rate is influenced by the chemical reactivity of the char, the external diffusion rate of oxygen from the bulk stream, and the internal diffusion of oxygen into the porous char matrix. Char ignition is likely to occur in regime I or II when a large proportion of the internal surface is available for reaction. The final burnout is likely to occur in regime II and III, when external diffusion may have a significant influence on the combustion rate of large particles. The time for the char to burn out is proportional to the square of the initial size of the char particles from the coarse end of the grind [17].

The processes controlling the rate of char combustion in fluidized-bed systems differ slightly from pulverized coal-fired systems. In fluidized-bed combustion, particle sizes are larger, the processes by which oxygen is brought to the coal surface differ because of the presence of the surrounding bed particles, and the heat transfer processes also differ from those in a pulverized coal-fired furnace [17].

All three regimes illustrated earlier in Figure 5.8 are important in fluidized-bed combustion: regime I during ignition and for the smaller particles burning in the bed and freeboard; regime II for medium-sized particles; and regime III for large particles in the bed. The surface reaction rate q (mass of carbon oxidized per unit area of particle outer surface per second), for the region separating regimes II and III, which are of special interest in coal combustion, is defined as

$$q = \frac{P_g}{(1/k_d + 1/k_c)} = kP_g \tag{5.15}$$

where P_g is the oxygen partial pressure in the gas outside of the boundary layer (kN/m^2). The mass transfer coefficient for the oxygen diffusion to the particles is

$$h_m = \frac{ShD}{d} \tag{5.16}$$

where h_m is the mass flux of oxygen per unit area of surface per unit of concentration difference between that at the surface and that in the gas outside the boundary layer (m/s), Sh is the Sherwood number (dimensionless), and D is the diffusion coefficient of oxygen through the gas mixture surrounding the particle (m^2/s) [17].

The combustion rate q, in units of mass of carbon oxidized per unit time per unit area of particle outer surface can be calculated by:

$$q = k_d(P_g - P_s) + \frac{12\phi ShD}{dRT_m}(P_g - P_s) \tag{5.17}$$

where D is evaluated at a mean temperature T_m in the diffusion layer (K), P_s is the oxygen partial pressure at the particle surface (kN/m^2), and the universal gas constant R has the units 8.31 J/mol K.

The chemical kinetic rate coefficient, k_c, is expressed by an Arrhenius type equation

$$k_c = A_a \exp(-E_a/RT_p) \tag{5.18}$$

where A_a is the apparent rate constant based on the particle outer surface area (kg/m^2 s per kN/m^2 of partial pressure of oxygen), and E_a is the apparent activation energy for regime II combustion (kJ kg/mol).

5.1.4 Coal Combustion Systems

The manner in which coal is burned and the devices in which it is burned are primarily determined by the desired unit size or capacity (i.e., required hourly steam production or electricity generation) and coal type and quality. The combustion methods, fixed bed (i.e., stokers), fluidized bed, and suspension firing, are discussed in the following sections.

Fixed-Bed Combustion

Fixed-bed combustion covers a wide variety of applications, including domestic space heaters, underground gasification, and industrial stokers. The latter are of interest for steam and power generation. Stokers were used to burn coal as early as the 1700s [7, 16]. Stokers have evolved over the years from simple design to quite sophisticated devices used to burn a variety of fuels, including coal. Stokers are generally divided into three general groups, depending on how the fuel reaches the grate (i.e., surface that contains the coal and allows combustion air to be introduced into the fuel bed) of the stoker for burning: underfeed, overfeed, and spreader designs.

Three patterns of feeding coal and combustion air have been developed and are used singly or in combination in commercial equipment [2]. These patterns, which are illustrated in Figure 5.9, are called overfeed (i.e., fuel is fed onto the top of the bed and flows down as it is consumed, while combustion air flows up through successive layers of ash, incandescent coke, and fresh coal), underfeed (i.e., the flow of coal and combustion is parallel and usually upward), or crossfeed (i.e., the fuel moves horizontally, with the combustion air moving upward at right angles to the fuel).

Unfortunately, because there are many examples of commercial equipment with more than one pattern of combustion air and fuel feed, disagreement is sometimes found in the literature over which type predominates in a given stoker application that also leads to differences in system classification. For this discussion, underfeed stokers include single- and multiple-retort ones; overfeed stokers include chain grates, traveling grates, and water-cooled vibrating grate ones; and spreader

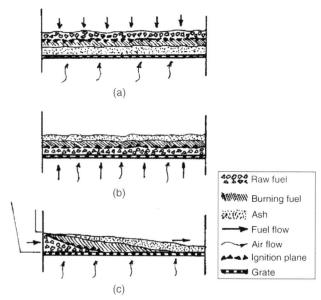

Figure 5.9 Patterns of feeding coal and combustion air to stokers: (a) overfeed, (b) underfeed, and (c) crossfeed.
Source: From Elliot (1981).

stokers are classified into several groups—depending on the type of grate selected—including stationary, dumping, reciprocating, vibrating, traveling, and water-cooled vibrating grates. Some characteristics of the stokers and their fuel requirements and capacities are given in Tables 5.3 through 5.5. Figure 5.10 illustrates the working principles of the three groups of stokers [21].

Underfeed Stokers

Underfeed stokers are used primarily for burning coal in small boilers serving relatively constant steam loads of less than 30,000 lb/h [7]. Coal is introduced beneath

Table 5.3 Characteristics of Various Types of Stokers

Ability of the Unit to:	Spreader	Overfeed	Underfeed
Increase load rapidly	Excellent	Fair	Fair
Minimize carbon loss	Fair	Fair	Fair
Overcome coal segregation	Fair	Poor	Poor
Accept a wide variety of coals	Excellent— traveling grate; fair– vibrating grate	Poor	Poor
Burn extremely fine coal	Poor	Poor	Poor
Permit smokeless combustion at all loads	Poor	Good	Good
Minimize fly ash discharge to stack	Poor	Good	Good
Maintain steam load under poor operating conditions	Good	Poor	Poor
Minimize maintenance	Good	Good	Fair
Minimize power consumption (stoker and boiler auxiliaries)	Good	Good	Good
Handle ash and cinders easily	Excellent	Good	Fair

Source: From DOE (2003).

Table 5.4 Stoker Fuel Requirements

Stoker	Coal Types	Coal Sizes
Underfeed	Bituminous coals or anthracite	Nut (i.e., 2 × 3/4 in.) or prepared stoker (large) (i.e., 2 × 1/4 in.); smaller sizes acceptable if <50% minus 1/4 in.
Overfeed	All coal ranks	Nut (i.e., 2 × 3/4 in.) or prepared stoker (large or intermediate) (2 × ¼ in. or 1 × 1/8 in., respectively), <20% minus 1/4 in.
Spreader	All coal ranks	Prepared stoker (small) (i.e., 3/4 × 1/16 in.), <30% minus 1/4 in.

Source: Modified from Berkowitz (1979).

Table 5.5 Capacities of Coal-Fired Stokers

Stoker Type	Recommended Grate Heat-Release Rates (1,000 × Btu/h ft^2 of grate)	Steam Generation (1,000 lb/h)
Underfeed Stokers		
Single retort (ram feed)	250–475	2–50
Multiple retort	450–500	40–300
Spreader Stokers		<10–400
Stationary and dumping		<50
grates	400–450	
Reciprocating and		up to 75 (reciprocating grate)
vibrating grates	600–650	up to 100 (vibrating grate)
Traveling grate	650–750	up to 400
Overfeed Stokers	400–425	10–300

Source: Modified from Elliot (1989).

the active fuel bed and is moved from a storage hopper by means of a screw or ram into a retort. As coal is fed into the retort, the force of the incoming fuel causes the coal to rise in the retort and spill over onto the fuel bed or grate surface on either side of the retort. No air is supplied in the retort proper; it comes through openings, called tuyeres, in the grate section adjoining the trough. Underfeed stokers are either of the single- or multiple-retort design, and water-cooled furnaces are preferred with underfeed stokers.

A relatively wide range of bituminous coals, as well as anthracite, can be burned on single- or multiple-retort stokers, but typical specifications call for ¾ × 1¼-inch coal with less than 50 percent of the fines passing through a ¼-inch screen [16]. The free swelling index of the coal should be limited to 5 with single-retort stokers equipped with stationary tuyeres and up to 7 on single-retort stokers with moving tuyeres and on multiple-retort stokers. It is normally recommended that the iron content in the ash be less than 20 percent as Fe_2O_3 with an ash fusion temperature above 2,400°F and below 15 percent, with coals having a lower ash fusion temperature.

Overfeed Stokers

Overfeed, or mass-burning stokers, convey coal from the fuel hopper located at the front of the stoker. The depth of the fuel bed conveyed into the furnace is regulated by a vertically, adjustable feed gate across the width of the unit [7, 16, 23]. The fuel is conveyed into and through the furnace and passes over several combustion air zones. The ash is continuously discharged into a storage hopper at the rear end of the grate. Overfeed mass-burning stokers consist of three designs: chain grate, traveling grate, and water-cooled vibrating grate. Water-cooled furnaces are preferred with all moving-grate stokers to prevent slag formation on the furnace walls.

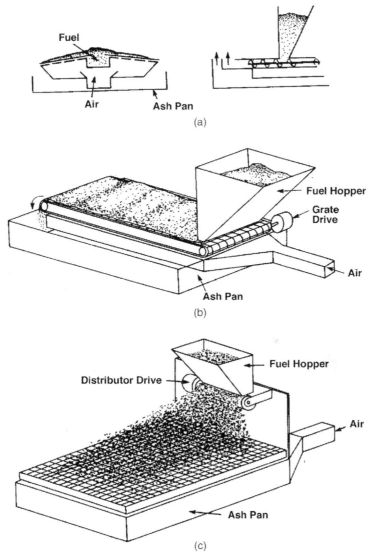

Figure 5.10 Working principles of mechanical stokers: (a) underfeed, (b) overfeed (traveling grate stoker), and (c) spreader.
Source: From Berkowitz (1979).

Chain grates consist of a wide chain with grate bars forming the links. The links are staggered and connected by rods extending across the stoker width. This chain assembly is continuously pulled or pushed through the furnace by an electric or hydraulic drive. The traveling grate has a chain drive (powered electrically or hydraulically) at the side of the grate with crossbars at intervals. Fingers, keys, or clips that form the grate surface are attached to these crossbars in an overlapping fashion to prevent ash from sifting through. The water-cooled, vibrating grate stoker consists of a grate surface mounted on, and in contact with, a grid of watertubes. These tubes are connected to the boiler circulatory system to ensure positive cooling. The structure is supported by a number of flexing plates, allowing the water-cooled grid and grate surface to move freely in a vibratory mode as the fuel bed moves through the furnace.

Chain grate stokers originally were developed for bituminous coal and traveling grate stokers for small sizes of anthracite [23]. However, almost any type of solid fossil fuel can be burned on the three stoker designs, including peat, lignite, subbituminous coal, noncaking bituminous coal, anthracite, and coke breeze [7]. Strongly caking bituminous coals may have a tendency to coke and prevent proper passage of combustion air through the fuel bed in the chain grate and traveling grate designs. In these designs, tempering (i.e., the addition of water or steam) of the fuel bed is done in the fuel hopper to make the bed more porous, although the coal's heating value is decreased. The vibrating action of the grate in the water-cooled vibrating grate design, however, keeps the fuel bed uniform and porous without the addition of water or steam.

Coal size ranges for overfeed mass-burning stokers are listed earlier in Table 5.4. These stokers are quite sensitive to segregation of coal sizes or distribution of the coal. If the fuel size is not uniform across the width of the stoker, the fuel bed will not burn uniformly, resulting in unburned carbon being discharged into the ash hopper.

Spreader Stokers

Spreader stokers are the most popular of the three types. One reason for this is that they are capable of burning all ranks of coal as well as many waste fuels [23]. In addition, they can accommodate a wide range of boiler sizes. Spreader stokers take fuel from feeders located across the front of the furnace and distribute it uniformly over the grate surface. The objective is to release an equal amount of energy from each square foot of active grate surface [7]. As the coal is spread over the grate, fines in the incoming coal stream burn in suspension, while the large pieces fall to the grate, forming a fuel bed; thus, to a limited extent, spreader firing has characteristics similar to pulverized coal combustion. Primary air for combustion is admitted evenly throughout the active grate area with an overfire air system providing secondary air and turbulence above the grate.

The fuel bed is normally thin and there is rarely more than a few minutes' worth of coal on the grate. This, coupled with 25 to 50 percent of the coal burned in suspension, allows the spreader stoker to respond quickly to load swings. This makes the spreader stoker well suited for industrial applications where process loads fluctuate rapidly.

The most common types of grates used today for spreader-stoker firing are the vibrating (or oscillating), traveling, and water-cooled vibrating grates. The water-cooled vibrating grate stoker is designed primarily for refuse burning (although conceivably could be used for coal) and is not discussed here. Stationary, dumping, and reciprocating grates see limited service. Not all of these grates are suited for coal firing.

The intermittent cleaning types of grates are stationary and dumping [16]. The stationary grate is seldom used because of hazards to the operator when removing ash through an open fire door. The dumping grate is seldom used for coal because the cleaning process results in high opacity in the stack. When it is used, it is for capacities of under 50,000 lb steam/h and a heat release rate from the grate of no more than 450,000 Btu/h-ft^2.

The reciprocating grate discharges ash by a slow back-and-forth motion of moving grates alternating with stationary grates, which causes the fuel bed to move forward, dumping the ash into a pit at the front of the boiler. The grate can be used on boilers from 5,000 to 75,000 lb steam/h and can accommodate a wide range of bituminous coals or lignite without preparation other than sizing. Because of the stepped nature of the reciprocating grate, it is only used for fuels with sufficient ash quantity to provide an adequate ash depth for insulation on the top of the grates.

The vibrating or oscillating grate is suspended on flexing plates with an eccentric drive or weights used to impart a vibrating action to the grate surface, which conveys the ash to the front of the stoker and discharges it into an ash pit [16]. This grate type is well suited for coal. The traveling grate spreader stoker is the most popular type. The endless grate moves at speeds between 4 and 20 feet per hour, depending on the steam demand, toward the front of the boiler, discharging ash continuously into an ash pit. The return grate then passes underneath in the air chamber. Traveling grate spreader stokers are designed to handle a wide range of coals.

Fluidized-Bed Combustion

Fluidized-bed combustion (FBC) is a mature technology for the combustion of fossil and other fuels and is attractive because of its many inherent advantages over conventional combustion systems. The FBC technology is introduced in this section and discussed in more detail in Chapter 7. These advantages include fuel flexibility, low NO_x emissions, and in situ control of SO_2 emissions. The fluidized-bed concept was first used around 1940 in the chemical industry to promote catalytic reactions. In the 1950s, the pioneering work on coal-fired FBC started in Great Britain, particularly by the National Coal Board and the Central Electricity Generating Board [2, 24].

The U.S. Department of Interior's Office of Coal Research, one of the predecessors of the current Department of Energy (DOE), began studying the fluidized-bed combustion concept in the early 1960s (and still continues sponsoring research into advanced fluidized-bed combustion systems) because it recognized that the fluidized-bed boiler represented a potentially cheaper, more effective, and cleaner way to burn coal [25]. Around 1990, atmospheric fluidized bed combustion crossed the commercial threshold, and every major boiler manufacturer in the United States currently offers fluidized-bed boilers as a standard package. Fluidized-bed coal combustors have been called "the commercial success story of the last decade in the power-generation business" and are perhaps the most significant advance in coal-fired boiler technology in a half century.

FBC technology is used in both the utility and nonutility sectors and comprises approximately 1 percent of fossil fuel–fired capacity. Approximately half of the facilities using FBC technologies are utilities or independent power producers. Facilities in the food products and pulp and paper industries, along with educational institutions, make up most of the nonutility FBC facilities [11].

FBC technology accounts for a small proportion of capacity, but the technology has increased dramatically over the last 20 years [11]. In 1978, four U.S. plants had four FBC boilers. As of December 1996, 84 facilities had with a total of 123 FBC boilers, representing 4,951 MW of equivalent electrical generating capacity. Because of the fuel flexibility, efficiency, and emissions characteristics of the FBC boilers, this technology is predicted to increase in the future, and additional units are being installed, both commercial units and advanced concepts through cofunded DOE programs (which are discussed in more detail in Chapter 11).

Figure 5.11 shows the geographic distribution of the FBC facilities [11]. These facilities are distributed throughout the Unites States, but Pennsylvania and California account for the largest numbers of plants. Pennsylvania accounts for more than 20 percent of capacity and California more than 10 percent. Pennsylvania is a leader in utilizing coal wastes, anthracite culm, and bituminous coal gob in FBC boilers.

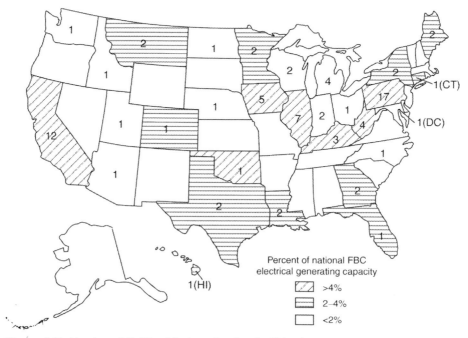

Figure 5.11 Number of fluidized-bed combustion facilities by state.
Source: EPA (1999).

In a typical FBC, solid, liquid, or gaseous fuel (or fuels), an inert material such as sand or ash (referred to as bed material), and limestone are kept suspended through the action of combustion air distributed below the combustor floor [16]. The primary functions of the inert material are to disperse the incoming fuel particles throughout the bed, heat the fuel particles quickly to the ignition temperature, act as a flywheel for the combustion process by storing a large amount of thermal energy, and provide sufficient residence time for complete combustion. The FBC concept is attractive because it increases turbulence and permits lower combustion temperatures. Turbulence is promoted by fluidization making the entire mass of solids behave much like a liquid. Improved mixing (and hence enhanced heat transfer to the bed material) permits the generation of heat at a substantially lower and more uniformly distributed temperature than occurs in conventional systems such as stoker-fired units or pulverized coal-fired boilers.

The bed temperature in an fluidized-bed combustion boiler is typically 1,450 to 1,650°F. This operating temperature range is well below that at which significant thermally induced NO_x production occurs. Staged combustion can be applied to minimize fuel-bound NO_x formation as well. With regard to SO_2 emissions, the operating temperature range is where the reactions of SO_2 with a suitable sorbent, commonly limestone, are thermodynamically and kinetically balanced [26]. The percent capture for a given sorbent addition rate drops significantly outside the 1,450 to 1,650°F range. Another reason the bed temperature must be kept above 1,400°F is that carbon utilization decreases with decreasing temperature, thereby reducing combustion efficiency.

Role of Sorbents in an FBC Process

In an FBC system, the sorbent, usually limestone but sometimes dolomite (a double carbonate of calcium and magnesium), undergoes a thermal decomposition commonly known as calcination. When using limestone, the calcination reaction is

$$CaCO_3 + heat \rightarrow CaO + CO_2 \qquad (5.19)$$

(limestone + heat $\rightarrow$ lime + carbon dioxide)

Calcination of limestone is an endothermic reaction, which occurs when limestone is heated above 1,400°F. Calcination is thought to be necessary before the limestone can absorb and react with gaseous sulfur dioxide.

Capture of the gaseous sulfur dioxide is accomplished via Eq. (5.20) to produce a solid product, calcium sulfate:

$$CaO + SO_2 + \tfrac{1}{2}O_2 \rightarrow CaSO_4 \qquad (5.20)$$

lime + sulfur dioxide + oxygen $\rightarrow$ calcium sulfate

The limestone is continuously reacted, and therefore it is necessary to continuously feed limestone with the fuel. The sulfation reaction requires that there always be an excess amount of limestone present. The amount of excess limestone that is required is dependent on a number of factors, such as the amount of sulfur in the fuel, the temperature of the bed, and the physical and chemical characteristics of the limestone.

The primary role of the sorbent in an FBC process is to maintain air quality compliance; however, the sorbent is also important in bed inventory maintenance, which affects the heat transfer characteristics and the quality and handling characteristics of the ash. Depending on the sulfur content of the fuel, the limestone can comprise up to 50 percent of the bed inventory, with the remaining portion being fuel ash or other inert material. This is especially true of FBCs firing refuse from bituminous coal cleaning plants that contains high levels of sulfur. When the bed is comprised of a large quantity of calcium (oxide or carbonate), there is the potential for ash disposal concerns, since the pH of the ash can become very high.

Comparison of Bubbling and Circulating Fluidized-Bed Combustion Boilers

The principle of FBC systems can be explained by examining Figure 5.12. The fundamental distinguishing feature of all FBC units is the velocity of the air through the unit, as illustrated in Figure 5.12 (modified from [7 and 16]). Bubbling beds have lower fluidization velocities, where the concept is to prevent solids elutriating (i.e., carrying over) from the bed into the convective passes. Circulating fluidized-bed units apply higher velocities to promote solids elutriation.

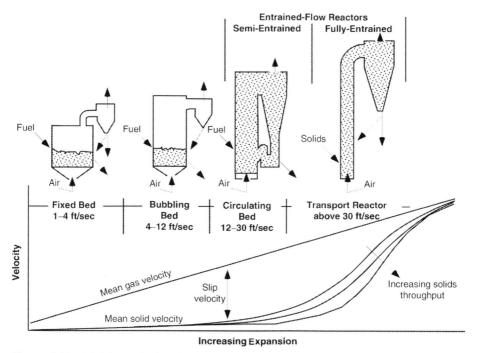

Figure 5.12 Fluidizing velocity of air for various bed systems.
Source: Modified from Power, Special Edition (June 1988) and Elliot (1989).

Bubbling fluidized-bed units characteristically operate with a mean particle size of between 1,000 and 1,200 µm and fluidizing velocities between the minimum fluidizing velocity and the entraining velocity of the fluidized solid particles (i.e., 4 to 12 ft/s). Under these conditions, a defined bed surface separates the high solids loaded bed and the low solids loaded freeboard region. Most bubbling-bed units, however, utilize reinjection of the solids escaping the bed to obtain satisfactory performance. Some bubbling-bed units have the fuel and air distribution configured so a high degree of internal circulation occurs within the bed [27]. A generalized schematic of a bubbling fluidized-bed boiler is shown in Figure 5.13 [28].

Circulating fluidized-bed (CFB) units operate with a mean particle size between 100 and 300 µm and fluidizing velocities up to about 30 ft/s. A generalized schematic of a CFB boiler is given in Figure 5.14 [28]. Since CFBs promote elutriation and the solids are entrained at a high rate by the gas, bed inventory can be maintained only by recirculation of solids separated by the off-gas by a high-efficiency process cyclone. Notwithstanding the high gas velocity, the mean solids velocity in the combustor is lowered due to the aggregate behavior of the solids. Clusters of solids are continuously formed, flow downward against the gas stream, are dispersed, are reentrained, and form clusters again. The solids thus flow upward in the combustor at a much

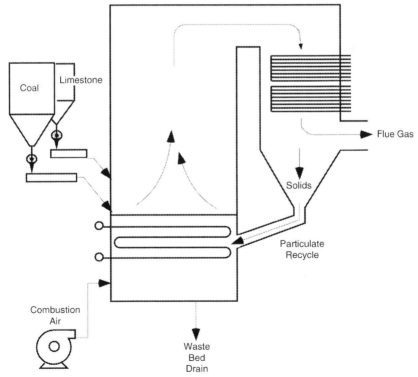

Figure 5.13 Generalized schematic diagram of a bubbling fluidized-bed boiler. *Source:* Gaglia and Hall (1987).

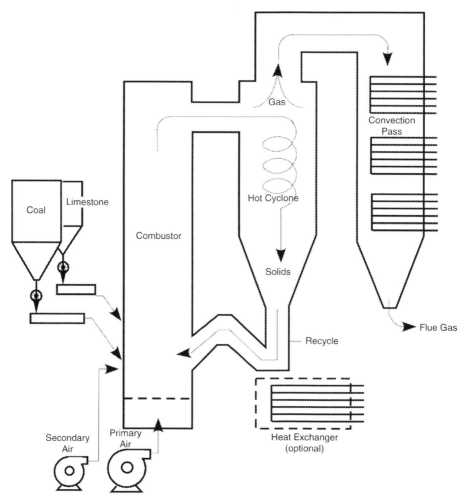

Figure 5.14 Generalized schematic diagram of a circulating fluidized-bed boiler.
Source: From Gaglia and Hall (1987).

lower mean velocity than the gas. The slip velocity between gas and solids is very
high, with corresponding high heat and mass transfer. This is further illustrated in
Figure 5.15, which shows that CFBs achieve higher rates of heat transfer from the
solids to the boiler tubes, as do bubbling fluidized-bed units [16].

Circulating fluidized-bed units include a refractory-lined combustor bottom section with fluidizing nozzles on the floor above the windbox; an upper combustor section, usually with waterwalls; a transition pipe, including a hot-solids separator and
reentry downcomer; a convective boiler section; and in some designs, an external heat
exchanger [16]. An external heat exchanger is a refractory-lined box containing an air
distribution grid and an immersed tube bundle designed to cool material from the hot-
solids separator; it is used to compensate for variations in the heat absorption rate

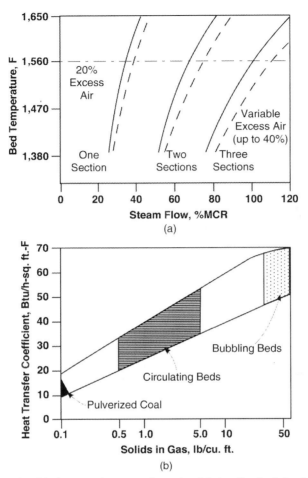

Figure 5.15 Relationship between heat transfer and solids loading/bed density.
Source: From Elliot (1989).

caused by changes in fuel properties and load conditions. The solids separators are refractory-lined cyclones that are used to keep the solids circulating. The solids reinjection device, called an L-valve, a J-valve, a loop seal, a Fluoseal, or a Sealpot, depending on the manufacturer and configuration, provides a simple, nonmechanical hydraulic barometric seal against the combustor shell.

Suspension Firing

Pulverized coal firing is the method of choice for large industrial boilers (e.g., >250,000 lb steam/h) and coal-fired electric utility generators, since pulverized coal-fired units can be constructed to very large sizes (i.e., up to ≈1,300 MW or ≈9.5 million lb steam/h), and unlike stoker units, where some designs have coal restrictions, it can accommodate virtually any coal with proper design provisions.

The coal size distribution for pulverized coal-fired units is typically less than 2 percent by weight greater than 50 mesh (300 μm), with 65 to 70 percent less than 200 mesh (74 μm) for lignites and subbituminous coals and 80 to 85 percent less than 200 mesh for bituminous coals [21]. After the coal is pulverized, it is pneumatically transported to the burners using a portion of the combustion air, typically 10 percent (the remaining combustion air is introduced at or near the burner), in a manner that permits stable ignition, effective control of flame shape and travel, and thorough and complete mixing of fuel and air.

Pulverized coal-fired units are typically classified into two types, depending on the furnace design for ash removal. Dry-bottom furnaces are units where the ash is removed from the system in dry form, whereas wet-bottom or slag-tap furnaces are units where the ash is removed in molten form. Dry-bottom furnaces are the more common of the two types and are now almost the only type sold in the United States. Dry-bottom furnaces are simpler to operate, more flexible with respect to fuel properties, and more reliable than slag-tap furnaces [2]. Dry-bottom furnaces are larger and thus more expensive than wet-bottom furnaces, since they must be sized to accommodate the ash where most of it (>80 percent) remains entrained in the flue gas and must be removed by particulate control devices at the back end of the system. Slag-tap units were developed to reduce the amount of fine fly ash that had to be handled by producing a heavier, granular ash and retaining most of the ash (up to 80 percent) in the furnace.

Dry-Bottom Firing

The most frequently used dry-bottom furnace and burner configurations are shown in Figure 5.16 [2]. These arrangements cover firing systems suitable for all ranks of coal and coal qualities, including high ash or moisture content, low heating value, low ash fusion temperature, and high potential for ash deposition.

Dry-bottom furnaces are designed to remove the ash as a solid; therefore, the rate of heat transfer and temperature in the furnace must be controlled. The dry-bottom furnaces are designed such that the heat release rates are much lower than wet-bottom and cyclone furnaces, and this, coupled with maintaining the furnace exit gas temperature below the ash fusion temperature, results in larger furnace designs. Flame temperatures in the pulverized coal-fired units are typically around 2,750°F. Heat is lost primarily by radiation in the furnace to the waterwalls and superheater/reheater tubes suspended in the furnace, and the temperature of the flue gas exiting the furnace is typically 1,850°F.

Horizontal and opposed horizontal furnaces (Figure 5.16a and b) are usually fired by circular burners spaced uniformly across the width of the furnace on the front, rear, or both front and rear walls. Each burner has its own flame envelope, and the firing system can be designed so an individual burner may be placed in service, adjusted, or removed from service independently of the other burners. In front or rear wall firing, the burners are arranged in such a way as to promote turbulence. In opposed firing, the burners in opposite walls of the furnace impinge their flames against each other to increase turbulence [7].

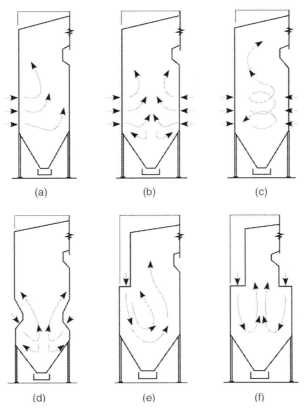

Figure 5.16 Dry-bottom furnace and burner configurations: (a) horizontal (front or rear), (b) opposed horizontal, (c) tangential (or corner firing), (d) opposed inclined, (e) single U-flame, and (f) double U-flame.
Source: From Elliot (1981).

In tangential, and to a lesser extent opposed inclined, furnaces (Figure 5.16c and d), the burner turbulence is replaced by the overall furnace turbulence. In these furnaces, there is a single flame envelope that promotes combustion stability and avoids the high flame temperatures that tend to favor NO_x formation. In addition, the burners in the tangential furnace, where the fuel and air are admitted at all four corners and at different levels of the furnace and the burners, can be tilted upward or downward by 20 degrees from the horizontal, thereby changing the temperature of the flue gas by as much as 150°F [7]. This allows for changing the combustion volume of the furnace to control superheat and reheat temperatures.

Single and double U-flame furnaces (Figure 5.16e and f) are used for firing hard to ignite and slowly burning fuels such as anthracite and coke. In these designs, the fuel is fired downward, and radiation from the rising portion of the flames and from the burners in the opposite arch (in the double U-flame units) assists in maintaining a stable flame over a wide load range.

Wet-Bottom Firing

Early wet-bottom furnaces (around the 1920s) were simply open, single-stage furnaces with burners located close together and near the furnace floor to achieve the high temperatures necessary for melting the ash [2]. The furnace type used was usually one of those shown in Figure 5.16, which was modified to accommodate the molten ash and satisfactorily used favorable coals where limited turndown was required. For coal ash that is difficult to melt and a larger turndown range is required, two-stage designs have been developed. Examples of two-stage slag-tap firing are shown in Figure 5.17 [2].

The primary advantage of wet-bottom furnaces is easier ash handling and disposal. However, the disadvantages of using wet-bottom furnaces have led to its decline in the United States. These disadvantages include lower boiler efficiency through sensible heat loss of the slag, less fuel flexibility, higher incidences of ash fouling and external corrosion of pressure parts, lower average steam generator availability, and higher levels of NO_x emissions [2].

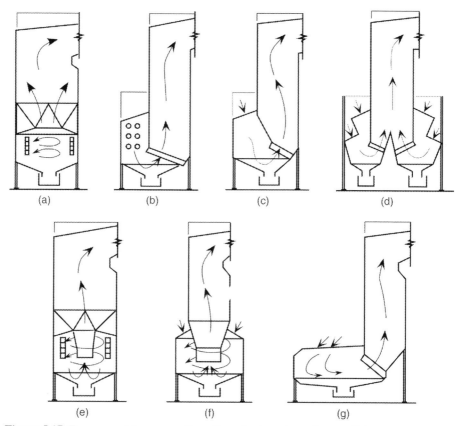

Figure 5.17 Furnace and burner configurations for two-stage slag-tap firing.
Source: From Elliot (1981).

Cyclone Furnaces

Cyclone-furnace firing, shown in Figure 5.18, is a form of two-stage, wet-bottom design, although some do not classify it as suspension firing, since a large portion of the fuel is burned on the surface of a moving slag layer [2]. In cyclone firing, one or more combustors are mounted on the wall of the main furnace. Most cyclone furnaces in the United States are fired with coal crushed to about 1/4-inch top size, while foreign practice uses partially pulverized coal (e.g., 25 percent finer than 200 mesh—74 μm).

In the screened furnace type, the gases exiting the cyclone pass through a small chamber and slag screen before entering the main furnace. This design has been largely replaced by the open-furnace arrangements as larger units have been developed. The development of the horizontal cyclone furnace occurred rapidly in the United States in the mid-1940s, with B&W the leader in this technology development. The interest in the cyclone furnace is due to its several good features: a very high rate of heat production (i.e., up to 500,000 Btu/h ft^2 compared to 150,000 and 400,000 Btu/h ft^2 in dry-bottom and slag-tap furnaces, respectively); high flame temperatures ($\approx$3,000°F) to sufficiently melt the ash; the ability to utilize coarser particles than pulverized coal-fired units, which results in lower system costs, since pulverizers are not required; and the ability to be designed to use almost any coal type as well as opportunity fuels such as tires, petroleum coke, and others.

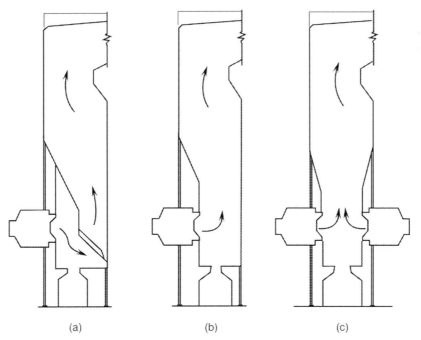

(a)	(b)	(c)

Figure 5.18 Horizontal cyclone-furnace arrangements: (a) screened furnace, (b) one-wall open furnace, and (c) opposed open furnace.
Source: From Elliot (1981).

The fuel characteristics of greatest interest in cyclone firing are the ash fusibility and viscosity of the ash lining the walls of the cyclone. The composition of the ash must be such that the ash will melt, coat the walls of the cyclone, and tap (i.e., be fluid and exit steadily) from the cyclone. In addition, the moisture content of the fuel, such as in lignites, is important because high-moisture fuels will consume heat while the moisture is being evaporated, which can affect the temperature of the cyclone and thus the fluid behavior of the slag.

As previously mentioned, the elevated temperatures produced in wet-bottom furnaces results in the generation of high levels of NO_x. Because of this, the use of cyclone furnaces in future installations is unlikely, and more attractive alternatives are pulverized-coal and fluidized-bed systems.

5.1.5   Influence of Coal Properties on Utility Boiler Design

The design of a utility steam-generating plant requires a technical and economic evaluation. Parameters that must be considered to arrive at a final design include the heat release rate, fuel properties (e.g., ash fusion temperatures, volatile matter, ash content), percentage of excess air, production of emissions (e.g., NO_x), boiler efficiency, and steam temperature [16], with the most important item to consider being the fuel burned [8]. This section discusses the influence of coal properties on boiler design, specifically as it relates to suspension firing (primarily pulverized coal), since this is the primary combustion technique used by the electric-generating industry today.

The coal properties that influence the design of the overall boiler system include, but are not limited to, coal and ash and handling, coal pulverizing, boiler size and configuration, burner details, amount of heat recovery surface and its placement, types and sizing of pollution control devices, and auxiliary components such as forced and induced-draft fan sizes, water treatment, and preheaters. The discussion in this section focuses on the influence of coal properties on furnace design.

Furnace Design

Furnaces for burning coal are more liberally sized than those for gas or fuel oil firing, as illustrated in Figure 5.19 (modified from [8]). This is necessary to complete combustion within the furnace and to prevent the formation of fouling or slagging deposits. A furnace is designed to take advantage of the high radiant heat flux near the burners [16]. Since the flue gas temperature at the exit of the boiler (i.e., the entrance to the convective section) must be at least 100°F below the ash-softening temperature (which varies from ≈2,000–2,500°F), the radiant heat transfer surface in a coal-fired boiler must be increased by 15 percent or more in order to achieve a steep reduction in temperature from the flame temperature of about 2,750°F.

The convective section of the boiler is designed to extract the maximum amount of heat from the partially cooled flue gas exiting the boiler. The flue gas velocity should not exceed about 60 ft/s in the convective section, when firing coal, to

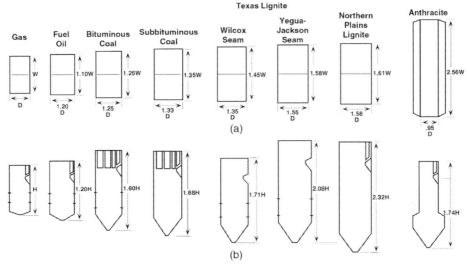

Figure 5.19 Effect of fuel type on furnace sizing assuming a constant heat input: (a) plan view and (b) elevation view. (H = distance between centerlines of lowest hopper headers and furnace roof tubes; h = distance between centerline of top burners and furnace nose apex; W = width; and D = depth)
Source: Modified from Singer (1981).

minimize erosion of the tubes from the fly ash [16]. Sootblowers are required in coal-fired power plants to keep the heat transfer surface clean.

The furnace size and shape must allow for adequate coal residence time within the furnace to achieve complete combustion. Sufficient heat must be contained in the flue gas exiting the boiler to enable efficient designs of superheater, reheater, economizer, and air heater heat transfer surfaces. The flue gas exiting the stack must be low to minimize heat losses from the system but must be above the dew point of the acid gases so metal corrosion is not experienced.

In general, ignition stability in a pulverized coal-fired furnace varies directly with the ratio of volatile matter to fixed carbon [16], and therefore coals like anthracite are typically fired in U-flame furnaces. Coals with higher volatile matter can be more easily burned in suspension, which allows for a lower furnace temperature but requires a larger furnace heat-release area. Also, coals with higher volatile matter tend to have lower ash fusion temperatures (i.e., higher fouling and slagging tendencies), thereby requiring lower furnace temperatures [16].

Figure 5.19 illustrates the relationship between furnace exit gas temperature requirements and heat release rates for typical coals as compared to natural gas and fuel oil. For a coal having a relatively low ash fusion temperature, a lower heat release rate is required than for a coal having a higher ash fusion temperature to avoid slagging problems. The furnace exit gas temperature is primarily a function of the heat release rate, which is the available heat divided by the equivalent water-cooled furnace surface [16].

Ash Characteristics

Coal is a very heterogeneous substance, and the mineral matter distributed throughout the coal comes in various forms, compositions, and associations and cannot be simply represented by composition (i.e., elemental oxides) and a single set of melting temperatures (i.e., ash fusion temperatures). Empirical indices have been developed using coal ash chemistry, but they are successfully used only part of the time and are not applicable to ranks of coal beyond that for which they were developed. Some of these historical indices that are important when evaluating coal ash behavior as they affect furnace slagging and fouling on both the furnace walls and convective surfaces (which vary among boiler manufacturers) include, but are not limited to, ash fusibility temperatures, dolomite concentrations, total ash concentration and composition, and base-to-acid, iron-to-calcium, and silica-to-aluminum ratios [8, 29, 30]. These parameters indicate an ash's slagging and fouling potential. In addition, slag viscosity is an important parameter for cyclone furnace operation.

Slag Viscosity

The viscosity of the slag formed from the coal ash is an important parameter for cyclone-fired furnaces. Slag will just flow on a horizontal surface at a viscosity of 250 poises [6]. The temperature at which this viscosity occurs (T_{250}) is used as the criterion to determine the suitability of a coal for a cyclone furnace. The T_{250} can be either calculated from a chemical analysis of the coal ash or, more preferably, determined experimentally using a high-temperature viscometer, and a value of 2,600°F is considered maximum. Coals with a slag viscosity of 250 poises at 2,600°F or lower are considered candidates for cyclone furnaces, provided the ash analysis does not indicate excessive formation of iron or iron pyrites [6]. The T_{250} index is used for all ranks of coal.

In dry-bottom furnaces, the formation of slag must be avoided so as not to adversely affect the unit's operation. The following table shows the relationship between furnace slagging and T_{250}.

Slagging Rating	T_{250}
Low	>2,325°F
Medium	2,550–2,100°F
High	2,275–2,050°F
Severe	<2,200°F

Slagging and Fouling Potential

The potential for slagging—fused slag deposits that form on furnace walls and other surfaces exposed to predominately radiant heat—is temperature and ash composition related. Slagging potential affects furnace sizing, the arrangement of radiant and convective heating surfaces, and the number of sootblowers required [30].

Fouling deposits form primarily in the lower-temperature regions of the furnace and convective section and affect the design and maintenance of superheaters, reheaters, furnace waterwalls, air heaters, and the number of sootblowers required.

Potential for fouling is linked to the alkaline content of the ash, specifically the active alkalis.

Ash fusibility has long been recognized as a tool for measuring the performance of coals related to slagging and deposit buildup. ASTM Standards D1857 gives the experimental procedure for determining the ash fusion temperatures. The test is based on the gradual thermal deformation of a pyramid-shaped ash sample in either an oxidizing or reducing atmosphere. Four temperatures are obtained during the test:

1. The *initial deformation temperature,* which is the temperature where the tip of the pyramid begins to show evidence of deformation
2. The *softening or fusion temperature,* which is the temperature where the ash sample has fused and the height equals the width
3. The *hemispherical temperature,* which is the temperature where the sample has fused into a hemispherical shape of the height equaling half of the width at the base
4. The *fluid temperature,* which is the temperature where the sample has fused down into a nearly flat layer

The ash-softening temperature is related to the type and ease of deposit removal from the heat transfer surfaces. If ash particles arrive at heat-absorbing surfaces at temperatures below their softening temperature, they will not form a bonded structure, and ash removal is relatively easy. If the ash particles arrive at these surfaces after they have been subjected to temperatures above their softening temperature and have become plastic or liquid, the resulting deposit will be tightly bonded and more difficult to remove [8]. Also, the temperature difference between the initial deformation and fluid temperatures provides information on the type of deposit to expect on furnace tube surfaces [8]. A small temperature difference indicates a thin, running, tenacious, difficult to remove slag, whereas wider temperature differences indicate less-adhesive deposits.

Another measure of ash viscosity that is used to predict furnace slagging is the temperature of critical viscosity, T_{cv} [31]. This is the temperature at which the viscosity properties of the molten slag change on cooling from those of a Newtonian fluid to those of a Bingham plastic and is believed to be the temperature where solid phases start to crystallize from the melt.

Several slagging and fouling indices have been developed based on the ash composition. Many of these indices have been developed for eastern U.S.–type coals. These are characterized as high iron and low alkali and alkaline-earth content coals. The ash composition of such coals reflects a Fe_2O_3 level exceeding the combined CaO, MgO, Na_2O, and K_2O percentage. In addition, sulfur percentages are higher than western U.S. coals.

The acidic ash constituents (reported as weight percent on an oxide basis) SiO_2, Al_2O_3, and TiO_2 are generally considered to produce high melting temperature ashes. Temperatures will be lowered by the relative amounts of basic oxides, Fe_2O_3, CaO, MgO, Na_2O, and K_2O available in the ash. A base/acid ratio defined as

$$\frac{B}{A} = \frac{Fe_2O_3 + CaO + MgO + Na_2O + K_2O}{SiO_2 + Al_2O_3 + TiO_2} \qquad (5.21)$$

has been developed as an indictor to predict the relative performance of coal ash in the furnace. This index can be used for all ranks of coal. A base/acid ratio in the range of 0.4 to 0.7 manifests low-fusibility temperatures and a higher slagging potential [8].

The base/acid ratio has also been used to define a slagging index expressed as $R_S = B/A \times S$, where S is the weight percent sulfur in the dry coal. The following table shows the slagging index that has been used with success to identify four types of slagging coals [32].

Slagging Type	Slagging Index, R_s
Low	<0.6
Medium	0.6–2.0
High	2.0–2.6
Severe	>2.6

The influence of alkalis, notably sodium (Na_2O) and potassium (K_2O), on fusibility and slagging are proportional to the quantity in the coal ash. With sodium-containing coals, the rate of buildup is a function of the sodium concentration. The alkalis can be present in the coal in various forms. Alkalis that vaporize during combustion are classified as active or mobile alkalis and are free to react or condense in the boiler and consist primarily of simple inorganic salts and organically bound alkalis. More stable forms of alkalis exist in impurities such as clays and shales and remain inert during combustion. The mode of occurrence of the alkalis in coal is determined through a coal leaching process via sequential washings using water, ammonium acetate, and hydrochloric acid. The most active alkalis are those soluble in water and ammonium acetate. The fouling potential of coals is directly related to the soluble concentration of sodium and has been shown to vary from about 0.001 lb soluble sodium/lb ash/million Btu fired for a low-fouling coal to about 0.044 lb soluble sodium/lb ash/million Btu fired for high/severe-fouling coals [8].

Two fouling indices related mainly to sodium have been proposed to predict the extent of fouling of convective heat transfer surfaces. Both apply to eastern U.S coals, rather than to the lignites, in which the CaO + MgO content of the ash may be greater than the Fe_2O_3 content [31, 32]. The indices are R_F = Base/Acid × Na_2O (ASTM ash) and = Base/Acid × water-soluble Na_2O (LTA; i.e., low-temperature ash). Sodium is determined conventionally on the ASTM ash and on the water-soluble portion of LTA. The fouling characteristics of coals are divided into the four categories in the table [31, 32].

Fouling Tendency	R_F	R'_F
Low	<0.2	<0.1
Medium	0.2–0.5	0.1–0.25
High	0.5–1.0	0.25–0.7
Severe	>1.0	>0.7

Numerous fouling studies, especially those using low-rank coals, have shown a relationship between the sodium in the ash and the fouling rate. This is particularly true of coals with lignitic-type ash (low iron, high alkali, and high alkaline–earth metals), as found primarily in the United States west of the Mississippi River, although sodium does contribute to deposition in higher-rank coals as well. A fouling index based on the percent of Na_2O in coal ash is as follows [29].

Lignitic Ash		Bituminous Ash	
Percent Na_2O	Fouling Potential	Percent Na_2O	Fouling Potential
<2.0	Low	<0.5	Low
2–6	Medium	0.5–1.0	Medium
6–8	High	1.0–2.5	High
8	Severe	>2.5	Severe

Chlorine content of the coal is also used by some to predict fouling [31]. There is some doubt as to the validity of this parameter, but the following values are recognized as representative of those used by industry.

Total Chlorine in Coal (%)	Fouling Rating
<0.2	Low
0.2–0.3	Medium
0.3–0.5	High
>0.5	Severe

Other fouling and slagging parameters have been developed, but many are guides to be used in conjunction with other parameters and operating experience of the various coals and boiler units. Here are some of them:

1. The silica/alumina ratio (SiO_2/Al_2O_3) can provide additional information relating to ash fusibility [8]. The general range of values is between 0.8 and 4.0, and for two coals having similar base/acid ratios, the one with a higher silica/alumina ratio should have lower fusibility temperatures.
2. The iron/calcium ratio (Fe_2O_3/CaO) indicates the fluxing (the lowering of the ash fusion temperature) potential of the iron and calcium in the ash. Iron/calcium ratios between 10.0 and 0.2 have a marked effect on lowering the fusibility temperatures of coal ash, and extreme effects are evident between ratios of 3.0 and 0.3 [8].
3. A dolomite percentage, DP, defined as

$$DP = \left(\frac{CaO + MgO}{Fe_2O_3 + CaO + MgO + Na_2O + K_2O}\right) x100 \qquad (5.22)$$

which is used primarily for coal ashes with a basic oxide content over 40 percent (i.e., western U.S. coals). It has been empirically related to the viscosity of coal ash slags, and, at a given basic concentration, a higher DP usually results in higher fusion temperatures and higher slag viscosities [8].

4. A silica percentage, SP, defined as

$$SP = \left(\frac{SiO_2}{SiO_2 + Equiv.Fe_2O_3 + CaO + MgO} \right) x100 \tag{5.23}$$

which has been empirically correlated with the viscosity of coal ash slags. As SP increases, the slag viscosity increases [8].

The development of a reliable coal-screening tool has long been a goal of the utility industry. The indices based on ASTM coal and ash analysis have provided useful information to boiler designers and operators but are not refined enough to be applied to all coals and all boilers. Also, since many of these indices were developed for bituminous coals, they are poor indicators of performance when applied to low-rank coals. Consequently, different analytical techniques are being applied, and new indices have been and are continually being developed. Many of these new indices rely heavily on computer-controlled scanning electron microscopy coupled with coal and coal ash analyses.

The technique of automated or computer-controlled scanning electron microscopy has enabled the identification and sizing of coal mineral matter in situ and is also used to determine the inherent or extraneous nature of the mineral matter in coal. The direct analysis of mineral matter in coal, size determinations, and the observation of the association of mineral matter with coal particles are essential to determine the behavior of mineral matter during combustion of pulverized coal. Examples of indices include wall slagging, convective pass fouling based on either sulfates or silicates being the primary bonding component, cyclone slagging, and deposit strength [33]. Many of the indices developed by the industry are proprietary.

As emphasized in this section, a major factor affecting furnace performance in large coal-fired utility boilers is the inorganic matter of the fuel. Most problems with ash are associated with its effect on heat transfer by thermally insulating furnace wall tubes and convective pass tube banks. Accumulation of ash deposits can decrease the heat transfer rate to the tube surface, resulting in high flue gas temperatures.

Boiler manufacturers each have their own empirical relationships to predict furnace performance and the effect of ash slagging and fouling. Also, testing coals in pilot-scale slagging and fouling combustors or boilers is routinely performed by boiler manufacturers, universities, and other test facilities to assess deposition performance. Each boiler manufacturer has its own criteria to allow for the effects of the ash characteristics; these are based on sound engineering judgment and years of experience with boilers of similar sizes.

5.2 Carbonization

Carbonization is the process by which coal is heated and volatile products—both gaseous and liquid—are driven off, leaving a solid residue called char or coke. Coal carbonization processes are classified into high-temperature operations if they are performed at temperatures greater than 1,650°F or low-temperature operations if they are conducted below 1,350°F. These temperatures are somewhat arbitrary, since they reflect pronounced physical changes that coal undergoes at temperatures between 1,110 and 1,470°F [21]. Sometimes carbonization processes that reach into the 1,350 to 1,650°F range are termed medium-temperature processes.

Coals of a very definite range of rank soften on heating, swell on decomposition, and resolidify on continued degasification [1]. Devolatilization is a continuous process, but a distinction can be made between the primary carbonization stage, in which mainly tar is evolved, and the secondary carbonization stage, in which only gas is split off. The characteristic temperatures and stages of the carbonization process are illustrated in Figure 5.20 [1].

The main purpose of high-temperature carbonization is the production of metallurgical coke for use in blast furnaces and foundries. Some coke is used for the manufacture of calcium carbide and electrode carbons, as reductant in certain ferrous and nonferrous open-hearth operations, and in foundries to produce cast iron; however, more than 90 percent of the coke produced is used in blast furnaces to smelt iron ore and produce pig iron, and modern coke-making practices are virtually dictated by the coke quality in this market. Low-temperature carbonization was mainly used to provide town gas for residential and street lighting, tars for use in chemical production, and smokeless fuels for domestic and industrial heating.

To put the use of coal used for carbonization into perspective with the electrical power-generation industry (i.e., using coal in combustion technologies), of the approximately 1,065.8 million short tons of coal consumed in the United States in 2002, approximately 22.5 million short tons, or about 2 percent, were used in coke plants as compared to approximately 975.9 million short tons (or ~92 percent) consumed by the electric power sector. This chapter provides a brief history of high-temperature carbonization, specifically as it was developed for the iron and steel industry, reviews coking processes, discusses coal properties of interest for

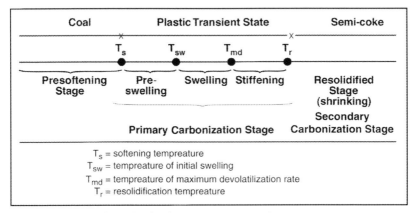

Figure 5.20 Characteristic carbonization temperatures and stages.
Source: From Van Krevelen (1993).

producing coking coals and the uses of coke, and concludes with a review of low-temperature carbonization.

5.2.1   Brief History of Carbonization (High-Temperature)

The carbonization of coal has its historical roots in the iron and steel industries. The iron-making process developed around the Mediterranean Sea, spreading northward through Europe [34]. Historians state that the Phoenicians, Celts, and Romans all helped spread iron-making technology, with one of the iron-making techniques spread by the Romans as far north as Great Britain. Originally, charcoal produced from wood was the fuel used to melt the iron ore. A tremendous amount of wood was needed for this industry. For example, one type of furnace, the Stuckofen, used in fourteenth-century Germany, could produce 4,000 pounds of iron per day with a fuel rate of 250 pounds of charcoal per 100 pounds of iron produced [34]. This was an early version of the charcoal blast furnace, and these furnaces that developed in Continental Europe soon spread to Great Britain.

By 1615, over 800 furnaces, forges, or ion mills existed in Great Britain, with 300 of them blast furnaces. The rate of growth in the number of these furnaces was so great that during the 1600s parliament passed laws to protect the forests. Consequently, many blast furnaces were shut down, alternative fuels were sought, and England encouraged the production of iron in its North American colonies, which had abundant supplies of wood and iron ore. The first successful charcoal blast furnace in the New World was constructed in Saugus, Massachusetts, in 1645.

As a result of the depletion of virgin forests in Great Britain to sustain the charcoal iron, the ironmasters were forced to consider alternative fuel sources. The alternative fuels included bituminous coal, anthracite, coke, and even peat [34]. The development of coke and anthracite iron-making paralleled each other and coexisted with charcoal production during the 1700s and 1800s, while the use of bituminous coal and peat never became a major iron-making fuel. The wide use of coke in place of charcoal came about in the early 1700s when Abraham Darby and his son demonstrated that coke burned more cleanly and with a hotter flame than coal [21]. Up until 1750, the only ironworks using coke on a regular basis were two furnaces operated by the Darby family [34]. However, during the period 1750 to 1771, the use of coke spread, with eventually 27 coke furnaces in production. The use of coke increased iron production because it was stronger than charcoal and could support the weight of more raw materials, and thus furnace size was increased.

The use of coke then spread to Continental Europe: Creussot, France, in 1785; Gewitz, Silesia, in 1796; Seraing, Belgium, in 1826; Mulhiem, Germany, in 1849; Donete, Russia, in 1871; and Bilbao, Spain, in 1880 [34]. In North America, the first attempt to use coke as 100 percent fuel was in the Mary Ann furnace in Huntington, Pennsylvania, although coke was mixed with other fuels as early as 1797 in U.S. blast furnaces.

The efficient use of coke and anthracite in producing iron was accelerated by the use of steam-driven equipment, the invention of equipment to preheat air entering the blast furnace, and the design of the tuyeres and the tuyere composition [34].

The evolution of both coke and anthracite iron-making paralleled each other in the United States during the 1800s, and by 1856 there were 121 anthracite furnaces in operation. With coke being the strongest and most available fuel, the evolution of 100 percent coke furnaces continued, with major steps being made in the Pittsburgh area between 1872 and 1913. The Carnegie Steel Company and its predecessor firms developed technological process improvements at its Monongahela Valley iron-making furnaces that ultimately made it possible for the United States to take over worldwide leadership in iron production. This is not true today, however, as much of the steel production has shifted overseas beginning in the 1960s and early 1970s.

5.2.2 Coking Processes

Early processes for the production of coke were similar to those employed for the production of wood charcoal. Bituminous coal was built up into piles and ignited in such a way that only the outside layers actually burned while the central portion was carbonized [2]. Piles, also called kilns, appeared for the first time in England in 1657 and spread from there to other European bituminous coal-producing regions.

Around 1850, half-open brick kilns (i.e., the Schaumburg kiln) were constructed, from which circular mounds of coal emitted tar-containing volatilized gases directly into the atmosphere [2]. The next development was the closed beehive oven, which in its original form discharged the distillation and flue gases through a chimney at a greater height. Beehive ovens were used in England until the end of the 1800s, although some are still in operation today in South Africa, South America, Australia, and the United States.

The beehive oven is a simple domed brick structure into which coal can be charged through an opening at the top and then leveled through a side door to form a bed about 2 feet thick [21]. Heat is supplied by burning the volatile matter released from the coal, and carbonization progresses from the top down through the charge. Approximately 5.6 short tons of coal can be charged, and 48 to 76 hours are required for carbonization. Some beehive ovens are still in operation because of system improvements and the addition of waste heat boilers to recover heat from the combustion products.

Similarly, the heat required for coking in the pile and the Schaumburg kiln is produced by partial combustion of the coal. This results in a substantial loss of material by combustion, with a coke yield in these ovens (including the beehive oven) at most 55 percent of the coal. Flame ovens, in which the coal was coked in chambers heated from the outside, were developed in 1850 in Belgium and the Saar District in Germany [2]. The high-heating value volatilized gases were burned in flues in the walls of the ovens, producing coke yields of about 75 percent, with coking times of 48 hours.

The first coke ovens that produced satisfactory blast furnace or foundary coke as the main product, and tar, ammonia, and later benzene as by-products, were built around 1856 and known as by-product recovery ovens [2]. Modifications to the design continued, but the basic design of these ovens, essentially the modern coke oven, was completed by the 1940s [1].

The horizontal slot-type coke (by-product recovery) oven, in which higher temperatures can be attained and better control over coke quality can be exercised, has superseded other designs and is used for coking bituminous coal [21]. Modern slot-type coke ovens are comprised of chambers 50 to 55 feet long, 20 to 22 feet high, and about 18 inches wide. A number of these chambers, from 20 to 100, alternating with similar cells that accommodate heating flues, are built as a battery. Coal, crushed to 80 percent minus 1/8 inch with a top size of 1 inch, is loaded along the top of the ovens using a charging car on rails and is leveled by a retractable bar. Coking takes place in completely sealed ovens, and when carbonization is completed (after 15–20 hours), the oven doors are opened, and a ram on one side pushes the red-hot coke into a quenching car or onto a quenching platform. Coke yield is about 75 percent. By-product gas and tar vapors are removed from the oven to collector mains for further processing or for use in the battery.

A block flow diagram of the recovery of by-products from a coke oven is shown in Figure 5.21 (modified from [24]). From a ton of coal, a modern by-product coke oven yields about 1,500 lb of coke, 11,000 ft^3 of gas, 8 to 10 gallons of light oil, and 25 lb of chemicals, mostly ammonium compounds.

The by-product gas and tar vapors leaving the coke oven undergo a separation process to remove the tars from the gas. The gas is then treated to recover ammonia, as ammonium sulfate or phosphate, while the tars are fractionated by distillation into three oil cuts, which are designated as light oil, middle (or tar acid) oil, and heavy oil. The gas, mainly a mixture of hydrogen and methane, has about one-half the heating value of natural gas and is used on site as fuel in the flue chambers in the coke ovens or in the furnaces used for heat-treating finished steel [24].

The light oil cut (boiling point <430°F) from the distillation of the coal tar, which consists of mostly benzene (45–72 percent), toluene (11–19 percent), xylenes (3–8 percent), styrene (1–1.5 percent), and indene (1–1.5 percent) is either processed into gasoline and aviation fuels or fractionated to provide solvents and feedstocks for chemical industries [21]. In either case, sulfur compounds, nitrogen bases, and undesirable unsaturates are removed.

Middle oils are usually cut to boil between 430°F and 710°F, and after sequential extraction of tar acids, tar bases, and naphthalene, they are processed to meet specifications for diesel fuels, kerosene, or creosote. The tar acids are mostly comprised of phenols, while the compounds produced from the tar bases include pyridine, picolines, lutidines, anilines, quinoline, isoquinoline, and methylquinolines [21].

The temperature at which distillation of the heavy oils is performed depends on what type of pitch residue is desired, but it is usually between 840°F and 1,040°F [21]. The distillate is a rich source of hydrocarbons—mainly anthracene, phenanthrene, carbazole, acenaphthene, fluorine, and chrysene. The remaining heavy oils are marketed as fuel oils or blended with pitches to meet specifications for various grades of road tar.

The residual coal tar pitches are complex mixtures of over 5,000 compounds. They have economic importance because of their resistance to water and weathering and are used as briquetting binders and binders in the preparation of carbon electrodes and other carbon artifacts.

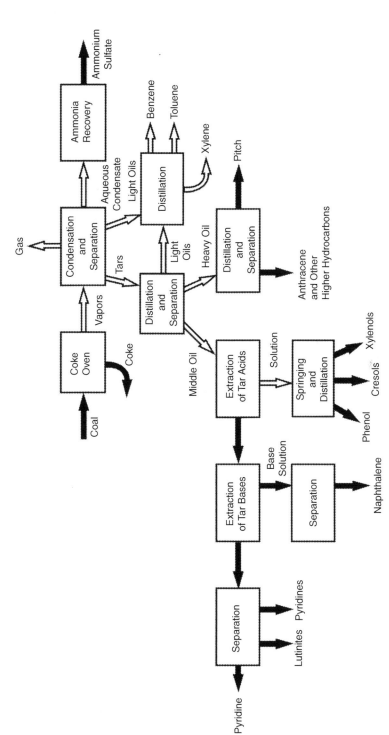

Figure 5.21 Simplified block diagram illustrating the by-products recovered and processing performed to produce useful chemicals.
Source: Modified from Schobert (1987).

5.2.3   Coal Properties for Coke Production

The properties of the coke are influenced by the coal or coal blend from which the coke is made, as well as by the carbonizing conditions used. The rank of the coal, or the average rank of a coal blend used, affects the properties of the coke product. The choice of a suitable coal is crucial to quality control of the coke, and coke producers blend coals to manipulate coke characteristics. The practice of blending coals originally developed to eliminate difficulties that were encountered when carbonizing highly fluid coals, but it was later used to stretch supplies of scarce or costly ideal caking coals, and it soon became the principal method for manipulating coke characteristics. It is common practice to blend high-volatile with low-volatile and/or medium-volatile coals to improve the strength of the coke.

The ability of a coal to melt upon heating and to form a coherent residue on cooling is termed *caking*, and it is an essential prerequisite for a coking coal that it should cake or fuse when heated. Coals that are low in rank, such as lignites, or high in rank, such as anthracites, do not cake and are therefore not capable of forming coke. Several properties of coals are measured to identify appropriate coking coals: swelling, fluidity, composition, maceral analyses, and vitrinite reflectance. These coal characteristics are listed in Table 5.6 along with typical values [1, 21, 35]. The mineral matter, or ash, content of the coal is of interest in coke production because the ash dilutes the coal and affects its caking properties. The composition of the ash is important as well because iron and steel quality is affected by sulfur and phosphorus content.

For coke to form, some of the coal's organic constituents or macerals must melt when the coal is heated. A caking coal behaves as if it were a pseudo-liquid when this occurs, and the viscosity of this material at various temperatures plays an important role in coking operations. The relative proportions of reactive and inert materials in a coking coal affect the strength of the final carbonized product.

Table 5.6 Coal Properties for Coke Production

Coal Parameter	Typical Values
Carbon content, wt.% (dry, ash free)	≈85
Hydrogen content, wt.% (dry, ash free)	≈5.25
Volatile matter content, wt.% (dry, ash free)	24–28
Ash content, wt.%	<10
Sulfur content, wt.%	≈0.5
H/C ratio	0.725
O/C ratio	0.04
Heating value, Btu/lb (moist, mineral matter free)	≈15,500
Vitrinite reflectance, %	≈1.25
Vitrinite/inertinite/exinite, % (dry basis)	≈55/35/10
Free swelling index	6.5.8
Maximum fluidity, dial divisions per minute	≈1,000
Roga index	≈45
Gray-King assay	≥G4

5.2.4 Coking Conditions

In addition to the coal properties listed in the previous section, the carbonization conditions influence the properties of the resultant coke. Carbonization conditions of interests include the particle size of the coal charged to the coke oven, charge density, rate of heating, oven design, and special conditions such as preheating and partial briquetting.

It is important to pulverize the coals (measured as the quantity of coal passing through a 1/8-inch screen, which is typically 80 percent but can vary from 50 to 100 percent) to reduce the inert particles of the coal as well as to reduce the size of coal particles that exhibit low fluidities [2]. Higher levels of pulverization tend to make a more homogenous mixture of the reactive and inert components of the coals blended. The bulk density of the coal charged to the oven is also adjusted in order to produce denser or more homogenous cokes. In addition to varying the coal particle size, this is accomplished by adding small amounts of water or oil to the blend.

Much effort has been made to increase the productivity of coke ovens by improved oven design and operating practice. The rate of heating the coal charge has been shown to be important in coke yield and properties [2]. As coking rate is increased, the coke size, shatter index, and stability decrease, while the hardness factor increases. Coke size becomes more uniform with increased heating rate. Faster coking rates and the resultant in situ crushing is an advantage to the iron makers, who crush coke to produce a uniform-sized burden for the furnace [2].

Preheating high-oxygen coals or marginally coking coals can substantially improve the quality of coke from these coals and reduce the required coking times [2]. Coking times for these coals can be reduced by 30 to 45 percent, while preheating strengthens the resultant cokes. In an effort to reduce the cost of blast furnace coke and to extend the range of ranks and the types of coals that can be incorporated into blast furnace coke blends, "formed coke" processes have been developed. Formed coke processes involve carbonizing coal or a blend of coals (that may contain low-rank or high-volatile coals) that have been compressed into shaped briquettes.

5.2.5 Low-Temperature Carbonization

Low-temperature carbonization was originally developed to provide town gas for residential and street lighting and to manufacture a smokeless fuel for domestic and industrial heating. The by-product tars were economically important and were often essential feedstocks for the chemical industry or were refined to gasoline, heating oils, and lubricants [21]. Low-temperature carbonization evolved and was used extensively in industrialized European countries but was eventually abandoned after 1945 as oil and natural gas became widely available. These early processes used fixed- and moving-bed technology, operated in batch or continuous mode, that consisted of vertical or horizontal retorts with direct or indirect heating [1]. In the 1970s, interest in low-temperature carbonization was resurrected after the oil crisis.

Most of the techniques utilized now, however, are different and mainly consist of using fluidized-bed or entrained-flow pyrolyzers.

Some of the low-temperature carbonization technologies are the FMC Coke, COED, U.S. Steel Coke, Occidental Pyrolysis, Lurgi-Ruhrgas, Coalite, Phurnacite, and Home Fire processes [2, 36]. In addition, technologies, such as ENCOAL's Liquids-From-Coal (LFC) process and Rosebud SynCoal Partnership's Advanced Coal Conversion Process (ACCP), are examples of commercial and/or near-commercial technologies to upgrade low-rank coals [37–39]. Similarly, ConvertCoal, Inc.'s (CCI) Rinker-England-Skov (RES) coal conversion process, which is a newer generation of the LFC process, upgrades low-rank coal and produces liquids, although this process is more of a medium-temperature carbonization process [40].

The preferred coals for low-temperature carbonization are typically lignites, subbituminous coal, or high-volatile bituminous coal that, when pyrolyzed at temperatures between 1,100°F and 1,300°F, yield a porous char with reactivities that are typically not much lower than those of their parent coals [21]. These reactive chars are easily ignited and are used as smokeless fuels or as feedstocks to gasification processes, are blended with coals to make coke-oven feed, or are used as a power plant fuel [2, 21, 24, 38, 39]. The tars that are produced during low-temperature carbonization are much different from those from high-temperature carbonization. High-temperature carbonization tends to produce mainly aromatic compounds, whereas those produced during low-temperature carbonization are predominately aliphatic compounds—thus, the different end-use applications of the tar by-products. Gas yield and composition are also different during low-temperature carbonization, with gas yields being about 25 percent of that produced during high-temperature carbonization, although the gas contains more methane and less hydrogen, giving it a higher heating value [24].

Smokeless Fuel Commercial Processes

The primary application of low-temperature carbonization is to make smokeless fuels for use in homes and small industrial boilers in areas that have high population density and rely on coal as a fuel, particularly coal that has a high volatile matter content. This is especially true in Great Britain, which has regulated smoke-controlled areas, and several commercial plants are producing smokeless fuels for open fires, room heaters, multifuel stoves, cookers, and independent boilers [41]. These fuels are marketed under names such as Coalite, Sunbrite, Phurnacite, Taybrite, and Home Fire.

Two processes used in Great Britain are the Coalite process and the Home Fire process. The Coalite process uses moving-bed, vertical retort technology and is a continuous process using indirect heating [1]. The Coalite works are located at Bolsover in Derbyshire and started operation in 1937. In the Coalite process, deep-mined British bituminous coal is carbonized in several batteries, each battery consisting of 40 metal retorts, assembled in two rows of 20, at a temperature of approximately 1,200°F [42]. The coal charge to each retort, which is about 660 lbs, remains in the retort for four hours, after which a ram pushes the Coalite into a cooler.

Typically, one metric ton of coal blend will produce 1,100 to 1,870 lbs of smokeless coal or semicoke; 5,300 to 6,350 standard cubic feet (scf) of gas; 18 to 20 gallons of coal oil; 3 to 5 gallons of light oil; and 45 to 48 gallons of aqueous liquor. The gas is recycled for on-site use to heat the batteries, generate steam, and general heating. Oils are distilled to yield pitch (for use as a boiler fuel), heavy oil (to produce creosotes and disinfectants), and middle and light oils (to produce phenols, cresols, and xylenols). The liquor contains dissolved chemicals—principally ammonia, monohydric and dihydric phenols. Through extraction and fractionation, a wide range of chemicals are produced, including catechol, resorcinol, and methyl resorcinol. Coalite is the leading manufacturer of smokeless fuel in the United Kingdom.

The Home Fire process uses a blend of bituminous coals and fluidized-bed technology. At the Home Fire plant, located near Coventry, the coal is crushed to 1/4-inch particles, dried, and devolatilized for 20 minutes at 800°F in a fluidized-bed reactor [24, 36]. The hot char is extruded into hexagonal briquettes, cooled, and quenched.

Low-Rank Coal Upgrading

The U.S. Department of Energy cofunded two programs through its Clean Coal Technology (CCT) program (which is discussed in detail in Chapter 11), where low-rank coals were upgraded using low-temperature pyrolysis/thermal upgrading technologies as part of the process. Western U.S. low-rank coals, primarily subbituminous coals and lignite, are generally low in sulfur, making them attractive (specifically the subbituminous coals) as power plant fuels in place of high-sulfur eastern U.S. coals; however, the disadvantages of the low-rank coals include high moisture content and low heating value. Consequently, two processes, ENCOAL's LFC process and Rosebud SynCoal Partnership's ACCP, have been developed and successfully demonstrated, and are now ready for commercialization. A third-generation process, ConvertCoal, Inc.'s RES process, which is based on the LFC process, is another technique for upgrading low-rank coals.

The LFC technology uses a mild pyrolysis or mild gasification process to produce a low-sulfur, high-heating value fuel and a coal-derived liquid [37–39, 43]. In the process, coal that has undergone some coal drying is fed into a pyrolyzer that is operated near 1,000°F to remove any remaining moisture and release volatile gases. The solid fuel, called process-derived fuel (PDF), is used as a boiler fuel. The pyrolysis gas stream is sent through a cyclone to remove entrained particles and then cooled to condense the desired hydrocarbons, called coal-derived liquids (CDL), and stop any secondary reactions. The CDL were utilized at seven industrial fuel users and one steel mill blast furnace during the five-year demonstration; however, studies were performed on upgrading the CDL to produce cresylic acids, petroleum refinery feedstock for producing transportation fuels, oxygenated liquids, and pitch.

The process was demonstrated for approximately five years in a plant feeding 1,000 short tons/day coal located near Gillette, Wyoming's, Triton Coal Company's Buckskin Mine, and it produced over 83,000 short tons of solid fuel product and

4.9 million gallons of liquid product. The process is considered commercial and is actively being marketed in the United States (i.e., in the Powder River Basin, Alaska's Beluga field, and the lignite fields of North Dakota and Texas) and abroad (i.e., in China, Indonesia, and Russia) [38, 43], and five detailed commercial feasibility studies have been completed [44]. A large-scale commercial plant has been designed, with participation from Mitsubishi Heavy Industries, to utilize 15,000 metric tons of coal feed. The plant, located near Gillette, Wyoming, has received the Industrial Siting Permit and an Air Quality Construction Permit, but it is on hold due to lack of funding [44].

The ACCP process is an advanced thermal conversion process coupled with physical cleaning techniques to upgrade high-moisture, low-rank coals to produce a high-quality, low-sulfur fuel. In this process, coal is fed into a vibratory fluidized-bed reactor to remove surface moisture and then flows to a second vibratory reactor, where the coal is heated to 600°F to remove chemically bound water, carboxyl groups, volatile sulfur compounds, and a small amount of tar. In this process, the volatiles are not collected but are used in a process heater. The technology was demonstrated from 1993 to 2001 (longer than the planned five-year period) in a 45 short ton/hour facility located adjacent to a unit train load-out facility at Western Energy Company's Rosebud coal mine near Colstrip, Montana, in which 2.8 million short tons of raw coal were processed. Nearly 1.9 million short tons of SynCoal were produced and shipped to various customers, including cement and lime kilns and utility boilers, and was used as a betonite additive in the foundary industry.

Three different feedstocks were tested at the ACCP facility—two North Dakota lignites and a subbituminous coal—and the products were fired in a utility boiler in North Dakota and three utility boilers in Montana [45]. The technology is being marketed and promoted worldwide, and a project was actively pursued for Minnkota's Milton R. Young Power Station, which had test-fired SynCoal produced from one of the North Dakota lignites; however, the project was suspended due to a lack of equity investors.

CCI's RES process has been developed for converting lignite and subbituminous low-rank coals into a synthetic crude oil (SCO) suitable for petroleum refining and into a low-emission coal-char fuel (CCF) for power generation and iron ore reduction. The primary processes are coal drying, pyrolysis, coal-tar oil recovery, and conversion of recovered coal-tar oil into SCO. Coal is dried to remove moisture and then undergoes two-stage pyrolysis. In the pyrolysis stages, coal is pyrolyzed at temperatures up to 1,650°F. The pyrolysis gases are hydroprocessed to refinery-grade SCO, which is similar to a low-sulfur West Texas Intermediate crude oil but with higher aromatics content similar to Alaska North Slope crude oil. When using a PRB subbituminous coal, the SCO product will be about 30 API gravity (0.975 g/ml), 60 wt.% aromatics, less than 7 wt.% fraction with a boiling point above 824°F (440°C), and less than 0.1 wt.% sulfur and 0.2 wt.% nitrogen [40]. The CCF product remaining after oil recovery is an upgraded fuel with a heating value that has been increased by 40 to 50 percent, which results in a 5 to 10 percent increase in boiler efficiency. Emissions are lower due to the removal of sulfur, nitrogen, and mercury from the coal.

5.3   Gasification

Gasification is a process for upgrading solid feedstocks, which are difficult to handle, by removing undesirable impurities and converting them into a gaseous form that can be purified and used directly as a fuel or further reacted to produce other gaseous or liquid fuels or chemicals. There are many reasons for the interest in gasification as a process for utilizing coal. Liquid and gaseous fuels are easier to handle and use than coal, whether the fuel is used for heating, cooking, transportation, or power production. Shipping coal can be difficult and labor intensive, and it can have negative environmental impacts. Impurities in coal can be more readily removed through gasification than when utilized directly. Synthetic fuels burn more cleanly than coal, and less sulfur and nitrogen oxides are formed during combustion. Carbon dioxide, which is being considered for regulation as a pollutant, can be more readily removed from the gas stream before gasification products are combusted.

Gasification of coal is especially attractive to nations that have coal reserves and lack reserves of oil and gas or are depleting them. And finally, gasification of coal reduces the concerns of volatile swings in the availability and cost of gaseous or liquid fuels that are experienced with petroleum or natural gas. Although the use of coal gasification is currently rather limited, this technology is poised to be the technology of the future for the production of electricity, steam, chemicals, and fuels such as hydrogen.

5.3.1   Brief History of Coal Gasification

The discovery of gases from coal, insofar as verified by written records, dates back to the early 1600s. In 1609, a Belgian chemist, Jan van Helmont, observed that gas was evolved from coal when it was heated [24]. At the end of the 1600s, John Clayton, a clergyman from Yorkshire, England, experimented with collecting gas from coal. In 1792, William Murdock, a Scottish engineer, pioneered the commercial gasification of coal using the technique of heating coal in a retort in the absence of air to convert coal to gas and coke. The gas produced in this manner (i.e., carbonization) has several names: coal gas, town gas, city gas, and illuminating gas. By 1798, gas-lighting systems were being installed in factories and mills in England, which was followed with their installation as street lighting systems beginning in 1807. By 1816, most of London was lit by gas. In 1816, the Baltimore Gas Company, the first coal gasification company in the United States, was established. Gas-lighting spread throughout the east coast: Boston in 1821, New York in 1823, and Philadelphia in 1841 [24].

By the mid-1920s, about 20 percent of the gas supply in the United States was being produced from coal [24]. Prior to World War II, there were at least 20,000 gasifiers operating in the United States. In the 1940s, the increasing availability of low-cost natural gas led to its substitution for gases derived from coal and the demise of the gasifiers. In the 1950s and 1960s, petroleum dominated the market; no new gasification processes emerged, and few coal gasification plants were installed in the United States [2]. It was not until the latter 1960s and early 1970s, when the United States began to experience natural gas shortages and the oil embargo of the 1970s occurred, that the significance of its coal reserves was recognized. This led to a tremendous surge in interest in coal utilization, primarily in the areas of gasification and liquefaction (i.e., production of coal to liquid products).

5.3.2 Principles of Coal Gasification

Carbonization of coal to produce coal gas is a relatively simple process to perform and is done in a retort in the absence of air. The composition of the gas being produced varies depending on the coal being used, but it is typically comprised of hydrogen (40–50 percent) and methane (30–40 percent) with minor amounts (2–10 percent) of nitrogen, carbon monoxide, ethylene, and carbon dioxide. The gas yield is approximately 10,000 scf per short ton of coal carbonized with a heating value of 550 to 700 Btu/scf. When carbonizing a bituminous coal, about 20 percent of the weight of the coal is converted to gas [24]. This gas is used as a fuel in coking operations.

Although carbonization of coal is a simple process, only a small fraction of the coal is converted to gas. Consequently, processes to convert all of the carbon in the coal to gas were developed. In one of these processes, air is slowly passed through a hot bed of coal, converting most of the carbon to carbon monoxide, with some carbon dioxide being formed. Some of the carbon dioxide is then converted to carbon monoxide by reacting with hot fuel carbon. The following reactions occur:

$$C + O_2 \rightarrow CO_2 \tag{5.4}$$

(combustion of carbon; $\Delta H = -170.0 \times 10^3$ Btu/lb mole of carbon gasified)

and

$$C + CO_2 \rightarrow 2CO \tag{5.3}$$

(Boudouard reaction; $\Delta H = +72.19 \times 10^3$ Btu/lb mole of carbon gasified)

resulting in

$$2C + O_2 \rightarrow 2CO \tag{5.1}$$

($\Delta H = -97.81 \times 10^3$ Btu/lb mole of carbon gasified)

Equations (5.4) and (5.3) are collectively sufficiently exothermic to sustain reaction with the reactants fed at ambient conditions.

The gas produced by this method is called producer gas, and when a bituminous coal is used, the gas composition is typically 20 to 25 percent carbon monoxide, 55 to 60 percent nitrogen, 2 to 8 percent carbon dioxide, and 3 to 5 percent hydrocarbons [24]. Unfortunately, the producer gas is diluted with nitrogen, and the heating value of the gas is only about 100 to 150 Btu/scf. The yield of producer gas is 150,000 to 170,000 scf per short ton of coal. Producer gas was used in a variety of industrial applications such as open-hearth furnace steel mills, glass-making furnaces, and pottery kilns. However, the demand for producer gas has been reduced with the demise of open-hearth furnaces in the steel industry and the development of natural gas and electric furnaces.

The temperatures developed in the fuel bed during Eqs. (5.4) and (5.3) can be very high, and when the ash in the bed is fusible, the endothermic carbon-steam reaction must be imposed by adding steam to the air:

$$C + H_2O \rightarrow CO + H_2 \tag{5.7}$$

(carbon-steam reaction; $\Delta H = +58.35 \times 10^3 \text{Btu/lb mole of carbon gasified})$

This reaction moderates the temperature and yields hydrogen in the product gas. This mixture of carbon monoxide and hydrogen is also called water gas. The water gas process is cyclic, involving a gas-making period during which the fuel bed is blown with steam to produce carbon monoxide and hydrogen, followed by an air-blowing period during which the heat is generated in the fuel bed.

A typical water gas contains 50 percent hydrogen, 40 percent carbon monoxide, and small amounts of carbon dioxide and nitrogen with a heating value of 300 Btu/scf. When a water-gas generator is being blown with air to reheat the bed, producer gas is made from the reaction of the hot carbon with oxygen, yielding about 35,000 scf of water gas and 80,000 scf of producer gas from one short ton of coal [24]. Water gas is a useful starting material for synthesizing chemicals or liquid fuels and is a good source of hydrogen. Treating the water gas with steam oxidizes the carbon monoxide to carbon dioxide and increases the amount of hydrogen by the equation

$$CO + H_2O \rightarrow CO_2 + H_2 \tag{5.8}$$

(water-gas shift reaction; $\Delta H = -3.83 \times 10^3 \text{Btu/lb mole of carbon gasified})$

The carbon dioxide can be removed from the product stream, leaving reasonably pure hydrogen.

Hydrogen can also be reacted with carbon at elevated pressures by the carbon hydrogenation or hydrogasification reaction

$$C + 2H_2 \rightarrow CH_4 \tag{5.24}$$

($\Delta H = -39.38 \times 10^3 \text{Btu/lb mole of carbon gasified})$

and whenever the carbon source generates volatile matter, further quantities of methane will form by thermal cracking.

5.3.3 Gasifier Types

Gasification processes are classified on the basis of the method used to bring the coal into contact with the gasifying medium (air or oxygen). The three principal commercial modes are fixed-bed, fluidized-bed, and entrained-flow systems, and

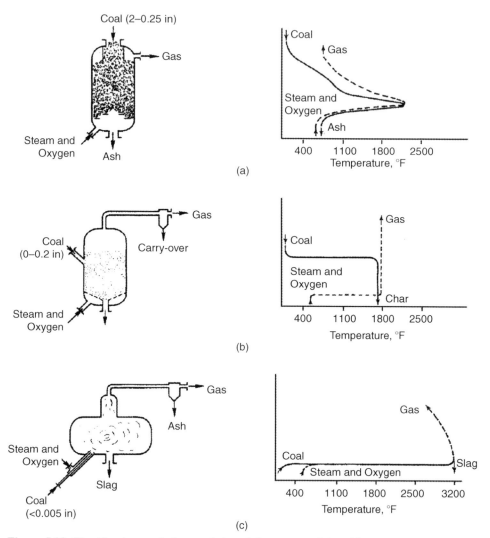

Figure 5.22 Classifications and characteristics of the commercial gasification systems:
(a) fixed bed (nonslagging), (b) char fluidized bed, and (c) entrained flow.
Source: From Elliot (1981).

their principal features are illustrated in Figure 5.22 [2], with important characteristics listed in Table 5.7 [46].

Fixed-Bed Gasifiers

In a fixed-bed gasifier, ¼- to 2-inch-sized coal is supplied countercurrent to the
gasifying medium. Coal moves slowly down (sometimes this type of gasifier is
called a moving-bed gasifier), ideally in plug flow against an ascending stream of

Table 5-7 Characteristics of Generic Gasifier Types

Gasifier Type	Fixed-Bed		Fluidized-Bed		Entrained-Flow
Ash Conditions	Dry Ash	Slagging	Dry Ash	Agglomerating	Slagging
Fuel Characteristics					
Fuel size limits	1/4–2 inches	1/4–2 inches	<1/4 inch	<1/4 inch	<0.005 inches
Acceptability of caking coal	Yes (with modifications)	Yes	Possibly	No, noncaking only	Yes
Preferred feedstock	Lignite, reactive bituminous coal, anthracite, wastes	Bituminous coal, anthracite, petcoke, wastes	Lignite, reactive bituminous coal, anthracite, wastes	Lignite, bituminous coal, anthracite, cokes, biomass, wastes	Lignite, reactive bituminous coal, anthracite, petcokes
Ash content limits	No limitation	<25% preferred	No limitation	No limitation	<25% preferred
Preferred ash melting temperature, °F	>2,200	<2,370	>2,000	>2,000	<2,372
Operating Characteristics					
Exit gas temperature, °F	Low[a] (800–1,200)	Low (800–1,200)	Moderate (1,700–1,900)	Moderate (1,700–1,900)	High (>2,300)
Gasification pressure, psig	435+	435+	15	15–435	<725
Oxidant requirement	Low	Low	Moderate	Moderate	High
Steam requirement	High	Low	Moderate	Moderate	Low
Unit capacities, MWth equivalent	10–350	10–350	100–700	20–150	Up to 700
Key Distinguishing Characteristics	Hydrocarbon liquids in raw gas	Hydrocarbon liquids in raw gas	Large char recycle	Large char recycle	Large amount of sensible heat energy in the hot raw gas
Key Technical Issue	Utilization of fines and hydrocarbon liquids	Utilization of fines and hydrocarbon liquids	Carbon conversion	Carbon conversion	Raw gas cooling

[a]Fixed-bed gasifiers operating on low-rank coals have exit temperatures lower than 800°F.

Source: From Ratafia-Brown et al. (2002).

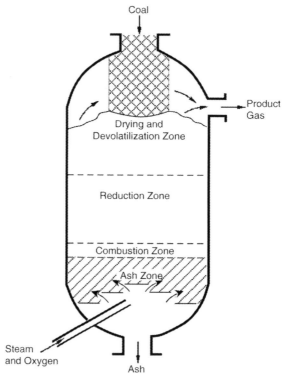

Figure 5.23 Reaction zones in a fixed-bed gasifier.
Source: From Elliot (1981).

gasifying medium. Reaction zones, shown in Figure 5.23, typically consist of drying and devolatilization, reduction, combustion, and ash zones. In the drying and devolatilization zone, located at the top of the gasifier, the entering coal is heated and dried, and devolatilization occurs. In the reduction/gasification zone, the devolatilized coal is gasified by reactions with steam and carbon dioxide. In the combustion zone, oxygen reacts with the remaining char, and this zone is characteristic of high temperatures. The ash is removed from the bottom of the gasifier either in dry form if the temperature in the gasifier is controlled with excess steam to maintain the temperature below the ash fusion point or as liquid slag. Both the ash and the product gas leave at modest temperature as a result of heat exchange with the entering gasifying medium and fuel, respectively. Fixed-bed gasifiers have the following characteristics [46]:

- Low oxidant requirements
- Design modifications are required for handling caking coals
- High cold-gas thermal efficiency when the heating value of the hydrocarbon liquids are included
- Limited ability to handle fines

Fluidized-Bed Gasifiers

In a fluidized-bed gasifier, coal that has been crushed to smaller than 1/8 to 1/4 inch in size enters the side of the reactor and is kept suspended by the gasifying medium. Similar to a fluidized-bed combustor, mixing and heat transfer are rapid, resulting in uniform composition and temperature throughout the bed. The temperature is sustained below the ash fusion temperature, which avoids clinker formation and possible slumping (i.e., defluidization of the bed). Some char particles are entrained in the product gas as it leaves the gasifier, but they are recovered and recycled back into the gasifier via a cyclone. The ash is discharged with the char, and the product gas and char temperatures are high, with some heat transfer occurring with the incoming steam and recycle gas. Fluidized-bed gasifiers have the following characteristics [46]:

- A wide range of solid feedstock (including solid waste, wood, and high ash content coals) accepted
- Uniform temperature
- Moderate oxygen and steam requirements
- Extensive char recycling

Entrained-Flow Gasifiers

In the entrained-flow gasifier, pulverized coal (<0.005 inches) is entrained with the gasifying medium to react in cocurrent flow in a high-temperature flame. Residence time in this type of gasifier is very short. Entrained-flow gasifiers generally use oxygen as the oxidant and operate at high temperatures, well above ash-slagging conditions, to ensure high carbon conversion. The ash exits the system as a slag. The product gas and slag exit close to the reaction temperature. Entrained-flow gasifiers have the following characteristics [46]:

- Ability to gasify all coals regardless of coal rank, caking characteristics, or amount of coal fines, although feedstocks with lower ash contents are favored
- Uniform temperatures
- Very short fuel residence times in the gasifier
- Solid fuel must be very finely sized and homogenous
- Relatively large oxidant requirements
- Large amount of sensible heat in the raw gas
- High-temperature slagging operation
- Entrainment of some molten slag in the raw gas

5.3.4 Influence of Coal Properties on Gasification

Coal properties have a major influence on the process and gasifier design; they include moisture, ash, volatile matter, and fixed carbon content, caking tendencies, reactivity, ash fusion characteristics, and particle size distribution. Some of these properties are listed earlier in Table 5.7, and each is briefly discussed in the following sections.

Moisture

Fixed-bed gasifiers can accommodate moisture contents of up to 35 percent, provided the ash content is not in excess of about 10 percent [2]. Predrying may be performed if the moisture and ash contents are above these amounts. Entrained-flow or fluidized-bed gasifiers require the moisture content to be reduced to less than about 5 percent by drying to improve coal handleability. In the entrained-flow system, the residual moisture contributes to the gasification steam but requires heat to evaporate it.

Ash

Ash should be kept at a minimum, since provisions must be made for introducing it to, and withdrawing it from the system, provisions that add to the complexity and cost of the overall system. Ash can be used as a heat transfer medium, either by its flow countercurrent to the products of gasification and gasifying agents in fixed-bed systems or by provisions in entrained-flow systems [2]. In fixed-bed systems, ash accumulates at the base of the fuel bed and is withdrawn by a mechanical grate if unfused or through a taphole if it is a liquid slag. In the enrained-flow system, it is removed as a liquid slag. In fluidized-bed systems, the ash is mixed with the char, and the ash is separated either by sintering and agglomeration of the ash or circulation from the bed through a fully entrained combustor to melt and separate the ash as a liquid slag. Fluidized-bed and entrained flow gasifiers tend to have higher losses of carbon in the ash than the fixed-bed systems.

Ash constituents are important in the selection of materials of construction, particularly in slagging combustors. In addition, proper ash composition, or its chemical manipulation through the addition of fluxing agents, is necessary for desirable slagging operations.

Volatile Matter

The volatile matter from the coal can add to the products of gasification without incurring steam decomposition or oxygen consumption. The volatile matter, which can vary from less than 5 percent (on a moist, ash-free basis) for anthracite to over 50 percent for subbituminous coal or lignites, can consist of carbon oxides, hydrogen, and traces of nitrogen compounds [2]. The volatile matter composition, the type of coal, and the conditions under which the volatile matter is driven off all affect the nature of the residual fixed carbon or char that remains.

Fixed Carbon

The nature of the fixed carbon, which is the major component of the char after the moisture and volatile matter is driven off, is important to the performance of the gasifier and can vary physically and chemically. Properties such as density, structure, friability/strength, and reactivity depend primarily on the original coal, but they are influenced by the pressure, the rate at which the coal is heated, and its final temperature [2].

Caking Tendencies

The coal's caking tendencies—strongly caking and swelling, weakly caking, and noncaking—must be considered during the design of the gasifier process [2]. Some gasifiers can be designed to handle caking and swelling coals, but others will require the coal to be pretreated. Table 5.7 listed the acceptability of caking coal to the generic types of gasifiers [46].

Ash Fusion

The ash fusion temperature is a measure of when the ash will melt and transform from a solid to liquid state. This temperature is an important parameter for the design and operation of gasification systems, for those that operate below the ash fusion temperature so as not to incur fusion, sintering, or clinkering of the ash, as well as those gasification systems that operate above the ash fusion temperature to promote slag production. Typical preferred ash melting temperatures for the generic types of gasifiers are listed in Table 5.7.

Another important ash characteristic is the relationship between temperature and ash viscosity, since it is the flow characteristics of the slag that are critical. The importance of slag viscosity, discussed earlier in the chapter, is as relevant to slagging gasifiers as it is to slagging combustors.

Reactivity

Coals vary in their reactivity to steam and, to a lesser extent, to hydrogen [2]. Reactive coals, or their chars, decompose steam more rapidly and sustain that decomposition down to a lower temperature than do less reactive coals. Reactivity has three important influences: (1) it favorably influences methane formation; (2) it reduces oxygen consumption by allowing steam decomposition down to a lower temperature; and (3) it allows less steam to be used per volume of oxygen or air than with less reactive coals without incurring ash clinkering.

Coal Size Distribution

The coal size limits are important gasifier system design considerations. In fixed-bed gasification systems, provisions have to be made for the fines that are generated from mining, transportation, and processing. This may include steam and power generation or briquetting, extrusion, or injection to allow them to be supplied to the fixed-bed gasifier [2]. The size distribution is less critical with fluidized-bed and entrained-flow gasification systems.

5.3.5 Regional Distribution of Gasification Systems

There is much interest in gasification technology, and the growth of this industry has been followed closely for the last decade. In 1999, the first World Gasification Survey was conducted by SFA Pacific, Inc., with support from the U.S. Department of Energy (DOE) and in cooperation with the member countries of the Gasification

Technologies Council [47, 48]. The survey was updated in 2001 and 2004 [49] and was followed by the DOE-sponsored 2007 World Gasification Survey [50]. The recent survey identified 144 commercial gasification plants, with a total of 427 gasifiers in operation in 27 countries in Africa/Middle East, Asia/Australia, Europe, North America, and Central and South America.

The gasification capacity of these units has grown to 56,238 megawatts thermal (MW_{th}) of syngas output. An additional 10 plants with 34 gasifiers have been announced and are forecast to become optional by 2010. The additional capacity from these new plants is 17,135 MW_{th}, an expected increase of over 30 percent. Worldwide capacity by 2010 is projected at 73,373 MW_{th} of syngas output from 154 plants and 461 gasifiers. Figure 5.24 shows the distribution of gasifier projects by geographic region. A summary of the survey listing plant owner, location, technology vendor, number of gasifiers, syngas capacity, feedstock, and application (chemicals, gaseous fuels, and power), is provided in Appendix B [50].

Because of the versatility and flexibility of gasification plants, they offer a wide range of products, including Fischer-Tropsch liquids, chemicals, fertilizers, power, steam, gaseous fuels, and various other products (e.g., ammonia and hydrogen). The distribution of gasification applications is illustrated in Figure 5.25 on a megawatt thermal (MW_{th}) syngas basis and show that for operating plants identified in the 2007 survey, chemicals and Fischer-Tropsch liquids represent the leading products, with 45 percent and 28 percent, respectively, of the world's gasification capacity [50].

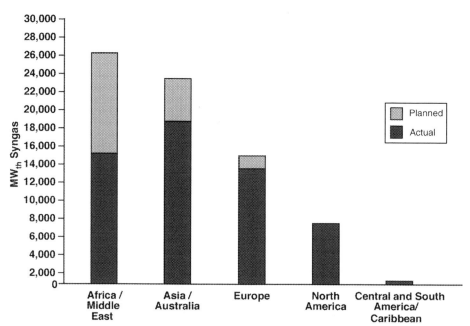

Figure 5.24 Distribution of gasification projects by geographic region.
Source: From DOE (2007).

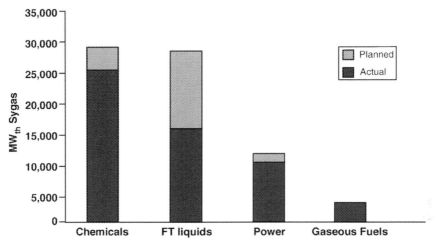

Figure 5.25 Distribution of gasification applications.
Source: From DOE (2007).

Gasification facilities utilize a variety of feedstocks, including natural gas, coal, petroleum, petcoke, biomass, and industrial wastes. Coal now dominates as the feedstock in 55 percent or 30,825 MW_{th} of syngas capacity, representing 45 plants. Petroleum, including fuel oil, refinery residue, and naphtha, is the second leading feedstock, with 18,454 MW_{th} or 33 percent of the total gasification capacity with 57 plants. Gasification capacity by feedstock is illustrated in Figure 5.26 on a MW_{th} syngas basis [50].

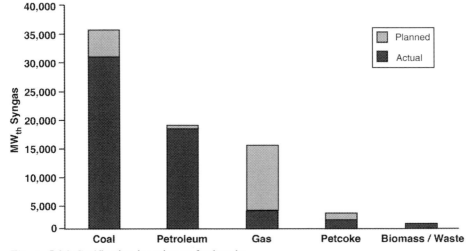

Figure 5.26 Gasification by primary feedstocks.
Source: DOE (2007).

5.3.6   Commercial Gasification Systems

Gasification of coal provides an alternative to coal-fired combustion systems, since it is more efficient and environmentally friendly. Coal gasification is a well-proven technology that started with the production of coal gas for towns, progressed to the production of fuels, such as oil and synthetic natural gas (SNG), chemicals, and more recently, to large-scale Integrated Gasification Combined Cycle (IGCC) power generation. Currently, the number of operating IGCC power plants is small; however, several power projects are currently under construction or in planning and design stages, and gasification technology is the center of future U.S. energy complexes that are under development and discussed in detail in Chapter 11. A description of an IGCC power plant and details of the IGCC power plants in operation and proposed for the future are given in Chapter 7 and 11.

The commercially well-proven GE Energy (formerly ChevronTexaco), Shell, and Sasol Lurgi (formerly Lurgi dry ash) dry-ash gasification technologies represent a major portion of the worldwide gasification capacity, as illustrated in Figure 5.27 [50]. These technologies currently hold 93 percent of the world market, with GE Energy, Shell, and Sasol Lurgi accounting for 31, 28, and 34 percent of the market share, respectively. In addition, these three technologies represent 27 of the 30 largest commercial projects (as of 2004) in the world (Table 5.8) [47].

Fixed-Bed Gasifiers

Fixed-bed gasifiers differ in exit gas conditions and in special design configurations. There are two main commercial fixed-bed gasifier technologies. The Sasol-Lurgi dry-ash gasifier was originally developed in the 1930s and has been used extensively

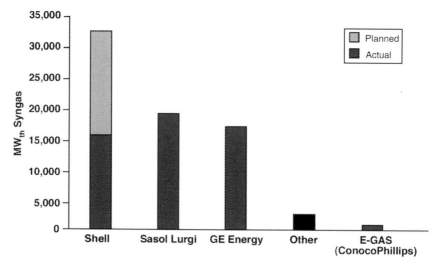

Figure 5.27 Gasification by technology.
Source: DOE (2007).

Table 5.8 World's 30 Largest Commercial Gasification Projects

Gasification Plant Owner	Location	Gasification Technology	MW$_{th}$ Output	Startup Year	Feed/Product
Sasol-II	South Africa	Sasol-Lurgi	4,130	1977	Subbituminous coal/F-T liquids
Sasol-III	South Africa	Sasol-Lurgi	4,130	1982	Subbituminous coal/F-T liquids
Respol/ Iberdrola	Spain	GE Energy	1,654	2004	Vacuum residue/ electricity
Dakota Gasification Company	United States	Sasol-Lurgi	1,545	1984	Lignite and refinery residue/ SNG
SARLUX srl	Italy	GE Energy	1,067	2000	Visbreaker residue/ electricity and H$_2$
Shell MDA Sdn. Bhd.	Malaysia	Shell	1,032	1993	Natural gas/ middistillates
Linde AG	Germany	Shell	984	1997	Visbreaker residue/H$_2$ and methanol
ISAB Energy	Italy	GE Energy	982	1999	ROSE asphalt/ electricity and H$_2$
Sasol-I	South Africa	Sasol-Lurgi	911	1955	Subbituminous coal/F-T liquids
Total France/ EdF/Texaco	France	GE Energy	895	2003	Fuel oil/electricity and H$_2$
Unspecified Owner	United States	GE Energy	656	1979	Natural gas/ methanol and CO
Shell Nederland Raffinaderij BV	Netherlands	Shell	637	1997	Visbreaker residue/H$_2$ and electricity
SUV/EGT	Czech. Republic	Sasol-Lurgi	636	1996	Coal/electricity and steam
Chinese Petroleum Corporation	Taiwan	GE Energy	621	1984	Bitumen/H$_2$ and CO
Hydro Agri Brunsbüttel	Germany	Shell	615	1978	Heavy vacuum residue/ammonia
Public Service of Indiana	United States	E-Gas (Destec)	591	1995	Bituminous coal/ electricity
VEBA Chemie AG	Germany	Shell	588	1973	Vacuum residue/ ammonia and methanol
Elcogas SA	Spain	Prenflow	588	1997	Coal and petcoke/ electricity

Continued

Table 5.8 World's 30 Largest Commercial Gasification Projects—Cont'd

Gasification Plant Owner	Location	Gasification Technology	MW_{th} Output	Startup Year	Feed/Product
Motiva Enterprises LLC	United States	GE Energy	558	1999	Fluid petcoke/ electricity and steam
API Raffineria di Anocona S.p.A.	Italy	GE Energy	496	1999	Visbreaker residue/ electricity
Chempoetrol a.s.	Czech Republic	Shell	492	1971	Vacuum residue/ ammonia and methanol
Demkolec BV	Netherlands	Shell	466	1994	Bituminous coal/ electricity
Tampa Electric Company	United States	GE Energy	455	1996	Coal/electricity
Ultrafertil S.A.	Brazil	Shell	451	1979	Asphalt residue/ ammonia
Shanghai Pacific Chemical Corp.	China	GE Energy	439	1995	Anthracite/ methanol and town gas
Exxon USA Inc.	United States	GE Energy	436	2000	Petcoke/electricity and syngas
Shanghai Pacific Chemical Corp	China	IGT U-GAS	410	1994	Bituminous coal/ fuel gas and town gas
Gujarat National Fertilizer Co.	India	GE Energy	405	1982	Refinery residue/ ammonia and methanol
Esso Singapore Pty. Ltd.	Singapore	GE Energy	364	2000	Residual oil/ electricity and H_2
Quimigal Adubos	Portugal	Shell	328	1984	Vacuum residue/ ammonia

Notes: Gasifier production is reported as MW_{th} equivalent; F-T is Fischer-Tropsch synthesis; SNG is synthetic natural gas.
Source: From SFA Pacific, Inc., and U.S. DOE (2000).

for town gas production and in South Africa, by Sasol (South African Coal Oil and Gas Corporation), for chemicals from coal. In fact, of the chemicals produced world-wide from syngas, Sasol produces nearly a third of them and uses Lurgi gasifiers to do so. Also, Sasol-Lurgi gasifiers are used in Dakota Gasification's Great Plains Synfuels Plant located near Beulah, North Dakota, which is the only facility in the United States that manufactures a high-Btu synthetic natural gas (SNG) from lignite, where 17,000 short tons of lignite are processed daily to produce 148 million scf per day of SNG along with by-product chemicals. In the dry-ash Sasol-Lurgi gasifier, the temperature at the bottom of the bed is kept below the ash fusion point so the ash is removed as a solid. In the 1970s, Lurgi and the British Gas Corporation developed a slagging version of the gasifier, referred to as the BGL gasifier, in which the temperature at the bottom of the gasifier is sufficient for the ash to melt.

Sasol-Lurgi Gasifiers

The most successful fixed-bed gasifier is the Sasol-Lurgi gasifier. It was developed in Germany during the 1930s as a means to produce town gas. The first commercial plant was built in 1936. Initially used for lignites, process developments in the 1950s allowed for the use of bituminous coals as well. The Sasol-Lurgi gasification process has been used extensively worldwide.

The Sasol-Lurgi dry-ash gasifier, which is shown schematically in Figure 5.28, is a pressurized gasifier typically operating at 30 to 35 atmospheres [21]. Sized coal enters the top of the gasifier through a lock hopper and moves down through the bed. Steam and oxygen enter at the bottom and react with the coal as the gases move up the bed. Ash is removed at the bottom of the gasifier by a rotating grate and lock hopper and is kept in a dry state through the injection of steam to cool the bed below the ash fusion point. As the coal moves down the gasifier, it goes through sequential stages of drying and devolatilization, with the resultant char undergoing gasification and combustion. The countercurrent operation results in a temperature drop in the gasifier. Gas temperatures are approximately 500 to 1,000°F in the drying and devolatilization zone, 1,800°F in the gasification zone, and 2,000°F in the combustion zone [24, 46]. The raw syngas, a mixture of carbon monoxide and hydrogen, which also contains tar, exits the gasifier at 570 to 930°F.

BGL Gasifiers

The BGL fixed-bed gasifier was developed in the 1970s to provide a syngas with a high methane content so SNG could be manufactured from coal more efficiently. The BGL fixed-bed gasifier, shown schematically in Figure 5.29, is a dry-feed, pressurized, slagging gasifier. The operational concept is similar to the Lurgi dry-ash gasifier, with two notable differences. The BGL fixed-bed gasifier is more fuel flexible in that it can use run-of-mine coal (rather than sized coal) and the gasifier is operated at temperatures above the ash fusion point, forming a liquid slag. Slag is withdrawn from the slag pool through an opening in the grate. The slag flows into a quench chamber and lock hopper in series.

Syngas exits the gasifier at about 1,040°F and passes into a water quench vessel and a boiler feedwater preheater designed to lower the gas temperature to approximately 300°F. Soluble hydrocarbons, such as tars, oils, and naphtha, are recovered from the aqueous liquor, are poured into a gas-liquor separation unit, and are then recycled to the gasifier [46].

Fluidized-Bed Gasifiers

Fluidized-bed gasifiers may differ in ash conditions, dry or agglomerating, and in the design configurations for improving char use. Commercial versions of this

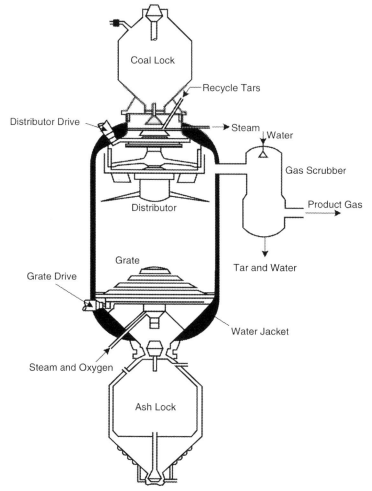

Figure 5.28 A modern Lurgi dry-ash gasifier.
Source: From Berkowitz (1979).

type of gasifier include the High-Temperature Winkler (HTW) and Kellogg-Rust-Westinghouse (KRW) designs.

HTW Gasifier
The High-Temperature Winkler gasifier, shown schematically in Figure 5.30, is a dry-feed, pressurized, fluidized-bed, dry-ash gasifier. The HTW process was developed by Rheinbraun in Germany during the 1920s to utilize coal with a small particle size and that was too friable for use in existing fixed-bed gasifiers. The HTW technology is capable of gasifying a variety of feedstocks, including

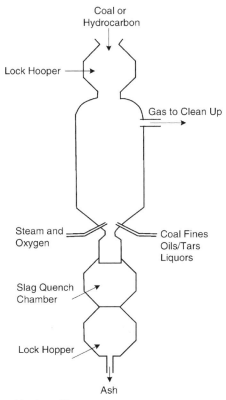

Coal or
Hydrocarbon

Lock Hooper

Gas to Clean Up

Steam and
Oxygen

Coal Fines
Oils/Tars
Liquors

Slag Quench
Chamber

Lock Hopper

Ash

Figure 5.29 A BGL fixed-bed gasifier.
Source: Modified from Ratafia-Brown et al. (2002).

reactive low-rank coals with a higher ash-softening temperature and reactive caking and noncaking bituminous coals.

Coal, with a particle size of less than 1/8 inch, is fed into the gasifier using a
screw feeder. The upward flow of the gasifying medium, air or oxygen, keeps the
particles of coal, ash, and semicoke/char in a fluidized state. Gas and elutriated
solids flow up the gasifier, with additional air or oxygen added in this region to
complete the gasification reactions. Fine ash particles and char that are entrained
in the gas are removed in a cyclone and recycled to the gasifier. Ash is removed
from the base of the gasifier by means of an ash screw. The syngas exiting the gasifier is at a high temperature so it does not contain any high molecular weight
hydrocarbons such as tars, phenols, and other substituted aromatic compounds [46].

The gasifier fluid bed is operated at about 1,470 to 1,650°F, and the temperature
is controlled to ensure that it does not exceed the ash softening point. The temperature in the freeboard can be significantly higher, up to 2,000°F. The operating pressure can vary between 145 psig for syngas manufacture and 360 to 435 psig for an
IGCC application [46].

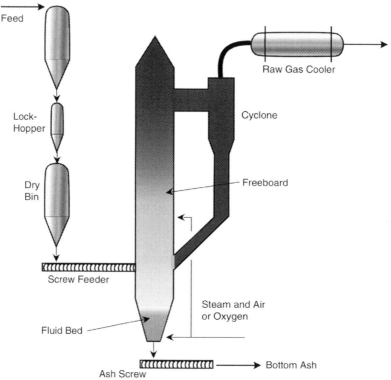

Figure 5.30 An HTW fluidized-bed gasifier.
Source: From Ratafia-Brown et al. (2002).

KRW Gasifiers

The KRW gasifier, shown schematically in Figure 5.31, is a pressurized, dry-feed, fluidized-bed slagging gasifier developed by M.W. Kellogg Company. The KRW gasifier is capable of gasifying all types of coals, including high-sulfur, high-ash, low-rank, and high-swelling coals.

Coal and limestone that have been crushed to less than 1/4 inch are fed into the bottom of the gasifier, with the air or oxygen entering through concentric high-velocity jets [46]. This ensures thorough mixing of the fuel and air/oxygen. The coal immediately releases its volatile matter upon entering the gasifier, which oxidizes rapidly, producing the heat for the gasification reactions. An internal recirculation zone is established, with the coal/char moving down the sides of the gasifier and back into the central jet. Steam is introduced with the air/oxygen and through jets in the side of the gasifier and reacts with the char to form the syngas.

Fine ash particles that are carried out of the bed are captured in a high-efficiency cyclone and reinjected into the gasifier. The internal recycling of the larger char particles results in the char becoming enriched in ash, and the low-melting components of the ash cause the ash particles to agglomerate. As the ash particles become larger, they begin to migrate toward the bottom of the gasifier, where they are removed

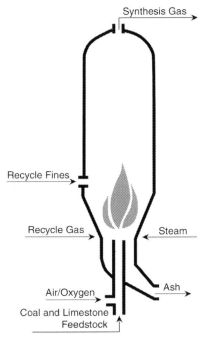

Figure 5.31 A KRW gasifier.
Source: From Ratafia-Brown et al. (2002).

along with spent sorbent (i.e., limestone that has reacted with sulfur to form calcium sulfide, CaS) and some unreacted char. The ash, char, and spent sorbent flow into a fluidized-bed sulfator, where the char and calcium sulfide are oxidized. The calcium sulfide forms calcium sulfate, $CaSO_4$, which is chemically stable and can be disposed in a landfill.

Entrained-Flow Gasifiers

Differences among entrained-flow gasifiers include the coal feed systems (coal-water slurry or dry coal), internal design to handle the very hot reaction mixture, and heat recovery configuration. Entrained-flow gasifiers have been selected for nearly all the coal-based IGCCs currently in operation or under construction [46]. The major commercial entrained-flow gasifiers include the GE Energy, Shell, Prenflo, and E-Gas gasifiers. Of these, both the GE Energy gasifier and the Shell gasifier have over 100 units worldwide [46].

GE Energy Gasifiers

The GE Energy gasifier, which is shown schematically in Figure 5.32, is a single-stage, down-fired, entrained-flow gasifier [46]. Fuel-water slurry (e.g., 60–70 percent coal) and 95 percent pure oxygen are fed into the pressurized gasifier. The coal

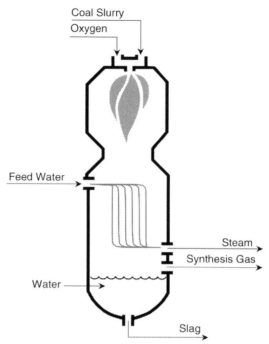

Figure 5.32 A GE Energy gasifier.
Source: From Ratafia-Brown et al. (2002).

and oxygen react exothermally at a temperature ranging from 2,200 to 2,700°F and a pressure greater than 20 atmospheres to produce syngas and molten ash. Operation at the high pressures eliminates the production of hydrocarbon gases and liquids in the syngas. The hot gases are cooled using either a radiant syngas cooler located inside the gasifier to produce high-pressure steam or an exit gas quench. Slag drops into the water pool at the bottom of the gasifier, is quenched, and is separated from the blackwater and finally removed through a lockhopper.

The GE Energy technology has operated commercially for over 40 years with feedstocks such as natural gas, heavy oil, coal, and petroleum coke. More than 60 commercial plants are in operation that use coke and coal, oil, and natural gas as feedstock [46, 51].

Shell Gasifiers

The Shell gasifier, which is the successor to the Koppers-Totzek process and then the Shell-Koppers process, is shown schematically in Figure 5.33 and is a dry-feed, pressurized, entrained-flow slagging gasifier that can operate on a wide variety of feedstocks. Pulverized coal is pressurized in lock hoppers and fed into the gasifier through dense-phase conveying with transport gas, which can be nitrogen or syngas [46]. Preheated oxygen is mixed with steam and used as a temperature moderator prior to feeding to the fuel injector. Temperatures and pressures in the gasifier are

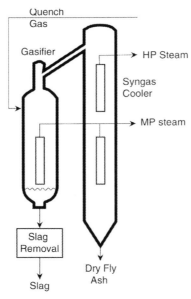

Figure 5.33 A Shell gasifier.
Source: From Ratafia-Brown et al. (2002).

2,700 to 2,900°F and 350 to 650 psig, respectively. A syngas is produced that contains mainly hydrogen and carbon monoxide with a little carbon dioxide. Elevated temperatures eliminate the production of hydrocarbon gases and liquids in the product gas.

The high temperature converts the ash into molten slag, which runs down the refractory walls into a water bath, where it is quenched and the ash/water slurry is removed through a lock hopper. The raw gas that leaves the gasifier, at 2,500 to 3,000°F, contains a small quantity of char and about half of the molten ash. The hot gas is partially cooled to temperatures below the ash fusion point by quenching it after it leaves the gasifier. The syngas undergoes further cooling before the particles are removed in a wet scrubber. The first Shell Gasification Process units were commissioned in the 1950s. Shell started development work with coal in 1972.

Prenflo Gasifiers

The Prenflo gasifier, developed by Uhde (formerly Krupp Uhde) and shown schematically in Figure 5.34, is a pressurized, dry-feed, entrained-flow slagging gasifier [46]. Coal, ground to about 100 μm, is pneumatically conveyed by nitrogen to the gasifier. The coal is fed through injectors located in the lower part of the gasifier with oxygen and steam. Syngas, produced at temperatures of approximately 2,900°F, is quenched with recycled cleaned syngas to reduce its temperature to about 1,500°F in an internal syngas cooler. The syngas is further cooled to about 700°F through evaporator stages before exiting the gasifier. The molten slag flows

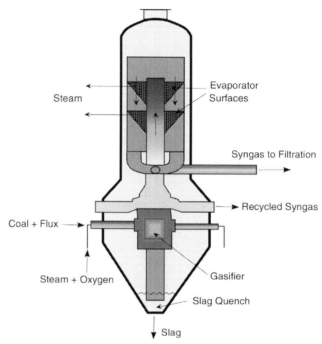

Figure 5.34 A Prenflo gasifier.
Source: From Ratafia-Brown et al. (2002).

down the walls into a water bath, where it is quenched and granulated before removal through a lock hopper system.

E-Gas Gasifiers

The E-Gas gasifier, shown schematically in Figure 5.35, is a slurry-feed, pressurized, entrained-flow gasifier. It is an upward flow gasifier with two-stage operation. The coal is slurried via wet crushing with coal concentrations ranging from 50 to 70 wt.%, and about 75 percent of the total slurry feed is fed into the first stage of the gasifier, which operates at 2,600°F and 400 psig. Oxygen is combined with the slurry and injected into the first stage of the gasifier, where the highly exothermic gasification/oxidation reactions occur. Operation at the elevated temperatures eliminates the production of hydrocarbon gases and liquids in the product gas. The molten ash flows into a water bath, where it is quenched and removed.

The raw gas from the first stage enters the second stage, where the remaining 25 percent of coal slurry is injected. The endothermic gasification/devolatilization reactions occur in this stage where the temperature is approximately 1,900°F, and some hydrocarbons are added to the product gas. Char is produced in the second stage and is recycled to the first stage, where it is gasified. The syngas exits the gasifier and undergos further cooling and cleaning.

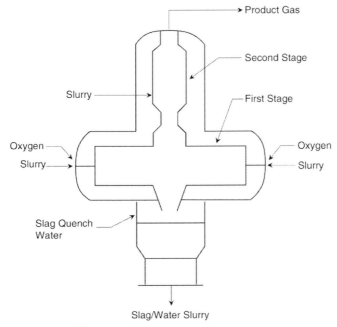

Figure 5.35 An E-Gas gasifier.
Source: From Ratafia-Brown et al. (2002).

5.4   Liquefaction

Liquefaction is the conversion of coals into liquid products. The three methods used to derive liquids from coals are pyrolysis, indirect liquefaction, and direct liquefaction. In pyrolysis processes, the liquids are a by-product during coke production. The term *liquefaction* refers to the conversion of the coal, where the primary product is a liquid.

In indirect liquefaction, the coal is gasified into a mixture of carbon monoxide and hydrogen—that is, syngas. The syngas is then processed into liquid products using Fischer-Tropsch synthesis. In direct liquefaction, also referred to as coal hydrogenation, coal is mixed with a hydrogen-donor solvent and reacted with hydrogen or syngas under elevated pressures and temperatures to produce a liquid fuel. Indirect liquefaction is used quite extensively throughout the world, as illustrated in Table 5.8, while direct liquefaction has not been able to compete with other liquid or gaseous fuels. Liquefaction processes are technically feasible, and Germany and South Africa, both countries with abundant coal supplies but little or no petroleum resources, have demonstrated that a country can meet part (Germany) or much (South Africa) of its liquid fuel needs through liquefaction. Germany did it during World War II, while South Africa became self-reliant during its years of apartheid, which made it susceptible to oil embargoes.

As discussed earlier in this chapter (and will be discussed in more detail in Chapters 7 and 11), gasification integrated with chemicals and fuels production is the technology anticipated to be utilized for future power complexes. Direct liquefaction is a much-researched technology, with proven processes operated at the pilot and demonstration scales; however, it is not a widely accepted commercial technology, although facilities are under construction and in operation in China. Hence, the next section presents a brief review of key direct liquefaction processes.

5.4.1   The Beginning of the Synthetic Fuel Industry

Coal was hydrogenated in the laboratory by Berthelot as early as 1869 [52]. The reaction was carried out with hydriodic acid at 520°F for 24 hours, and a 67 percent yield of oil containing aromatics and naphthenes was obtained.

Major advances were made by the Germans in the early 1900s. Germany, a country with abundant coal reserves but virtually no petroleum deposits, was becoming increasingly dependent on gasoline and diesel engines. To ensure that Germany would never lack a plentiful supply of liquid fuels, German scientists and engineers invented and developed two processes that enabled them to produce synthetic petroleum from their coal supplies and to establish the world's first technologically successful synthetic liquid fuel industry [53].

In 1911, Friedrich Bergius obtained oil by hydrogenating coal without a catalyst under hydrogen pressure at 570 to 660°F. In 1913 he applied for the first patent on coal hydrogenation, and in 1931 he was awarded the Nobel Prize in Chemistry [21, 52]. Bergius also observed that coal paste could be injected readily into a vessel under pressure. The role of catalysts in the hydrogenation of coal was not realized until later.

At the end of 1925, I.G. Farben, a chemical company, hydrogenated coal using a molybdenum oxide catalyst. The presence of the catalyst allowed the hydrogenation of coal in the presence of excess hydrogen at low pressure and at temperatures of 750 to 840°F. In the following year, I.G. Farben conducted the liquefaction process in two steps because high molecular weight materials in the intermediate hydrogenation product fouled the catalyst. Coal was mixed with the catalyst and hydrogenated in the liquid phase to middle oil, which was further hydrogenated to gasoline in vapor-phase over a fixed bed of catalyst [52].

About ten years after Bergius began his work, Franz Fischer and Hans Tropsch at the Kaiser-Wilhelm Institute for Coal Research in Mülheim invented a second process for the synthesis of liquid fuel from coal [53]. By the mid-1930s, I.G. Farben and other chemical companies like Ruhrchemie had started to industrialize synthetic liquid fuel production, resulting in the construction of 12 coal hydrogenation and 9 Fischer-Tropsch plants by the end of World War II. The processes were complementary, with coal hydrogenation producing high-quality aviation and motor gasoline, and Fischer-Tropsch syntheses producing diesel and lubricating oil, waxes, and some lower-quality motor gasolines [53].

Germany's technologically successful synthetic fuel industry continued to grow through the 1930s and in the period 1939 to 1945 produced 18 million metric tons of liquids from coal and tar, and another 3 million metric tons of liquids from

Fischer-Tropsch synthesis [53]. Neither coal-to-oil process could produce a synthetic liquid fuel at a cost competitive with natural petroleum, but they persevered because they provided the only path Germany could follow in its search for petroleum independence.

Similarly, South Africa, fearing a boycott as a result of their racial policies, decided to proceed with a synthetic fuels plant in 1951, although this process would produce liquid fuels that were more expensive than petroleum refined products. The South Africans selected Fischer-Tropsch technology because the Germans had successfully used it and because direct liquefaction had not been used on a scale that was as large as the South Africans were planning [24]. This led to the creation of Sasol and government subsidies, which exist to this day [54]. However, now that Sasol has the equipment in place, crude oil is readily available, and apartheid has been abolished, political and economic considerations have forced the South African government to phase out the subsidy for transportation fuels. Sasol anticipated losing its subsidies and has gradually shifted its emphasis from producing only transportation fuels to deriving a significant fraction of its profits from the sale of chemicals and petrochemical feedstocks [54]. Plants were constructed in South Africa in three stages—1955, 1982, and 1992 (refer to Table 5.8)—and are discussed later in this chapter. All three of these plants are currently operating.

In the United States, interest in converting coal to liquid products has been cyclic and affected by the cost and availability of petroleum. In the beginning of the industrial revolution, coal was the major source of energy in the United Stated and continued to dominate the energy supply in the United States for the next hundred years. Petroleum, however, quickly became the preferred energy source after its discovery in Pennsylvania in 1859 and rapid commercial production in the early 1900s. By the early 1920s, worries that oil supplies were becoming depleted along with an expanding automobile industry resulted in interest in coal liquefaction. But this was short-lived; when oil was discovered in Texas in the mid-1920s, further interest in coal liquefaction ceased.

After World War II, consumption of petroleum and natural gas in the United States exceeded that of coal, and the country experienced petroleum shortages; as a result, coal liquefaction was again considered as an alternative. A sizeable research effort resulted. However, the discovery of massive petroleum reserves in the Middle East in the mid-1940s once again made coal liquefaction uneconomical. In 1972, petroleum production in the United States began to decline, and unrest developed in the Middle East. The limited availability of domestic supplies of natural gas and crude oil and the desire to reduce the country's dependence on foreign sources of energy led to significant liquefaction research by the federal government, private industry, and universities.

5.4.2   Indirect Liquefaction—Fischer-Tropsch Synthesis

Fischer-Tropsch synthesis is the conversion of carbon monoxide and hydrogen to liquid hydrocarbons and related oxygenated compounds over variously promoted Group-VIII catalysts and is defined by the reaction

$$CO + 2H_2 \rightarrow (-CH_2-) + H_2O \qquad (5.25)$$

Fischer-Tropsch synthesis can, in principle, provide almost all hydrocarbons conventionally obtained from petroleum. The actual product mix depends on the temperature, pressure, CO/H_2 ratio, and catalysts used, and there are several variants of the process. Table 5.9 illustrates how the product composition can be influenced [21], and Table 5.10 lists some technical information on three industrial processes [1].

Medium-pressure synthesis, conducted at 430 to 640°F and 5 to 50 atmospheres (note that 1 standard atmosphere = 14.7 psia) over iron catalysts, yields mainly gasoline, diesel oils, and heavier paraffins [21]. The proportion of gasoline in the product mix increases and gasoline quality improves as the hydrogen-to-carbon monoxide ratio in the feed gas is increased, and the overall product composition is influenced by the type of reactor—fixed catalyst beds or in a fluid-bed reactor.

High-pressure synthesis, conducted at 210 to 300°F and 50–1,000 atmospheres over ruthenium catalysts, furnishes mainly straight-chain paraffin waxes with molecular weights up to 105,000 and melting ranges up to 270 to 273°F. However, the formation of lower molecular weight hydrocarbons can be increased by increasing the proportion of hydrogen in the feed gas, raising the reaction temperature and decreasing the pressure.

Iso-synthesis is a reaction that converts carbon monoxide and hydrogen to branched-chain hydrocarbons and is usually conducted at 750 to 930°F and 100 to 1,000 atmospheres over thoria or K_2CO_3-promoted thoria-alumina catalysts and yields predominately low molecular weight (C_4 and C_5) isoparaffins [21]. Iso-synthesis at temperatures much above 750°F promotes coproduction of aromatics, while oxygenated compounds are formed at temperatures below 750°F.

The synthol process is used for the production of oxygenated compounds. In this process, the feed gas is reacted at 750 to 840°F and almost 140 atmospheres over an alkalized iron catalyst [21]. There are two variations of this process: The synol synthesis involves the interaction of carbon monoxide and hydrogen at 355 to 390°F and 5 to 50 atmospheres in the presence of highly reduced ammonia catalysts to

Table 5.9 Influence of Operating Parameters on the Composition of the Product of Medium-Pressure Fischer-Tropsch Synthesis

		Yield[a]	
Increase in	**Mean Molecular Weight of Products**	**Oxygenated Products**	**Olefins**
Pressure	+	+	−
Temperature	−	−	+
CO conversion	+	−	−
Flow rate	−	+	+
Gas recycle ratio	−	+	+
H_2:CO ratio	−	−	−

[a]+, increasing; −, decreasing

Source: From Berkowitz (1979).

Table 5.10 Technical Information on Three Industrial Fischer-Tropsch Processes

Process	Cobalt F-T (normal/medium pressure)	ARGE[a] (Sasol I)	Synthol[b] (Sasol I and II)
Reactor stages	Fixed Bed 2–3	Fixed Bed 1	Entrained 1
Pressure, atm.[c]	1–12	23–25	24–28
Temperature, °F	380	430–480	610–640
Catalyst	Co	Fe (Cu)[d]	Fused Fe
Promotor	MgO, ThO_2	K_2O	K_2O
Flow rate, h^{-1}	100	500–700	NA[e]
Recycle ratio	0	2.5	2.0–2.4
CO conversion	90–95	73	77–85
Yield, wt.%			
C_1–C_2	NA	7	20
C_3–C_4	7	14 (of which 45% are olefins)	23 (of which 87% are olefins)
C_5–C_{12}	30	24.8 (of which 50% are olefins)	39 (of which 70% are olefins)
C_{13}–C_{20}	10	14.7 (of which 40% are olefins)	5 (of which 60% are olefins)
C_{20+}	11	36.2 (of which 15% are olefins)	5
Oxygen comp.	NA	4	8
Alcohols/ketones	NA	2.3	7.8
Acids	NA	NA	1.0

[a]ARGE—Arbeitsgermeinschaft Ruhrchemie-Lurgi
[b]Circulating entrained-flow reactor
[c]1 standard atmosphere = 14.7 psia (pounds per square inch absolute)
[d](Cu) = about 5% Copper
[e]NA—not available
Source: From Van Krevelen (1993).

produce a product with 40 to 50 percent oxygenated straight-chain compounds. The oxyl synthesis process is carried out at 355 to 390°F and 20 to 50 atmospheres over a precipitated iron catalyst to produce a product with over 30 percent oxygenated straight-chain compounds.

The oxo-synthesis process is used for producing aldehydes and occurs with the reaction between an olefin and syngas at 210 to 390°F and 100 to 500 atmospheres over cobalt carbonyl catalysts. The oxo-synthesis process has become the most common industrial method for producing C_3 to C_{16} aldehydes.

5.4.3   Direct Liquefaction

Direct liquefaction was developed in the 1930s by Bergius and collaborators and involved reacting pulverized coal or coal–oil slurries with gaseous hydrogen at high pressures and temperatures. It was used extensively in Germany during World War II and in the former Soviet Union and Czechoslovakia for several years after the war. Plants were also constructed in Japan and Great Britain but on a much smaller scale. These plants were either destroyed during the war or shut down shortly after because they were uneconomical to operate. Direct liquefaction research and development continued after the war and increased significantly after the oil crisis in the 1970s. Direct liquefaction research funding essentially ended in the mid- to late 1980s. The funding for the research came from a host of agencies, companies, and states, and examples of funding sources include (but the list is obviously not inclusive) the U.S. DOE (and its predecessors the Office of Coal Research and Energy Research and Development Administration (ERDA)); Electric Power Research Institute (EPRI); Edison Electric Institute (EEI); oil companies such as Amoco, Ashland, Conoco, Mobil, Exxon, Phillips, and Gulf; major utilities such as Southern Services, Inc.; companies such as the Great Northern Railroad; and major coal-producing states like Kentucky. With the exception of the coal-to-liquids plants that Hydrocarbon Technologies, Inc. (HTI) has constructed and started operating in China, development of direct liquefaction technologies only achieved demonstration-scale status. This was primarily due to the availability of lower-priced petroleum-refined fuels.

From a chemical standpoint, coal has a lower hydrogen-to-carbon (H/C) ratio than petroleum: approximately 0.7 to more than 1.2. Direct liquefaction transforms coal into liquid hydrocarbons by directly adding hydrogen to the coal. Examples of some of the operating parameters of the primary processes are summarized in Table 5.11 [1, 2, 55, 56].

The Bergius/I.G. Farben Process

The Bergius process was put into commercial practice by I.G. Farben in Leuna, Germany, in 1927, with additional plants erected in the 1930s. The process operated in two stages, with liquid-phase hydrogenation first transforming the coal into middle oils with boiling points between 300 and 615°F, and subsequent vapor phase hydrogenation then converting these oils to gasoline, diesel fuel, and other relatively light hydrocarbons [21]. The Bergius process converts 1 short ton of coal to 40 to 45 gallons of gasoline, 50 gallons of diesel fuel, and 35 gallons of fuel oil [24]. The gasoline fraction contains 75 to 80 percent paraffins and olefins and 20 to 25 percent aromatic compounds.

Liquid-phase hydrogenation of low-rank coals was usually accomplished at 890 to 905°F under pressures of 250 to 300 atmospheres using an iron oxide hydrogenation catalyst. For liquid-phase hydrogenation of bituminous coals, temperatures were similar, but pressures were in the range of 350 to 700 atmospheres.

Solvent Refining Processes

The solvent-refined coal (SRC) process is considered the least complex of the various process schemes. Hydrogenation of the coal in the SRC process occurs at

Table 5-11 Information on Select Liquefaction Processes

Process	I.G. Farben/Bergius	SRC-I	SRC-II	H-Coal	EDS	Costeam	HTI DCL
Temperature (°F)							
Stage 1	900	840	860	840	700–900	750–840	800
Stage 2	750	NA[a]	NA	NA	500–840		?[b]
Pressure (atm)							
Stage 1	350–700	140	140	210	20–170	270	170
Stage 2	300	NA	NA	NA	80–210[c]	NA	?
Carbon content of coal (wt.%)	70–83	70–85	≈75	≈78	75–80	60–70	?
H₂ consumption Stage 1							
(wt.% on coal) Stage 2							
Stage 1	6	2.4	4.7	5.5	—	6.5+	?
Stage 2	4	NA	NA	NA	6[c]	NA	?
Scale of operation (short tons per day)	150–1,800	50	50	600	250	(5–10 lb/h)[d]	4,300[e]
Sulfur in coal (wt.%)	<3.0	3	3	1–3	1–3	1	?
Sulfur in product (wt.%)	0.1	0.7	0.3–0.7	0.3–0.7	0.3	low	low

[a]Not applicable
[b]Unknown
[c]Recycle oil hydrogenation
[d]Small-scale, lb/h
[e]Per train, 3 trains planned

Source: Modified from Van Krevelen (1993), Elliot (1981), HTI (2004), and Miller (1982).

elevated temperatures and pressures in the presence of hydrogen. Catalysts other than the minerals contained in the coal are not used [2].

The SRC-I Process

In the SRC-I process, ground and dried coal is fed to the reactor as a slurry. The transport liquid, or process solvent, is a distillate fraction recovered from the coal hydrogenation product and serves as the hydrogen donor. The coal slurry and hydrogen are introduced into a dissolver at 840°F and 1,500 psig. The process produces a low-ash ($\approx$0.1 percent), low-sulfur ($\approx$0.3 percent), solid fuel with a heating value of about 16,000 Btu/lb. The SRC-I fuel was envisioned as a boiler fuel to replace natural gas and fuel oil.

Testing of this process started with Gulf Oil during the 1960s. Much of the developmental work on this process was performed at Wilsonville, Alabama, in a 6-short-ton/day plant with funding from EEI, Southern Services, DOE, and EPRI. A 50-ton/day plant was built at Fort Lewis, Washington, and testing was performed by Gulf Oil.

The SRC-II Process

The SRC-I process was modified to eliminate a number of processing steps and to produce an all-distillate, low-sulfur fuel oil from coal rather than a solid fuel. In the SRC-II process, pulverized and dried coal is mixed with recycled slurry solvent from the process. The slurry mixture is pumped to reaction pressure, about 140 atmospheres, preheated to about 700 to 750°F, and fed into the dissolver, which is operated at 820 to 870°F [2]. The dissolver effluent is separated into vapor and liquid phase fractions. The overhead vapor stream undergoes several stages of cooling and separations, with the condensed liquid distilled to produce naphtha and a middle distillate oil, which are converted to gasoline and diesel fuel, respectively. The gaseous products are purified to remove hydrogen sulfide and carbon dioxide, and the hydrogen-rich gas is then recycled to the reactor with makeup hydrogen. The liquid phase product acts as the solvent for the SRC-II process. SRC-II testing was performed at the Tacoma, Washington, pilot-plant by Gulf Oil. Funding for this work was provided by DOE.

Costeam Processes

The Costeam process, investigated at the Pittsburgh Energy Research Center (DOE) and the University of North Dakota/Grand Forks Energy Research Center was intended to produce low-sulfur liquid products from lignites and subbituminous coals. This process uses crude syngas containing about 50 to 60 percent carbon monoxide and 30 to 50 percent hydrogen, rather than pure hydrogen. The high moisture content of the low-rank coals provides the water that is converted to steam.

Ground coal is slurried with recycled product oil from the process. The slurried coal is pumped to unit pressure (4,000 psig), mixed with syngas, heated to 750 to 840°F, and maintained at conditions for periods of one to two hours to liquefy the coal. Severe operating conditions are required for the dissolution of the lignite. These operating conditions were a drawback for the process, since long residence times and high pressures are costly and present major engineering problems.

This process, although a potential candidate for liquefying low-rank coals, was only tested in small-scale equipment. It is included in this discussion because it was considered a potential process for utilizing low-rank coals.

Catalytic Processes

The most important process of this group is the H-Coal process, developed by Hydrocarbon Research, Inc. (HRI) as an outgrowth of previous work on the hydrogenation of petroleum fractions. The development of this process was sponsored by ERDA and a large group of oil companies [1].

H-Coal Processes

In the H-Coal process, pulverized coal is slurried with recycled oil and, along with hydrogen, fed into an ebullated-bed reactor, a feature that distinguishes this process from others. Reactor conditions are normally in the range of 825 to 875°F and 2,500 to 3,500 psig. The reactor contains a bed of catalytic particles, cobalt molybdate on alumina oxide. The products from the reactor include hydrocarbon gases, light and heavy distillate oils, and bottoms slurry. Variations of the processing scheme can produce fuel oil, naphtha, synthetic crude, ammonia, and fuel gas. Further processing can produce gasoline and jet fuel. HRI operated several small-scale reactors in their Trenton, New Jersey, test facility, and an 800-short-tons coal/day pilot-plant was constructed in Cattletsburg, Kentucky.

In 1995, HRI changed its name to HTI (Hydrocarbon Technologies, Inc.) when it became an employee-owned company. HTI has modified its process (now known as HTI's Direct Coal Liquefaction (DCL) process and shown schematically in Figure 5.36), and it has now gone from a single- to a double-stage reactor system [57]. In the HTI DCL process, pulverized coal is dissolved in recycled, coal-derived, heavy process liquid at about 2,500 psig and 800°F while hydrogen is added. Most of the coal structure is broken down in the first-stage reactor. Liquefaction is completed in the second-stage reactor at a slightly higher temperature and lower pressure. A proprietary GelCat catalyst is dispersed in the slurry for both stages. The process produces diesel fuel and gasoline.

HTI has entered into an agreement with Shenhua Group Corporation, Ltd, for a direct coal liquefaction plant that has been constructed in the People's Republic of China, and start-up began in late 2008 [57]. The plant is located approximately 80 miles south of Baotou, at Majata, Inner Mongolia in China. The plant has an ultimate capacity of 50,000 barrels per day of ultraclean, low-sulfur, diesel fuel and gasoline produced from Chinese coal. Shenhua Group intends to construct three additional plants in the Shengdong Coalfield of China, which spans Shaanxi Province and Inner Mongolia. These would be the only commercial, large-scale direct coal liquefaction plants operational in the world since the World War II–vintage plants constructed by the Germans.

Donor Solvent Processes

The main representative of this process is the Exxon Donor Solvent (EDS) process. In a donor solvent process, the solubilization of the coal is done by the hydrogen

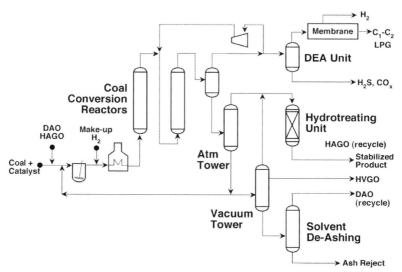

Figure 5.36 HTI's direct coal liquefaction process scheme.
Source: From Hydrocarbon Technologies, Inc. (2002).

donor liquid, and hydrogen molecules come from the donor liquid and not gaseous hydrogen. The donor liquid is then recycled and hydrotreated to add hydrogen back into the donor liquid.

Exxon Donor Solvent Processes

In the EDS process, finely ground coal is mixed with the donor solvent, and the coal is liquefied in a noncatalytic tubular plug-flow reactor in the presence of molecular hydrogen and a hydrogen-rich donor solvent [2]. The liquefaction reactor operates at 800 to 880°F and 1,700 to 2,300 psig. The products from the liquefaction reactor are separated by distillation into light hydrocarbon gases, ranging from methane to propane and methylpropane, a naphtha fraction, a heavy distillate, and a bottoms fraction [24]. The naphtha and heavy distillate fractions are treated by conventional petroleum-refining technology. About 85 percent of the naphtha is recovered as gasoline, and about 50 percent of the heavy distillate is recovered as a mixture of benzene, toluene, and xylenes. Further processing of the heavy distillate produces fractions comparable to jet fuel and heating fuel [24]. A portion of the heavy distillate is hydrotreated and recycled to slurry the coal. In addition, the bottoms are either coked in a fluid coking plant or recycled to the liquefaction reactor. Recycling to the liquefaction unit results in a dramatic increase in the conversion of the coal to liquid products.

In addition to small-scale test reactors, Exxon, DOE, EPRI, and an international group of industrial sponsors installed a 250-short-tons coal/day pilot plant located in Baytown, Texas. This was the minimum size needed for confident scale-up of the critical process and mechanical features of the EDS process.

Conclusion

Several direct liquefaction processes were introduced in this chapter. Although these were some of the most important and, in most cases, the most successful processes, it must be noted, however, that there were many other processes conceived and researched. As a result of economics, however, none were able to compete with available liquid and gaseous fuels, especially in the industrialized countries. The work of the Germans during the 1930s and 1940s and the facilities planned for China illustrate that large-scale, direct coal liquefaction commercial plants are technically feasible. They can be constructed and operated when necessary, if and/or when dictated by political mandates, whether it is isolation during wartime or a decision to be self-sufficient and use indigenous resources.

REFERENCES

[1] D.W. Van Krevelen, Coal: Typology—Physics—Chemistry—Constitution, Third Edition, Elsevier Science, 1993.

[2] M.A. Elliot (Ed.), Chemistry of Coal Utilization, Secondary Supplementary Volume, John Wiley & Sons, 1981.

[3] E.S. Moore, Coal: Its Properties, Analysis, Classification, Geology, Extraction, Uses, and Distribution, John Wiley & Sons, 1922.

[4] S.E. Kuehn, Power for the Industrial Age: A brief history of boilers, Power Engineering 100 (2) (1996) 5–19.

[5] J.G. Landels, Engineering in the Ancient World, University of California Press, 1978.

[6] J.B. Kitto, S.C. Stultz, Steam: Its Generation and Use, Babcock & Wilcox Company, 1978.

[7] Power, Boilers and Auxiliary Equipment, Special Edition, June 1988.

[8] J.G. Singer (Ed.), Combustion: Fossil Power Systems, Combustion Engineering, Inc., 1981.

[9] CIBO (Council of Industrial Boiler Owners), Industrial Combustion Boiler & Process Heater MACT Summary Sheet, July 3, 2002.

[10] CIBO, Energy Efficiency & Industrial Boiler Efficiency—An Industry Perspective, March 2003.

[11] EPA (U.S. Environmental Protection Agency), Report to Congress: Wastes from Combustion of Fossil Fuels, Volume 2—Methods, Findings, and Recommendations, U.S. Government Printing Office, March 1999 (Chapter 3).

[12] B.G. Miller, S. Falcone Miller, A.W. Scaroni, Utilizing Agricultural By-Products in Industrial Boilers: Penn State's Experience and Coal's Role in Providing Security for our Nation's Food Supply, Nineteenth Annual International Pittsburgh Coal Conference, University of Pittsburgh, September 23–27, 2002a.

[13] B.G. Miller, S. Falcone Miller, R.E. Cooper, N. Raskin, J.J. Battista, A Feasibility Study for Cofiring Agricultural and Other Wastes with Coal at Penn State University, Nineteenth Annual International Pittsburgh Coal Conference, University of Pittsburgh, September 23–27, 2002b.

[14] B.G. Miller, S. Falcone Miller, Utilizing Biomass in Industrial Boilers: The Role of Biomass and Industrial Boilers in Providing Energy/National Security, The First CIBO Industrial Renewable Energy & Biomass Conference, Washington, D.C., April 7–9, 2003.

[15] B.G. Miller, S. Falcone Miller, R.T. Wincek, R.S. Wasco, Fuel Flexibility in Boilers for Fuel Cost Reduction and Enhanced Food Supply Security Final Report, Prepared for Pennsylvania Energy Development Authority, Agreement No. PG050020, June 30, 2006, 509 pages.

[16] T.C. Elliot (Ed.), Standard Handbook of Powerplant Engineering, McGraw-Hill, 1989.

[17] C.J. Lawn (Ed.), Principles of Combustion Engineering for Boilers, Academic Press, 1987.

[18] D.A. Tillman, B.G. Miller, D.K. Johnson, D.J. Clifford, Structure, Reactivity, and Nitrogen Evolution Characteristics of a Suite of Coals and Solid Fuels, 29th International Technical Conference on Coal Utilization & Fuel Systems, Coal Technology Association, April 2004.

[19] J.B. Howard, R.H. Essenhigh, Mechanism of Solid-Particle Combustion with Simultaneous Gas Phase Volatiles Combustion, in: 11th Symposium (Int.) on Combustion, The Combustion Institute, Pittsburgh, 1967, pp. 399–408.

[20] M.A. Field, D.W. Gill, B.B. Morgan, P.G.W. Hawksley, Combustion of Pulverized Coal, The British Coal Utilisation Research Association, Cheney & Sons Ltd, 1967.

[21] N. Berkowitz, An Introduction to Coal Technology, Academic Press, 1979.

[22] P.L. Walker, F. Rusinko, L.G. Austin, Gas Reactions of Carbon, Advances in Catalysis and Related Subjects, vol. XI, Academic Press, 1959.

[23] Power, Power from Coal, Special Report, February 1974.

[24] H.H. Schobert, Coal: The Energy Source of the Past and Future, American Chemical Society, 1987.

[25] DOE (U.S. Department of Energy), Fluidized-Bed Combustion: An R&D Success Story, www.fe.doe.gov/programs/powersystems/combustion/fluidizedbed_success.shtml, September 17, 2003.

[26] J.T. Tang, F. Engstrom, Technical Assessment on the Ahlstrom Pyroflow Circulating and Conventional Bubbling Fluidized Bed Combustion Systems, in: Proceedings of the International Conference on Fluidized Bed Combustion, May 3–7, 1987.

[27] M.J. Virr, The Development of a Modular System to Burn Farm Animal Waste to Generate Heat and Power, in: Proceedings of the Seventeenth Annual International Pittsburgh Coal Conference, University of Pittsburgh, September 11–14, 2000.

[28] B.N. Gaglia, A. Hall, Comparison of Bubbling and Circulating Fluidized Bed Industrial Steam Generation, in: Proceedings of the International Conference on Fluidized Bed Combustion, May 3–7, 1987.

[29] R.W. Bryers, Fireside Behavior of Minerals in Coal, in: Symposium on Slagging and Fouling in Steam Generators, January 31, 1987, p. 63.

[30] K.S. Kumar, R.E. Sommerland, P.L. Feldman, Know the Impacts from Switching Coals for CAAA Compliance, Power, vol. 135, No. 5, May, 1991, pp. 31–38.

[31] E.C. Winegartner, Coal Fouling and Slagging Parameters, Research Department, ASME, 1974.

[32] R.C. Attig, A.F. Duzy, Coal Ash Deposition Studies and Application to Boiler Design, in: Proceedings American Power Conference, vol. 31, 1969, pp. 290–300.

[33] M.I. Microbeam, Introduces New Indices to Assist Prediction of Ash Deposition, vol. 14, Winter, 2003, pp. 1–3.

[34] D.H. Wakelin (Ed.), The Making, Shaping and Treating of Steel, eleventh ed., Ironmaking Volume, The AISE Steel Foundation, 1999.

[35] C.R. Ward (Ed.), Coal Geology and Coal Technology, Blackwell Scientific, 1984.

[36] UK Coal (United Kingdom Coal), Section 20: Coking and Carbonisation Plant, www.open.voa.gov.uk/Instructions/chapters/Rating/vol15/sect270/s270.htm (accessed March 26, 2009).

[37] DOE, Clean Coal Technology, Upgrading Low-Rank Coals, Technical Report Number 10 Office of Fossil Energy, August 1977.

[38] ENCOAL Corporation, ENCOAL Mild Coal Gasification Project: ENCOAL Project Final Report, prepared for U.S. Department of Energy, No. DE-FC21-90MC27339, September 1997.

[39] DOE, Clean Coal Technology Demonstration Program: Program Update 2001 as of September 2001, Office of Fossil Energy, July 2002.

[40] ConvertCoal, Inc., New Domestic Oil Resource—Efficient, Cleaner Energy Supply, *www.ConvertCoal.com* (accessed November 3, 2009).

[41] Coal Merchants, Listing of Economy Fuels, *www.coalmerchantsfederatoin.co.uk/Authorised%20Fuels.htm* (accessed March 26, 2009).

[42] Coalite Smokeless Fuels, The Process, *www.coalite.co.uk/page4.html* (accessed March 26, 2009).

[43] J.P. Frederick, B.A. Knottnerus, Role of the Liquids From Coal Process in the World Energy Picture, Fifth Annual Clean Coal Technology Conference, January 8, 1997.

[44] Fact Sheet, *www.lanl.gov/projects/cctc/factsheets/encol/encoaldemo.html*, last modified January 13, 2003.

[45] DOE, Advanced Coal Conversion Process Demonstration—Project Fact Sheet, *www.lanl.gov/projects/cctc/factsheets/rsbud/adcconvdemo.html*, last modified December 2, 2002.

[46] J. Ratafia-Brown, L. Manfredo, J. Hoffman, M. Ramezan, Major Environmental Aspects of Gasification-Based Power Generation Technologies, Office of Fossil Energy, December 2002.

[47] SFA Pacific, Inc., and DOE, Gasification: Worldwide Use and Acceptance, Office of Fossil Energy, January 2000.

[48] Gasification Technology Council, Gasification: A Growing, Worldwide Industry, *www.gasification.org.story.worldwide/worldwide.html* (accessed February 10, 2003).

[49] DOE, Gasification World Survey Results 2004, Office of Fossil Energy, September 2005.

[50] DOE, Gasification World Database 2007, Office of Fossil Energy, October 2007.

[51] R. Breault, Gasification Tutorial, 33rd International Technical Conference on Coal Utilization & Fuel Systems, Coal Technologies Association, June 1–5, 2008.

[52] W.R.K. Wu, H.H. Storch, Hydrogenation of Coal and Tar, U.S. Bureau of Mines Bulletin, No. 633, 1968, pp. 1–10.

[53] A.N. Stranges, Germany's Synthetic Fuel Industry 1927–1945, Energia 12 (5) (2001) 2.

[54] B. Davis, Fischer-Tropsch Synthesis, The CAER Perspective, Energia 8 (3) (1997) 1.

[55] Hydrogen Technologies, Inc. (HTI), Coal-to-Liquids (CTL), *www.covol.com/coalLiquids.asp*, 2004.

[56] B.G. Miller, Autoclave Studies of Lignite Liquefaction, M.S. Thesis, University of North Dakota, 1982.

[57] HTI, Headwater's HTI Subsidiary Signs License Agreement with Shenhua Group—China's Largest Coal Company, *www.htinj.com/news/061802shenhualicagment.html*, June 2002.

6 Anatomy of a Coal-Fired Power Plant

This chapter discusses a coal-fired power plant from coal receipt to gases emitted from the stack, along with water usage and treatment and ash handling. Subsequent chapters will focus on combustion and gasification as they relate to clean coal usage. This chapter, however, focuses on the details of a conventional coal combustion power plant to illustrate the importance of this technology. This is especially true, since developing countries such as China and India are constructing new coal-fired power plants at record numbers, and clean coal or emissions control have not necessarily been a priority.

Figure 6.1 shows the major subsystems of a modern coal-fired power plant [1]. Key subsystems include coal receiving and preparation, coal combustion and steam generation, turbine generator, heat rejection, emissions control, water treatment, and ash and by-product handling. These are discussed in the subsequent sections.

Coal-fired power plants have two main "paths": the fuel and products of the combustion path and the steam-water path. In the first path, the fuel handling system stores the coal, prepares the coal for combustion in the steam generator, and transports it to the boiler, which is the steam generator. The associated air system supplies air to the burners through a forced-draft fan. The steam generator subsystem, which includes the air preheater, burns the fuel/air mixture, recovers the heat, and generates the steam. The flue gas leaves the boiler; passes through a series of pollution control equipment to remove fly ash, oxides, and nitrogen and sulfur (and in some cases mercury); passes through an induced-draft fan; and exits the stack. In the steam-water path, the boiler evaporates the water and supplies high-temperature, high-pressure steam to a turbine-generator set that produces electricity. The steam may also be reheated in the boiler after passing through part of a multistage turbine system and sent to lower-temperature and lower-pressure sections of the turbine. Ultimately, the steam is passed from the turbine to the condenser, where the remaining heat is extracted. The water then passes through several pumps and heat exchangers (i.e., feedwater heaters) to increase its pressure and temperature before being returned to the boiler. The heat absorbed by the condenser is rejected to the atmosphere by cooling towers.

6.1 Coal Transport to the Power Plant

Coal is delivered to the power plant by rail, barge, truck, conveyor, pipelines, or ocean transport [2]. The mode of delivery to the plant is determined by the transport infrastructure available, the location of the mine (or mines) and power plant with

Clean Coal Engineering Technology. DOI: 10.1016/B978-1-85617-710-8.00006-6
Copyright © 2011 by Elsevier Inc. All rights of reproduction in any form reserved.

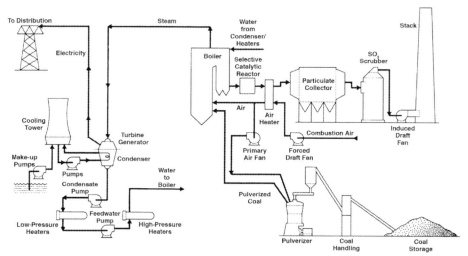

Figure 6.1 Coal-fired utility power plant schematic.
Source: Modified from Kitto and Stultz (2005).

respect to each other, quantities of coal needed at the power plant, and capital and operating costs. For example, power plants located near a coal mine will receive their coal via truck or overland conveyor. Some power plants located near the coast-line or near shipping terminals receive their coal via ocean transport. Much of the subbituminous coal that is mined in the Powder River Basin, which is located in the interior of the United States, is shipped via rail throughout the country. Some power plants located near large rivers receive their coal via barges.

The volume and distribution of coal transported by the various methods are shown in Table 6.1 [3]. Combinations of these are often used.

Trucks used to deliver coal typically fall into two categories. They can be of massive size, with capacities up to 300 tons, that operate on mine roads and deliver to

Table 6.1 Distribution of Coal Transportation Methods for Electricity
Generation in the United States in 2007

Method	1,000 Short Tons/Year	Percent of Total
Rail	751.69	73.2
Barge	96.10	9.4
Truck	96.77	9.4
Conveyor/pipeline/tramway	73.68	7.2
Other water shipping methods	8.82	0.8
Unknown	0.19	<0.02
U.S. Totals	**1,027.25**	**100.0**

Source: Energy Information Administration (2008).

Figure 6.2 Coal-hauling truck with 100-short-ton capacity.

mine-mouth power plants. An example of a larger-sized type of truck is shown in Figure 6.2, which has a 100-ton coal capacity. This truck is used to haul lignite from Dakota Westmoreland Company's Beulah mine south of Beulah, North Dakota, to a coal sizing and transfer station that either loads coal onto railcars or transports the coal to the Coyote Station, a 427 MW_e pulverized coal-fired plant operated by Otter Tail Power Company [4]. Trucks can also be much smaller so they can operate on existing highways such as the triaxle truck shown in Figure 6.3, which can haul up to 25 tons of coal. As an example, for many years, this was the primary mode of transportation for Edison International Company's Homer City Power Station, located near Homer City, Pennsylvania, which received 17,000 tons/day of coal (i.e., about 30 trucks/hour) for its three units generating 1,884 MW of electricity from mines in about a 50-mile radius [5]. In 2006, a rail line was reactivated because many local mines shut down and about a third of the coal is now delivered by rail.

Figure 6.3 Coal-hauling truck with 25-short-ton capacity.

Truck transportation has high operating costs per ton-mile relative to rail or barge transport. Truck transportation has limited haulage distances (up to 50 miles), can cause congestion around power plants, and requires a high degree of monitoring at the plant, including weighing all the trucks. Trucking, however, is the most flexible mode of coal transportation because schedules can be adjusted to meet the generating plant's supply needs, existing highway infrastructure can be used (for the smaller trucks), it is the least capital intensive, and it has a high degree of competition.

Many power plants receive their coal via rail in a "unit train," an example of which is shown in Figure 6.4. This is a large coal train that typically contains 100 railcars with 100 short tons of coal in each one, for a total load of 10,000 short tons. A large power plant can require at least one coal delivery of this size every day, with some plants receiving as many as three to five trains a day. The 3,100 MW_e Monroe Power Plant, located near Detroit, Michigan, operated by DTE Energy, and composed of four units each with 775 to 795 MW_e capacity, consumes 25,000 to 30,000 short tons/day of coal, or two or three unit trains of coal/day [6].

In 2007, railroads delivered nearly 75 percent of the coal transported to electric generating stations. Rail transport is used for distances ranging from 10 to 1,500 miles. In general, rail transport has the second lowest unit cost per ton-mile, and it is extremely efficient.

Coal is also conveyed over land from the mines to the power plants. Conveyors with capacities of up to 5,000 tons/h are used [7]. An example of a conveying system is shown in Figure 6.5 [8]. This is a 5.5-mile overland conveyor belt system that transfers coal from Jim Walter Resources' Mine No. 7, located in central Alabama, to Mine No. 7's coal preparation plant, where it is washed and processed for power generation.

Figure 6.4 Coal transportation via railcar.

Figure 6.5 Overland coal conveyor.
Source: Courtesy of Walter Energy, Inc.

Transportation by water is done using either barges (on rivers) or ships (on the ocean). Barges typically carry 1,000 to 1,500 short tons [2]. Cargo ships carrying coal typically have capacities of up to 40,000 short tons. Barge transport of coal is the most cost effective method compared to rail or truck transport. The capacity of one barge is equivalent to 15 railcars or 58 trucks. In addition, less energy is expended to move cargo by barge than by railcar or truck. Barges, however, are not self-unloading. The coal delivery costs by barge, and the capital investment required for unloading facilities, must be factored into a decision when identifying the method of coal delivery.

6.2 Coal Handling, Storage, and Processing

When it is received at the power plant, there are several steps the coal undergoes before it is used in the boiler: coal handling, storage, and processing.

6.2.1 Coal Handling

Coal handling equipment is based on the method of coal delivery to the power plant, the boiler type, and the required coal capacity. Portable conveyors may be used in small plants to unload railcars, reclaim coal from yard storage piles, and fill bunkers while dedicated handling facilities are installed in larger plants to meet the demand for fuel supply. The coal handling components are determined by the design and requirements of the boiler. The coal handling system capacity is determined by the boiler's rate of coal use, the frequency of coal deliveries to the plant, and the time available for unloading.

Railcars can be of the rotary dump or bottom discharge type. In automatic rotary car dumping systems, the railcars are hydraulically or mechanically clamped in a cradle, and the cradle is rotated about a rotating track section so the coal falls into the hopper below the tracks. These railcars have a special coupling that allows the dumping to be performed without uncoupling the cars. This type of system has the advantages of fast unloading and thus higher capacity per period of time; however, its main disadvantage is high capital cost. Bottom dump cars are characterized by simple unloading unless the coal has high moisture content and sticks to the sides of the car or is frozen in the car.

Barge unloading is done using either a clamshell bucket mounted on a fixed tower or a bucket wheel or elevator unloader. With the clamshell bucket unloader, the barge is located below the bucket and is moved as necessary to unload. A shore-mounted bucket wheel or elevator unloader is more efficient and has a higher capacity than a clamshell bucket unloader. Large vessels used for ocean transport are equipped with a bucket wheel for self-unloading.

Truck unloading is relatively simple and usually consists of the dumped coal passing through a grid into a storage hopper. At smaller facilities or in cases where coal blending occurs, the coal may be dumped into temporary storage facilities.

Other types of handling equipment consist of a variety of belt conveyors, bucket elevators, screw feeders, chutes, feeders, and reclaiming equipment. This auxiliary equipment is used from coal unloading, coal yard storage (and in some cases blending), transferring from the coal yard to the crushers, and, in the case of pulverized coal-fired boilers, to the pulverizers. The most common type of conveyor is the belt conveyor, which consists of a rubber belt running over spaced troughed idler rollers and terminating at both ends over pulleys. Belt conveyors can handle up to 5,000 tons/h of coal.

Bucket elevators are the simplest and most dependable units for making vertical lifts. They are available in a wide range of capacities and may operate entirely in the open or be totally enclosed. Chutes are utilized in the process where gravity can be used to transfer coal from one location to another. These come in a variety of designs (steeper slopes for materials with high friction) and materials of construction.

Feeders are used to control flow from the storage bunker (discussed in the next section) at a uniform rate. Feeder selection is based on the analysis of the material properties, such as moisture level, maximum particle size, particle size distribution, bulk density, and abrasiveness, the desired flow rate, and the degree of control required. Feeders can generally be classified as volumetric or gravimetric. Volumetric feeders provide a controlled volume of fuel to the downstream components.

Typical examples of feeders include screw, apron, belt, table, rotary valve, or vibrating feeders. Gravimetric feeders compensate for variations in bulk density due to moisture, coal size, and other factors. They provide a precise weight of coal to the downstream components. In the most common gravimetric feeder system, coal is carried over a load cell that monitors the coal weight on a belt.

6.2.2 Coal Storage

Bulk storage of coal at the power plant is necessary to ensure a continuous supply of fuel. The amount of coal stored at the site depends on many factors, including the size of the power plant; the amount that may be required by law, such as in the case of public utilities that must maintain minimum supplies; available storage space; the weathering characteristics of the coal used; dependability of coal delivery; among others.

Storage facilities may take the form of open storage piles or enclosed bins and silos. The storage piles may take one of several shapes, while the bins and silos may be round, square, or rectangular and be made from a variety of materials such as steel, concrete, and so on. The type of storage facilities depends on the following factors:

- The required sizes of the process or emergency storage
- The maximum allowable moisture content of the coal
- The necessity for dust control
- The potential effects of freezing when stored in cold weather
- The availability of capital for investing

In addition, a basic choice may be made between bin or ground storage. Usually volumes of 26,000 yd^3 or more are stored in open storage, while piles of less than 13,000 yd^3 are stored in bins. Volumes from 13,000 to 26,000 yd^3 can be stored in either manner [9].

Open storage piles take many shapes, including a cone, wedge, or kidney, or variations thereof. Figure 6.6 provides some practical ranges of storage volumes as a function of pile height. The shape of the stockpile is generally dependent on the type of equipment used for pile construction and for reclaiming coal from the pile. Conical piles are generally associated with a fixed stacker, while a radial stacker generates a kidney-shaped pile. A rail-mounted traveling stacker can be used to form a rectangular pile. Angles of repose of the piles are shallow, usually 40 degrees. The lower table in Figure 6.6 shows the approximate area in square feet per ton of 50 lb/ft^3 coal.

Many different issues must be taken into consideration for long-term storage of coal. These include coal oxidation and spontaneous combustion, fugitive dust emissions, and coal pile runoff. To address these, coal compaction is used to reduce airflow through the open stockpile and avoid undue oxidation problems. For typical utility storage pile applications, a loose coal bulk density of 50 lb/ft^3 is typical; however, for well-compacted piles, a bulk density of 65 to 72 lb/ft^3 is more appropriate. Sometimes inactive coal piles are sprayed with petroleum products or aqueous emulsions of hydrocarbons, soaps, fatty acids, or milk of lime to seal the outer surface [10].

Covered stockpiles or silos are used where severe winter conditions or limited land for storage can justify the added cost. Dome-shaped buildings typically can vary

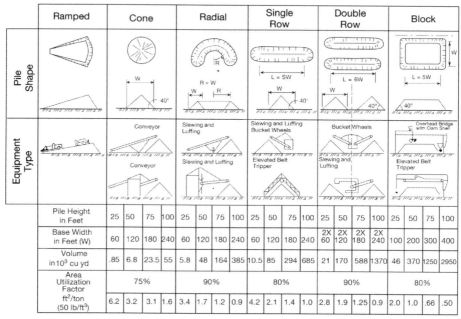

	Ramped				Cone				Radial				Single Row				Double Row				Block			
Pile Height in Feet	25	50	75	100	25	50	75	100	25	50	75	100	25	50	75	100	25	50	75	100	25	50	75	100
Base Width in Feet (W)	60	120	180	240	60	120	180	240	60	120	180	240	60	120	180	240	2X 60	2X 120	2X 180	2X 240	100	200	300	400
Volume in 10³ cu yd	.85	6.8	23.5	55	5.8	48	164	385	10.5	85	294	685	21	170	588	1370	46	370	1250	2950				
Area Utilization Factor	75%								90%				80%				90%				80%			
ft²/ton (50 lb/ft³)	6.2	3.2	3.1	1.6	3.4	1.7	1.2	0.9	4.2	2.1	1.4	1.0	2.8	1.9	1.25	0.9	2.0	1.0	.66	.50				

Figure 6.6 Shapes and dimensions of storage piles.

from 50 to 200 feet in height and diameters from 200 to 300 feet. Tent-shaped buildings typically have lengths of 100 to 1,000 feet, widths of 100 to 250 feet, and heights of 50 to 100 feet. Bins and silos are usually cylindrical and can be made from steel or concrete. Large silos at power stations can store 10,000 to 20,000 short tons of coal, and multiple silos can be installed for the necessary storage requirements.

Coal bunkers provide intermediate storage ahead of other coal processing equipment such as pulverizers. The bunkers are normally sized for 8 to 24 hours of fuel supply. Normally there is one bunker for each pulverizer and boiler coal feed point [1].

6.2.3 Coal Processing/Size Reduction

Coal received at the power station is typically crushed to 2 inch × 0 and must be further reduced in size for utilization in the boiler. Coal is then crushed for use in stoker boilers, fluidized-bed boilers, and cyclone-fired boilers. Crushed coal is ground finer for use in pulverized coal-fired boilers. Equipment used for coal size reduction is usually grouped into three broad categories: primary crushing, secondary crushing, and grinding, which is also referred to as pulverizing. Tables 6.2 through 6.4 list important characteristics in each category, respectively. Sometimes the grinding category is subdivided into coarse and fine grinding. Table 6.5 lists combustion process fuel sizing requirements, and Table 6.6 provides a list of fuel equipment descriptions used for various combustion applications [6].

Table 6.2 Characteristics of Equipment for Primary Crushing

Characteristics	Equipment			
	Rotary Breaker	Single-Roll Crusher	Double-Roll Crusher	Single-Rotor Impact Crusher
Influence of feed rate variations on performance	Not sensitive	Sensitive	Sensitive	Sensitive
Influence of foreign material in feed	Not sensitive	Sensitive	Sensitive	Not sensitive
Relative reduction ratio[a]	High	Low	Low	High moderate
Relative degree of fines produced	Low	Moderate	High	High moderate
Mode of crushing	Impact	Compression, shear	Compression, shear	Impact
Custom built	Yes	No	No	No
Relative equipment availability	High	Moderate	Low	—[b]
Relative maintenance cost	Low	Moderate	High	—
Advantages	Provides primary beneficiation by rejecting hard foreign material; provides size calibrated product top size	Effective for crushing hard or resilient coal; also used for crushing of refuse	Effective for crushing hard or resilient coal	Effective for crushing hard or resilient coal
Disadvantages	Not effective in crushing very hard or resilient coal; not applicable for crushing refuse	Can produce oversize material	Can produce oversize material	Tramp material can cause downtime by shear pin release on breaker bar

[a]Reduction ratio is defined as the ratio between the top size of the feed material and the top size of the crushed product.
[b]Not available

Table 6.3 Characteristics of Equipment for Secondary Crushing

Characteristics	Equipment		
	Double-Roll Crusher	Impact Crushers (hammermill, ring mill)	Impact Crushers (cage mill)
Most common application	Crushing clean coal	Liberation of impurities	Pulverizing clean coal
Relative capacity per unit	Low	High	High
Constraints on maximum top size of feed	Restricted	Less restricted	Less restricted
Relative degree of fines produced	Low	High	High
Relative reduction ratio	Low	High	Medium
Mode of crushing	Compression, shear	Impact, attrition, shear	Impact, attrition, shear
Relative availability	High	Moderate	Low
Relative maintenance cost	Low	Moderate	High
Miscellaneous comments	Capable of producing a crushed product with a low quantity of fines	Capable of uniform product size with minimum fines production	Requires soft, friable coal to avoid excessive wear

Bradford Breakers are normally used for crushing, sizing, and cleaning run of mine coals at the mine or plant site. They produce a relatively coarse product with minimum fines that are 100 percent sized. The Bradford Breaker, which is shown in Figure 6.7 (see page 231), is a large cylinder made of a perforated plate fitted with internal shelves. As the cylinder rotates, the shelves lift the feed, and the feed slides off the shelves and drops onto the screen plates below, where it shatters along its natural cleavage lines.

Table 6.4 Characteristics of Equipment for Grinding

		Equipment		
		Vertical Spindle Mill (roller mill)		
Characteristics	Ball Mill	Ball-and-Race Mill	Roll-and-Race Mill	Impact/Attrition Mill
Pulverizing method	Impact, attrition	Compression	Compression	Impact, attrition
Metal-to-metal contact in grinding zone	Point contact of balls on liner, balls on balls	In contact	None	None
Drive	Horizontal, pinion and large gear	Vertical, bevel gears	Vertical, worm gears	Horizontal, direct coupling from motor
Speed, rpm	15–25	90–200	20–105	900–1,800
Tramp material	Assists grinding until ground down	Promptly rejected after initial contact with grinding elements		Unitype, magnet in feeder and tramp pocket in mill. Attrita, magnet in feeder
Capacity, tons/h	5–160	2–20	2–105	3–30
Coal top-size input, inches	¾	1½	1–2	2½
Fineness through 200 mesh (74 μm) screen, %	70–95	70	65–85	70–95
Loss of fineness with wear of grinding elements	None but requires ball and liner replacement	None	Adjustment provides for constant fineness	Largest
Noise	Highest	High	Quiet	Least
Pulverizer fan	Handles coal and air	Handles clean air only	Handles coal and air	Handles coal and air
Normal type of coal	Anthracite, bituminous, subbituminous	Bituminous, subbituminous, lignite		Bituminous, subbituminous

Continued

Table 6.4 Characteristics of Equipment for Grinding—Cont'd

		Equipment		
		Vertical Spindle Mill (roller mill)		
Characteristics	**Ball Mill**	**Ball-and-Race Mill**	**Roll-and-Race Mill**	**Impact/Attrition Mill**
Main advantages	Low maintenance, high reliability	Compact, low capital cost, accommodates many coals	Low maintenance, low vibration, capacity over wear cycle	Compact, rapid response to load changes, quite operation
Main disadvantages	Large space requirements, noisy, high power consumption	High maintenance cost, low capacity	Outlet temperature 210°F maximum	High maintenance costs with abrasive coals

Table 6.5 Combustion Process Fuel Sizing Requirements

Fuel Type	As-Received Coal Size	Pulverized Coal	Circulating Fluidized-Bed	Stoker	Cyclone
		Process Distribution (200/50 mesh)	*Process Top Size*	*Process Top Size*	*Process Top Size*
Anthracite	<2"	80%/99%	¼"	5/16"–1½"	¼"
Low volatile bituminous coal	<2"	75%/99%	3/8"	3/4"–1½"	¼"
High volatile bituminous coal	<2"	75%/99%	½"	3/4"–1½"	¼"
Subbituminous coal	<2"	65%/98%	½"	3/4"–1½"	¼"
Lignite	<2"	65%/98%	1"	3/4"–1½"	¼"

Source: From Miller and Tillman (2008).

The size of the screen plate perforations determines product size. Sized product flows through the screen, but oversized pieces are relifted and redropped until they are able to pass through. Tramp iron, lumber, or other debris is transported to the discharge end of the cylinder, where it is scooped out for disposal by a refuse plow [11].

Cage mills are used for sizing of friable, dry materials. A cage mill, shown in Figure 6.8, is comprised of a series of concentric sleeves that are driven. The feed

Table 6.6 Fuel Sizing Equipment Storage Summary

Technology	Grinding Type	Use	Output	System
Bradford breakers	Gravity	Primary crushing	Relatively coarse, minimum fines, 100% sized	Mine or plant primary sizing and tramp material cleaning
Cage mills	Impact	Primary crushing, final sizing	Fine-to-coarse	CFB[a], stoker, cyclone
Granulators	Impact, compression	Primary crushing, final sizing	Medium-to-coarse, minimum fines	CFB, stoker, cyclone
Hammermills	Impact, attrition, shear	Primary crushing, final sizing	Controlled top size, medium fineness, high fines	CFB, stoker, cyclone
Impactors	Impact	Final sizing	Controlled top size, minimum fines	CFB, stoker
Double roll crushers	Compression	Final sizing	Controlled top size, minimum fines	CFB, stoker
Ball mills	Gravity impact	Final sizing	Controlled top size, maximum fines	PC[b] fired, arch fired
Vertical spindle mills	Compression	Final sizing	Controlled top size, maximum fines	PC fired, arch fired

[a]Circulating fluidized-bed boiler
[b]Pulverized coal-fired boiler
Source: From Miller and Tillman (2008).

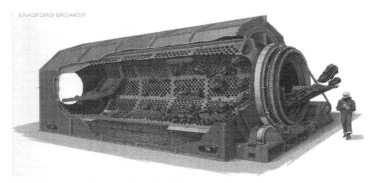

Figure 6.7 Pennsylvania Crusher Bradford Breaker.
Source: From Miller and Tillman (2008).

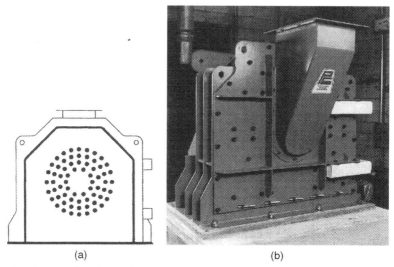

Figure 6.8 A Pennsylvania Crusher cage mill: (a) view of inside of mill and (b) actual mill. *Source:* From Miller and Tillman (2008).

enters the center of the mill and is ground through impaction by being struck by the first row of sleeves. After impact, the shattered coal is passed to the second row of sleeves, where the sizing process continues. Size reduction continues through the subsequent rows until the fuel is thrown against an impact plate and then finally exits the mill. The product size is controlled by varying the spacing between the sleeves and adjusting the speed of the rotating cage [11].

Granulators crush by a combination of impact and rolling compression, using rows of ring hammers that crush with a slow, positive rolling action that produces a granular product with minimum fines. Product size is determined by screen openings and is adjusted by changing the clearance between the cage and the path of the ring hammers [11]. A granulator is shown in Figure 6.9 [6].

Hammermills reduce the material sizing in a two-step process. The material is first reduced by dynamic impact from the rotating hammers and continues with attrition and shear in the small clearance between the hammers and the screen bars. Hammermill crushing produces great product uniformity with a minimum of oversized product and will produce a high throughput capacity. Modern-day systems use a reversible hammer to increase the time between maintenance shutdowns and to maximize the hammer material use [11]. An example of a hammermill is shown in Figure 6.10.

Impactors are used to produce an optimum percentage of products below 1/8-inch size and with a minimum amount of fines. The modern impactors can maintain their throughput with wet coals and are able to pass through noncrushables. Their rotors can usually be run in either direction to provide equal wear on both hammer faces [6]. An impactor is shown in Figure 6.11.

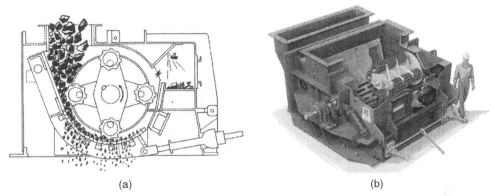

(a)   (b)

Figure 6.9 A Pennsylvania Crusher granulator: (a) view of inside of mill and (b) actual mill. *Source:* From Miller and Tillman (2008).

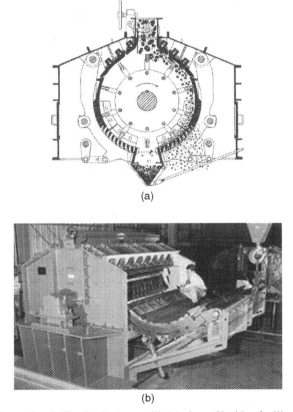

(a)

(b)

Figure 6.10 A Pennsylvania Crusher hammermill: (a) view of inside of mill and (b) actual mill. *Source:* From Miller and Tillman (2008).

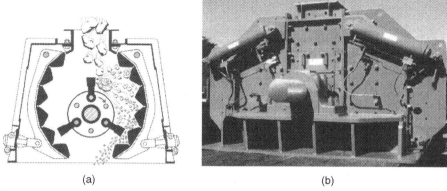

(a)   (b)

Figure 6.11 Pennsylvania Crusher impactor and Coalpactor mills: (a) view of inside of mill and (b) actual mill.
Source: From Miller and Tillman (2008).

Double-roll crushers grind the fuel between two counterrotating rolls. As the two rolls rotate toward each other, the material is pulled down into the crushing zone, where it is grabbed and compressed by the rolls. Product size is determined by the size of the gap between the rolls [11]. An example of a double-roll crusher is shown in Figure 6.12.

Ball mill pulverizers use an air sweeping action for impaction and gravity grinding of the fuel in a large rotating drum filled with hardened steel balls. The main components are a large drum lined with a wear-resistant material, hollow trunnions at each end, and a classifier that sizes the final product. The drum is mounted horizontally and supported by trunnions bearings. As the ball mill rotates, the profiled liners lift and tumble the fuel and steel balls at a low speed, pulverizing the fuel. The ground product is sized by an integral or externally mounted classifier that passes the correct-size fuel and rejects the oversized fuel back into the mill for further grinding. An example of a ball mill is shown in Figure 6.13. Hot air is used to transport and dry the fuel during the grinding process. The inlet air temperature is varied to maintain a set air/coal temperature leaving the mill [6].

Figure 6.12 A Pennsylvania Crusher double-roll crushing mill.
Source: From Miller and Tillman (2008).

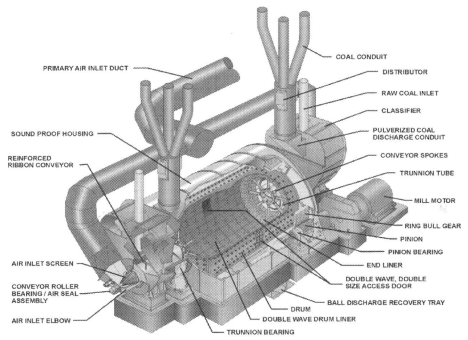

Figure 6.13 A Foster Wheeler ball mill pulverizer.
Source: From Miller and Tillman (2008).

Ball mill pulverizers produce a high fineness at the cost of high power usage. They have historically been used for anthracite and petcoke grinding. Their lower moisture levels and unique combustion characteristics make both of these fuels ideal candidates for ball mills.

Vertical spindle mills use an air sweeping action for attrition and compression grinding of the fuel between a rotating table and large grinding rollers. The rollers use an external loading mechanism to increase the grinding capability on the system. The mill operates by maintaining a bed of fuel between the grinding tables and the grinding rollers. The rollers impart pressure on the fuel bed, causing the coal particles to rub against each other and to break apart. Hot air is used to transport and dry the fuel. The inlet air temperature is varied to maintain the set air/coal temperature when leaving the mill. Figure 6.14 labels the key components of a Foster Wheeler MBF pulverizer and shows the coal flow path through the system.

Vertical spindle pulverizers are the most common grinding mill on pulverized coal-fired utility steam generators. They provide a reasonable fuel fineness capability with a fast fuel change response, low-horsepower utilization, and a relatively long time between replacement of wear parts. The coal mills have undergone continual improvements by all the manufacturers as new materials are being used and more stringent combustion and emissions requirements have come into play.

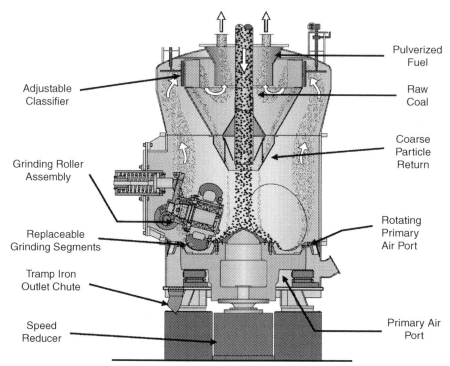

Figure 6.14 Foster Wheeler MBF coal mill.
Source: From Miller and Tillman (2008).

6.3   Steam Generation

Steam generation in a pulverized coal-fired boiler is accomplished in a configuration of thermal-steam/water sections, which preheat and evaporate water, and superheat and reheat steam. An example of a coal-fired utility boiler was shown in Figure 5.5 (the fundamentals of steam production and energy flow were discussed in Chapter 5). The major components in the steam generating and heat recovery system of a utility boiler include the following [1]:

- Furnace and convection pass
- Steam superheaters
- Steam reheater
- Economizer
- Steam drum
- Attemperator and steam temperature control system
- Air heater

6.3.1   Furnace and Convection Pass

The furnace is a large chamber in which the coal is combusted. The geometry and dimensions of the furnace are influenced by the fuel and type of combustion equipment, as discussed in Chapter 5. The furnace walls, which are comprised of tubes containing water and steam (hence the term *waterwalls*), are designed to extract the appropriate quantity of heat to lower the gas temperature, leaving the furnace to minimize ash deposition on the tubes in the convection section and ensure that excessive tube metal temperatures are not encountered. When firing pulverized coal, the products of combustion rise through the upper furnace and pass through the superheater, reheater, and economizer sections, as shown in Figure 5.5. The heat transfer surfaces located in the flue gas horizontal and vertical downflow sections of the boiler are the convection pass. This section is designed to extract the maximum amount of heat from the partially cooled flue gas [2]. The section should be as compact as possible in order to obtain the most efficient use of construction material and plant space. Design factors include the flue gas velocities between tubes, the tube spacing, and the use of finned tubes.

The furnace and convection pass walls are constructed of water- or steam-cooled carbon steel or low-alloy tubes to maintain metal temperatures within acceptable limits so they maintain their integrity [1]. The tubes are connected at the top and bottom by headers, which distribute or collect the water, steam, or steam-water mixture. An example of a tube wall/header arrangement is shown on a smaller boiler system in Figure 6.15, which is a D-type industrial boiler. The tubes are welded together with steel bars to provide membrane wall panels that are gas-tight, continuous, and rigid.

6.3.2   Steam Superheaters and Reheaters

The superheaters and reheaters (refer to Figure 5.5 for typical locations) are tube bundles located in the flow gas stream that increase the temperature of the saturated steam, which is produced in the furnace waterwalls. These tubes are constructed from steel alloy material because of their high operating temperature. The main difference between superheaters and reheaters is the steam pressure [1]. In a typical drum boiler, the superheater outlet pressure might be 2,700 psig, while the reheater outlet pressure might be only 580 psig [1]. They are used to help control steam temperatures, control steam flow pressure loss, and optimize heat recovery.

6.3.3   Economizers

The economizer (refer to Figure 5.5) is a heat exchanger for recovering energy from the flue gas downstream of the superheater and reheater. It is used to increase the temperature of the feedwater entering the steam drum. Steam is not generated inside these tubes.

6.3.4   Steam Drums

The steam drum (shown earlier in the schematic in Figure 5.5 and in Figure 6.15) serves two purposes in a boiler. It acts as a header for distributing water to the downcomers (tubes that contain cold water, which is circulated to the lower portion

Figure 6.15 *Top:* membrane wall arrangement with header—outside view; *middle:* membrane wall arrangement—inside furnace view; *bottom:* cutaway tube construction. *Source:* Modified from Kitto and Stultz (2005) and Keeler Dorr Oliver [12].

of the furnace), and it separates the steam from the water after the steam/water mixture rises to the top of the boiler through the boiler tubes (also known as risers) [1, 2]. The steam drum is a large cylindrical vessel (3 to 6 feet in diameter) that runs the entire length of the boiler, which can approach lengths of 100 feet. Steam drums are constructed from thick steel plates that have been rolled into cylinders with hemispherical heads. They house connections for the downcomers, boiler tubes (risers), continuous blowdown, chemical addition (to remove oxygen and modify the pH), feedwater inlet, gauge glass, pressure taps, and steam outlets to the superheaters. In addition, they contain baffles and separators to remove moisture from the steam before the steam is sent to the superheaters.

6.3.5 Steam Temperature Controls

Steam temperature control is an important aspect of steam generation. It can be performed by several methods, including the addition of water or low-temperature steam to the high-temperature steam flow (i.e., attemperation). In large utility units, spray attemperation is most frequently used for dynamic control because of their rapid response [1]. These systems are located at the inlet to the superheater section or between individual superheater sections to control the superheater metal temperature.

6.3.6 Air Heaters (Preheaters)

The air heater is not part of the steam-water circulation system, but it is an important component in the steam generator system heat transfer and efficiency. The flue gas that exits from the steam generator is often as hot as 600°F (315°C) and represents a major loss of heat and a source of inefficiency in the power plant. To address this, power plants design air heaters to reduce the flue gas temperature to about 300°F (150°C) by heating the incoming combustion air. Air heater designs typically include tubular, flat plate, and regenerative heat exchangers.

6.4 Steam Turbines and Electricity Generation

In a power plant, steam turbines convert the thermal energy of the steam into mechanical energy, or shaft torque, which is then converted to electrical energy. The conversion of thermal energy into mechanical energy is accomplished by controlled expansion of the steam through stationary nozzles and rotating blades. The shaft torque drives electric generators to produce electricity. The process of imparting heat to pressurized water to produce high potential energy steam, and translation of the potential energy to work through interaction of steam and turbine blades, is called a Rankine cycle (shown in Figure 6.16). In the Rankine cycle, the turbine exhaust (steam), now at low pressure and temperature, is condensed and recycled back to the boiler.

6.4.1 Steam Turbines

Steam turbines can be classified in a number of ways [2]. They can be classified by cycle and steam conditions, such as Rankine cycle, Rankine regenerative cycle, reheat cycle, condensing or noncondensing, and by the number and arrangement of turbine

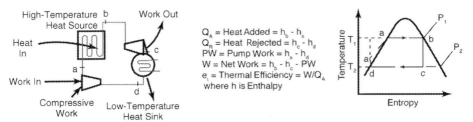

Figure 6.16 Simple Rankine cycle without superheat.
Source: Modified from Elliot (1989) and Power, Special Edition (June 1989).

Figure 6.17 A cross-compound steam turbine.
Source: From U.S. Department of Energy (2009).

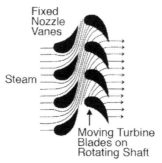

Figure 6.18 A stationary nozzle and moving blade illustrating steam flow in what is called a reaction stage.
Source: From U.S. Department of Energy (2005).

shaft shafts and casings. When two or more casings are involved, the designs are either tandem-compound (all casings on the same shaft) or cross-compound (casings on two or more shafts) [2]. Figure 6.17 shows a cross-compound steam turbine [14].

Turbines are designed with multiple stages to accommodate the volume expansion of steam as the pressure drops. As steam moves through the system and loses pressure and thermal energy, it expands in volume, requiring a larger diameter and longer blades in each succeeding stage to extract the remaining energy. Every stage of a turbine has two basic design elements: the stationary nozzle and the moving blade or bucket. This is illustrated in Figure 6.18 [15], and a close-up view of a blade is shown in Figure 6.19

Figure 6.19 Close-up view of a turbine blade.
Source: From U.S. Department of Energy (2009).

[14]. The design of these parts depends on factors such as entering steam conditions, exhaust steam pressure, shaft speed, rated capacity, and steam flow [2].

As the turbine size increases (i.e., the electricity generation increases), the last-stage annulus area requires that there be two parallel-flow paths, or double flow, resulting in separate low-pressure and high-pressure casings. For reheat turbines, the expansion of the steam occurs in three sections, with the addition of an inter-mediate pressure stage [2]. Figure 6.20 shows a high-pressure section of a turbine with the casing removed [14]. Figure 6.21 shows two turbine sections [14].

Figure 6.20 A high-pressure turbine section.
Source: From U.S. Department of Energy (2009).

Figure 6.21 A turbine with two sections.
Source: From U.S. Department of Energy (2009).

6.4.2 Electric Generators

Electric generators convert mechanical energy to electrical energy. The most common type of generator used by electric utilities is the synchronous alternator. It consists of a stationary stator with the output windings arranged around its periphery and a rotating, driven rotor with direct current windings acting as an electromagnet [2]. When the electromagnet rotates, it generates a current in the stator. In operation, it generates up to 24,000 volts of current. The power that is generated is alternating current, which is transformed to the voltage required (e.g., 115, 230, 500, or 765 kV), rectified (converted) to direct current (because direct current can be transported over large distances efficiently), transmitted over transmission lines as direct current, and then inverted back to alternating current at the point of application.

6.5 Steam Condensers and Cooling Towers

Steam condensers and cooling towers are part of the power plant's water circuitry, as shown earlier in Figure 6.1. Most systems are closed-loop cooling in which water is taken from a pump basin, circulated through the condenser and a cooling tower, and then returned to the basin. Water is not discharged from the system, except for a small amount used in cooling tower blowdown and lost through evaporation in the cooling tower. This water is replenished with makeup water, as shown in Figure 6.1.

6.5.1 Steam Condensers

The steam condenser's function is to condense the turbine exhaust steam using cooling water. In condensing the steam, a vacuum is created, which reduces the backpressure on the turbine, thereby maximizing plant efficiency. In addition, quality feedwater is recovered and recycled back to the economizer and steam generation unit.

The most common type of condenser is the surface condenser, in which the cooling water and steam remain separate, with indirect contact only. This condenser is usually a shell and tube heat exchanger, where the cooling water is circulated through the tubes and the low-pressure steam exhaust is condensed by flowing over the tubes on the shell side of the heat exchanger.

6.5.2 Cooling Towers

The heat absorbed by the cooling water in the condenser must be removed to maintain overall system efficiency. This is accomplished by pumping the warm cooling water from the condenser to a cooling tower, where the waste heat is dissipated. Cooling towers are characterized by shape, flow relationship of air and water (counterflow or crossflow), manner of producing airflow (natural or induced), and heat transfer methodology. The most common types of cooling towers used at power plants are of the wet design [2, 13]. Wet towers are divided into natural draft (hyperbolic) and induced draft or mechanical [2]. These can be further divided into counterflow and crossflow designs.

Even though natural-draft towers are typically more expensive than mechanical types (which use motor-driven fans to generate air movement), they are commonly used in the power-generation sector [2, 13] under the following circumstances:

- Long amortization periods allow sufficient time to recoup the differential cost of not requiring fan power.
- Operating conditions couple low wet-bulb temperature and high relative humidity.
- A combination of low wet-bulb temperature and high inlet- and exit-water temperatures exists.
- A heavy winter load is possible.

An example of a hyperbolic tower is shown schematically in Figure 6.22 [13] and in Figure 6.23.

Mechanical-draft towers can be designed to be more compact than natural-draft towers and are designed in cells. They can be designed to accommodate a broad range of water flows by boosting cell size and number of cells. Although the wet-flow cooling towers have been the traditional choice, there is considerable research underway by the U.S. Department of Energy and the Electric Power Research Institute to improve cooling tower designs to conserve the water and make power plants zero-discharge.

6.6 Water Treatment

The water used in a boiler system must be of high purity, and water treatment is performed to prevent or to control contamination of the water by species that may corrode metals or form scale on heat-transfer surfaces. Feedwater (to the boiler) consists of makeup water that is added to the condensate at the condenser outlet (called the condenser hot well). This makeup water has undergone some treatment before

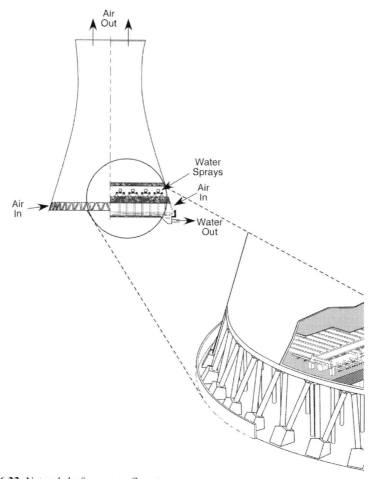

Figure 6.22 Natural-draft, counterflow tower.

being introduced into the system. Makeup water treatment involves demineralization to remove dissolved matter and pretreatment steps to remove suspended solids [16]. Standard pretreatment equipment consists of clarifiers, softeners, and media and pre-coat filters to remove suspended solids, calcium, and magnesium (i.e., hardness). Table 6.7 lists the effect of treatment techniques on makeup water contaminants [16].

This water is then passed through a series of feedwater heaters and introduced into the deaerator. Deaerating heaters are direct-contact devices that use steam as a stripping agent to remove noncondensable gases from feedwater. These devices remove dissolved gases from the water, with the removal efficiency of dissolved oxygen the highest. The oxygen concentration in the feedwater is reduced to 5 parts per billion (ppb) to reduce the corrosivity of the water. Carbon dioxide and other volatile substances such as ammonia and hydrazine, if present, will be only partially removed because of their solubility in water [2, 16].

Figure 6.23 A counterflow tower with water sprays shown in insert.
Source: Photograph courtesy of Anna F. Miller.

Table 6.7 Effects of Treatment Techniques on Makeup Water Contaminants

Treatment Technique	Suspended Solids	Alkalinity	Hardness	Dissolved Solids	Total Dissolved Solids
Filtration	Reduction close to 100%	No change	No change	No change	No change
Cold process softening	Reduction close to 90%	Reduction 30–50%	Reduced to 50–75 ppm	Reduced to 3–5 ppm	Reduction of 10–20%
Hot process softening	Reduction close to 95%	Reduction 50–70%	Reduced to 10–30 ppm	Reduced to 1–3 ppm	Reduction of 20–30%
Sodium cycle cation exchange	Prefiltration removes 100%	No change	100% removal	No change	No change
Demineralization	Prefiltration removes 100%	100% removal	100% removal	Reduction of 80–90%	100% removal
Reverse osmosis	Removes 100%	At least 90% removal	At least 90% removal	Reduction of 90%	Reduction of 90%

Source: From Power, Special Section (June 1988)

Table 6.8 Feedwater Treatment Techniques

Treatment Technique	Effect of Treatment
Blend of neutralizing amines	Low pH protection throughout steam and condensate sections
Neutralizing-amine blend plus filing amine	Low pH protection throughout steam and condensate sections O_2 corrosion protection Better distribution of filming amines
Neutralizing amines plus O_2 scavenger	Low pH protection throughout steam and condensate sections O_2 corrosion protection Passivation of metal surfaces
Filming amine plus O_2 scavenger	Low pH protection throughout steam and condensate sections O_2 corrosion protection Passivation of metal surfaces Surface preparation for filming amine
Neutralizing amines plus O_2 scavenger and filming amine	Low pH protection throughout steam and condensate sections O_2 corrosion protection Passivation of metal surfaces Surface preparation for filming amine Backup protection against film disruption

Source: From Power, Special Section (June 1988).

Under normal operating conditions, minimal chemical treatment of feedwater is required assuming good control of condensate purity. The main chemicals added to feedwater are an oxygen scavenger, such as hydrazine, and ammonia or an amine for pH control. These and other boiler water treatment chemicals, such as chelants and iron dispersants, are mainly introduced to the feedwater via condensate entering the deaerating heater [2]. Solids such as sodium phosphates and caustic soda are most often added directly to the boiler water in the steam drum to minimize boiler water pH in the alkaline range and provide buffering capacity against boiler water pH excursions as the result of cycle contamination. These techniques are summarized in Table 6.8.

6.7 Environmental Protection

An important element of a coal-fired boiler system is environmental protection. There are numerous federal, state, and local regulations that limit gaseous, solid, and liquid emissions from a power plant. This section will not discuss regulations or emissions control strategies in detail because they are discussed in more detail in Chapters 8 and 9, respectively.

Primary air pollutant emissions that are controlled include airborne particulate or fly ash, sulfur dioxide (SO_2), and nitrogen oxides (NO_x). Federal legislation to control airborne mercury emissions from power plants are close to promulgation and are anticipated to be enacted within the next few years, while some state regulatory agencies have started to impose regulations. Currently, carbon dioxide emissions are a topic of discussion, and legislation to control CO_2 emissions is also expected in the near future.

Regulations are in place to control water discharge quality, which can contain trace chemicals used to control corrosion and fouling as well as waste heat rejected from the condenser. Solid wastes such as residual ash from the fuel and spent sorbent from pollution control systems are also regulated.

6.8   Ash and By-Product Handling

The ash handling systems of a coal-fired power plant collect the ash and residue from different locations along the boiler flue gas streams. For a pulverized coal-fired system, a typical ash distribution is boiler bottom ash (15–40 percent of the total ash), convective pass/economizer ash ($\approx$5 percent), air heater ash ($\approx$5 percent), and fly ash from the particulate collection device (50–75 percent) [1]. The characteristics of the ash differ throughout the boiler system, so the collection, transport, and storage systems are usually separate for the furnace and collection points downstream.

6.8.1   Bottom Ash Systems

Bottom ash is most often conveyed in a hydraulic system in which the ash is entrained in a high-flow, circulating water system and delivered to an ash pond or to dewatering storage bins. A simplified schematic of a hydraulic system is shown in Figure 6.24 [16]. In this system, the bottom ash hoppers are filled with water to quench the hot ash, which may be as hot as 2,400°F, and are usually lined with refractory. The hoppers are emptied periodically by hydraulic sluicing. When the boiler is a continuously slagging wet bottom design, the ash hopper is replaced with a slag tank.

Some boilers are equipped with a mechanical conveying system, which uses less energy than the hydraulic system. Mechanical systems have submerged scrapers/conveyors where flight bars attached to chains at intervals push the ash that falls between them to a quenching pool. Ultimately the bottom ash is dumped into a storage bin or dewatering bin.

6.8.2   Convective Pass/Economizer Ash Systems

The fly ash that is removed from the hoppers located under the economizer, selective catalytic reactor (shown earlier in Figure 6.1 and discussed in Chapter 9), and air heater contains a mixture of larger particles (which may contain unburned carbon) and fine fly ash. The ash from this part of the boiler system can be handled in several ways. The coarser ash, which is the typical size of the economizer ash,

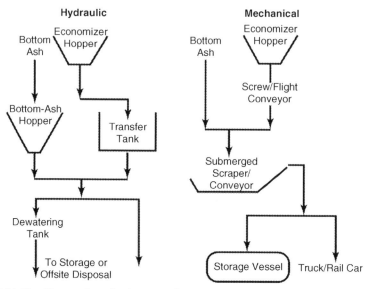

Figure 6.24 Handling options for bottom ash.
Source: Modified from Power, Special Section (June 1988).

is often crushed to reduce the particles to about 0.25 inch in diameter for pneumatic transport. The finer ash sizes are generally combined with the fly ash transport system, which can be vacuum or pressurized pneumatic. Economizer ash is mixed with a hydraulic bottom ash system (see Figure 6.24) when the bottom ash is being disposed of in a landfill because this ash is difficult to dewater [1]. Mechanical drag systems are also used for conveying economizer ash.

6.8.3 Fly Ash Systems

Most of the ash from a pulverized coal-fired boiler is entrained by the flue gas and is carried through the boiler system. As much as 70 percent of the ash generated in the boiler is removed from the flue gas by the particulate removal device, which is either an electrostatic precipitator or a baghouse (see Chapter 9 for a detailed discussion of fly ash removal). The fly ash that is removed by the particulate control devices is of low density and therefore not conveyed in a water slurry. This ash is almost always transported pneumatically, with some mechanical conveyance used. Pneumatic transport systems generally fall into three general types: vacuum, pressure, or combination vacuum/pressure. A simplified diagram of fly ash transport systems is shown in Figure 6.25 [16].

A vacuum system uses air, below atmospheric pressure, to entrain and convey ash particles. This system provides the lowest cost for short transport distances [1]. Pressurized conveying of fly ash offers some advantages over vacuum systems, including the ability to convey over longer distances, and simplified ash/air-separation

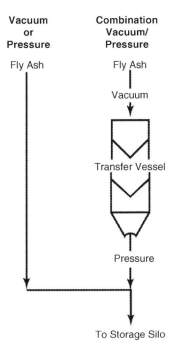

Figure 6.25 Handling options for fly ash.
Source: Modified from Power, Special Section (June 1988).

equipment at the storage silo. A disadvantage is that equipment sizes are larger. Combination vacuum/pressure systems are used to transport ash longer distances than vacuum systems alone can accommodate, while retaining some of the advantages of the vacuum system, such as its lower space requirement than a pressurized system. In the combination system, the initial withdrawal of the ash from the hoppers is by a vacuum system to an intermediate collection vessel, where it is transferred to a pressure conveying system for delivery to a distant storage vessel.

6.8.4 Scrubber Sludge Systems

Boiler systems that contain wet flue gas desulfurization systems (shown in Figure 6.1 and discussed in detail in Chapter 9) must handle the solid effluent from the scrubber (also called the scrubber sludge) which contains the reaction products from limestone and SO_2. The sludge can be disposed of in either slurry ponds or landfills, or used for wallboard ($CaSO_4$) manufacture.

The gypsum/water slurry from the scrubber system is first passed through a primary hydrocyclone to reduce the moisture content of the slurry. The concentrated slurry stream is then sent to a vacuum filter to reduce the moisture content to less than 10 percent. The gypsum is then sent to the wallboard manufacturing plant. Sometimes the gypsum is sent to the cement industry or fertilizer company.

References

[1] J.B. Kitto, S.C. Stultz (Eds.), Steam: Its generation and use, fourty-first ed., The Babcock & Wilcox Company, 2005.

[2] T.C. Elliot (Ed.), Standard Handbook of Power Plant Engineering, McGraw-Hill, 1989.

[3] EIA (Energy Information Administration), Domestic Distribution of U.S. Coal by Destination, State, Consumer, Origin and Method of Transportation, 2007, U.S. Department of Energy, December 2008.

[4] Lignite Energy Council, *www.lignite.com*, (accessed 16.06.09).

[5] Edison International Company, Homer City Generating System Fact Sheet, 2007.

[6] B.G. Miller, D.A. Tillman (Eds.), Combustion Engineering Issues for Solid Fuel Systems, Academic Press, 2008.

[7] Roberts & Schaefer Company, Coal Production Overland Conveying Fact Sheet, *www. r-s.com*, 2007.

[8] Walter Energy Downloaded Operational Photos, *www.sloss.com* (accessed 7.12.09).

[9] Institute of Gas Technology, Coal Conversion Systems Technical Data Book, Prepared for U.S. Department of Energy, Contract No. No. AC01-81-FE05157, June 1982.

[10] J.W. Leonard III, B.C. Hardinge (Eds.), Coal Preparation, fifth ed., Port City Press, 1991.

[11] Pennsylvania Crusher Corporation, Bulletin 4050-D 00-3-03R-IM: Handbook of Crushing, 2003.

[12] Keeler Dorr Oliver, Packaged Steam Generator Brochure, Bulletin No. 95B187.

[13] Power, Steam Turbines and Auxiliaries, Special Section, June 1989.

[14] DOE (U.S. Department of Energy), National Energy Technology Laboratory, Steam Turbine Designs for Coal Fired Generation, *www.netl.doe.gov/publications/proceedings/ 02/combustion/boss.pdf*, June 19, 2009.

[15] DOE, National Energy Technology Laboratory, Turbine Program, Enabling Near-Zero Emission Coal-Based Power Generation, June 2005.

[16] Power, Boilers and Auxiliary Equipment, Special Section, June 1988.

7 Clean Coal Technologies for Advanced Power Generation

One approach to reducing emissions of air pollutants and greenhouse gases, specifically carbon dioxide (CO_2), is to increase the energy efficiency of existing thermal cycles and to develop advanced, higher-efficiency cycles. An increase in energy efficiency reduces the amount of fuel consumed and consequently the quantity of pollutants and greenhouse gases emitted. Increasing the thermal efficiency of a power plant is one of three primary techniques to reduce pollutants and CO_2 emissions. A second method is to change to less carbon-intensive sources such as natural gas, nuclear power, or renewable sources. The third method is the capture and storage of CO_2 from fossil fuel–fired (i.e., coal in particular) power plants. There are three technological pathways for CO_2 capture: postcombustion capture, oxy-fuel combustion, and precombustion capture—that is, integrated gasification combined cycle (IGCC).

In this chapter, technologies to increase thermal efficiency and environmental performance are discussed. These include supercritical pulverized coal combustion; fluidized-bed combustion, both atmospheric and pressurized; and IGCC, which is a technology that both improves cycle efficiency and is a CO_2 capture technique. Similarly, oxy-fuel combustion, which is an advanced power-generation technology, is also presented in this chapter for both pulverized coal and fluidized-bed combustion systems. Postcombustion CO_2 capture is presented in Chapter 10. This chapter begins with an overview of thermodynamic cycles because of the major efforts underway worldwide to increase their efficiency in power-generation facilities.

7.1 Power Cycles

Thermal efficiency is a measure of the performance of a power plant. The two main thermodynamic cycles used widely in power generation are the Rankine and Brayton cycles. These cycles are also referred to as power cycles because thermal energy is converted into mechanical energy, which is then converted to electrical energy. When these two cycles are combined, as in an IGCC system, it is referred to as a Combined cycle.

7.1.1 Rankine Cycle

The Rankine cycle is the basis of all large steam power plants, as briefly discussed in Chapter 6 (refer to Figure 6.16). In coal-fired power plants, high-temperature, high-pressure steam is produced by converting the chemical energy stored in the

Clean Coal Engineering Technology. DOI: 10.1016/B978-1-85617-710-8.00007-8
Copyright © 2011 by Elsevier Inc. All rights of reproduction in any form reserved.

coal into thermal energy and transferring the energy to the working fluid (e.g., water), which passes through the boiler to produce the steam. The steam from the boiler is expanded in a series of high- and low-pressure steam turbines, which convert the energy into mechanical shaft work to drive an electric generator and to produce electricity. After the last turbine stage, the steam is routed to a condenser, and the condensate is then pumped back into the boiler, repeating the cycle.

There are variations of the Rankine cycle, but for our discussion purposes the basic Rankine cycle will be presented in this section for simplicity. The Rankine cycle is shown in Figure 7.1 [1]. The working fluid is compressed from low to high pressure (Step 1-2). At this stage, the pump requires little input energy because the fluid is a liquid. The high-pressure liquid enters a boiler, where it is heated at constant pressure to the compressed liquid (Step 2-3) and two-phase (i.e., water and steam) mixture (Step 3-4). The steam is then passed through the superheaters, where it superheated (Step 4-5). The steam is then expanded through a turbine to generate power (Step 5-6). The expanded vapor, which is decreased in temperature and pressure, enters a condenser, where it is condensed at a constant pressure and temperature to become a saturated liquid and the unavailable heat is rejected to the atmospheric sink (Step 6-1).

The thermodynamic efficiency depends on the temperatures at which heat addition and rejection occur:

$$\eta = \frac{T_1 - T_2}{T_1} \tag{7.1}$$

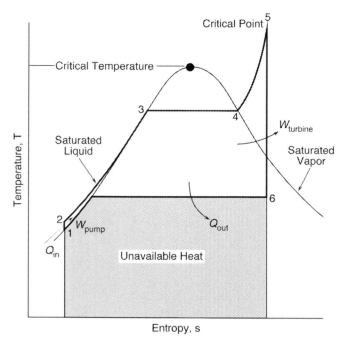

Figure 7.1 Temperature-entropy diagram of the ideal Rankine cycle.
Source: Modified from Kitto and Stultz (2005).

where η is thermal efficiency of conversion from heat into work; T_1 is the absolute temperature of the heat source ($^\circ$R or $^\circ$K); and T_2 is the absolute temperature of the heat sink ($^\circ$R or $^\circ$K). From Equation (7.1), there are three ways to increase ideal cycle efficiency: decrease T_2, increase T_1, or both. Little can be done to reduce T_2 in the Rankine cycle because of the limitations imposed by the temperatures of available rejected heat sinks in the environment [1]. Consequently, much of the effort in improving cycle efficiency is directed toward increasing T_1, which is discussed in more detail later in this chapter.

7.1.2   The Brayton Cycle

The Rankine cycle efficiency limit is dictated by the ratio of the maximum and minimum cycle temperatures. The maximum temperature of the steam Rankine cycle is about 1,150°F ($\approx$620°C), which is set primarily by material constraints at the elevated pressures of the steam cycles [1, 2]. Gas turbines, however, can operate at much higher temperatures than steam turbines, with gas turbine inlet temperatures as high as 2,350°F (1,288°C), thereby making them more thermodynamically efficient.

The conventional Brayton cycle is a gas turbine cycle in which air is compressed and burned with fuel in a combustor. The hot gases are then expanded in a turbine coupled to an electric generator. A simple gas turbine system is shown in Figure 7.2 [1]. In this system, a portion of the work produced by the turbine is used to drive the compressor, and the remainder is used to produce power. The Brayton cycle in Figure 7.2 consists of compressing the combustion air (Step 1-2), adding heat through the combustion process (Step 2-3), expanding the gases through the turbine and performing work (Step 3-4), and rejecting heat to the atmosphere (Step 4-1).

7.1.3   Combined Cycle

The gas turbine (Brayton cycle) efficiently uses high-temperature gases from a combustion process but discharges its exhaust gas at a relatively high temperature, which is wasted heat/energy. The steam turbine (Rankine cycle) is not able to achieve the high temperatures of the gas turbine due to materials of construction. Combining the two cycles (high-input temperatures and low-output temperatures) allows one to improve the overall efficiency of the power plant. The combined cycle power plant utilizes a gas turbine (Brayton cycle) to generate electricity, while the waste heat is used in a heat recovery boiler to produce steam (Rankine cycle) to generate additional electricity using a steam turbine. The combined cycle efficiency is a sum of both cycle efficiencies, since they are both powered by the same fuel source. When using coal as the fuel source, one approach to coal firing is the indirect-fired gas turbine cycle. In this cycle, coal combustion gases are passed over tubes that are cooled with clean, high-pressure air; the tubes heat the air to temperatures above 2,000°F (1,093°C), and the hot air is used to drive the gas turbine. Another approach is to gasify the coal and burn the gases that are produced in the gas turbine. This is the concept behind IGCC, which is discussed in more detail later.

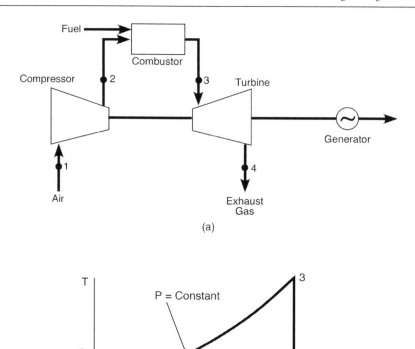

(a)

(b)

Figure 7.2 Simple gas turbine system (a) and temperature-entropy diagram (b) for a Brayton cycle.
Source: Modified from Kitto and Stultz (2005).

7.2   Pulverized Coal-Fired Power Plants

Pulverized coal technology is mature, well understood, and discussed in detail in Chapters 5 and 6. The United States has more than 1,100 pulverized coal units, and about 5,000 more units are in operation worldwide [3]. This technology is the most widely used in power generation and is based on decades of experience. The main developments in pulverized coal combustion have involved increasing plant thermal efficiencies by raising the steam pressure and temperature used at the boiler outlet/steam turbine inlet, while ensuring that the units can operate reliably and follow load requirements satisfactorily [4].

7.2.1 Advanced Pulverized Coal-Fired Plants

The thermodynamic efficiency of a Rankine steam cycle increases with the increasing temperature and pressure of the superheated steam entering the turbine. It is possible to increase the mean temperature of heat addition by taking partially expanded (and reduced temperature) steam from the turbine and sending it back to the boiler, reheating it, and reintroducing it to the turbine. In the usual designation of steam parameters, the first temperature is the main steam temperature, and the second and third temperatures refer to single and double reheat—that is, 4,484 psia/1,100°F/1,100°F/1,100°F (30.9 MPa/594°C/594°C/594°C). Boilers are referred to as subcritical, supercritical, and ultra-supercritical, depending on the steam's state. The term *subcritical* refers to conditions below the critical point of water, which is 3,208 psia (22.1 MPa) and 706°F (374°C). Water's critical pressure is the maximum pressure in which liquid and vapor can coexist in equilibrium.

At this critical point, the density of steam and water are equal and there is no distinction between the two states. In subcritical boilers, which have two-phase flow, a drum-type boiler is used to separate the steam from the water (as discussed in Chapter 6) before it is superheated and sent into the turbine. As steam pressure and superheat temperature are increased above 3,208 psia (22.1 MPa) and 706°F (374°C), the steam becomes supercritical—that is, it undergoes a gradual transition from water to vapor with corresponding changes in physical properties. In coal-fired boilers, above an operating pressure of 3,208 psia (22.1 MPa) in the evaporator part of the boiler, the cycle is termed *supercritical*. The cycle medium is a single-phase fluid with homogenous properties, and it is not necessary to separate steam from water in a drum. Therefore, once-through boilers are used because they do not contain drums.

Figure 7.1 shows the critical point on a temperature-entropy Rankine cycle diagram. *Ultra-supercritical* steam generally refers to supercritical steam at more than 1,100°F (593°C), although the definitions of supercritical and ultra-supercritical boiler pressure and temperature profiles differ from one country to another. Table 7.1

Table 7.1 Characteristics of Subcritical, Supercritical, and Ultra-supercritical Power Plants

Type of Unit	Main Steam Pressure (psia/MPa)	Main Steam Temperature (°F/°C)	Reheat Steam Temperature (°F/°C)	Efficiency (%, based on HHV[a] and bituminous coal)
Subcritical	<3,201/<22.1	Up to 1,050/565	Up to 1,050/565	33–39
Supercritical	3,207–3,628/ 22.1–25	1,000–1,075/ 540–580	1,000–1,075/ 540–580	38–42
Ultra-supercritical	>3,628/>25	>1,075/>580	>1,075/>580	>42

[a]Based on the coal's higher heating value
Source: From Nalbandian (2008).

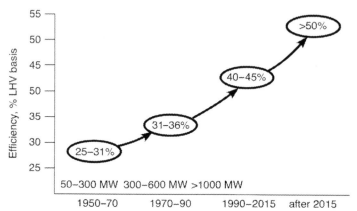

Figure 7.3 Progression of coal-fired power plant efficiencies.
Source: Modified from Baruya (2008).

shows one definition of these plants, and it lists approximate pressure, temperature, and efficiency ranges for the subcritical, supercritical, and ultra-supercritical pulverized coal-fired power plants [4].

Note that efficiencies are expressed on the basis of either lower heating value (LHV) or higher heating value (HHV) of the fuel. Europe bases its efficiencies on the LHV, which excludes the latent heat of vaporization of the water/moisture in the combustion process (net heat of combustion). In the United States, efficiencies are expressed in terms of the HHV, which includes the latent heat of vaporization of the water/moisture formed in the combustion process (gross heat of combustion). Since the LHV does not take into account the energy used to vaporize the water, efficiencies expressed in LHV are usually higher than those based on HHV. The result is that reported U.S. efficiencies are generally 2 to 4 percent lower than European efficiencies. This chapter discusses efficiencies based on both heating values and notes which basis is being used.

More than 80 percent of the existing coal-fired power plants around the world are subcritical units, and the majority of them are based on a conventional single reheat thermal cycle with subcritical steam pressure in the range of 2,320 to 2,610 psia (16–18 MPa) and the main/reheat steam temperatures both in the range of 1,000 to 1,050°F (535–565°C) [5]. Under these conditions, the efficiencies (LHV basis) range from 33 to 39 percent but can be as low as 20 to 25 percent (LHV basis) in developing nations. Figure 7.3 illustrates the progression of the efficiency of new coal-fired power plants since the 1950s with efficiency targets in 2015 and beyond [5]. The average efficiency of the existing U.S. coal-fired generating fleet is about 32 percent (HHV basis).

Supercritical units first came into operation in the early 1960s and experienced material failures [4]. Problems that were encountered in the early/older supercritical units included [4]:

- The erosion of start-up valves due to high differential pressure because of operating at a constant pressure
- Longer start-up times due to a complicated start-up system and operation
- Low ramp rates due to turbine thermal stresses caused by temperature change in the high-pressure turbine during load changing
- High minimum stable operation load due to by-pass operation and pressure ramp-up operation
- Slagging due to an undersized furnace and inadequate soot blower systems
- Waterwall tube cracking due to metal temperature rise because of inner scale deposits and fireside wastage
- Frequent acid cleaning required due to inappropriate water chemistry
- Lower efficiency than expected due to high air leakage because of the pressurized furnace and complications from reheat temperature control
- Low availability due to the preceding problems

Over the last few decades, supercritical technology has improved with better materials of construction developed and countermeasures to the problems just listed implemented such that supercritical pulverized coal combustion is considered a well-established technology and is expected to play a significant role in reducing emissions from coal-fired power plants, especially with the application of CO_2 capture and storage [6, 7]. Efficiencies of 43 to 45 percent (LHV basis) in supercritical pulverized coal-fired power plants built in the last decade are commonly being achieved [7]. State-of-the-art supercritical plants currently have steam temperatures of 1,112°F (600°C) and reheat temperatures of 1,121 to 1,148°F (605–620°C), which, by some definitions, is considered ultra-supercritical [8]. These plants achieve efficiencies of 43 to 44 percent (LHV basis) for lignite-fired plants and 45 to 46 percent (LHV basis) for bituminous coal-fired plants. These plants have a long history of excellent availability.

In an effort to increase efficiencies further, major development programs are underway to achieve higher ultra-supercritical steam conditions. These programs are being funded by the U.S. Department of Energy's National Energy Technology Laboratory, the Electric Power Research Institute, the Ohio Coal Development Office, the European Commission, and others. Two notable programs are the European Commission's Thermie Project and the Ultra-Supercritical Materials Consortium in the United States.

The power-generating industry is moving toward the use of advanced, ultra-supercritical steam cycles, with steam temperatures approaching 1,400°F (760°C) and operating pressures of 5,000 psia (35 MPa) [9, 10]. Increasing the steam temperature is key, since the plant efficiency increases by about one percentage point for every 20°C rise in superheat and reheat temperature [11]. Figure 7.4 shows a generalized relationship between steam temperature and pressure and power plant thermal efficiency [12]. Figure 7.5 shows the extent of improvement possible, as well as the relative effects of increasing steam temperature and pressure [11]. Figure 7.6 shows efficiency improvements for the various power plant steam cycles along with future efficiency targets for ultra-supercritical plants [13].

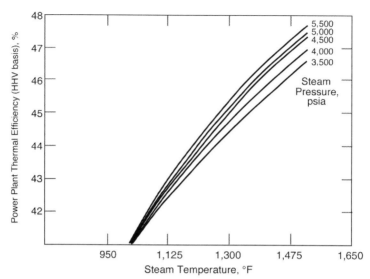

Figure 7.4 The effect of steam temperature and pressure on the thermal efficiency of steam cycle power plants.
Source: Modified from Ruth (1998).

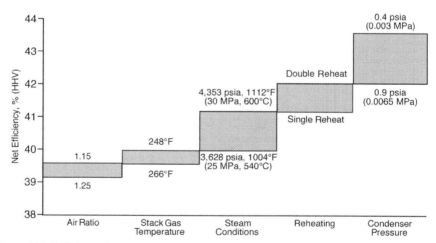

Figure 7.5 Efficiency improvement measures for a pulverized coal-fired plant.
Source: From National Coal Council (2007).

Supercritical pulverized coal-fired power plants are currently routinely chosen in many countries with unit sizes up to 1,000 MW$_e$ [4]. There are approximately 500 units operating, under construction, and planned with supercritical and, to a lesser extent, ultra-supercritical, steam parameters [4]. The mature nature of this technology is such that further development of the combustion process itself will consist

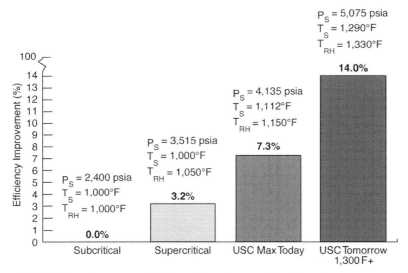

Figure 7.6 Efficiency improvements for various steam cycles. P_S = steam pressure, T_S = steam temperature, T_{RH} = reheat steam temperature.
Source: Modified from Kloth, Davis, and Pickering (2009).

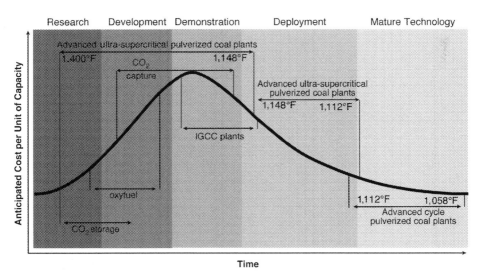

Figure 7.7 New technology development curve for coal.
Source: Modified from National Coal Council (2007).

of refinements rather than major breakthroughs. Emphasis is mainly on improving efficiencies of these facilities. The main area of activity is the further development of steam conditions beyond the current state-of-the-art supercritical and ultra-supercritical technologies in order to raise thermal efficiency even further [4]. Figure 7.7 shows a new technology development curve (often called "the mountain of death") for coal depicting the approximate stage of development/commercialization for the various technologies discussed in this chapter [11].

7.2.2 Advanced Ultra-Supercritical Research and Development

In the development of ultra-supercritical steam power plants, currently several areas are undergoing research and development to achieve higher steam pressures and temperatures. This includes materials development, improved fabrication technology, and steam turbine improvements [8, 14]. One of the major limitations in designing advanced ultra-supercritical plants is the creep damage of the materials used in the boiler and in the steam turbine [9]. Also, the boiler materials of conventional power plants do not possess the requisite high-temperature mechanical properties and corrosion/oxidation resistance to meet the requirements of advanced ultra-supercritical plants [10].

Materials of Construction

At present the materials are based on ferritic/martensitic alloys that permit steam temperatures up to around $1,112°F$ ($600°C$) in state-of-the-art supercritical plants. Although iron-based alloys could be further developed to achieve even higher conditions, it was recognized during the early 1990s that there would be more opportunities for advancing efficiency improvements by using alloys based on nickel [4]. New nickel-based superalloys for long-term operation at steam temperatures in the range of $1,400°F$ ($760°C$) are under development for use in thin-walled outlet headers and steam piping, as well as for castings and forgings of turbine parts [8–10, 14]. In order to reduce the costs that are associated with nickel alloys, further development of existing ferritic-martensitic materials is also necessary. These can be used in areas of lower temperatures (e.g., some furnace walls and turbine stages) to reduce the need for expensive nickel-based superalloys. Corrosion continues to be a problem in high-temperature areas. Hence, development of enhanced coating systems to reduce oxidation, corrosion, and erosion in the high-temperature areas of power plants with higher steam conditions is underway [8, 10].

Improved Fabrication Technology

Improved fabrication technology is another necessary area of research and development for advanced supercritical units. This includes developing fabrication methods for large cast and forged components of nickel-based materials (e.g., turbine rotor and casing). Manufacturing methods for large-diameter pipes composed of nickel-based alloys that transport the steam (under high pressure and temperature) from the last superheater to the turbine inlet need to be developed. Work is underway to develop technologies for welding thick-walled components such as headers and main steam lines. Similarly, work is underway on developing welded joints between materials with different thermal stress behavior—in other words, between superalloys and ferritic steel.

Steam Turbine Improvements

Improvement of the sealing technology for steam turbines is a continuing area of research and development. This becomes more important as the steam pressure is further increased. Also, the development of long-term measurement equipment for static strains in turbine components, as well as sensors for measuring the radial clearance in steam turbines, is necessary.

7.2.3 Oxy-Fuel Firing in Pulverized Coal-Fired Boilers

Oxy-fuel combustion is an emerging approach to postcombustion CO_2 capture. Oxy-fuel combustion is not technically a capture technology but rather a process in which coal combustion occurs in an oxygen-enriched (i.e., nitrogen-depleted) environment, thereby producing a flue gas comprised mainly of CO_2 (up to 89 vol.%) and water. The water is easily separated and the CO_2 is ready for sequestration.

The most common oxy-fuel process involves the combustion of pulverized coal in an atmosphere of nearly pure oxygen (> 95 percent) mixed with recycled flue gas. A schematic of this system is provided in Figure 7.8. Almost pure oxygen for combustion is produced in a cryogenic air separation unit (ASU) and is mixed with recycled flue gas (about 2/3 of the flue gas flow from the boiler) prior to combustion in the boiler to maintain combustion conditions similar to an air-fired configuration. Mixing the recycled flue gas with the pure oxygen is necessary because materials of construction currently available cannot withstand the high temperature resulting from coal combustion in pure oxygen. Research has shown that in order to maintain the oxygen/recirculated flue gas flame so the pulverized coal oxy-combustion flame has heat transfer characteristics similar to that of an air-fired system, an oxygen level of about 30 to 35 vol.% is required in the gas entering the boiler [15].

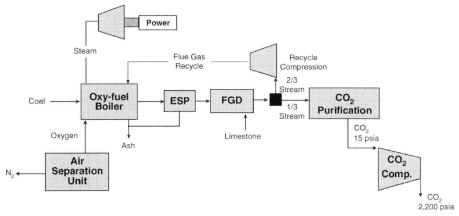

Figure 7.8 A pulverized coal oxy-fuel combustion system.
Source: Modified from Figueroa et al. (2008) [16] and Kather et al. (2008).

Power plants that are retrofitted to oxy-combustion would likely operate within the range of oxygen levels that will maintain combustion conditions near those of air-fired systems. New oxy-combustion plants or repowered plants, however, will not necessarily be constrained, and smaller boiler equipment will be able to be used, which will increase efficiency. The critical parameter of oxygen content may be adjusted to increase radiative heat transfer at the expense of convective heat transfer zones further from the flame. This is in contrast to air-fired systems, where approximately half of the heat transfer occurs in the boiler and the balance of the heat transfer is convective in the backpasses. In both retrofit and new systems, the changes inherent in oxy-combustion affect many parameters, including flame behavior, heat and mass transfer, combustion gas chemistry and behavior, char burnout, and slag development, chemistry, and deposition.

Even though oxy-combustion technology is very promising, with a significant amount of research and development focused on it and serious interest in designing and constructing new power plants with it [18], many technical (and economic) issues must be addressed. These can be categorized into combustion technology, steam generator design, CO_2 purification, and overall process issues [17].

Combustion Technology Issues

The following are some combustion technology issues:

- *Identification of optimum oxygen concentration when firing.* The challenge is to distribute a homogenous mixture of gas and coal to the burner to balance high oxygen levels, and consequently high compressor demands, with low oxygen levels where corrosive atmospheres in the furnace form and incomplete burnout might occur.
- *Combustion behavior of coals in atmospheres comprised of O_2, CO_2, and H_2O.* There are many experimental data published, but more information is needed on oxy-fuel operation of large-scale boilers because not all data are readily scaleable to full-scale systems.
- *Formation mechanisms of pollutants.* The formation of pollutants such as sulfur oxides, NO_x, and CO needs to be examined in more detail. A thorough understanding of the formation mechanisms is necessary to develop cost-effective strategies for the control of these pollutants.
- *Reliable oxygen addition to the recycled flue gas.* Oxygen is not introduced into the system where coal drying, transportation, or milling occurs due to the concern for explosions; consequently, it is mixed with recycled flue gas directly to the boiler. In doing so, the oxygen must be reliably mixed to avoid substoichiometric areas near the boiler walls to ensure good coal burnout and minimize corrosion. Handling of high concentrations of oxygen also requires appropriate safety measures, which are not typical of conventional coal-fired boilers.

Steam Generator Design Issues

The following are some steam generator design issues:

- *Effect of flue gas composition on heat transfer.* Several heat transfer issues must be addressed, especially since radiation is the predominant heat transfer mechanism in the lower furnace. The atmosphere of the oxy-fuel fired boiler chamber is different from that from an air-fired boiler, which leads to different flame characteristics and radiation heat transfer behavior. The heat capacity and density of CO_2 is higher than N_2, which results in a smaller furnace cross section to maintain similar flue gas velocities. CO_2 and H_2O will increase the radiative heat transfer, but more information is needed to determine how significant this is compared to the radiative contribution from coal and fly ash particles.

- *Optimum temperature for flue gas recirculation.* The temperature of the flue gas dictates the cross section and materials of construction of the flue gas ducting, length of flue gas ducting, and power requirements for recirculating the flue gas. Defining the optimum temperature is dependent on the overall plant configuration and is crucial in defining the overall economics.
- *Fouling and corrosion under oxy-fuel environments.* Fouling and corrosion in oxy-fuel combustion environments is expected to be more severe due to the higher loadings of CO_2 and sulfur oxides and, in future designs, higher temperatures. This is a critical issue that needs to be addressed.

CO_2 Purification Issues

The following are some issues related to CO_2 purification by liquefaction:

- *Determining phase equilibria of flue gas compositions as basis for liquefaction equipment design.* Knowing the phase equilibria of flue gas mixtures is important in the development of high-efficiency liquefaction equipment, developing simulation tools, and designing CO_2 transport and storage systems.
- *Influence of kinetic parameters on the concentration of liquid CO_2 and the behavior of contaminants such as sulfur oxides, NO_x, and CO during humidification.* A better understanding of gas-liquid solution kinetics can result in the development of improved liquefaction and flue gas cleaning equipment. Similarly, the fate of contaminants during humidification must be better understood because many will be dissolved during humidification, and knowing their effect on effective CO_2 capture is important.
- *Pumps for transporting liquid CO_2.* The knowledge base for pumping CO_2 must be expanded. Large pumps need to be developed for the transportation of liquid CO_2.

Overall Process Issues

The following are some overall process issues:

- *Minimizing auxiliary power by integrating key elements.* The oxy-fuel process has large energy penalties from the ASU and liquefaction equipment. Efficiency improvements must be realized through process integration in order to make the economics more favorable.
- *Long-term stability of materials under flue gas humidification operation.* Corrosion of materials during flue gas humidification is a major concern, and the development or identification of suitable materials is needed.
- *Development of ASU technology.* The development of highly efficient ASU technology for power plants with capacities greater than 400 MW_e is necessary. The largest current ASUs are able to produce approximately half of the oxygen necessary for a 400 MW_e plant.
- *Developing new pollution control process schemes.* When high-purity CO_2 is required, it is necessary to remove sulfur and nitrogen oxides upstream of the recycle point. This in turn will influence the size of the equipment, power requirements, efficiency, and investment cost. In addition, if a distillation process must be installed for high-purity CO_2, increases in energy demand and capital investment will also increase.

- *Developing a new mill design.* Current mill designs use air for coal drying and transport, which is not acceptable for oxy-fuel systems. Recycled flue gas can be used, but this may require the flue gas to be dried and stripped of sulfur compounds to reduce corrosion in the mill and to achieve proper coal drying, which in turn results in increased process costs. A preferred option is the development of a mill that can use the recycled flue gas without prior removal of the sulfur compounds and water vapor.
- *Control of air in-leakage.* The most important source of impurities in the flue gas stream is from uncontrolled air in-leakage, which is typical for boilers operating at slightly sub-atmospheric pressure. The largest concentration of impurities is nitrogen, which in turn is converted to NO_x. In conventional power plants, air in-leakage is in the range of 4 to 10 percent of the flue gas volume, which is problematic when pure CO_2 streams are the target. It has been reported that a 75 percent reduction in air-leakage should be achievable for an oxy-fuel retrofit and an air in-leakage rate of 2 percent of the flue gas should be feasible for new boilers being designed for oxy-fuel operation [17].

7.3   Fluidized-Bed Combustion

Fluidized-bed combustion (FBC) is a leading technology for the combustion of a range of fuels, fossil and others, because of the many inherent advantages it has over conventional combustion systems, including fuel flexibility, low NO_x emissions, in situ control of SO_2 emissions, excellent heat transfer, high combustion efficiency, and good system availability. In addition, it is recognized as an early CO_2-mitigation technology by cofiring biomass with coal. General FBC operating principles are presented in Chapter 5, specifically atmospheric bubbling fluidized-bed combustion (BFBC) and atmospheric circulating fluidized-bed combustion (CFBC) systems. This section discusses fluidized-bed combustion units as advanced combustion systems, specifically CFB boilers, pressurized FBC technology, and oxy-fuel firing in CFB boilers. As part of this discussion, some fundamental information on FBC technology is presented as well.

7.3.1   Introduction

Circulating fluidized-bed combustion technology has a number of advantages over BFBC technology including [19]:

- Improved combustion efficiency and sulfur retention due to the use of finer particles, turbulent gas-particle mixing, and a high recycle rate
- Smaller bed area due to the use of high fluidizing velocities
- Reduced number of fuel feed points due to the smaller combustor size and turbulent mixing
- Reduced erosion and corrosion of heat transfer tubes because tubes immersed in the fluidized-bed cooler are subjected to significantly lower gas and particle velocities than in a BFBC, and the cooler is subjected to oxidizing conditions, whereas reducing conditions occur near the fuel feed points in the BFBC
- Increased convective heat transfer coefficients

While CFBC technology has several advantages over BFBC technology, there are also some areas of concern and disadvantages:

- Increased height of the CFBC boiler compared to the BFBC boiler
- Additional height and size of the fluidized-bed ash cooler
- Greater pressure drop across the CFBC boiler resulting in increased fan power requirements
- Need for high cyclone efficiencies for bed material recovery for solids recycle
- Higher erosion in the combustor, cyclone, and associated ducting due to the high gas velocities

CFB boilers have evolved into the utility boiler size range, with a number of units as large as 250 to 300 MW$_e$ in operation, and they are poised to enter into the realm of larger once-through, supercritical units [20]. However, several opportunities remain to support further development of supercritical circulating fluidized-bed boilers and the pressurized version. Further development in efficiency improvement, fuel flexibility, effective scale-up, and reducing capital cost are underway [21].

As in atmospheric designs, two types of pressurized fluid beds have been applied to power generation: bubbling and circulating. Bubbling bed designs (PBFBC) have been developed, and a few are already in operation, while circulating designs (PCFBC) are mainly in the pilot and conceptual design stages. An example of a PBFBC boiler is shown in Figure 7.9, which is the Tidd PFBC demonstration, the United States' first PFBC combined-cycle demonstration [22].

Combustion occurs in a large pressure vessel at pressures of 10 to 15 atmospheres and temperatures similar to atmospheric FBC boilers. Because the

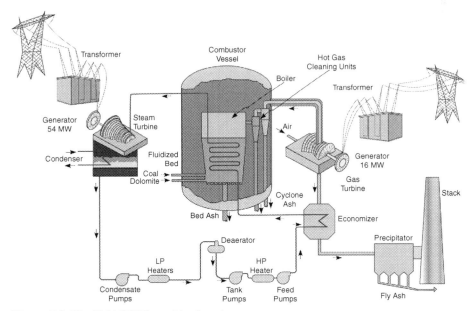

Figure 7.9 The Tidd PFBC combined-cycle process.
Source: From U.S. DOE and American Electric Power Service Corporation.

system is operated under pressure, provisions must be made to feed the fuel and sorbent to the combustor and remove the ash from the combustor across pressure boundaries. Also, primary particulate removal is performed under pressure using high-efficiency cyclones or high-temperature filters. The advantages of PFBC include improved cycle efficiency, reduced emissions, reduced boiler size, improved combustion, and reduced tube erosion [23].

Elevated pressures, and in some cases temperatures, produce a high-pressure gas stream that can drive a gas turbine, and steam generated from the heat in the fluidized bed is sent to a steam turbine, creating a highly efficient combined cycle steam. About 80 percent of the electricity is generated in a conventional steam turbine-generator set, while the balance of the electricity is generated in a gas turbine [21]. Cycle efficiencies of 40 percent are achieved in PBFBC boilers, with efficiencies greater than 50 percent targeted in second-generation PCFBC boilers [24, 25].

The increased pressure of PFBC systems and corresponding air/gas density allows much lower superficial fluidizing air velocities. For a PBFBC boiler, this is approximately 3 ft/sec compared to 10 ft/sec in a BFBC boiler. This reduces erosion of submersed boiler tubes and also permits the use of much deeper beds. The combined effect of lower velocity and deeper bed results in an increased in-bed gas residence time and smoother fluidization, resulting in better gas-solids contact and ultimately better SO_2 capture and improved combustion efficiency. Another advantage of PFBC is reduced boiler size. The high gas density results in a smaller required bed plan area. The lower velocity reduces the total height required for the bed and freeboard [25].

7.3.2 Heat Transfer

In fluidized beds, good mixing is usually achieved, which gives good heat distribution and a uniform temperature distribution. This, in turn, results in effective gas-solids contact, giving a high rate of heat transfer from the burning fuel to the waterwalls or immersed tubes. In conventional furnaces (stoker-fired or pulverized coal-fired), the solids loading in the gas stream is low (i.e., approximately 10 lb per 1,000 lb of gas), and heat transfer from the gas to the waterwalls is primarily by radiation, with a lesser contribution from convective heat transfer [23]. In contrast, in a CFB boiler, the gas leaving the furnace contains a high solids concentration, which can be greater than 5,000 lb per 1,000 lb of gas, and thus convective heat transfer dominates over radiative heat transfer. For equal temperatures, the heat transfer coefficients in an FBC boiler are considerably higher than those in a conventional furnace [23]. However, because the temperatures are lower in an FBC boiler, the overall heat fluxes between the two systems are similar.

What differs between the two systems are the types of heat transfer surfaces employed between the combustion systems. In a conventional system, heat is transferred to the waterwalls (i.e., containment structure) and tubes in the convective pass. In an FBC boiler, three zones of heat transfer must be considered: in-bed, splash zone (interface between the bed and freeboard), and above-bed. In addition, heat transfer areas outside of the furnace must also be considered. In a BFB boiler, heat transfer surfaces include tube banks in the dense bed, waterwalls in the dense bed, and tubes in the convective pass. CFB boilers do not incorporate a tube bank surface in the bed and rely on heat absorption of the containment walls, internal partitions such as division walls and wingwalls, external ash coolers, and tubes in the convective pass.

7.3.3 Combustion Efficiency

Combustion efficiency, defined as the ratio of heat released by the fuel to the heat input by the fuel, is generally high in FBC systems. The combustion efficiency is typically higher than stoker-fired systems and is comparable to pulverized coal-fired systems. It is generally higher in a CFB boiler than in a BFB boiler because of the use of finer particles, more turbulent environment, and a high solids recycle rate [23]. Similarly, PCFBC boilers achieve higher efficiencies due to smaller and more frequent bubbles, which results in better gas-solid contact.

Combustion efficiency is affected by fuel type, bed temperature, gas velocity, and excess air levels. Combustion efficiency increases with fuel volatile matter content and bed temperature. Combustion efficiency decreases with increasing superficial gas velocity. Combustion efficiency initially increases with increasing excess air level and then decreases. This is believed to be due to an increase in CO and hydrocarbon emissions as the excess air level increases to higher levels [23].

7.3.4 Fuel Flexibility

The flexibility in fuel utilization—mainly the ability to use low-quality fuels— makes FBC boilers an attractive technology. These fuels can be fired solely, in combination with other low-grade fuels, or cofired with coal. The lower combustion temperatures permit burning high-fouling and high-slagging fuels at temperatures below their ash fusion temperature. This greatly reduces operating problems associated with these fuels; however, care is still required, since these fuels contain significant concentrations of alkali and alkaline earth metals. In addition, fuels with low heating values, due to high moisture and/or ash contents, or low volatile matter contents can be successfully burned using an FBC boiler because of the large mass of hot bed material and long residence time that the fuel spends in the bed. Examples of these include fuels such as brown coal, peat, and sludge with moisture contents up to 60 percent; waste coals with ash contents up to 76 percent and higher heating values as low as 2,600 Btu/lb; and petroleum coke with volatile matter content less than 10 percent. The fact that many of these low-grade fuels are difficult to reduce to fine size, due to high ash contents or fibrous structures (as in the case of biomass), makes them candidates for FBC technology, since the fuel does not need to be pulverized but can be processed to sizes 0.25 inch × 0.

Although a main advantage of FBC boilers is that they can be designed to burn a wide variety of low-grade fuels, once an FBC boiler has been designed, there are limitations in deviating from the design values so as not to exceed design limits [23]. As of 2005, biomass is the main fuel type used in BFB boilers, with more than 2,000 MW$_e$ of installed capacity [21]. This is followed by peat, various ranks of coal (bituminous coal, subbituminous coal, and lignite), coal wastes, and other wastes, each contributing less than 500 MW_e in installed capacity. In contrast, the main fuel type used in CFBC boilers is bituminous coal ($\approx$10,000 MW_e) followed by lignite (about 4,000 MW_e), with smaller contributions from coal wastes, petroleum coke, biomass, other ranks of coal (anthracite, subbituminous coal, brown coal), peat, wastes, and RDF pellets, where each category contributes 2,000 MW_e or less [21].

Figure 7.10 illustrates the wide range of fuels that have been used/tested in FBCs, but the range of fuels is endless. Candidate fuels include, not inclusive, the various ranks of coal (anthracite, bituminous coal, subbituminous coal, lignite, brown coal), waste coal from coal cleaning operations, petroleum coke, oil shale, refinery bottoms, peat, woody biomass, herbaceous biomass, manure and litter, animal-tissue biomass, tires, paper mill sludge, sewage sludge, refuse-derived fuel, pellet-derived fuel, plastics, industrial wastes, and more.

The fuel characteristics have a minor impact on a CFB utility boiler design compared to that of a pulverized coal-fired boiler. Parameters that must be considered when arriving at a final pulverized coal-fired design include the heat release rate, fuel properties (e.g., ash fusion temperatures, volatile matter content, ash content), percentage of excess air, production of emissions, boiler efficiency, and steam temperature [27], with the most important item to consider being the fuel burned. Furnaces for burning coal are more liberally sized than those for gas or fuel oil firing. This is necessary to complete combustion in the furnace and to prevent formation of fouling or slagging deposits. This is depicted in Figure 7.11, where the impact of coal type on furnace size is shown [28]. By contrast, Figure 7.12 shows the impact of coal type on a CFBC design, which is small [28].

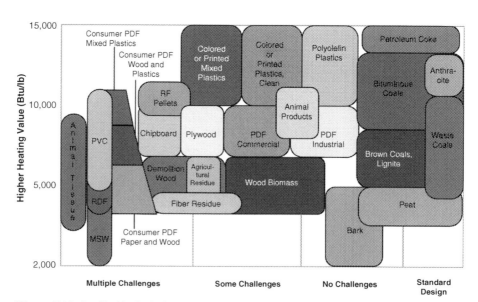

Figure 7.10 Applicable fuels for FBC technology.
Source: Modified from Foster Wheeler (2006) [26].

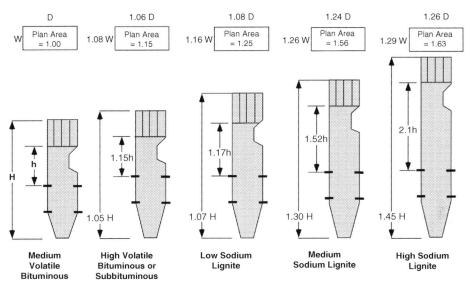

Figure 7.11 Impact of coal type on pulverized coal-fired furnace design.
Source: From Foster Wheeler (2007).

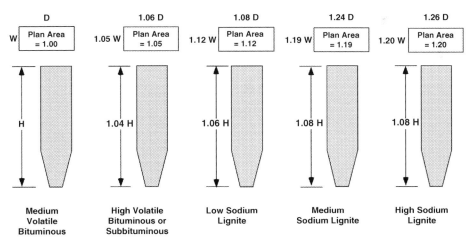

Figure 7.12 Impact of coal type on circulating fluidized-bed furnace design.
Source: From Foster Wheeler (2007).

7.3.5 Pollutant Formation and Control

Fluidized-bed coal combustors have been called the "commercial success story of the last decade in the power-generation business" and are perhaps the most significant advance in coal-fired boiler technology in half a century. Originally, development of the technology was focused on manufacturing a compact, package boiler that could be preassembled at the factory and shipped to a plant site, thereby providing a lower-cost alternative to onsite assembly of conventional boilers. By the mid-1960s, however, it became apparent that the fluidized-bed boiler not only represented a potentially lower cost and more efficient way to burn coal, but it also generated less emissions than conventional boilers. Because the technology can control sulfur dioxide and nitrogen oxides at lower cost than conventional boilers, FBC technology has been developed from the "package boiler" concept to the utility boiler concept. This section discusses pollutant formation and control, with an emphasis on sulfur and nitrogen oxides as it pertains to FBC systems. In Chapter 9, additional pollutant formation and control information are presented.

Sulfur Dioxide

Sulfur in the fuel is oxidized to SO_2 during the combustion process. In an FBC boiler, the SO_2 is captured in situ by adding a sorbent material, which most commonly is limestone but sometimes can be dolomite. Sulfur retention can be greater than 95 percent in an FBC boiler, but sorbent utilization levels are relatively low (e.g., typically only about 40 percent of the calcium is utilized in the capture of sulfur). The effect of operating parameters and sorbent characteristics on sulfur capture are discussed in the following sections.

Transformation of Sorbents in the FBC Process

In an FBC system, limestone and dolomite will undergo thermal decomposition, a process commonly known as calcination. The decomposition of limestone proceeds according to the following equation:

$$CaCO_3 \rightarrow CaO + CO_2 \tag{7.2}$$

Calcination of limestone is an endothermic reaction, which occurs when limestone is heated above 1,400°F (760°C). Calcination is necessary before the limestone can absorb and react with gaseous sulfur dioxide. Calcined limestone is porous due to the loss of carbon dioxide.

Capture of the gaseous sulfur dioxide is accomplished via the following equation to produce a solid product—calcium sulfate:

$$CaO + SO_2 + \frac{1}{2}O_2 \rightarrow CaSO_4 \tag{7.3}$$

The reaction of porous calcium oxide with sulfur dioxide involves a continuous variation in the physical structure of the reacting solid with conversion. One of the major factors responsible for these changes is the formation of calcium sulfate, which has a higher molar volume than calcium oxide (52 cm^3/mole for CaSO$_4$ compared to 17 cm^3/mole for CaO). Because of this expansion in the solid volume from the reactant to the product, the pore network within the reactant will be progressively blocked as conversion increases. As a result, the reaction rate will decay rapidly as soon as a shell of the product layer is formed on the outside of the reacting solid. For pure CaO prepared by the calcination of reagent grade $CaCO_3$, the theoretical maximum conversion of $CaCO_3$ to $CaSO_4$ has been calculated to be 57 percent.

At this conversion, the void spaces in the solid are completely filled by the $CaSO_4$ formed. In practice, the actual conversion obtained using natural limestones is much lower. Calcium utilization as low as 15 to 20 percent has been reported in some cases [29] and is typically 30 to 40 percent. Low utilization efficiency can be explained by early pore blockage making it more difficult for SO_2 to reach the reactive CaO in the particle interior. Sulfur dioxide can usually penetrate only a small distance (50–100 μm) into a particle until pore plugging inhibits further sulfation [30]. Therefore, fine particles can be sulfated more completely than coarse particles, but they also tend to be elutriated from the system before they have had time to become fully sulfated.

Physically, the limestone undergoes structural changes such as porosity, pore volume, and surface area development; changes in pore size distribution; and sintering. The particles also undergo degradation, which is commonly referred to as attrition, with the resultant small particles removed from the system with the offgas. The effect of each of these physical properties on sulfur capture is discussed later in this chapter.

The decomposition of dolomite, $MgCa(CO_3)_2$, which is a double carbonate of magnesium and calcium, is similar in many respects to the decomposition of limestone. The reaction proceeds either in a single step or in stages, depending on the operating conditions and the chemical composition of the sample. For a single-step decomposition, the overall process can be described by the equation

$$MgCa(CO_3)_2 \rightarrow MgO + CaO + CO_2 \tag{7.4}$$

When decomposition occurs in stages, $MgCO_3$ will decompose first. The following reactions will take place in sequence:

$$MgCa(CO_3)_2 \rightarrow MgO + CaCO_3 + CO_2 \tag{7.5}$$

$$CaCO_3 \rightarrow CaO + CO_2 \tag{7.6}$$

MgO will not react with sulfur dioxide in the reaction gas at temperatures above 1,400°F (760°C); therefore, the sulfation reaction of dolomite is basically the reaction of sulfur dioxide with calcium oxide (Eq. 7.3).

In PFBC systems, however, the partial pressure of CO_2 (1,400°F; 760°C) is so high that calcination of limestone does not proceed because of thermodynamic restrictions. For example, at 1,560°F (850°C), calcium carbonate does not calcine if the CO_2 partial pressure exceeds 0.5 atmospheres. In these conditions, the sulfation reaction for limestone is

$$CaCO_3 + SO_2 + \frac{1}{2}O_2 \rightarrow CaSO_4 + CO_2 \tag{7.7}$$

Increasing the pressure from 1 to 5 atmospheres significantly increases the sulfation rate and calcium utilization [31, 32]. Under pressure, dolomite may partially calcine, and sulfation is represented by the following reaction:

$$CaCO_3 \bullet MgCO_3 + SO_2 + 1/2O_2 \rightarrow CaSO_4 \bullet MgO + 2CO_2 \qquad (7.8)$$

Bed Temperature
The effect of bed temperature on sulfur capture is well known. The peak sulfur retention in an atmospheric fluidized-bed combustion boiler is between 1,450°F and 1,650°F (790–900°C), which coincides with the increasing extent of calcinations. Plants with insufficient heat transfer area have exhibited high bed temperature and poor limestone utilization. In addition to the level of the bed temperature, the uniformity of the bed temperature is important because hot and cold locations in the bed must be avoided.

An optimum temperature exists that is sensitive to sorbent type, particle size, calcination conditions, and pressure. This temperature factor is due to an irreversible effect—the generation and loss of pores by sintering [33, 34]—and decomposition of $CaSO_4$ formed during oxidizing conditions [35].

For PFBC boilers, however, there is no pronounced maximum for sulfur retention as a function of temperature. Within the normal operation range, sulfur capture is found to increase slightly with temperature [19]. When dolomite is used as a sorbent, the increase in sulfur retention with an increase in bed temperature can continue to as high as 1,750°F (950°C) [19].

Particle Residence Time
The residence time of the sorbent particles in an FBC boiler is a function of the particle size (and the fluidizing gas velocity). The particle size distribution is a continuum, but a variety of particle sizes are found in the system. Sorbent particle size (and thus residence time) is important because the rate of sulfation is proportional to the particle surface area, which is inversely proportional to the particle diameter. In other words, the smaller the particle, the faster the rate of sulfation. Unfortunately, the smaller particles, which may have a fast sulfation rate, have the shortest residence time in the combustor (≈ 2 seconds, similar to that of the gas residence time). The larger particles may have slower rates of sulfur capture, but they are contained in the system for a significantly longer period of time.

Bed Quality
Bed quality, which includes limestone distribution, mixing, and fluidization, is important in capturing sulfur in the bed. Obviously, the better the contact between the limestone, sulfur dioxide, and oxygen, the greater the quantity of sulfur that can be captured. Also, increasing the fluidizing gas velocity decreases the residence time of the gas in the bed. This, in turn, reduces the effective contact between SO_2 and the sorbent and, consequently, reduces sulfur capture.

Gaseous Environment

Another important parameter is the gaseous atmosphere in the combustor. Oxygen is necessary for the sulfation reaction. FBC boilers normally operate with about 4 percent. vol O_2 in the bed, which is far in excess of the SO_2 concentration. Under this condition it is commonly assumed that the sulfation reaction has zero dependency on O_2 concentration. Oxygen can indirectly reduce the rate of CaO sintering, however. Sintering is a complex process that results in a reduction in the total surface area and thus a change in the pore size. Carbon dioxide enhances sintering, which can be detrimental to sulfur capture by causing CaO particle shrinkage and densification (which occurs during the later stages of sintering). Excess O_2 will reduce the CO_2 partial pressure and thus impede sintering.

The initial rate of sulfation is approximately proportional to the SO_2 concentration, but the reaction is terminated by pore blockage occurring more quickly as the SO_2 concentration is raised. In addition, it appears that the ultimate extent of sulfation attainable is slightly dependent on the amount of SO_2 present in the bulk gas.

Combustor Pressure

Combustor pressure also affects sulfur capture in an FBC boiler. No peak in the sulfur retention versus temperature curve is observed at high pressure (≈ 12 atmospheres). Limestone performance is reduced, while dolomite performance improves. Dolomite is the preferred sorbent in pressurized fluidized-bed combustors.

Chemical Composition

Sorbent performance in FBC boilers has been well documented in literature. It is generally accepted that sorbent performance has no relationship to chemical composition [19, 29, 30, 36, 37]. A high-purity limestone (>95 percent $CaCO_3$) will not necessarily perform better than a low-purity limestone (<70 percent $CaCO_3$).

Porosity

The total porosity, a dimensionless quantity, is defined as the ratio of the sum of the pore volume to the total volume. It is an important physical property because it provides insight into whether the solid has sufficient interior vacancy in which the sulfation reaction can occur. Porosities of raw stones generally vary between 0.3 and 12.0 percent and that of calcines from 20 to 50 percent, which is less than the theoretical porosity of 54 percent for CaO derived from carbonates [38].

The porosity of a stone increases upon heating because of the release of CO_2. The rate of porosity increase depends on the calcination conditions. Since this porosity increase is so condition-specific, it is impossible to predict a priori. However, porosity increases for different stones between a factor of 5 and 180 for the same calcination conditions have been reported [39]. This increase in porosity can be quickly negated by four processes: sulfation, thermal sintering (more than 1,470°F; less than 950°C), moisture-activated sintering, which is believed to be most prevalent at temperatures around 1,110°F (600°C), and CO_2-activated sintering, which is most intense at approximately 1,650°F (900°C) [40].

The importance of using an initially porous stone can be seen from the following equation [41]:

$$x_s = \varepsilon_o / (z - 1)(1 - \varepsilon_o) \tag{7.9}$$

where x_s = the maximum conversion at the surface of CaO to $CaSO_4$ before pore closure occurs; ε_o = the initial porosity of the calcine; and z = molar volume of $CaSO_4$/molar volume of CaO.

Using 0.53 for the initial porosity of CaO, x_s is maximized at 56 percent utilization. However, stone utilizations typically range between 30 and 40 percent. The 16 to 26 percent difference is partly because the porosity of the stone is quickly reduced by the formation of $CaSO_4$, and only a 30 to 40 mol% fraction of the outermost region of the stone reacts with the gaseous SO_2. This phenomena is often referred to as pore mouth plugging and is the dominant cause for only partial sulfation of sorbent particles in the size range between 18 and 140 mesh (1,000 and 105 μm) [40]. There are four main ways to increase stone utilization via optimizing calcine structure:

- Calcining with high CO_2 partial pressures to produce a wide-mouth pore size distribution
- Using slow calcination rates
- Calcining at temperatures between 1,380°F and 1,470°F (750–950°C)
- Using impure stones

Particle porosity is believed to be the controlling parameter for larger particles. Porosity accounts for larger product molar volumes, which in turn can result in larger sorbent utilizations at long residence times (i.e. when pore plugging becomes rate limiting) [42]. The pore size distribution plays only a minor role in controlling sulfation, whereas increasing the sorbent porosity appears to be the single most promising method to enhance sulfation, especially if the stone is initially 15 percent porous [43]. However, is has been shown that large-mouth pores greatly enhance stone utilizations and identified pore volume distribution and grain size as the determining factors for stone utilization [44].

Surface Area

Surface area is related to porosity. As with porosity, the surface area of the resulting calcine increases substantially upon heating. The surface area development is controlled by the particle size and calcining conditions: temperature, heating rate, and environment. Surface areas as large as 90 m^2/g have been obtained by using low temperatures (1,110–1,650°F; 600–900°C), high-heating rates, small particles, and CO_2-free environments.

As with porosity, internal surface area is the limiting parameter for smaller particles, smaller product molar volumes, lower conversions, and short residence times (i.e., when kinetics may be rate determining) [42].

Like chemical composition and porosity, there is no experimental correlation between surface area and stone performance. However, for an irreversible first-order gas-solid reaction under chemical control, the time required to reach a given conversion is inversely proportional to the active surface area and inversely proportional to the square of the surface under product layer diffusion control. For fluidized-bed applications, the latter relation is of most interest [45].

Particle Size

Conversion of CaO to CaSO$_4$ increases as particle size decreases over the entire particle size range of 1 to 1,000 μm. From a fundamental viewpoint, reducing the particle size is beneficial because less CaO will be inaccessible (on a weight basis). The ideal particle size distribution for an FBC power plant is site specific. That is, it depends on the manufacturer and specifications of the combustor and cyclone, the economics of producing and feeding fines, the ash content of the fuel, and the propensity for the stone to attrit. These are only a few of the many considerations that must be included in the evaluation process. Laboratory results indicate that reducing the particle size will be beneficial in capturing more sulfur per unit mass of stone. However, in the field, operating conditions and economics will dictate the specifications of the particle size distribution.

Nitrogen Oxides

The nitrogen oxides (mainly NO, NO$_2$, and N$_2$O, but referred to as NO$_x$) that are formed in an FBC derive from two sources: oxidation of fuel nitrogen, which is referred to as fuel NO$_x$, and reactions of oxygen and nitrogen in the air, which is referred to as thermal NO$_x$. Since thermal NO$_x$ is a product of a high-temperature process, mainly above 2,700°F (1,480°C), it makes only a minor contribution to the overall NO$_x$ emissions, and fuel NO$_x$ is the primary contributor. Typically, more than 90 percent of the NO$_x$ is in the form of NO, approximately 20 to 300 ppm as N$_2$O, with the balance NO$_2$ [19]. A discussion of NO$_x$ formation mechanisms, the effect of fuel characteristics and operating parameters on NO$_x$ formation, and NO$_x$ reduction techniques is presented in the following sections.

NO$_x$ Formation

The formation of NO$_x$ compounds is complex. Upon heating the fuel, the organically bound nitrogen volatilizes, forming volatile-nitrogen and char-nitrogen. The volatile-nitrogen compounds are primarily released as HCN and NH$_3$. In the char, the nitrogen is bound in aromatic structures [19].

HCN and NH$_3$ undergo the following reactions during combustion [19]:

$$HCN + \frac{5}{4}O_2 \rightarrow NO + CO + \frac{1}{2}H_2O \tag{7.10}$$

$$HCN + \frac{3}{2}O_2 + NO \rightarrow N_2O + CO + \frac{1}{2}H_2O \tag{7.11}$$

$$HCN + \frac{3}{4}O_2 \rightarrow \frac{1}{2}N_2 + CO_2 + \frac{1}{2}H_2O \tag{7.12}$$

$$NH_3 + \frac{5}{4}O_2 \rightarrow NO + \frac{3}{2}H_2O \tag{7.13}$$

$$NH_3 + \frac{3}{4}O_2 \rightarrow \frac{1}{2}N_2 + \frac{3}{2}H_2O \tag{7.14}$$

During char oxidation, the nitrogen is mainly oxidized to NO and N_2O. These are, in turn, partially reduced to N_2. NO is reduced through reactions with NH_3, carbon in the char, and CO. N_2O can be reduced by temperature effects or reactions with char and CO. NO and N_2O reduction reactions are [19]

$$NO + NH_3 + \frac{1}{4}O_2 \rightarrow N_2 + \frac{3}{2}H_2O \tag{7.15}$$

$$NO + \frac{2}{3}NH_3 \rightarrow \frac{5}{6}N_2 + H_2O \tag{7.16}$$

$$NO + C(char) \rightarrow \frac{1}{2}N_2 + CO \tag{7.17}$$

$$N_2O \rightarrow N_2 + \frac{1}{2}O_2 \tag{7.18}$$

$$N_2O + C(char) \rightarrow N_2 + CO \tag{7.19}$$

$$N_2O + CO \rightarrow N_2 + CO_2 \tag{7.20}$$

Fuel Nitrogen and Volatile Matter Content, and Fuel Rank

In general, NO and N_2O emissions increase with increasing nitrogen content in the fuel, whereas with increasing fuel volatile matter content, NO emissions usually increase, but N_2O emissions decrease. N_2O emissions are primarily dependent on the type of fuel (as well as combustion temperature, as discussed later) with low-rank fuels generating the lower amounts of N_2O than higher-rank fuels—that is, N_2O emissions from lowest to highest—biomass, peat, oil shale, brown coal/lignite, bituminous coal [46, 47]. This is attributed to the amount of NH_3 released from the fuel as lower-rank fuels (i.e., higher-volatile-matter fuels) release more NH_3, and their NH_3 to HCN ratio is greater than for higher-rank fuels. Oxidation of NH_3 results in NO formation, while HCN produces both NO and N_2O.

Combustion Temperature

Combustion temperature has a significant effect on both NO and N_2O emissions [46, 47]. NO emissions increase, but N_2O emissions decrease. Higher temperatures promote the oxidation of nitrogen radicals to NO and reduce the char and CO concentrations, which in turn decreases the reduction of NO to N_2 on the char surface. The reduction reactions of N_2O with hydrogen radicals (H, OH) are significantly enhanced via the following reactions [19]:

$$N_2O + H \rightarrow N_2 + OH \tag{7.21}$$

$$N_2O + OH \rightarrow N_2 + HO_2 \tag{7.22}$$

Excess Air

An increase in excess air leads to increases in NO and N_2O emissions. As excess air is increased, the combustion rate increases. However, excess air has a minor effect on N_2O emissions compared to fuel type and combustion temperature [47]. This is especially true in pressurized units [19].

Gas Velocity/Residence Time

The superficial gas velocity—that is, the residence time in the bed—has an effect, although minor, on N_2O emissions but not on NO emissions. With increasing gas velocity, the contact time between the gas and particles decreases, thereby resulting in higher N_2O emissions because there is less time for the carbon-N_2O reduction reaction to occur [47].

Limestone Effects

The effect of limestone addition on NO_x emissions is complex. In a BFBC boiler, the limestone is primarily in the dense phase of the bed where CO concentration is high and O_2 concentration is low, and the catalyzed reduction of NO by CO may dominate over oxidation of NH_3, thereby leading to lower NO emissions [19]. In a CFBC boiler, however, the limestone is distributed through the entire combustor, and oxidation of volatiles to NO may occur in the upper zone where CO concentrations are lower, thereby increasing NO emissions. The addition of limestone is usually found to decrease N_2O emissions, although the effect is minor [19, 47]. The decrease is attributed to the limestone catalyzing the decomposition of N_2O.

NO_x Reduction Techniques

The FBC process inherently produces lower NO_x emissions due to its lower operating temperature. However, where necessary, additional combustion modifications or flue gas treatment for NO_x control can also be employed. Techniques currently used for FBC systems include reducing the peak temperature by flue gas recirculation, natural gas reburning, overfire air/air staging, fuel reburning, low excess air, and reduced air preheat. Postcombustion control is also used, including selective noncatalytic reduction (SNCR) and selective catalytic reduction (SCR), which achieve 35 to 90 percent NO_x reductions. Air staging, SNCR, and SCR will be briefly discussed here but are presented in more detail in Chapter 9.

Air staging can reduce both NO and N_2O emissions from FBC boilers. This is accomplished by introducing less than the theoretical amount of combustion air through the distributor plate and adding the remainder of the combustion air above the dense bed or, in the case of CFBC boilers, injecting some of the secondary air into the cyclone. As a result, some of the fuel nitrogen compounds decompose into molecular nitrogen rather than forming NO_x due to the reducing atmosphere in the bed. Char and CO concentrations in the bed increase, thereby enhancing the rates of NO and N_2O reduction on the char.

SNCR is a technology that involves injecting nitrogen-containing chemicals into the FBC boiler within a specific temperature window without the expensive use of catalysts. The chemicals, with the two most common being ammonia and urea, selectively react with NO in the presence of oxygen to form molecular nitrogen and water. The main reactions when using ammonia or urea are, respectively,

$$4NO + 4NH_3 + O_2 \rightarrow 4N_2 + 6H_2O \qquad (7.23)$$

$$4NO + 2CO(NH_2)_2 + O_2 \rightarrow 4N_2 + 2CO_2 + 4H_2O \qquad (7.24)$$

The optimum temperature window for ammonia is 1,560 to 1,920°F (850–1,050°C) and for urea is 1,830 to 2,100°F (1,000–1,150°C). While both have been used, ammonia is favored due to its optimum temperature window and that fact that ammonia produces less N_2O than urea [47]. SNCR operation is very effective for reducing NO, but it does increase N_2O emissions. Fortunately, the overall concentration of N_2O is much less than NO. Where it is necessary to reduce N_2O emissions even further, a combination SNCR/SCR can be employed for overall NO_x control [47].

Particulate Matter

Particulate matter is generated from two sources: the fuel and the sorbent. The mineral matter in the fuel is released during combustion, and this residue, along with calcined sorbent and reacted sorbent, is referred to as ash. In addition to the ash, there can be unburned carbon in the residue, although this amount should be minimal provided the system is properly operating. Some of the ash remains in the fluidized bed and is discharged by the bed material drain system. This ash is normally larger than 105 μm and is relatively easy to handle and transport [23]. The remaining ash is fine and leaves the boiler in the flue gas. This material is typically less than 44 μm and requires a high-efficiency collection device. Normally this is a fabric filter but can also be an electrostatic precipitator (see Chapter 9 for a discussion of particle removal devices). Ash from an FBC system contains a significant amount of $CaSO_4$, CaO, and $CaCO_3$ and is more alkaline that that from conventional combustion systems.

Carbon Monoxide/Hydrocarbons

Carbon monoxide is the product of incomplete combustion of carbon. It is formed when the oxygen supplied is less than the amount required for stoichiometric combustion of carbon to CO_2, when there is inadequate fuel/air mixing or insufficient residence time for combustion. Fuel reactivity and bed temperature also influence CO emissions. Hydrocarbon emissions are produced under similar conditions. Typical flue gas concentrations are less than 200 ppm for CO and 20 ppm for hydrocarbons in a CFB boiler burning coal [23]. Overfire air—air staging—is a common technique used for controlling CO and hydrocarbon emissions.

Trace Elements

All solid fuels contain small concentrations of trace elements, usually measured in parts per million. Trace elements enter the atmosphere through natural processes, and sources of trace elements include soil, seawater, and volcanic eruptions. Human activities such as power generation and combustion of fuels in the industrial and commercial sectors also lead to emissions of some elements.

Title III of the U.S. Clean Air Act Amendments of 1990 designates 188 hazardous air pollutants. Eleven trace elements are in the list: antimony (Sb), arsenic (As), beryllium (Be), cadmium (Cd), chromium (Cr), cobalt (Co), lead (Pb), manganese (Mn), mercury (Hg), nickel (Ni), and selenium (Se). In addition, barium (Ba) is regulated by the Resources Conservation and Recovery Act, and boron (B) and molybdenum (Mo) are regulated by Irrigation Water Standards. Vanadium (V) is regulated based on its oxidation state, and vanadium pentoxide (V_2O_5) is a highly toxic regulated compound. Other elements, such as fluorine (F) and chlorine (Cl), which produced acid gases (i.e., HF and HCl) on combustion, and radionuclides, such as radon (Rn), thorium (Th), and uranium (U), are also of interest.

The distribution of trace elements in the bottom ash, ash collected in the particulate control devices, and fly ash and gaseous constituents emitted into the atmosphere depend on many factors, including the volatility of the elements, temperature profiles across the system, pollution control devices, and operating conditions [48, 49]. Numerous studies have shown that trace elements can be classified into three broad categories based on their partitioning during coal combustion. A summary of these studies is presented by Clarke and Sloss [49], and Figure 7.13 illustrates the classification scheme for selected elements [50].

Class I elements are the least volatile and are concentrated in the coarse residues (i.e., bottom ash) or are equally divided between coarse residues and finer particles (i.e., fly ash). Class II elements will volatilize in the boiler but condense

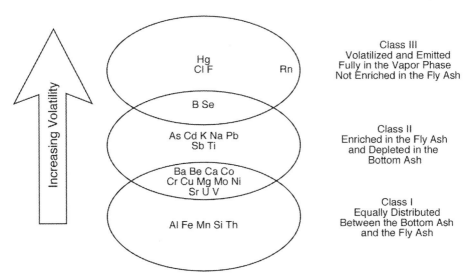

Figure 7.13 Classification scheme for selected trace elements relative to their volatility and partitioning in power plants.
Source: Adapted from Clarke and Sloss (1992) and Miller et al. (1996).

downstream and are concentrated in the finer-sized particles. Class III elements are the most volatile and exist entirely in the vapor phase. Overlap between the classifications exists and is a function of fuel, combustion system design, and operating conditions, especially temperature [49].

The operating temperature in an FBC boiler is lower than that from a conventional system (i.e., pulverized coal or stoker-fired), which may lead to reduced volatilization of some elements. However, this reduction may be offset to some degree by the longer residence times at a relatively high temperature in the FBC boiler, allowing more volatilization to occur [49].

Mercury, which has become an element of much interest due to its toxicity and recent regulation for coal-fired boilers [51], has been shown to be captured by the limestone that is used for sulfur control in CFBC boilers. The mercury is captured by the fine limestone particles and removed by the particulate control device. According to Hughes and Littlejohn [52], most trace elements associated with the particulates and emissions of the trace elements from the system depend primarily upon the efficiency of the particulate control device. Conventional baghouse filters have an overall removal efficiency of greater than 99 percent. Once a dust cake is formed, the efficiency for removal of even the smaller particles (down to 0.1 μm) can approach 100 percent [49, 53]. The following discussion is evidence of the low mercury emissions rate from CFBC boilers, where the mercury is associated with the limestone particles captured by bagfilters.

In 1999, the U.S. Environmental Protection Agency (EPA) approved an Information Collection Request (ICR) to study mercury emissions from power plants to provide some framework for comparison as a basis for implementing mercury control legislation. Part III of the ICR was to determine speciated mercury emissions from stationary sources. This included power plants of various sizes, systems, and firing configurations throughout the United States. The details of the ICR can be found at the U.S. EPA website [54]. Three CFBC units were included in the ICR: Stockton CoGen Plant, Stockton, California; PG&E Scrubgrass Generating Station, Unit 1, Kennerdell, Pennsylvania; and Tractebel Power Inc., Kline Township Cogeneration Plant Unit 1, New York.

The Stockton plant is a CFB boiler that fired bituminous coal and fluid coke, with mercury concentrations of 26 to 29 ppb and 12 to 45 ppb, respectively, at 55,000 lb/h during the test. Ammonia and limestone were also fed into the system for emissions control. The emissions tests showed no detection of mercury at the baghouse outlet. Detection limits were equivalent to 0.1 mg/dscm (milligram per dry, standard cubic meter), which translates to 0.00005 to 0.00008 lb/h. Mercury was detected in the particulate samples at the baghouse inlet; however, no oxidized or elemental mercury (gas phase) were detected.

The Scrubgrass plant fired waste bituminous coal in a CFB boiler with limestone injection. The fuel was fed at a rate of 75,000 lb/h (3.95 lb Hg/h based on 527 ppb (parts per billion) Hg in the waste coal). Emissions tests showed that 99.8 percent of the mercury was captured with the ash in the baghouse. This corresponds to emissions of 1.39 lb Hg/year. Capture efficiency across the baghouse resulted in a reduction of 99.99, 85.88, and 19.00 percent, respectively, in particle bound mercury, oxidized mercury, and elemental mercury.

Similarly, Tractebel's plant fired anthracite cleaning wastes (i.e., culm) in a CFB boiler. Test results showed that 99.80 percent of the mercury was captured with the ash in the baghouse. This amounts to approximately 0.53 lb Hg release per year. The average mercury in the fuel during the test was 330 ppb; however, the plant reported that the mercury concentration is normally about 170 ppb, which correlates to emissions of 0.25 to 0.30 lb per year. The baghouse removal efficiency of the particulate bound mercury was 99.995. Oxidized and elemental mercury removal through the baghouse was 43.69 and 91.16 percent, respectively.

Overall, certain trace elements are captured by the bed material, which is dependent on the type of element, bed temperature, and fuel characteristics. The remaining elements will exit the combustor by the flue gases and, as the temperature decreases, condense onto cooler surfaces and fly ash particles, which will be collected by the particulate collection devices. The highly volatile elements will remain in the flue gas or condense on the submicron particles that escape the particulate removal device.

Ash Chemistry and Agglomeration Issues

An important issue that must be assessed in an FBC system is the behavior of the inorganic elements toward bed agglomeration. As previously discussed, the lower bed temperatures are advantageous in that they can be kept below the ash-softening temperatures, especially when firing coals. However, because an FBC boiler can fire a wide range of fuels with varying ash chemistries, the interactions of the ash with the bed material can lead to agglomeration, which has a detrimental effect on fluidization. This is especially true with biomass materials, where care must be taken in the design of the system (e.g., incorporation of kaolin clay injection systems for agglomeration control) or selection of the feedstocks.

It has long been recognized that the mode of occurrence of inorganic elements in fossil fuels has a direct bearing on their behavior during combustion [55–58]. The occurrence of inorganic elements in bio-based fuels is also important, especially as cofiring coal and biomass is becoming increasingly more popular [59–64]. Inorganic species can occur as ion-exchangeable cations, as coordination complexes, and as discrete minerals. In the case of firing a single fuel, such as coal, it is possible to predict ash behavior to avoid system problems. However, it becomes more complex to predict ash behavior in the case of firing multiple fuels in proportions that vary with time—for example, seasonal changes with biomass, and are extremely heterogeneous.

Low-rank coals and biomass materials often contain significant amounts of alkali metals (potassium and sodium) and alkaline earth metals (calcium and magnesium), which are rapidly released into the gas phase and interact with other elements resulting in problems with fouling, slagging and corrosion. In general, potassium and sodium that are associated with the organic structure of the fuel tend to be problematic in that they can contribute to the formation of inorganic phases that have lower melting points. Studies conducted on ash formation during coal combustion show that the incorporation of moderate amounts of alkali and alkaline earth elements into silicates enhances the coalescence and agglomeration of inorganics due to formation of "sticky" molten phases [56, 57, 65–67].

The presence of low-melting-point phases in a fluidized-bed combustor results in the formation of clinkers that can compromise the bed fluidity. It is also important to recognize that the blending of biomass feedstocks and coal does not necessarily result in simply an additive effect of problematic elements. Changes in the feed blend may or may not have devastating effects on system operation. Predicting these effects is based on understanding the manner in which the inorganics in fuels interact during combustion and their effect on the chemical and physical properties of the ash and gas phases in the system. Their behavior can be predicted from detailed fuel analysis, such as chemical fractionation, to assess the mobility of the inorganic elements and models to predict sintering potential and viscosity behavior of ash produced during combustion.

7.3.6 Supercritical Fluidized-Bed Boilers and Oxy-Coal Firing in Fluidized-Bed Boilers

Power plant efficiency improvements through the use of advanced steam cycles are as big a concern to the FBC boiler manufacturers as to the pulverized coal-fired boiler vendors. Supercritical boiler technology is being implemented on CFB boilers by the major boiler manufacturers to realize efficiency improvements that lead to CO_2 mitigation, savings in fuel costs, reductions in power plant auxiliary energy consumption, and reducing heat losses of the boiler [68].

Similarly, there is growing interest in oxy-fuel combustion in circulating fluidized-bed (CFB) boilers to build upon their inherent advantages over conventional combustion systems discussed in the previous sections [68–70]. Many of the challenges for pulverized coal oxy-fuel boilers are valid for CFB boilers. To date, all oxy-fuel demonstrations (not including pilot-scale test units) use pulverized coal-fired boilers. Projects are underway to demonstrate oxy-fuel firing in CFBs [70].

7.4 Integrated Gasification Combined Cycle

Integrated gasification combined cycle (IGCC) is the most advanced precombustion carbon capture technology available [3, 6–8, 65]. IGCC is the integration of two different technologies: coal gasification from the chemical industry and combined-cycle power generation from the power industry. The characteristics of the IGCC technology make it attractive for CO_2 capture because syngas is generated in the gasifier at elevated pressures and is then chemically converted to enrich the CO_2 concentration, which are two of the driving forces for CO_2 separation. In addition, IGCC technology is able to take advantage of the efficient high-temperature gas turbines that are available. As a result, IGCC can offer high efficiencies, typically in the mid-40s (LHV basis), thereby producing less CO_2 per unit of fuel than lower efficiency systems. In addition to the carbon capture capability, IGCC has the advantages of using less water, about 33 percent, than a similar-size pulverized coal-fired plant and produce a usable by-product (i.e., a glassy slag used in the manufacture of cement or roofing shingles, or as asphalt filler or aggregate) [71].

7.4.1 Introduction

The basic layout of an IGCC plant is shown in Figure 7.14, which is a simplified flowsheet with precombustion CO_2 capture and cold-gas cleanup (modified from [6, 8]). Within this framework, however, there is considerable variation among systems, as is demonstrated by five existing coal-based IGCCs (as of 2009) of the 250 to 350 MW_e class listed in Table 7.2 and discussed later in this chapter. Note that there are 17 IGCC units in operation worldwide, but only five are coal-based [8]. The purpose of this section is to review the choices available within the basic structure and discuss the main systems in the IGCC plant, which are further summarized following. The IGCC process is often divided into blocks or islands, but the definitions of these blocks vary among manufacturers.

The first variation to be considered is the extent of air-side integration, which can range from 0 percent, as in Polk or Wabash, to 100 percent, as in Buggenum or Puertollano, where the degree of integration is defined as the percentage of air supplied to the Air Separation Unit (ASU) by extraction from the gas turbine. This difference in the first generation (1990s) of IGCCs can largely be attributed to the gas turbines available at the time. The 100 percent integration is clearly disadvantageous in today's economic environment, since a long start-up period using an

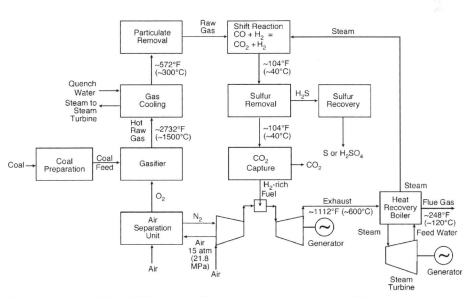

Figure 7.14 Simplified IGCC process flowsheet with precombustion CO_2 capture and coal gas cleanup.
Source: Modified from Henderson (2003) and Kather et al. (2008).

Table 7.2 Operating Coal-Based IGCCs Worldwide as of 2009

Facility	Company	Location	Feedstock	Gasifier Technology
Wilem-Alexander Centrale	Nuon	Buggeneum, Netherlands	Coal/biomass	Shell
Wabash River	SG Solutions/Duke Energy Indiana	W. Terre Haute, Indiana	Coal/coke	ConocoPhilips
Polk Power Station	Tampa Electric	Mulberry, Florida	Coal/coke	GE Energy
Puertollano	ELCOGAS	Puertollano, Spain	Coal/coke	Prenflo
Clean Coal Power R&D Co.	Consortium of Japanese utilities, MTI, CRIEPI	Nakoso, Japan	Coal	Mitsubishi Heavy Industries (MHI)

Source: From Neville (2009).

expensive backup fuel is required. On the other hand, zero air extraction from the gas turbine air compressor does not allow optimum use of the machine over a full range of ambient conditions. The optimum degree of integration is dependent on many factors, including the ambient temperature range and the turbine selection. A typical figure today might be around 30 percent.

The quality of the oxygen is another variable. Over the range 85 percent (Puertollano) to 95 percent (most other plants), the optimum curve for energy consumption is fairly flat. Additional energy is required in the ASU to raise the quality further. A purity of 99.5 percent O_2 is typically not attractive for a straight IGCC application, but for chemical applications it is generally the standard. Where production of power with one or more chemicals (often known as polygeneration) is to be considered, one would need to review the oxygen purity specification on an individual case. If the coproduct were to be ammonia for instance, 95 percent O_2 would be acceptable.

7.4.2 Gasification Island

Variations in the gasification island are almost entirely dependent on the choice of technology supplier. Important differences are discussed in this section. In feedstock preparation, rod mills are typically used for slurry preparation (e.g., GE Energy or ConocoPhillips) or roller mills and drying are used in dry-feed systems (e.g., Shell or Siemens). In both cases, the particle size is of the order of magnitude of less than 100 μm. In the case of fluid-bed processes, the particle size is much larger: about 6 mm.

More specifically, in a dry-feed system, the coal is ground and dried in a roller mill with a hot gas drying circuit, similar to those used in conventional pulverized coal units. The pulverized coal is then fed through a lock hopper system into the pressurized feed vessel. The coal is then transported to the burners from the feed vessel by pneumatic conveying in the dense phase. The carrier gas is typically pure nitrogen from the ASU. The lock hopper system, which relies on gravity flow to move the coal from the uppermost, atmospheric bunker through the lock hopper into the feed vessel requires a support structure that can be as tall as the gasifier itself, so there is great interest in the development of a "solids pump" that could reduce the cost of the feed system.

For wet-feed systems, the slurry is made in a rod mill into into which precrushed coal and water are fed. The coal is ground in a wet milling process to a size of about 100 μm. The product slurry is sieved to remove oversize material, which can be discarded or recycled according to need. The slurry is pumped to the reactor pressure typically with a membrane piston pump.

For an entrained-flow gasifier, the flow direction can be downflow (e.g., GE Energy or Siemens) or upflow (e.g., ConocoPhillips or MHI). The temperature containment can be with refractory (e.g., GE Energy or ConocoPhillips) or using a membrane wall (Shell or MHI). ConocoPhillips and MHI use a two-stage gasifier. The other technology suppliers use single-stage gasification. The oxidant can be oxygen (e.g., Shell ConocoPhilips, Prenflo, or GE Energy) or air (e.g., MHI).

Syngas cooling is available in a number of variations. Water quench (GEE or Siemens) is not currently used in the coal-based IGCC configuration, primarily because of the associated efficiency penalty, whereas it is used in chemical applications, particularly where CO shift for hydrogen manufacture is involved (Kingsport, Tennessee; Coffeyville, Kansas). It is also used in a number of refinery-based IGCC units. Should CO_2 capture be implemented, then this cooling technique would probably be favored also for coal-based IGCC applications.

Of the different steam-raising configurations, radiant cooling is only offered by GE Energy (e.g., Polk). ConocoPhillips uses a firetube convection cooler after the second stage of its E-Gas gasifier. GE Energy has also used a firetube convection cooler as a second cooling stage in Polk, but it has been deleted in current designs. Shell uses a gas quench and watertube syngas cooling.

A certain amount of syngas pretreatment is generally included in the scope of the gasification technology supplier. This includes removal of particulate matter and a number of trace components in the gas, particularly ammonia and chlorides. Shell and ConocoPhillips remove the particulates and the water-soluble gases in separate stages, using a candle filter (sinter metal for ConocoPhillips and ceramic for Shell) for particulate removal, and a water wash for ammonia and chlorides. GEE combines these steps in a single scrubber. Slag removal from the pressurized gasifier is achieved using a lock hopper arrangement in most processes. ConocoPhillips has a proprietary continuous letdown system.

7.4.3   Gas Treatment and Sulfur Recovery

Although sulfur species (primarily H_2S) are the principal targets of the gas treatment system, it is necessary to consider the full range of potential contaminants, which include COS (a minor sulfur species) and mercury. Depending on the selection of desulfurization technology, COS will probably need to be hydrolyzed to H_2S to achieve the required level of sulfur removal. Typical temperatures for COS hydrolysis are between 320°F and 390°F (160°C and 200°C).

Mercury removal is best performed at ambient temperatures upstream of the acid gas removal, so some of the gas treatment will need to be integrated with the low-temperature gas cooling. Mercury removal from syngas has only been practiced industrially at Kingsport, although it is a regular feature of natural gas pretreatment in LNG plants.

There is an extremely wide variety of acid gas removal (AGR) systems on the market. These can be classified as chemical washes (which include all amines such as methyldiethanolamine (MDEA)) and physical washes such as Selexol or Rectisol. In addition, it is possible to have a mixed characteristic solvent such as Sulfinol. All of these named processes have been used in IGCC or chemical plant gasification operations. Selection is based on requirements for high purity (Rectisol) versus low cost (MDEA), with Selexol and Sulfinol lying in between on both counts. All these processes have a long track record in industrial practice, all with high availability records.

Chemical Solvent Processes

Figure 7.15 shows the flow sheet of a typical MDEA wash, although this flow sheet is representative of many other chemical washing processes.

Amine Processes
The raw syngas is contacted in a wash column with lean MDEA solution, which absorbs the H_2S and some of the CO_2. MDEA is to some extent selective in that

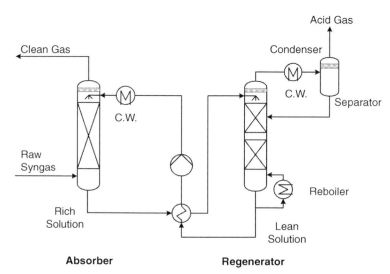

Figure 7.15 A typical MDEA flowsheet.

the bonding of the amine with H_2S takes places faster than with CO_2 and advantage can be taken of this in the design. The rich solution is preheated by heat exchange with the lean solution and enters the regenerator. Reboiling breaks the chemical bind and the acid gas components discharged at the top of the regenerator are cooled to condense out the water, which is recycled.

Physical Solvent Processes

Physical Washes
The following are important characteristics for any successful physical solvent:

- There should be good solubility for CO_2, H_2S, and COS in the operating range, preferably with significantly better absorption for H_2S and COS compared with CO_2 if selectivity is an important issue for the application of interest.
- Low viscosity at the lower end of the operating temperature range. Although lowering the operating temperature increases the solubility, the viscosity governs, in effect, the practical limit to lowering the operating temperature.
- A high boiling point reduces vapor losses when operating at ambient or near ambient temperatures.

Selexol
The Selexol process was originally developed by Allied Chemical Corporation and is now owned by UOP. It uses dimethyl ethers of polyethylene glycol (DMPEG). The physical properties of DMPEG are listed in Table 7.3. The typical operating temperature range is 15 to 100°F. The ability to operate in this temperature range offers substantially reduced costs by eliminating or minimizing refrigeration duty. On the other hand, for a chemical application such as ammonia, the residual sulfur in the treated gas may be 1 ppm H_2S and COS each after the CO_2 wash. This is, however, not an issue in power applications where the sulfur slip is less critical.

The ratio of absorption coefficients for H_2S, COS, and CO_2 is about 1:4:9 in descending order of solubility. A plant designed for 1 ppm COS in the clean gas would

Table 7.3 Properties of Physical Solvents

Process		Selexol	Rectisol
Solvent		DMPEG	Methanol
Formula		$CH_3O(C_2H_4O)_xCH_3$	CH_3OH
Molecular weight	lb/lb mol	178 to 442	32
Boiling point at 760 Torr	°F	415 to 870	147
Melting point	°F	–4 to –20	–137
Viscosity	cP	4.7 at 86°F	0.85 at 5°F
	cP	5.8 at 77°F	1.4 at –22°F
	cP	8.3 at 59°F	2.4 at –58°F
Specific weight	kg/m^3	1.031	790
Selectivity at working temperature	(H_2S:CO_2)	1:9	1:9.5

require about four times circulation rate of a plant for 1 ppm H_2S, together with all the associated capital and operating costs. In a gasification environment, it is therefore preferable to convert as much COS as possible to H_2S upstream of a Selexol wash. In a plant using raw gas shift for hydrogen or ammonia, this will take place simultaneously on the catalyst with the carbon monoxide shift. Where no CO shift is desired, COS hydrolysis upstream of the Selexol unit provides a cost-effective solution to the COS issue. Other characteristics that are favorable for gasification applications include high solubilities for HCN and NH_3 as well as for nickel and iron carbonyls.

The Selexol flowsheet in Figure 7.16 exhibits the typical characteristics of most physical absorption systems. The intermediate flash allows coabsorbed syngas components (H_2 and CO) to be recovered and recompressed back into the main stream. For other applications such as H_2S concentration in the acid gas or separate CO_2 recovery using staged flashing, techniques not shown here may be applied.

Rectisol

The Rectisol process, which uses cold methanol as a solvent, was originally developed to provide a treatment for gas from the Lurgi moving-bed gasifier, which in addition to H_2S and CO_2 contains hydrocarbons, ammonia, hydrogen cyanide, and other impurities. Figure 7.17 contains a flowsheet of the Rectisol process.

In the typical operating range of -20 to $-75°F$, the Henry's law absorption coefficients of methanol are extremely high, and the process can achieve gas purities

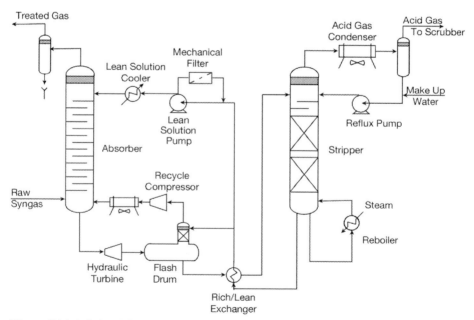

Figure 7.16 A Selexol flowsheet for selective H_2S removal.

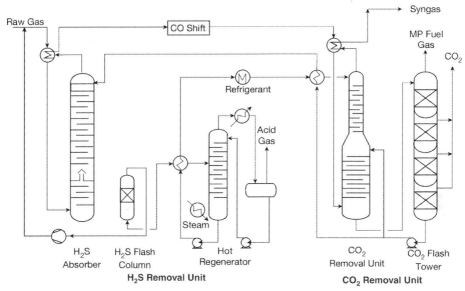

Figure 7.17 A flowsheet of the selective Rectisol process.

unmatched by other processes. This has made it a standard solution in chemical applications, such as ammonia, methanol, or methanation, where the synthesis catalysts require sulfur removal to less than 0.1 ppm. This performance has, however, a price in that the refrigeration duty required for operation at these temperatures involves considerable capital and operating expense.

Methanol as a solvent exhibits considerable selectivity, as can be seen earlier in Table 7.3. This allows substantial flexibility in the flowsheeting of the Rectisol process, and both standard (nonselective) and selective variants of the process are regularly applied according to circumstances.

As a physical wash, which uses at least in part flash regeneration, part of the CO_2 can be recovered under an intermediate pressure. Typically, with a raw gas pressure of 710 psig, about 60 to 75 percent of the CO_2 would be recoverable at 40 to 60 psig. Where CO_2 recovery is desired, whether for urea production in an ammonia application or for sequestration, this can provide significant compression savings.

Following the solvent circuit is an intermediate H_2S flash from which coabsorbed hydrogen and carbon monoxide are recovered and recompressed back into the raw gas. The flashed methanol is then reheated before entering the Hot Regenerator. Here the acid gas is driven out of the methanol by reboiling, and a Claus gas with an H_2S content of 25 to 30 percent (depending on the sulfur content of the feedstock) is recovered. Minor adaptations are possible to increase the H_2S content if desired.

Water entering the Rectisol unit with the syngas must be removed, and an additional small water-methanol distillation column is included in the process to cope with this. Typically the refrigerant is supplied at between −20°F and −40°F. Depending on application, different refrigerants can be used. In an ammonia plant, ammonia is used, and the refrigeration system is integrated with that of the synthesis. In a refinery environment, propane or propylene may be the media of choice.

The Rectisol technology is capable of removing not only conventional acid gas components but also, for example, HCN and hydrocarbons, and metallic mercury.

COS Hydrolysis

The chemical washes are generally not capable of absorbing COS, which must be converted to H_2S in a COS hydrolysis step upstream of the wash. Physical washes can absorb COS. In the case of Selexol, this capability is not very strong, and economics usually dictate the use of a COS hydrolysis as well (but not after a CO shift as in Coffeyville). Rectisol does not require any upstream COS hydrolysis.

Sulfur Recovery

Sulfur recovery is generally achieved using Claus technology, although Polk is an exception in that it manufactures sulfuric acid rather than elemental sulfur. Differences in the Claus technology itself are generally only of a detailed nature. Considerable variety is shown in the handling of the tailgas from the Claus plant, which in addition to H_2S also contains small quantities of SO_2, COS, CS_2, and elemental sulfur. In all plants, these are hydrogenated back to H_2S over a catalyst. In some plants, this gas is then treated separately in another washing stage and then incinerated and discharged to the atmosphere. In others it is recycled to a point and mixed with the raw gas upstream of the main AGR so that this remaining gas is treated there. The point at which the recycle is fed into the main gas stream varies.

In all plants, syngas dilution is used to reduce the NO_x emissions from the gas turbine. The dilution medium may be nitrogen only (e.g., Polk initially), steam only (e.g., Wabash), or a combination of the two (e.g., Buggenum). Steam is generally added by saturation using low-level heat to provide the necessary hot water. In some cases it is added by direct injection.

7.4.4   Combined Cycle Power Plant

The combined cycle power plant covers the typical scope of a natural gas combined cycle system complete with balance of plant. The principal difference lies in the use of syngas as a fuel. Note that to date little experience is available on the use of selective catalytic NO_x reduction (SCR) in IGCCs, where the residual sulfur content in the syngas could impact on the availability of the HRSG. Two units in Italy are equipped with SCRs, but the required NO_x emissions levels are not comparable with the values required of a NGCC plant. In one case, the SCR is only used when the gas turbine operates on the backup distillate fuel. The only significant experience with SCR in IGCC is at the Negishi plant of Nippon Oil. Values of less than 2.6 ppm NO_x and less than 2.0 ppm SO_x have been reported [72].

7.4.5   IGCC with Carbon Capture

Integration of carbon capture into an IGCC is relatively simple using techniques common in, for example, the fertilizer industry, where CO_2 is captured in the ammonia plant and compressed to typically 2,200 to 2,900 psig for urea manufacture. The principle changes to an IGCC power plant without carbon capture are the introduction of a CO shift reactor after the particulate removal and the modification to the acid gas removal to draw off the CO_2 stream separately from the H_2S. The configuration up to the outlet of the AGR is a regular feature of coal-based ammonia plants. The CO shift uses a sulfided chrome-molybdenum catalyst. The selective AGR is generally a physical wash. This has the added advantage that much of the CO_2 is available under pressure, which reduces compression costs. One well-documented example is the 1,000 t/d (100 MW_e equivalent) ammonia plant in Coffeyville, Kansas, which feeds petroleum coke into a GEE Quench Gasifier and uses Selexol for acid gas removal. It achieves on-steam times of 98 percent between annual turnarounds [73]. There are a number of basically similar plants in China using the Shell gasifier with Rectisol.

The main area where industrial experience is limited is on the use of hydrogen as fuel in a gas turbine, although there is more experience available than generally appreciated, much of it in industrial applications. One example described by GE as the "H2 Fleet Leader" is a frame 6B unit operating regularly on 85 to 97 percent hydrogen at "an availability of 96.5 percent+ running in uninterrupted operation, 24 hours a day over the year" since 1997 [74].

7.4.6   Benefits and Limits of IGCC

Several benefits and limits of IGCC warrant discussion. The most important ones are associated with system efficiency, environmental impact, pollutant emissions, system availability, and capital requirements.

Efficiency

As previously mentioned, one of the motivations for development of the IGCC power plant was to harness the high efficiency of combined cycle gas turbines for use with coal. The prototype 100 MW_e Cool Water IGCC operated between 1984 and 1988 with a heat rate of 10,950 HHV Btu/kWh net [75]. Wabash, which is representative of the 250 MW_e class of IGCC built in the mid-1990s and based on the GE 7FA gas turbine, had a heat rate of 8,900 HHV Btu/kWh net [76]. The 630 MW_e class currently being planned is based on two larger GE 7FB or Siemens SGT6-5000F gas turbines with a heat rate of about 8,500 HHV Btu/kWh net on bituminous coals. Dry-feed gasifiers achieve a similar heat rate when operating on high-moisture subbituminous coals such as those from the Powder River Basin. The heat rate of a slurry feed gasifier operating with PRB coals increases to 9,000 HHV Btu/kWh net or more.

IGCC efficiencies are typically between 40 and 50 percent with recent turbine technology [7]. Net efficiencies of the Buggenum and Puertollano plants are 43 percent and 45 percent (LHV basis), respectively. Further increases in efficiency (with projections of net efficiencies of 56 percent (LHV basis) are possible through the following [7]:

- Optimizing the integration of the gasification and power-generation blocks
- Advances in gas turbine technology, such as higher pressure ratios, higher turbine entry temperatures, and reheat
- Further development of the hot gas cleaning processes for the raw gas leaving the gasifier
- Advanced air separation technologies

Environmental Impact

Another motivation for development of the IGCC power plant was its potential for extremely low environmental emission rates compared with other coal-based technologies. The first-generation IGCC plants built during the 1990s have achieved the goals set at the time, which are discussed under the individual pollutants following. The technology to reduce these rates further is available, though its use will depend on a balance between cost and regulatory requirements.

Sulfur Emissions

An IGCC can readily reduce sulfur emissions to about 4 ppm SO_2 (wet basis, referred to 1 percent O_2) in the turbine exhaust. This is equivalent to about 30 ppm total sulfur (H_2S + COS) in the dry, undiluted gas leaving the acid gas removal, a value achievable with an MDEA or Selexol system. If a selective catalytic reactor (SCR) is required for Denox, then this would typically need to be reduced to about half this value, which would also be possible with a rather more elaborate version of Selexol. A further two orders of magnitude reduction would be possible using Rectisol instead of the currently used chemical or synthetic fuel applications. This is, however, considered to be unnecessarily expensive.

NO_x Emissions

The high hydrogen content of syngas (which would be even higher in the carbon capture scenario) prohibits the use of current dry low NO_x burners as developed for natural gas. Instead, diffusion burners are used. Typically one could achieve about 15 ppm (dry basis referred to 15 percent O_2) in the exhaust of a gas turbine without SCR. In fact, some IGCCs such as the plant in Buggenum regularly achieve values under 10 ppmv without SCR.

Some areas may, however, require SCR to achieve lower NO_x emission rates of 3 ppmv (dry basis referred to 15 percent O_2) comparable with those for natural gas fired turbines. In such a case, it is necessary to desulfurize the fuel gas further to about 15 ppmv to avoid formation of ammonium bisulfate from ammonia slip in the SCR, which can deposit on heat exchange surface downstream the SCR. The SCR in the oil-fired IGCC at Negishi in Japan is reported as meeting its permit level of 2.6 ppmv (dry basis referred to 15 percent O_2) [72].

Mercury

Mercury can be removed from the fuel gas with a fixed bed of activated carbon. Approximately 95 percent of the mercury leaving the gasifier in the fuel gas is captured.

Other Emissions

Typical values for other pollutants referred to the gross heat input to the gasifier, based on a heat rate of 8,500 Btu (HHV)/kW (net), are as follows:

- Particulate matter: 0.0145 lb/million Btu (including condensables)
- CO: 10–25 ppm (dry basis referred to 15 percent O_2)
- Unburned hydrocarbons: 7 ppmv (wet basis)
- VOC: 1.4 ppmv (wet basis)

Availability

It is generally acknowledged that while early IGCC plants met their efficiency and environmental goals, the availability results were not good [77]. This was also surprising to those in the industry who were accustomed to high availabilities such as those achieved by, say, Eastman at their Kingsport methanol plant, where 98 percent is regularly reported. An analysis of the causes of outage revealed some unexpected results [78]. Much of the lack of availability was due to fleet issues on early models of the gas turbines involved (in some cases up to 25 percent loss of annual availability), which had no relation to their utilization in an IGCC environment. This is contrasted with three refinery-based IGCC units built in Italy about five years later, which after a two- to three-year ramp-up period are reporting availabilities (and on stream times) of 90 to 95 percent. One of these plants achieved more than 90 percent availability in its second year of operation.

Capital Requirements

A discussion on capital costs of IGCC with numerical values is difficult in a period of rapid inflation in the capital plant industry. Estimates vary significantly, depending on a large number of factors, including gasification technology used, rank of coal, and level of built-in redundancy. It is, however, generally accepted that current IGCCs require between 10 and 20 percent additional investment when compared with conventional technology. Existing IGCCs are all tailor-made units with all the associated high costs of engineering, procurement, and construction. Vendors of IGCC systems have recognized the necessity of reducing costs through modularization and standardization and are all preparing so-called "reference designs" for two-train 630 MW net output plants. One vendor has estimated that current efforts could reduce the cost premium to about 10 percent [78].

Introduction of syngas operation to the next generation of gas turbines (H-class), which have now started operation with natural gas, will also reduce the specific costs per kW installed capacity by extracting a higher-power output from the same gas production facility.

It should also be noted that CO_2 capture from the high-pressure fuel gas is much less costly from flue gas and that in the event of carbon capture becoming a necessity, the overall cost of electricity is expected to be lower with IGCC than with conventional technologies and postcombustion capture. This is illustrated in Table 7.4, which lists CO_2 emissions and costs from an Electric Power Research Institute (EPRI) study for different technologies (modified from [79]).

Table 7.4 CO$_2$ Emissions, Efficiency, and Costs of Advanced Power-Generation Technologies with and without CCS

Technology: Carbon Capture and Sequestration	Supercritical Pulverized Coal		Ultra- Supercritical Pulverized Coal		Oxy- Fuel PC	IGCC	
	without CCS	with CCS	without CCS	with CCS	with CCS	without CCS	with CCS
CO$_2$ emitted, g/kWh	830	109	738	94	104	824	101
Efficiency, % (HHV)	38.5	29.3	43.4	34.1	30.6	38.4	31.2
Total capital requirement, $/kW	1,937	3,120	1,976	3,042	2,990	2,080	2,756
Cost of electricity, ¢/kWh	5.50	8.84	5.39	8.44	7.93	5.90	7.50

Source: Modified from Booras and Holt (2004).

7.4.7 Commercial Status

As previously mentioned and listed in Table 7.2, five coal-fueled IGCC units are in operation worldwide, with capacities of 250 to 350 MW$_e$ [6, 71]. Each of these units uses a different gasifier technology. Four of the five units use oxygen for gasification, and one uses air. Two of these are slurry-fed gasifiers. This section presents some of the key information of the units.

Peurtollano (Peurtollano, Spain)

The Prenflo gasifier-based IGCC (300 MW$_e$) at Puertollano, Spain, owned by ELCOGAS has been operating since 1996 [6, 8]. Oxygen is used for gasification. The plant has a Siemens V94.3 gas turbine. The raw product gas is cooled to 380°C in an integral syngas cooler above the gasification zone. Ceramic filters operating at about 240°C remove particulates before the scrubber. Quench gas is recycled to cool the raw gas to 800°C before the syngas cooler.

Initially, difficulties were encountered with fuel and slag flow and the gas turbine combustor, which resulted in intermittent operation (i.e., low availability). This was the result, in part, of the feed, which is a mixture of high-ash coal and high-sulfur petroleum coke. The combustor was modified based on experience from Buggenum, and since 2000, improved syngas availability has been achieved by restricting the proportion of coal fed to 38 percent by mass and the use of better offline cleaning cycles for the hot gas cleanup system, which are candle filters. Availability is approaching 75 percent, which is still below the 85 percent average availability of the natural gas combined cycle and the pulverized coal-fired power plants [8].

Polk Power Station (Tampa, Florida)

The Global Energy's E-Gas two-stage, upflow slurry feed gasifier-based IGCC (250 MW$_e$) at the Polk Power Station, owned by Tampa Electric, has been operating since 1996 [6, 8]. Oxygen is used for gasification. The feed is a mixture of coal and petroleum coke. The gas emerging from the radiant cooler is cooled in two parallel firetube boilers and by heat exchange with the cleaned gas after desulfurization in MDEA scrubbers. The plant has a General Electric STAG 107FA gas turbine combined cycle system.

The IGCC system availability is typically 60 to 70 percent. Carbon conversion has been lower than anticipated, low to mid-90s versus about 98 percent, as expected. The resultant heat rate penalty has been around 200 to 500 kJ/kWh [6].

Wilem-Alexander Cenrale (Buggenum, Netherlands)

The Shell entrained-flow gasifier-based IGCC (250 MW$_e$) at Beggenum, owned by NUON Power, has been operating commercially in The Netherlands since 1998 [6, 8]. Oxygen is used for gasification. The plant has a Siemens V94.2 gas turbine. The membrane-walled gasifier operates at a temperature of 1,500°C and pressure of 2.8 MPa. The raw gas is quenched to 900°C at the exit at the top of the gasifier by addition of a recycle stream of cooled, ash-free gas before being sent to a convective syngas cooler. The gas then passes to a cyclone then a ceramic filtration unit at 250 to 285°C. Final cleaning is by cold scrubbing to remove ammonia, chlorides, and sulfur gases, after which the gas is saturated with water for NO$_x$ control and sent to the gas turbine for combustion [6]. IGCC system availability is approximately 60 percent and has been operating at this level for several years [8].

Wabash River (W. Terre Haute, Indiana)

The ConocoPhilips E-Gas (formerly Dow Chemical then Dynergy then Global Energy) gasifier-based IGCC (260 MW$_e$) at Wabash River, Indiana, owned by SG Solutions/Duke Energy Indiana has been operating since 1999 [6, 8]. Oxygen is used for gasification. The feedstocks are a high-sulfur Illinois Basin bituminous coal and petroleum coke. The plant has a General Electric Frame 7FA gas turbine. Raw gas from the top of the gasifier is cooled to 370°C in a vertical firetube syngas cooler. A candle filter unit removes fly ash and char for recycle to the gasifier. The filtered gas is then cooled and cleaned of sulfur compounds in an MDEA scrubber. The clean flue gas is reheated and moisture added to it for control of NO$_x$ emissions before combustion in the gas turbine. IGCC system availability is approximately 70 percent, which is among the highest levels any of the five coal-fueled systems.

Clean Coal Power R&D Company (Nakoso, Japan)

The MHI air-blown, two-stage entrained-flow gasifier-based IGCC (250 MW$_e$) in Nakoso, owned by a consortium of Japanese utilities, corporations, and the Ministry of Economy, Trade, and Industry has been operating since September 2007 and as of April 2009 has accumulated more than 2000 hours of operation [80]. The plant uses filtration for particulates removal followed by scrubbing with MDEA for desulfurization [6]. The system is new with limited operational data available; however, target net efficiency is 42 percent (LHV basis) and target emissions are 8 ppm SO_2, 5 ppm NO_x, and 4 mg particulate/m^3.

7.5   IGCC Research Needs

Several areas of research and development are required to further the IGCC technology and the integration of CO_2 capture and storage [7]. These are generally divided into areas of availability and reliability, modeling and simulation, process and components, and precombustion capture. With respect to availability and reliability, both need to be increased; there must be an increase in robustness by preventing slagging, fouling, and corrosion; and dry-feed systems need optimizing or new ones must be developed. Research needs for modeling and simulation include buildup of databases for substance and process modeling, developing models for optimizing gasification, and modeling and simulation of gasifiers.

In the area of process and components, avenues of research include developing methods to use high-temperature heat without decreasing availability, improving hot gas cleanup technologies, optimizing gas turbines for synags, increasing gas turbine inlet temperatures, optimizing flowsheets, and reducing capital costs. Research needs related to precombustion capture in IGCC power plants include developing an optimized hydrogen-fueled gas turbine; optimizing the process by integrating the ASU, CO-shift, and CO_2 capture technologies; developing dynamic modeling of the entire process; and performing analysis of part-load ability and variable CO_2 capture.

References

[1] J.B. Kitto, S.C. Stultz (Eds.), Steam: its generation and use, fourty-first ed., The Babcock & Wilcox Company, 2005.
[2] A.M. Robertson, H. Gagliano, G. Hack, Stanko, Corrosion Tests of MA956 DOS Tube Specimens, in: Proc. of the 34th International Technical Conference on Clean Coal Utilization & Fuel Systems, May 31–June 4, 2009.
[3] T. Sarkus, Fossil Energy, Clean Coal Technology, and FutureGen, Coal Age 113 (7) (2008) 56–59.
[4] H. Nalbandian, Performance and risks of advanced pulverized coal plant, IEA Clean Coal Centre, 2008.
[5] P. Baruya, Competitiveness of coal-fired power generation, IEA Clean Coal Centre, 2008.
[6] C. Henderson, Clean coal technologies, IEA Clean Coal Centre, 2003.
[7] H. Kessels, S. Bakker, A. Clemens, Clean coal technologies for a carbon-constrained world, IEA Clean Coal Centre, 2007.
[8] A. Kather, S. Rafailidis, C. Hermsdorf, M. Kostermann, A. Maschmann, K. Mieske, et al., Research and development needs for clean coal deployment, IEA Clean Coal Centre, 2008.

[9] R. Viswanathan, J. Shingledecker, Materials for Advanced Ultra Supercritical Fossil Plant, in: Proc. of the 34th International Technical Conference on Clean Coal Utilization & Fuel Systems, May 31–June 4, 2009.

[10] M. Gagliano, H. Hack, G. Stanko, Update on the Fireside Corrosion Resistance of Proposed Advanced Ultra-supercritical Superheater and Reheater Materials: Laboratory and Field Test Results, in: Proc. of the 34th International Technical Conference on Clean Coal Utilization & Fuel Systems, Clearwater, FL, May 31 to June 4, 2009.

[11] National Coal Council, Technologies to Reduce or Capture and Store Carbon Dioxide Emissions, The National Coal Council, 2007.

[12] L.A. Ruth, Advanced Pulverized Coal Combustion, in: Proc. of Advanced Coal-Based Power and Environmental Systems Conference, U.S. Department of Energy, National Energy Technology Laboratory, 1998.

[13] H. Klotz, K. Davis, E. Pickering, Designing an Ultra-supercritical Steam Turbine, Power, July 2009, pp. 34–38.

[14] DOE (U.S. Department of Energy), New Turbine Materials for Advanced Ultra Supercritical Coal Power Plants, Clean Coal Today, 80 (Spring/Summer) (2009).

[15] DOE, Oxy-fuel Combustion Fact Sheet, *www.netl.doe.gov*, August 2008.

[16] J.D. Figueroa, T. Fout, S. Plasynski, H. McIlvried, R.D. Srivistava, Advances in CO_2 capture technology—U.S. Department of Energy's Carbon Sequestration Program, International Journal of Greenhouse Gas Control (2008) 9–20.

[17] A. Kather, S. Rafailidis, C. Hersforf, M. Klostermann, A. Maschmann, K. Mieske, et al., Research and development needs for clean coal deployment, International Energy Association Clean Coal Centre, 2008.

[18] H. Farzan, S. Vecci, D. McDonald, K. McCauley, P. Pranda, R. Varagani, et al., State-the-Art Oxy-Coal Combustion Technology for CO_2 Control from Coal-Fired Boilers: Are We Ready for Commercial Installation? in: Proc. of the 32nd International Technical Conference on Coal Utilization & Fuel Systems, Coal Technology Association, 2007.

[19] Z. Wu, Understanding Fluidized-Bed Combustion, IEA Coal Research, London, 2003.

[20] S.J. Goidich, S. Wu, Z. Fan, A.C. Bose, Design Aspects of the Ultra-Supercritical CFB Boiler, in: Proc. of the 22nd International Pittsburgh Coal Conference, University of Pittsburgh, September 12–15, 2005.

[21] J. Koornmeef, M. Junginger, A. Faaij, Development of Fluidized Bed Combustion—An Overview of Trends, Performance, and Cost, Progress in Energy and Combustion Science 33 (2007) 19–55.

[22] DOE and American Electric Power Service Corporation, Tidd: The Nation's First PFBC Combined-Cycle Demonstration, Clean Coal Technology Topical Report Number 1, March 1990.

[23] S.C. Stultz, J.B. Kitto (Eds.), Steam, Its Generation and Use, Babcock & Wilcox Company, 1992.

[24] DOE, Fluidized-Bed Combustion, *www.netl.doe.gov/technologies/coalpower/advresearch/ combustion/FBC/fbc-overview.html*.

[25] A. Robertson, Z. Fan, D. Horaazak, R. Newby, H. Goldstein, A.C. Bose, 2nd Generation PFB Plant with Supercritical Pressure Steam Turbine, in: Proc. of the 22nd International Pittsburgh Coal Conference, University of Pittsburgh, September 12–15, 2005.

[26] Foster Wheeler, Biomass Circulating Fluid Bed Boilers, Promotional Presentation Module 38 Biomass CFB 071106, 2006.

[27] T.C. Elliot (Ed.), Standard Handbook of Powerplant Engineering, McGraw-Hill, 1989.

[28] Foster Wheeler, Utility CFB Steam Generators, Promotional Presentation Module 31 Utility CFB Detail 042707, 2007.

[29] M.Z. Haji-Sulaiman, Sorbent Performance in Fluidized Bed Combustion, Ph.D. Thesis, The Pennsylvania State University, May 1988.

[30] W.S. Rickman, Modeling Sulfur Capture by Recycle Fines in Fluidized-Bed Combustion, EPRI CS-4442, Final Report, March 1986, pp. 2–3.

[31] B.G. Miller, D.E. Romans, A.W. Scaroni, Characterization of Limestones for FBC Systems, in: Proc. of The National Stone Association SO₂ Emission Control Conference, Pittsburgh, September 9–11, 1990.

[32] J.L. Morrison, B.G. Miller, D.E. Romans, S.V. Pisupati, A.W. Scaroni, Fluidized-Bed Boilers – SO2 Capture Aspects, in: Proc. of SO₂ Capture Seminar "Sorbent Options and Considerations", Cincinnati, September 19–21, 1993.

[33] P.G. Christman, T.G. Edgar, The Effect of Temperature on the Sulfation of Limestones, in: Proc. of 1982 American Institute of Chemical Engineers Meeting, Los Angeles, 1982.

[34] J.S. Dennis, A.N. Hayhurst, The Effect of Temperature on the Kinetics and Extent of SO₂ Uptake by Calcarious Materials during FBC of Coal, in: Proc. of the 20th Symp. (Int.) on Combustion, The Combustion Institute, 1984, pp. 1347–1355.

[35] P.F.G. Hansen, K.D. Johansen, L.H. Bank, K. Ostergaard, Sulfur Retention on Limestone under Fluidized-Bed Combustion Conditions: An Experimental Study, in: Proc. of the 11th International Conference on FBC, Montreal, April 21–24, 1991, pp. 73–81.

[36] S.V. Pisupati, J.L. Morrison, D.E. Romans, B.G. Miller, A.W. Scaroni, Importance of Calcium Carbonate Content on the Sulfur Capture Performance of Naturally Occurring Sorbens in a 30 MWₑ Circulation Fluidized-Bed Plant, in: Proc. of the 12th International FBC Conference, La Jolla, May 8–13, 1993.

[37] D.E. Romans, A.W. Scaroni, B.G. Miller, Evaluation of Sorbents for Capturing SO₂ in Fluidized-Bed Combustion Systems, in: Proc. of the 14th Annual Energy-Sources Technology Conference & Exhibition, American Society of Mechanical Engineers, Dallas, January 22, 1991.

[38] R.S. Boynton, Chemistry and Technology of Lime and Limestone, John Wiley & Sons, 1980.

[39] D.C. Fee, et al., Sulfur Control in Fluidized-Bed Combustors: Methodology for Predicting the Performance of Limestone and Dolomite Sorbents, ANL/FE-80-10, 1982.

[40] G.H. Newton, S.L. Chen, J.C. Kramlich, Role of Porosity Loss in Limiting SO₂ Capture by Calcium Based Sorbents, AIChE Journal 35 (6) (1989) 988–994.

[41] S.K. Bhatia, D.D. Perlmutter, The Effect of Pore Structure on Fluid-Solid Reactions: Applications to the SO₂-Lime Reaction, AIChE Journal 27 (2) (1981) 226–234.

[42] G.A. Simons, A.R. Garmen, Small Pore Closure and the Deactivation of the Limestone Sulfation Reaction, AIChE Journal 32 (9) (1986) 1491–1499.

[43] G.A. Simons, Parameters Limiting Sulfation By CaO, AIChE Journal 34 (1) (1988) 167–170.

[44] M.E. Ulerich, E.P. O'Neill, D.L. Keairns, Thermogravimetric Study of the Effect of Pore Volume-Pore Size Distribution on the Sulfation of Calcined Limestone, Thermochimica Acta 26 (1978) 269–282.

[45] R.H. Borgwardt, N.F. Roache, K.R. Bruce, Surface Area of Calcium Oxide and Kinetics of Calcium Sulfide Formation, Environmental Progress 3 (2) (1989) 129–135.

[46] M. Hiltunen, et al., N₂O Emissions from CFB Boilers: Experimental Results and Chemical Interpretation, in: Proc. of the 11th Conference on Fluidized-Bed Combustion, Montreal, April 21–24, 1991.

[47] M. Hiltunen, Nitrous Oxide (N₂O) Emissions in Fluidized-Bed Boilers, ECCP I Review: Non-CO₂ Gases, Brussells, January 30, 2006.

[48] W.T. Davis (Ed.), Air Pollution Engineering Manual, second ed., John Wiley & Sons, 2000, p. 9.

[49] L.E. Clarke, L.E. Sloss, Trace Elements: Emissions from Coal Combustion and Gasification, IEA Coal Research, 1992.

[50] S.J. Miller, S.R. Ness, G.F. Weber, T.A. Erickson, D.J. Hassett, S.B. Hawthorne, et al., A Comprehensive Assessment of Toxic Emissions from Coal-Fired Power Plants: Phase 1 Results from the U.S. Department of Energy Study Final Report, 1996. Contract No. DE-FC21-93MC30097 Subtask 2.3.3.

[51] Standards of Performance for New and Existing Stationary Sources: Electric Utility Steam Generating Unites, amended, Code of Federal Regulations, Part 60, 63, 72, and 75, Title 40, 2005.

[52] I.S.C. Hughes, R.F. Littlejohn, Emissions of Trace Elements Emissions from AFBC, in: Proc. of the International Conference on Fluidized-Bed Combustion: FBC Comes of Age, American Society of Mechanical Engineers, May 3, 1987.

[53] B.G. Miller, S. Falcone Miller, R.T. Wincek, A.W. Scaroni, A Demonstration of Fine Particulate and Mercury Removal in a Coal-Fired Industrial Boiler using Ceramic Membrane Filters and Conventional Fabric Filters, in: Proc. EPRI-DOE-EPA Combined Utility Air Pollutant Symposium, Atlanta, August 16–20, 1999.

[54] www.epa.gov/ttn/atw/combust/utiltox/mercury.html.

[55] J.D. Watt, The Physical and Chemical Behaviour of the Mineral Matter in Coal under the Conditions Met in Combustion Plant, Part II, BCUR, Leatherhead, 1969.

[56] E. Raask, Mineral Impurities in Coal Combustion, Hemisphere Publishing Corp, 1985.

[57] S.K. Falcone, H.H. Schobert, Mineral Transformations during Ashing of Selected Low-Rank Coals, in: K.S. Vorres (Ed.), Mineral Matter and Ash in Coal, ACS Symposium Series No. 301, American Chemical Society, 1986, pp. 114–127.

[58] H.H. Schobert, Lignites of North America, Coal Science and Technology vol. 23 Elsevier, 1995 (Chapter 5).

[59] B.G. Miller, S. Falcone Miller, R.E. Cooper Jr., N. Raskin, J.J. Battista, Biomass Cofiring: A Feasibility Study for Cofiring Agricultural and Other Wastes with Coal at Penn State University, in: Proc. of the 27th International Technical Conference on Coal Utilization & Fuel Systems, March 4–7, 2002.

[60] S. Falcone Miller, B.G. Miller, D. Tillman, The Propensity of Liquid Phases Forming during Coal-Opportunity Fuel (Biomass) Cofiring as a Function of Ash Chemistry and Temperature, in: Proc. of the 27th International Technical Conference on Coal Utilization & Fuel Systems, March 4–7, 2002.

[61] S. Falcone Miller, B.G. Miller, C.M. Jawdy, The Occurrence of Inorganic Elements in Various Biofuels and Its Effect on the Formation of Melt Phases during Combustion, in: Proc. of the 2002 International Joint Power Generation Conference, Phoenix, June 24–26, 2002.

[62] S. Falcone Miller, B.G. Miller, The Effect of Cofiring Coal and Biomass on Utilization of Coal Combustion Products: The U.S. Perspective, in: Proc. of the 11th International Conference on Ashes from Power Generation, Zakopane, October 13–16, 2004.

[63] S. Falcone Miller, B.G. Miller, The Occurrence of Inorganic Elements in Various Biofuels and its Effect on Ash Chemistry and Behavior and use in Combustion Products, Fuel Processing Technology 88 (2007) 1155–1164.

[64] J.D. Watt, The Physical and Chemical Behaviour of the Mineral Matter in Coal under the Conditions Met in Combustion Plant, Part II, BCUR, Leatherhead, 1969.

[65] B.G. Miller, D.A. Tillman (Eds.), Combustion Engineering Issues for Solid Fuel Systems, Academic Press, 2008.

[66] K.S. Vorres, S. Greenburg, R.B. Poeppel, Viscosity of Synthetic Coal Ash Slags, Argonne National Laboratory Report ANL/FE-85-10, 1985.

[67] B. Jung, Viscous Sintering of Coal Ashes in Combustion Systems, Ph.D. Thesis in Fuel Science, The Pennsylvania State University, 1990.

[68] S.N. Suranti, Nsakala, S.L. Darling, Alstom Oxyfuel CFB boilers: A promising option for CO_2 capture, Energy Procedia 1 (2009) 543–548.

[69] A.A. Lavasseur, P.J. Chapman, N. Nsakala, F. Kluger, Alstom's Oxy-Firing Technology Development and Demonstration—Near Term CO_2 Solutions, in: Proc. 34th International Technical Conference on Clean Coal & Fuel Systems, Clearwater, FL, May 31–June 4, 2009.

[70] T. Wall, J. Yu, Coal-Fired Oxyfuel Technology Status and Progress to Deployment, in: Proc. of the 34th International Technical Conference on Clean Coal & Fuel Systems, Clearwater, FL, May 31–June 4, 2009.

[71] A. Neville, IGCC Update: Are We There Yet? Power (August) (2009) 52–57.

[72] M. Yamaguchi, First Year of Operational Experience with the Negishi IGCC, Gasification Technologies Conference, Washington, DC, 2004.

[73] N.E. Barkley, Petroleum Coke Gasification Based Ammonia Plant, AIChE Ammonia Safety Conference, Vancouver, 2006.

[74] R.M. Jones, et al., Gas Turbine Requirements for a Carbon Constrained Environment, IChemE European Gasification Conference, Barcelona, 2006.

[75] EPRI (Electric Power Research Institute), Cool Water Coal Gasification Program: Final Report, Palo Alto, 1990.

[76] Wabash River Energy Ltd., Wabash River Coal Gasification Repowering Project: Final Technical Report, 2000.

[77] EPRI, Integrated Gasification Combined Cycle (IGCC) Design Considerations for High Availability, Vol. 1: Lessons from Existing Operations, (Product No. 1012226), Palo Alto, 2007.

[78] E. Lowe, IGCC reference Plant Inked, World-Generation (May/June), 2006.

[79] G. Booras, N. Holt, Pulverized Coal and IGCC Plant Cost and Performance Estimates, Gasification Technologies Conference, Washington, DC, 2004.

[80] Mitsubishi Heavy Industries, Ltd., Clean Coal Power R&D Co., Ltd., Winds the "Prime Minister Prize" of the Japan Industrial Technology Grand Prize with its IGCC Demonstration Plant, www.mhi.co (accessed October 8, 2009).

8 Coal-Fired Emissions and Legislative Action

The use of coal has a long history of emitting a variety of pollutants into the atmosphere and has been a contributor to several known health episodes in the past. Regulations on coal usage date back to medieval times when nobility became inconvenienced. It was the major health episodes in the 1940s and 1950s in both the United States and England, however, that resulted in severe impacts on human health that were the impetus for legislation in both countries to protect the health and welfare of the general public. In the United States, the major development of legislative and regulatory acts occurred from 1955 to 1970, and by the mid-1970s the basis for national regulation of air pollution was well developed. The regulations are continually changing as more information on the effect of emissions on health and the environment is learned, new control technologies are developed, and society demands a safe living environment. The use of coal for power generation is a highly regulated industry with more regulations soon to be implemented.

This chapter briefly discusses past health episodes, primarily their roles in the initiation of regulations on the use of coal. The emphasis of this chapter is on federal legislation and regulatory trends in the United States. A history of legislative action in the United States as it pertains to coal-fired power plants is presented. Impending legislation of emissions currently not regulated is also discussed. The types and quantities of emissions, with the emphasis from coal-fired power plants, are presented along with their trends over time as a consequence of legislative action. A brief discussion of emissions and legislation from other countries and how they compare to the United States is also provided.

8.1 Major Coal-Related Health Episodes

Air pollution is not a modern phenomenon, although legislative action to successfully regulate emissions and protect human health is. This section discusses the major health episodes caused by or contributed to by using coal, especially those that have raised public awareness of the ill effects of air pollution on health and have led to regulations, specifically with respect to the use of coal.

8.1.1 Pre-Industrial Revolution

Documentation of air pollution begins as early as Ancient Rome, when the statesman Seneca complained about "the stink, soot and heavy air" in the city [1]. As coal usage increased in England in medieval times and as coal became less expensive

Clean Coal Engineering Technology. DOI: 10.1016/B978-1-85617-710-8.00008-X
Copyright © 2011 by Elsevier Inc. All rights of reproduction in any form reserved.

and wood became more scarce, so came increased air pollution problems, mainly black smoke and fumes. It became so bad that in 1257 Queen Eleanor was driven from Nottingham Castle by the smoke and fumes rising from the city below [2]. In 1283 and 1288 there were complaints about air quality in London because coal, which was used by the various smithies and the general public for home heating, was now being used in lime kilns. By 1285 London's air was so polluted that King Edward I established the world's first air pollution commission, and 22 years later he made it illegal to burn coal [1]. This Royal Proclamation in 1307 forbade the use of coal in lime-burners in parts of South London; however, this proclamation did not work, and a later commission had instructions to punish offenders with fines and ransoms for a first offense and to demolish their furnaces for a second offense [2]. Eventually, economics (high-priced wood versus low-priced coal) and a change in government policy (laws to save the few remaining forests) won out over the populace's comfort, and London was to remain polluted by coal fumes for another 600 years [2], leading the poet Shelley to write in the early 1800s, "Hell must be much like London, a smoky and populous city" [1].

8.1.2 Post-Industrial Revolution

Two major air pollution episodes raised awareness of the effects of pollution on human health and were instrumental in passing the English Clean Air Act in the United Kingdom in 1956 and the Clean Air Act in the United States in 1970 (although federal air pollution control acts started being passed in 1955, as discussed later in this chapter) [3, 4]. These incidents occurred in Donora, Pennsylvania, in 1948 and London, England, in 1952 and illustrated that people will largely ignore pollution until the death toll is enormous. In fact, local officials in Donora and London initially placed blame on weather inversions, frail health, and flu epidemics [3].

Donora, Pennsylvania, which is located south of Pittsburgh in the Monongahela Valley, was a town of approximately 26,000 people in the 1940s and home to a steel mill and zinc works [3]. The inhabitants were accustomed to dreary days, dirty buildings, and barren ground where no vegetation would grow, but they ignored the effect that the steel and zinc mills had on the population and environment because about two-thirds of the workers were employed in the steel and zinc mills. This continued until the "killer smog" of October 1948. Nearly half of the town's inhabitants were sick by the end of the second day of the thick smog, 20 people died in three days, 50 more deaths than would be expected from other causes occurred in the month following the episode, and many people experienced breathing difficulties for the rest of their lives [3]. It was discovered later that the cause for many of the fatalities was not sulfur dioxide from firing coal in the steel and zinc mills, as initially thought and often reported [5], but was fluoride poisoning from the fluorspar used in the zinc mill [3]. Regardless of the cause, this episode brought increased public awareness of air pollution and helped in the passage of air pollution laws.

A similar, and even more severe, episode occurred in London. London has been renowned for centuries for its thick fog, the infamous London smog. Throughout the eighteenth century, London experienced about 20 foggy days per year, but by the end of the nineteenth century, this had increased to 60 days [1]. People were beginning to become aware of a connection between pollution and certain sicknesses, and bronchitis was initially known as the "British disease." A severe pollution episode occurred in London in December 1952 with substantial loss of life and was followed by a similar one in January 1956 [4]. The more severe episode, which occurred in 1952, happened at a time when London was economically depressed and still recovering from the Second World War [3]. The government was selling its cleaner burning anthracite reserves to help pay its war debt, while nearly all of its households were burning cheap, brown coal for heat. This became exacerbated in December 1952 when a cool spell (and temperature inversion) settled over London. London's 8 million residents burned the brown coal for warmth, and the week-long inversion kept the smoke at ground level. In seven days, an estimated 2,800 people died [3].

8.2 History of Legislative Action for Coal-Fired Power Plants

The major development of air pollution legislative and regulatory acts occurred from 1955 to 1970; however, the early acts were narrow in scope as the U.S. Congress was hesitant to grant the federal government a high degree of control, since air pollution problems were viewed as local or regional. This was found to be impractical, since some states were hesitant to regulate industry, and atmospheric transport of pollutants is not bounded by geographic lines. By the mid-1970s, the basis for national regulation of air pollution was developed, and the actual regulations are continually changing. Regulations on coal-fired emissions essentially started in 1970 with the passing of the Clean Air Act Amendments of 1970. There were a few regulatory changes in the 1980s, but the Clean Air Act Amendments of 1990 resulted in significant regulatory changes. On March 15, 2005, EPA issued the first-ever federal rule to permanently cap and reduce mercury emissions, making the United States the first country in the world to regulate mercury emissions from coal-fired power plants [6]. This rule, however, was vacated by the U.S. Court of Appeals for the District of Columbia Circuit on February 8, 2008. The EPA decided to develop mercury emissions standards for power plants under the Clean Air Act (Section 112). On March 10, 2005, the EPA issued the Clean Air Institute Rule (CAIR) to permanently cap emissions of sulfur dioxide and nitrogen oxides in the eastern United States [7]. The U.S. Court of Appeals for the D.C. Circuit determined that this rule violated the Clean Air Act, and the EPA is currently working on issuing a new rule to replace CAIR. Consequently, there is legislation under consideration to further reduce sulfur dioxide and nitrogen oxides. Legislation for controlling carbon dioxide is currently being debated.

8.2.1 Pre-1970 Legislation

The history of federally enacted air pollution legislation begins with the Air Pollution Control Act of 1955. The act was narrow in scope because of the federal government's hesitation to encroach on states' rights; however, it was the first step toward identifying air pollution sources and its effects, and setting the groundwork for effective legislation and enforcement by regulatory agencies that was developed over the next 15 years. The act initiated the following [4]:

- Research by the U.S. Public Health Service on the effects of air pollution
- Provision for technical assistance to the states by the federal government
- Training of individuals in the area of air pollution
- Research on air pollution control

The Air Pollution Control Act of 1955 was amended in 1960 and 1962 (i.e., Air Pollution Control Act Amendments of 1960 and 1962) because of worsening conditions in urban areas due to mobile sources. Through these acts, Congress directed the Surgeon General to study the effect of motor vehicle exhausts on human health. A more formal process for the continual review of the motor vehicle pollution problem was included in the Clean Air Act of 1963.

The Clean Air Act of 1963 provided, for the first time, federal financial aid for air pollution research and technical assistance [4]. The act encouraged state, regional, and local programs for the control and abatement of air pollution, while reserving federal authority to intervene in interstate conflicts, thereby preserving the classical three-tier system of government. The act provided for the following [4]:

- Acceleration in the research and training program
- Matching grants to state and local agencies for air pollution regulatory control programs
- Developing air quality criteria to be used as guides in setting air quality standards and emissions standards
- Initiating efforts to control air pollution from all federal facilities
- Federal authority to abate interstate air pollution
- Encouraging efforts by automotive companies and the fuel industries to prevent pollution.

The Clean Air Act of 1963 also provided for research authority to develop standards for sulfur removal from fuels and a formal process for reviewing the status of the motor vehicle pollution problem. This, in turn, led to the Motor Vehicle Air Pollution Control Act of 1965, which formally recognized the technical and economic feasibility of setting automotive emission standards. The act also gave the secretary of the Department of Health, Education, and Welfare (HEW) the authority to intervene in intrastate air pollution problems of "substantial significance."

The National Air Quality Control Act of 1967

The first federal legislation to impact stationary combustion sources was the National Air Quality Control Act of 1967. The act provided for a two-year study on the concept of national emissions standards for stationary sources and was the basis for the 1970 legislative action. Among the provisions of the National Air Quality Control Act of 1967 were the following [4]:

- Establishment of eight specific areas in the United States on the basis of common meteorology, topography, and climate
- Designation of air quality control regions (AQCR) within the United States, where evaluations were to be conducted to determine the nature and extent of the air pollution problem

- Development and issuance of air quality criteria for specific pollutants that have identifiable effects on human health and welfare
- Development and issuance of information on recommended air pollution control techniques, which would lead to recommended technologies to achieve the levels of air quality suggested in the AQC reports
- Requirement of a fixed-time schedule for state and local agencies to establish air quality standards consistent with air quality criteria. The states were allowed to set higher standards than recommended in the AQC reports; however, if a state did not act, the secretary of HEW had the authority to establish air quality standards for each air quality region. The states were given primary responsibility for action, but a very strong federal fallback authority was provided.

The federal program was not implemented within the required time schedule because federal surveillance of the overall program was understaffed and the process to set up the AQCRs proved to be too complex. Consequently, both President Nixon and Congress proposed new legislation in 1970.

8.2.2 The Clean Air Act Amendments of 1970

The Clean Air Act Amendments of 1970 extended the geographical coverage of the federal program aimed at the prevention, control, and abatement of air pollution from stationary and mobile sources. The act transferred administrative functions assigned to the secretary of HEW to the newly created Environmental Protection Agency (EPA). The act provided for the first time national ambient air quality standards and national emission standards for new stationary sources. It initiated the study of aircraft emissions and imposed carbon monoxide, hydrocarbons, and nitrogen oxide emissions control on automobiles.

The major goal of the act was the achievement of clean air throughout the United States by the middle of the decade. Two types of pollutants were to be regulated: criteria air pollutants and hazardous air pollutants (HAPs). The criteria air pollutants were to be regulated in order to achieve attainment of National Ambient Air Quality Standards by establishing emission standards (which were developed by state and local agencies) for existing sources and national emission standards for new sources through promulgation of New Source Performance Standards. HAPs were to be regulated under National Emission Standards for Hazardous Air Pollutants. These were the major provisions of the act [4]:

- The EPA was to establish National Ambient Air Quality Standards (NAAQS), including primary standards for the protection of public health and secondary standards for the protection of public welfare.
- New Source Performance Standards (NSPS) were to be required, with each state implementing and enforcing the standard of performance. Before a new stationary source could begin operation, state or federal inspectors were required to certify that the controls would function and the new stationary sources had to remain in compliance throughout the lifetime of the plant.
- National Emission Standards for Hazardous Air Pollutants (NESHAP) were to be established and would apply to existing as well as new plants.

- Funding was provided for fundamental air pollution studies, research on health and welfare effects of air pollutants, research on cause and effects of noise pollution, and research on fuels at stationary sources, including methods of cleaning fuels prior to combustion rather than flue gas cleaning techniques, improved combustion techniques, and methods for producing new or synthetic fuels with lower potential for creating polluted emissions.
- State and regional grant programs were authorized, and matching grants were established for implementing standards.
- The designation of air quality control regions was to be completed.
- Statewide implementation plans (SIPs) designed to achieve primary or public health standards had to be enacted within three years, and the overall plan had to be convincing as to its ability to meet and maintain the standards.
- Industry was required to monitor and maintain emission records and make these records available to EPA officials, and EPA was given the right of entry to examine records.
- Fines and criminal penalties were imposed for violation of implementation plans, emission standards, and performance standards that were stricter than those under earlier law.
- New automobile emission standards were set.
- Aircraft emission standards were to be developed by EPA.
- Citizens were permitted to sue those alleged to be in violation of emission standards, including the United States, and suits could be brought against the EPA administrator if he failed to act in cases where the law specified that he must.

Air Quality Criteria and National Ambient Air Quality Standards

The Air Quality Act of 1967 addressed the development and issuance of air quality criteria (AQC), and the need for such criteria was reaffirmed in the 1970 amendments. AQC indicates, both qualitatively and quantitatively, the relationship between various levels of exposure to pollutants and the short- and long-term effects on health and welfare [4]. Air quality criteria describe effects that can be expected to occur when pollutant levels reach or exceed specific values over a given time period and delineate the effects from combinations of contaminants as well as from individual pollutants. Economic and technical considerations are not relevant to the establishment of AQC.

The development of AQC is essential in providing a quantitative basis for air quality standards. Standards prescribe the pollutant levels that cannot be legally exceeded during a specific time period in a specific geographical region. The Clean Air Act Amendments of 1970 required federal promulgation of national primary and secondary standards that are to be established equitably in terms of the social, political, technological, and economic aspects of the problem. Standards are subject to revision as aspects change over time.

The purpose of the primary standards is immediate protection of the public health, including the health of sensitive populations such as asthmatics, children, and the elderly. Primary standards are to be achieved regardless of cost and within a specified time limit. Secondary standards are to protect the public welfare from known or anticipated adverse effects, including protection against decreased visibility, damage to animals, crops, vegetation, and buildings. Both standards have to be consistent with AQC, and in addition, the standards have to prevent the continuing deterioration of air quality in any portion of an air quality control region.

The Clean Air Act Amendments of 1970 defined the first six criteria pollutants as carbon monoxide, nitrogen dioxide, sulfur dioxide, total particulate matter, hydrocarbons, and photochemical oxidants, and NAAQS were established for these. Subsequently, the list has been revised with the following actions [4]:

- Lead was added to the list in 1976.
- The photochemical oxidant standard was revised and restated as ozone in 1979.
- The hydrocarbon standard was withdrawn in 1983.
- The total suspended particulate matter standard was revised in 1987 to include only particles with an aerodynamic particle size of less than or equal to 10 μm and referred to as the PM_{10} standard.
- The $PM_{2.5}$ (i.e., particles with an aerodynamic particle size of less than or equal to 2.5 μm) standard was added in 1997.
- In 1997, the EPA reviewed the air quality standard for ground-level ozone and established the primary and secondary eight-hour NAAQS for ozone at 0.08 ppm and the primary and secondary one-hour NAAQS for ozone at 0.12 ppm.

The current list of NAAQS is shown in Table 8.1 and includes carbon monoxide, nitrogen dioxide, ozone, lead, PM_{10}, $PM_{2.5}$, and sulfur dioxide [8].

National Emission Standards

Emission standards place a limit on the amount or concentration of a pollutant that may be emitted from a source. It is often necessary for certain industries to be regulated by emission standards promulgated by the federal or state government in order to maintain or improve ambient air quality within a region to comply with national or state air quality standards. Several factors must be considered when establishing emission standards [4]:

- The availability of technology that is appropriate for the cleanup of a given type of industry must be determined.
- Monitoring stations must be available to measure the actual industrial emissions for which control is considered, as well as the ambient air quality so the effectiveness of the standards can be determined.
- Regulatory agencies must be organized to cope with the measurement and enforcement of the standards.
- The synergistic effects of various pollutants must be determined.
- Models must be developed that reasonably predict the effects of reducing various emissions on the ambient air quality.
- Reasonable estimates of future emissions must be made based on the growth or decline of industry and population within a region.

Emission standards can be categorized into the following general types [4]:

- *Visible emissions standards*—the opacity of the plume from a stack or a point of fugitive emissions is not to equal or exceeds a specified opacity.

Table 8.1 National Ambient Air Quality Standards

Pollutant	Standard Value		Standard Type
Carbon Monoxide (CO)			
8-hour average	10 mg/m^3	9 ppm	Primary (no secondary)
1-hour average	40 mg/m^3	35 ppm	Primary (no secondary)
Nitrogen Dioxide (NO$_2$)			
Annual arithmetic mean	100 μg/m^3	0.053 ppm	Primary and secondary
Ozone (O$_3$)			
8-hour average[a] (1997 standard)	157 μg/m^3	0.08 ppm	Primary and secondary
8-hour average[b] (2008 standard)		0.075 ppm	Primary and secondary
1-hour average	235 μg/m^3	0.12 ppm	Primary and secondary
Lead (Pb)			
Rolling 3-month average	0.15 μg/m^3		Primary and secondary
Quarterly average	1.5 μg/m^3		Primary and secondary
Particulate (PM$_{10}$)			
24-hour average	150 μg/m^3		Primary and secondary
Particulate (PM$_{2.5}$)			
Annual arithmetic mean	15 μg/m^3		Primary and secondary
24-hour average	35 μg/m^3		Primary and secondary
Sulfur Dioxide (SO$_2$)			
Annual arithmetic mean	80 μg/m^3	0.03 ppm	Primary
24-hour average	365 μg/m^3	0.14 ppm	Primary
3-hour average	$1,300 \text{ μg/m}^3$	0.50 ppm	Secondary

- *Particulate concentration standards*—the maximum allowable emission rate is specified in mass/volume: grams per dry standard cubic meter (g/dscm) or grains per dry standard cubic foot (gr/dscf). For combustion processes, it is common to specify the concentration at a fixed oxygen (O_2) or carbon dioxide (CO_2) level to avoid dilution, thereby lowering the concentration.
- *Particulate process weight (or mass) standards*—the maximum allowable particulate emissions are tied to the actual mass of material being processed or used, and in combustion systems the standards are commonly reported in pounds of particulate matter per million Btu of fuel burned (lb/10^6 Btu or lb/MM Btu).
- *Gas concentration standards*—gas standards are typically reported in mass per volume or volume per volume (g/dscm or ppm).
- *Prohibition of emissions*—processes are banned outright.
- *Fuel regulations*—fuel standards may be specified for various fuel-burning equipment such as limiting sulfur concentration in a fuel.
- *Zoning restrictions*—emissions may be limited by passing zoning ordinances that dictate facilities that can be constructed.
- *Dispersion-based standards*—these standards limit the allowable emission of pollutants based on their contribution to the ambient air quality.

National emission or performance standards have been set for a number of industries, including fossil fuel-fired electric utility steam generating units. National standards are necessary for industries that are spread geographically across the country and provide a basic commodity that is essential for the development of the country. Unfair economic advantages might be gained by the states if the standards were set by individual states, with one state relaxing the standards to attract industries. These national emissions standards are referred to as New Source Performance Standards (NSPS) and apply to construction of new sources as well as sources that undergo operational and physical changes that either increase emission rates or initiate new emissions from the plant [4].

40CFR Part 60, Subpart D

On December 23, 1971, the first five final standards were published, and since then the EPA administrator has promulgated nearly 75 standards. The complete text for each NSPS is available in the Code of Federal Register at Title 40 (Protection of the Environment), Part 60 (Standards of Performance for New Stationary Sources) with steam electric plants found in Subpart D [9].

The 1971 regulations addressed coal use in utility and industrial steam-generation units, and the regulations were amended for lignite in March 7, 1978. The maximum SO_2 emissions allowable from electric utility steam generating units of more than 73 MW (megawatts) or 250 million (MM) Btu/h of heat input and under construction or modification after August 17, 1971, was 1.2 lb SO_2/MM Btu for solid fuels. Particulate matter was limited to 0.10 lb/MM Btu heat input, and opacity was not to exceed 20 percent for one 6-minute period per hour. NO_x standards were 0.70 lb NO_x/MM Btu heat input for solid fossil fuel or solid fossil fuel and wood residue (except for lignite); 0.60 lb NO_x/MM Btu heat input for lignite or lignite and wood residue; and 0.80 lb NO_x/MM Btu for lignite mined in North Dakota, South Dakota, or Montana and burned in a cyclone-fired unit (various combustion technologies are discussed in Chapter 5).

40 CFR Part 60 Subpart Da

The original regulations were significantly revised as of February 6, 1980. The EPA promulgated new regulations for control of SO_2, NO_x, and particulate matter from steam generating units with more than 73 MW or 250 MM Btu/h of heat input (40 CFR Part 60 Subpart Da) and under construction after September 18, 1978.

For a coal-fired unit, the 1980 standard requires at least a 90 percent reduction of potential SO_2 emissions and limits the rate to 1.2 lb SO_2/MM Btu heat input, or requires at least a 70 percent reduction and limits the emission rate to 0.6 lb SO_2/MM Btu heat input. In addition, the standard specifies a unique maximum allowable emission rate and unique minimum reduction of potential emission based on the sulfur content and heating value of the coal [10]. If the uncontrolled emission rate (UER) is determined, then the required efficiency is as shown in the table that follows.

UER of SO_2	Required Efficiency or Maximum Allowable Efficiency Rate
<2 lb SO_2/MM Btu	70%
2–6 lb SO_2/MM Btu	0.6 lb SO_2/MM Btu; efficiency (%) = [(UER – 0.6)/UER] × 100%
6–12 lb SO_2/MM Btu	90%
>12 lb SO_2/MM Btu	1.2 lb SO_2/MM Btu; efficiency (%) = [(UER – 1.2)/UER] × 100%

The NO_x standard varies according to fuel type and, with respect to coal, is 0.50 lb NO_x/MM Btu heat input from subbituminous coal, shale oil, or any solid, liquid, or gaseous fuel derived from coal; 0.80 lb NO_x/MM Btu heat input from the combustion in a slag tap furnace of any fuel containing more than 25 percent, by weight, lignite that has been mined in North Dakota, South Dakota, or Montana; and 0.60 lb NO_x/MM Btu heat input from anthracite or bituminous coal. The NO_x standard is based on a 30-day rolling average. Particulate emissions are limited to 0.03 lb/MM Btu heat input. In addition, opacity is limited to 20 percent for a 6-minute average.

The NO_x standards for Subparts Da and Db (see following for Subpart Db) were revised on September 16, 1998 [11]. Only those electric utility steam generating units for which construction, modification, or reconstruction is commenced after July 9, 1997, would be affected by these revisions; they changed the existing standards for NO_x emission limits to reflect the performance of best-demonstrated technology. The revisions also changed the format of the revised NO_x emission limit for new electric utility steam generating units to an output-based format to promote energy efficiency and pollution prevention.

The NO_x emission limit in Subpart Da is 1.6 lb NO_x/megawatt-hour (MWh) gross energy output regardless of fuel type for new utility boilers. For existing utility boilers that would become subject to the standards due to a modification or reconstruction, EPA revised the NO_x limit to be consistent with the requirements for new units but expressed the emission limits in an equivalent input-based format, which is 0.15 lb NO_x/MM Btu. This provision was withdrawn by the Environmental Protection Agency on August 7, 2001, however, after industry groups filed petitions for review and a motion to vacate the standards as applied to modified boilers in the U.S. Court of Appeals [12]. On September 21, 1999, the court issued an order granting the petitioner's motion, and as a result, owners and operators of electric utility boilers on which modification is commenced after July 9, 1997, will be required to comply with the applicable nitrogen oxides emission limits specified in the preexisting NSPS, which is 0.50 lb NO_x/MM Btu.

40 CFR Part 60 Subparts Db and Dc

Although the acid rain provisions of the 1990 Clean Air Act Amendments, which are discussed later in this chapter, place additional requirements on the electric utility industry, the NSPS, as revised in 1980, is still applicable for new sources. Smaller sources have been addressed over time, including industrial/commercial/institutional steam generators constructed after June 19, 1984 with heat inputs of 29 to 73 MW (40 CFR Part 60 Subpart Db) and small industrial/commercial/

institutional steam generating units with 2.9 to 29 MW of heat input constructed after June 9, 1989 (40 CFR Part 60 Subpart Dc).

On November 25, 1986, standards of performance for industrial/commercial/ institutional steam generation units were promulgated [13]. Coal-fired facilities that have a heat input capacity of between 29 and 73 MW (100 and 250 million Btu/h) are subject to the particulate matter and NO_x standards under 40 CFR Part 60 Subpart Db. Particulate matter (PM) standards for coal are more complicated than previous regulations and are as follows:

- 0.05 lb PM/MM Btu heat input if the facility combusts only coal or combusts coal and other fuels and has an annual capacity factor for the other fuels that is 10 percent less or
- 0.10 lb PM/MM Btu heat input if the facility combusts coal and other fuels and has an annual capacity factor for the other fuels greater than 10 percent and is subject to a federally enforceable requirement limiting operation of the affected facility to annual capacity factor greater than 10 percent for fuels other than coal
- 0.20 lb PM/MM Btu heat input if the affected facility combusts coal or coal and other fuels and has an annual capacity factor for coal or coal and other fuels of 30 percent or less; has a maximum heat input capacity of 73 MW (250 million Btu/h) or less; had a federally enforceable requirement limiting operation of the affected facility to an annual capacity factor of 30 percent or less for coal or coal and other solid fuels; and construction of the facility commenced after June 19, 1984, and before November 25, 1986

The NO_x standards for coal-fired facilities identified in 1986 in Subpart Db are as follows:

- 0.50 lb NO_x (expressed as NO_2)/MM Btu for mass-feed stokers
- 0.60 lb NO_x/MM Btu for spreader stoker and fluidized bed combustors
- 0.70 lb NO_x/MM Btu for pulverized coal-fired units
- 0.60 lb NO_x/MM Btu for lignite units except for lignite mined in North Dakota, South Dakota, or Montana and combusted in a slag tap furnace for which the emissions limits are 0.8 lb NO_x/MM Btu
- 0.50 lb NO_x/MM Btu for coal-derived synthetic fuels

As previously mentioned, the NO_x standard for Subpart Db was revised on September 16, 1998 [11]. Only those industrial steam generating units for which construction, modification, or reconstruction was commenced after July 9, 1997, would be affected by these revisions. For coal-fired Subpart Db units, the NO_x emission limit promulgated was 0.20 lb/MM Btu heat input. However, this provision was withdrawn on August 7, 2001, and owners and operators of industrial/commercial/ institutional boilers on which modification was commenced after July 9, 1997, would be required to comply with the applicable nitrogen oxides emission limits specified in the preexisting NSPS (i.e., 0.50 to 0.80 lb NO_x/MM Btu depending on fuel type and boiler configuration as just listed) [12].

On September 12, 1990, standards of performance for small industrial/commercial/ institutional steam generation units were promulgated (40 CFR Part 60 Subpart Dc) [14]. Under Subpart Dc, coal-fired facilities that have a heat input capacity of between 2.9 and 29 MW (10 and 100 million Btu/h) are not subject to NO_x standards, nor are they subject to SO_2 or particulate matter emissions limits during periods of combustion research. For nonresearch operations, SO_2 standards are as follows:

- When firing only coal, SO_2 emissions are limited to no more than 10 percent of the potential SO_2 emissions rate (i.e., 90 percent reduction) and less than 1.2 lb SO_2/MM Btu heat input.
- If coal is combusted with other fuels, the affected facility is subject to the 90 percent SO_2 reduction requirement, and the emission limit is determined by the equation

$$E_{SO_2} = (K_a H_a + K_b H_b + K_c H_c)/(H_a + H_b + H_c) \qquad (8.1)$$

where

E_{SO_2} is the SO_2 emission limit, expressed in lb/million Btu heat input
K_a is 1.2 lb/million Btu
K_b is 0.60 lb/million Btu
K_c is 0.50 lb/million Btu
H_a is the heat input from the combustion of coal, except coal combusted in a facility that uses an emerging technology for SO_2 control, in million Btu
H_b is the heat input from the combustion of coal in a facility that uses an emerging technology for SO_2 control, in million Btu
H_c is the heat input from the combustion of oil in million Btu

- When firing only coal in a facility that uses an emerging technology for SO_2 control, SO_2 emissions are limited to no more than 50 percent of the potential SO_2 emissions rate (i.e., a 50 percent reduction) and less than 0.60 lb SO_2/MM Btu heat input.
- If coal is combusted with other fuels, 50 percent SO_2 reduction is required, and the emission limit is determined using Eq. (8.1).
- Percent reduction requirements are not applicable for affected facilities that have input capacity of 22 MW (75 million Btu/h) or less; affected facilities that have an annual capacity for coal of 55 percent or less; affected facilities located in a noncontinental area; and affected facilities that combust coal in a duct burner as part of a combined cycle system where 30 percent or less of the heat entering the steam generating unit is from combustion of coal in the duct burner and 70 percent or more of the heat entering the steam generating unit is from exhaust gases entering the duct burner.
- Reduction of the potential SO_2 emission rate through fuel pretreatment is not credited toward the percent reduction requirement unless fuel pretreatment results in a 50 percent or greater reduction in the potential SO_2 emission rate and emissions from the pretreated fuel (without either combustion or post-combustion SO_2 control) are equal or less than 0.60 lb/MM Btu.

The particulate matter standards in Subpart Dc state that a facility that combusts coal or mixtures of coal with other fuels and has a heat input capacity of 8.7 MW (30 million Btu/h) or greater is subject to the following emission limits:

- 0.05 lb PM/MM Btu heat input if the facility combusts only coal, or combusts coal with other fuels and has an annual capacity factor for the other fuels of 10 percent or less; and
- 0.10 lb PM/MM Btu heat input if the facility combusts coal with other fuels, has an annual capacity factor for the other fuels greater than 10 percent, and is subject to a federally enforceable requirment limiting operation of the facility to an annual capacity factor greater than 10 percent for fuels other than coal.

Emission Factors

Once an NSPS has been established, it is necessary for new sources constructed after a defined date to meet the standards. It is not possible to sample new sources to determine required collection or removal efficiencies. In these cases, knowledge of the emission factors for the specific regulated pollutant from these sources is used to estimate the approximate level of control required to meet the NSPS. The U.S. EPA has published a document, *Compilation of Air Pollutant Emission Factors*, referred to as *AP-42*, since 1972. Supplements to AP-42 have been routinely published to add new emissions source categories and to update existing emission factors. This document is also provided on EPA's website on their CHIEF (Clearinghouse for Inventories and Emissions Factors; *www.epa.gov/ttn/ chief/ap42*) bulletin board.

The EPA routinely updates *AP-42* in order to respond to the new emission factor needs of state and local air pollution control programs, industry, and the agency itself. The current emission factors for bituminous and subbituminous coal, lignite, and anthracite firing are provided in Appendix C and are from Sections 1.1 ("Bituminous and Subbituminous Coal Combustion"), 1.2 ("Anthracite Coal Combustion"), and 1.7 ("Lignite Combustion") of *AP-42*, Fifth Edition, Volume I, Supplements A through G [16]. The emission factors, which are not inclusive, have been developed for the following:

*   Various fuel firing configurations
*   Uncontrolled and controlled emissions
*   Criteria gaseous pollutants (SO_x, NO_x, CO)
*   Filterable particulate matter and condensable particulate matter
*   Trace elements
*   Various polynuclear organic matter (POM), polynuclear aromatic hydrocarbons (PAH), and organic compounds
*   Acid gases (HCl and HF)
*   Other gaseous pollutants such as CO_2, N_2O, and CH_4
*   Cumulative ash particle size distribution and size-specific emissions

Emission factors and emissions inventories have long been fundamental tools for air quality management. Emission estimates are important for developing emission control strategies, determining the applicability of permitting and controlling programs, and ascertaining the effects of sources and appropriate mitigation strategies. Users include federal, state, and local agencies; consultants; and industry. Data from source-specific emission tests or continuous emission monitors are usually preferred for estimating a source's emissions because those data provide the best representation of the tested source's emissions. However, test data from individual sources are not always available, and they may not reflect the variability of actual emissions over time. Consequently, emission factors are often the best or only method available for estimating emissions.

The passage of the Clean Air Act Amendments of 1990 and the Emergency Planning and Community Right-to-Know Act of 1986 has increased the need for both criteria and hazardous air pollutant emission factors and inventories. The Emission

Factor and Inventory Group (EFIG), in the EPA's Office of Air Quality Planning and Standards develops and maintains emission-estimating tools. The *AP-42* series is the principal means by which EFIG can document its emission factors.

Emission factors may be appropriate for use in a number of situations, such as source-specific emission estimates for area-wide inventories. These inventories have many purposes including ambient dispersion modeling and analysis, control strategy development, screening sources for compliance investigations, and some permitting applications. The emission factors in *AP-42* are neither EPA-recommended emission limits (e.g., Best Available Control Technology, BACT, or Lowest Achievable Emission Rate, LEAR) nor standards (e.g., NSPS or NESHAP).

Figure 8.1 depicts various approaches to emission estimation in a hierarchy of requirements and levels of sophistication that need to be considered when analyzing the tradeoffs between cost of the estimates and the quality of the resulting estimates. More sophisticated and more costly emission determination methods may be neces-sary where risks of either adverse environmental effects or adverse regulatory out-comes are high. Less expensive estimation techniques, such as emission factors and emission models, may be appropriate and satisfactory where risks of using a poor estimate are low. Note that the reliability of the *AP-42* emission factors are rated from A through E, which is a general indication of the robustness of that fac-tor. This rating is assigned based on the estimated reliability of the tests used to develop the factor. In general, factors based on many observations, or on more widely accepted test procedures, are assigned higher rankings with A being the best.

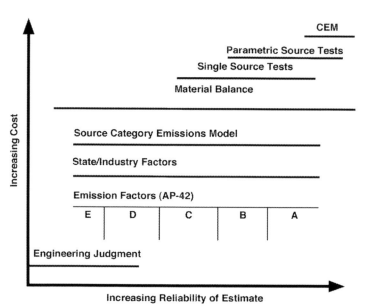

Figure 8.1 Approach to emission estimation.
Source: From *AP-42* (1998).

The emission factors rating is provided with the list of emission factors contained in Appendix C.

National Emission Standards for Hazardous Air Pollutants

The 1970 Clean Air Act Amendments also provided for national emission standards for hazardous air pollutants (NESHAPs). Only seven standards were established between 1970 and 1990, since the NESHAPs were the subject of numerous suits and court decisions regarding how to address emission limits on carcinogenic pollutants [4]. None of the original seven NESHAPs—asbestos, beryllium, mercury, vinyl chloride, benzene, radionuclides, or arsenic—specifically addressed coal-fired steam generating units. The Clean Air Act Amendments of 1990 made major changes in the approach taken to address HAPs, with amendments still being made as recently as the fall of 2009 [17]. HAPs are discussed in more detail later in this chapter.

8.2.3 Clean Air Act Amendments of 1977 and Prevention of Significant Deterioration

By 1977 most areas of the country had still not attained the National Ambient Air Quality Standards for at least one pollutant [4]. For those areas that had not attained an NAAQS (i.e., nonattainment areas), states were required to submit and have an approved state implementation plan (SIP) revision by July 1, 1979, that demonstrated how attainment would be achieved by December 31, 1982. This requirement was a precondition for the construction or modification of major emission sources in nonattainment areas after June 30, 1979. If a state could not attain primary standards for carbon monoxide or photochemical oxidants after implementation of all reasonably available measures, it was required to submit a second SIP revision by December 31, 1982, that would demonstrate how attainment would be achieved by December 31, 1987.

Prevention of Significant Deterioration

The concern over nonattainment areas and the controversy generated by the 1970 Clean Air Act Amendments' provision on standards preventing continuing deterioration of air quality led to a set of guidelines issued in 1974 by the EPA to prevent the significant deterioration of air quality in areas that were cleaner than that required by NAAQS—that is, that were in attainment of the NAAQS. This was necessary because some interpreted the 1970 act to mean that a region could not backslide in air quality even though the present air quality may be superior to the national standard. This interpretation would have stifled economic growth in a region (new industrial and commercial operations in the region could not contribute zero pollution), yet it would have failed to force sources in the region to decrease their contaminant emissions. This concern led to the passage of the regulations on prevention of significant deterioration (PSD). These regulations for attainment areas required that all areas be designated as Class I, II, or III, depending on the degree of deterioration to be allowed, and that incremental limits were placed on the amount of increase in deterioration allowed. These are the classifications [4]:

Class I—Pristine areas, including international parks, national parks, and national wilderness areas in which very little deterioration would be allowed

Class II—Areas where moderate change will be allowed but where stringent air quality constraints are desirable

Class III—Areas where major growth and industrialization would be allowed.

Congress specified which of the areas must be protected by the most stringent Class I designation, designated all other areas within the United States as Class II areas, and provided the option for redesignation of Class II areas to Class I or Class III areas by public referendum. Congress also specified the maximum allowable incremental increases in concentration of sulfur dioxide and particulate matter and charged the EPA to determine comparable increments for hydrocarbons, carbon monoxide, photochemical oxidants, and nitrogen oxides. The PSD increments are listed in Table 8.2 [4].

A requirement was placed on major sources in the preconstruction PSD review process that specified that each major new plant must install BACT, which was defined to be at least as stringent as NSPS, to limit its emissions. Major sources subject to PSD review are those with the potential to emit 100 short tons or more per year of any regulated pollutant under the Clean Air Act Amendment of 1977 [4]. All sources emitting greater than 250 short tons per year are subject to PSD review. For non-NSPS sources, a BACT review document is prepared in the preconstruction review.

Table 8.2 PSD Increments

Pollutant	Maximum Allowable Increase (μg per m^3)		
	Class I	Class II	Class III
Particulate Matter			
PM_{10}, annual geometric mean	5	19	37
PM_{10}, 24-hour maximum	10	37	75
Sulfur Dioxide			
Annual arithmetic mean	2	20	40
24-hour maximum	5	91	182
3-hour maximum	25	512	700
Nitrogen Dioxide			
Annual arithmetic mean	2.5	25	50

Source: From Wark, Warner, and Davis (1998).

The *1974 Guidelines on PSD* and the Clean Air Amendments of 1977 established the protocol for new sources and proposed modifications to major sources in attainment areas, where they would be subjected to a new source review and would have to meet certain PSD requirements. This review requires that dispersion modeling be conducted of the proposed emissions from the sources to ensure that the emissions from the proposed facility would not exceed the increments listed in Table 8.2 or cause an exceedance of the NAAQS. In concept, sources are allowed to use some fraction of the increment, as determined by the state or local agency and based on dispersion modeling.

The PSD increment can only be used to the extent that it does not cause the ambient concentration to exceed the NAAQS. Another important issue is the extent to which any single source would be allowed to use the available increment and the ramifications on long-term growth if a single source was allowed to use it all. Some state and local agencies address this on a case-by-case basis, while others have taken the approach of allowing only a certain fraction of the increment, or the remaining increment, to be used in a single PSD application [4].

Nonattainment Areas

It was feared that no industrial growth could occur in nonattainment areas, since these areas are in violation of one or more NAAQSs, and because PSD does not apply in these areas, there are no increments available. This was remedied on December 21, 1976, with an interpretive ruling, the Offsets Policy Interpretive Ruling, on the preconstruction review requirements for all new or modified stationary sources of air pollution in nonattainment areas [4]. This ruling, also known as the emission offset policy, requires that three conditions be met:

1. The source must meet the lowest achievable emission rate (LAER), defined as being more stringent as BACT.
2. All existing sources owned by the applicant in the same region must be in compliance or under an approved schedule to achieve compliance.
3. The source must provide an offset or reduction of emissions from other sources greater than the proposed emissions that the source contribute such that there is a net improvement in air quality.

The Clean Air Act Amendments of 1977 also provided for the banking of offsets. If the offsets achieved are considerably greater than the new source's emissions, a portion of this excess emission reduction (also known as emission reduction credits) can be banked by the source for use in future growth or traded to another source, depending on each state's offset/trading policy.

8.2.4   Clean Air Act Amendments of 1990

In June 1989, President George W. Bush proposed major revisions to the Clean Air Act. Both the House of Representatives and the Senate passed Clean Air bills by large votes that contained the major components of the president's proposals. After a joint conference committee met to work out the differences in the bills, Congress voted out the package recommended by the conferees, and President Bush signed the bill, the Clean Air Amendments of 1990 (1990 CAAA), on November 15, 1990. The 1990 CAAAs are the most substantive regulations adopted since the passage of the Clean Air Act Amendments of 1970. Specifically, the new law does the following:

- Encourages the use of market-based principles and other innovative approaches, like performance-based standards and emission banking and trading
- Provides a framework from which alternative clean fuels will be used by setting standards in the fleet and a California pilot program that can be met by the most cost-effective combination of fuels and technology
- Promotes the use of clean low-sulfur coal and natural gas, as well as innovative technologies to clean high-sulfur coal through the acid rain program
- Reduces enough energy waste and creates enough of a market for clean fuels derived from grain and natural gas to cut dependency on oil imports by 1 million barrels per day
- Promotes energy conservation through an acid rain program that gives utilities flexibility to obtain needed emission reductions through programs that encourage customers to conserve energy

The 1990 CAAA contains eleven major divisions, referred to as Titles I through XI, which either provided amendments to existing titles and sections of the Clean Air Act or provided new titles and sections. These are the titles for the 1990 CAAA:

Title I: Provisions for Attainment and Maintenance of National Ambient Air Quality
   Standards
Title II: Provisions Relating to Mobile Sources
Title III: Air Toxics
Title IV: Acid Deposition Control
Title V: Permits
Title VI: Stratospheric Ozone and Global Climate Protection
Title VII: Provisions Relating to Enforcement
Title VIII: Miscellaneous Provisions
Title IX: Clean Air Research
Title X: Disadvantaged Business Concerns
Title XI: Clean Air Deployment Transition Assistance

The titles that directly impact the use of coal—Titles I, III, IV, and V—are discussed in the following sections. The concepts of NAAQS, NSPS, and PSD, as defined in Title I, remained virtually unchanged; however, major changes have occurred in regulations and approaches used to address nonattainment areas in Title I, hazardous air pollutants in Title III, acid rain in Title IV, and permitting in Title V.

Title I: Provisions for Attainment and Maintenance of National Ambient Air Quality Standards

Although the Clean Air Act Amendments of 1970 and 1977 brought about significant improvements in air quality, urban air pollution persisted, and there were many cities out of attainment for ozone, carbon monoxide, and PM_{10}. Of these, the most widespread pollution problem is ozone (i.e., smog). One component of smog—hydrocarbons—comes from automobile emissions, petroleum refineries, chemical plants, dry cleaners, gasoline stations, house painting, and printing shops, while another key component—nitrogen oxides—comes from the combustion of fossil fuels for transportation, utilities, and industries [18].

The 1990 CAAA created a new, balanced strategy for the nation to address urban smog. The new law gave states more time to meet the air quality standard, but it also required states to make constant progress in reducing emissions. It required the federal government to reduce emissions from cars, trucks, and buses; from consumer products such as hair spray and window washing compounds; and from ships and barges during loading and unloading of petroleum products. The federal government must also develop guidance that states the need to control stationary sources.

The 1990 CAAA addresses the urban air pollution problems of ozone, carbon monoxide, and PM_{10}. Specifically, it clarifies how areas are designated, and it redefined attainment. The 1990 CAAA also allows EPA to define the boundaries of nonattainment areas and establishes provisions defining when and how the federal government can impose sanctions on areas of the country that have not met certain conditions.

For the pollutant ozone, the 1990 Clean Air Act Amendments establish nonattainment area classifications ranked according to the severity of the area's air pollution problem. These classifications are marginal, moderate, serious, severe, and extreme, and were designated based on the air quality of the nonattainment area during the period 1987 to 1989. EPA assigns each nonattainment area one of these categories, thus triggering varying requirements the area must comply with in order to meet the ozone standard. Table 8.3 lists the ozone design value, which determines the classification, attainment deadlines, minimum size of a new or modified source that would be affected, and offset requirements [4].

Table 8.3 Classifications for Nonattainment Areas

Pollutant	Classification	Design Value (ppm)	Attainment Deadline	Major Source (short tons VOCs/y for ozone; short tons CO/y for carbon monoxide)	Offset Ratio for Sources
Ozone	Marginal	0.121–0.138	11/15/1993	100	1.1 to 1
	Moderate	0.138–0.160	11/15/1996	100	1.15 to 1
	Serious	0.160–0.180	11/15/1999	50	1.2 to 1
	Severe	0.180–0.190	11/15/2005	25	1.3 to 1
	Severe	0.190–0.280	11/15/2007	25	1.3 to 1
	Extreme	>0.280	11/15/2007	10	1.5 to 1
Carbon monoxide	Moderate	9.1–16.4	12/31/1995	—	none
	Serious	>16.4	12/31/2000	50	none

Nonattaiment areas have to implement different control measures, depending on their classification. Nonattainment areas with worse air quality problems must implement more control measures. For example, in areas classified as extreme, boilers with emission rates greater than 25 short tons per year are required to burn clean fuels or install advanced control technologies.

The 1990 CAAA also establishes similar programs for areas that do not meet federal health standards for carbon monoxide and PM_{10}. Areas exceeding the standards for these pollutants are divided into moderate and serious classifications. The classifications for nonattainment areas for carbon monoxide are shown in Table 8.3.

Areas that were nonattainment for PM_{10} at the time the 1990 CAAA was passed were designated as moderate areas with an attainment deadline of December 31, 1994 [4]. Areas classified as nonattainment subsequent to passage of the 1990 CAAA are designated moderate with six years to achieve compliance. Major sources in moderate areas are those that emit 100 short tons or more of particulate matter per year. Moderate areas require the adoption of reasonably available control measures (RACM).

Moderate areas that fail to reach attainment are redesignated as serious areas and have ten years from the date of designation as nonattainment to achieve attainment. For serious areas, major sources include those that emit 70 short tons or more of particulate matter per year. Serious areas must also adopt best available control measures (BACM).

Title III: Air Toxics

Hazardous air pollutants (HAPs), also known as toxic air pollutants or air toxics, are those pollutants that cause or may cause cancer or other serious health effects, such as reproductive effects or birth defects, or adverse environmental and ecological effects, but are not specifically covered under another portion of the Clean Air Act. Most air toxics originate from human-made sources, including mobile sources (e.g., cars, trucks, buses), stationary sources (e.g., factories, refineries, power plants), and indoor sources (e.g., building materials and activities such as cleaning) [18]. The Clean Air Act Amendments of 1977 failed to result in substantial reductions of the emissions of these very threatening substances. Over the history of the air toxics program, only seven pollutants had been regulated. Title III established a list of 189 (later modified to 188) HAPs associated with approximately 300 major source categories. The list of HAPs is provided in Appendix D [19]. Under Title III, a major source is defined as any new or existing source with the potential to emit, after controls, 10 short tons or more per year of any of the 188 HAPs or 25 short tons or more per year of any combination of those pollutants. These sources may release air toxics from equipment leaks, when materials are transferred from one location to another, or during discharge through emissions stacks or vents.

The EPA must then issue maximum achievable control technology (MACT) standards for each listed source category according to a prescribed schedule. These standards will be based on the best demonstrated control technology or practices within the regulated industry, and the prescribed schedule dictated that the EPA must issue the standards for 40 source categories within two years, 25 percent of the source categories within five years, 50 percent of the source categories within seven years, and 100 percent of the source categories within ten years of passage of the new law. Eight years after MACT is installed on a source, EPA must examine the risk levels remaining at the regulated facilities and determine whether additional controls are necessary to reduce unacceptable residual risk [18].

The Bhopal, India, tragedy, where an accidental release of methyl isocyanate at a pesticide manufacturing plant in 1984 killed approximately 4,000 people and injured more than 200,000, inspired the 1990 CAAA requirement that factories and other businesses develop plans to prevent accidental releases of highly toxic chemicals. In addition, the Act established the Chemical Safety Board to investigate and report on accidental releases of HAPs from industrial plants.

Title III did not directly regulate air toxics from power plants but did state that regulation of air toxins from utility power plants will be based on scientific and engineering studies. Mercury is one pollutant that was identified for study and will be discussed in more detail later in this chapter. At power plants, compounds in the vapor phase (e.g., polycyclic organic matter) and those combined with or attached to particulate matter (e.g., arsenic) are subject to the Title III provisions [20].

Title IV: Acid Deposition Control

The Acid Rain Program was established under Title IV of the 1990 CAAA. The program required major reductions of sulfur dioxide (SO_2) and nitrogen oxides (NO_x) emissions, the pollutants that cause acid rain. Using an innovative market-based or "cap and trade" approach to environmental protection, the program sets a permanent cap on the total amount of SO_2 that may be emitted by electric power plants nationwide. The cap is set at about one-half the amount of SO_2 emitted in 1980, and the trading component allows for flexibility for individual fossil fuel–fired combustion units to select their own methods of compliance. The program also sets NO_x emission limitations for certain coal-fired electric utility boilers, representing about a 27 percent reduction from 1990 levels [21].

Under the Acid Rain Program, each unit must continuously measure and record its emissions of SO_2, NO_x, and CO_2, as well as volumetric flow and opacity [21]. In most cases, a continuous emissions monitoring (CEM) system must be used. Units report hourly emissions data to EPA on a quarterly basis. This data are then recorded in the Emissions Tracking System, which serves as a repository of emissions data for the utility industry. The emissions monitoring and reporting are critical to the program because they instill confidence in allowance transactions by certifying the existence and quantity of the commodity being traded and ensure that NO_x averaging plans are working. Monitoring also ensures, through accurate accounting, that the SO_2 and NO_x emissions reduction goals are met.

The SO_2 Program

Title IV of the 1990 CAAA called for a two-step program to reduce SO_2 emissions by 10 million short tons from 1980 levels and, when fully implemented in 2000, placed a cap of approximately 8.9 million short tons per year on SO_2 emissions, forcing all generators that burn fossil fuels after 2000 to possess an emissions allowance for each ton of SO_2 they emit. By January 1, 1995, the deadline for Phase I, half of the total SO_2 reductions were to occur by requiring 110 of the largest SO_2-emitting power plants (with 263 boilers or units) located in 21 eastern and midwestern states to cut their emissions to an annual average rate of 2.5 lb SO_2/million Btu. These stations, specifically identified in the 1990 CAAA (see Appendix E), consisted of boilers with output greater than or equal to 100 MW and sulfur emissions of greater than 2.5 lb SO_2/million Btu. Plants deciding to reduce SO_2 emissions by 90 percent were given to 1997 to meet the requirements. By the year 2000, the deadline for Phase II, virtually all power plants greater than 75 MW and discharging SO_2 at a rate more than 1.2 lb/million But were required to reduce emissions to that level. In 2001, there were 2,792 units were affected by the SO_2 provision of the Acid Rain Program [21].

The Phase I reductions were accomplished by issuing the utilities that operated these units emission allowances equivalent to what their annual emission would have been at these plants in the years 1985–1987 based on burning a coal with emissions of 2.5 lb SO_2/million Btu. One allowance is equivalent to the emission of one short ton of SO_2 per year. Utilities were allowed the flexibility of determining which control strategies to be used on existing plants as long as the total emissions from all plants listed in Phase I and owned by the utility did not exceed the available allowances. The law was designed to let the industry find the most cost-effective way to stay under the cap. This differs in approach from previous air quality regulations, such as NSPS and PSD, which are based on controlling emissions at their source and then monitoring to ensure compliance [22]. If a utility's emissions exceeded the available allowances, they were subject to fines assessed at $2,000/short ton of excess emissions with a requirement to offset the emissions in future years. Any emission reductions achieved that were in excess of those required could be banked by the utility for use at a later date or traded or sold to another utility.

The SO_2 component of the Acid Rain Program represents a dramatic departure from traditional regulatory approaches that establish source-specific emissions limitations; instead, the program uses an overall emissions cap for SO_2 that ensures emissions reductions are achieved and maintained and a trading system that facilitates lowest-cost emissions reductions. The program features tradable SO_2 emissions allowances, where one allowance is a limited authorization to emit one short ton of SO_2. A fixed number of allowances are issued by the government, and they may be bought, sold, or banked for future use by utilities, brokers, or anyone else interested in holding them. Existing units are allocated allowances for each year; new units do not receive allowances and must buy them. New coal-fired boilers are subject to the NSPS, which remains in effect (i.e., 70–90+ percent reduction) with the provision that they must acquire emission allowances to emit the residual SO_2 that is not controlled [4]. At the end of the year, all participants in the program are obliged to surrender to EPA the number of allowances that correspond to their annual SO_2 emissions [21].

The NOₓ Program

Title IV also required EPA development of a NO_x reduction program and a plan for reducing NO_x by 2 million short tons from 1980 levels. As with the SO_2 emission reduction requirements, the NO_x program was implemented in two phases, beginning in 1996 and 2000 [23]. The NO_x program embodies many of the same principles of the SO_2 trading program in its design: a results-orientation, flexibility in the method to achieve emission reductions, and program integrity through measurement of the emissions. However, it does not cap NO_x emissions as the SO_2 program does nor does it utilize an allowance trading system, although NO_x trading programs have been implemented.

Emission limitations for the NO_x boilers provide flexibility for utilities by focusing on the emission rate to be achieved, expressed in pounds of NO_x per million Btu of heat input. Two options for compliance with the emission limitations are: (1) compliance with an individual emission rate for a boiler or (2) averaging of emission rates over two or more units that have the same owner or operator to meet an overall emission rate limitation. These options give utilities the flexibility to meet the emissions limitations in the most cost-effective way and allow for the further development of technologies to reduce the cost of compliance. If a utility properly installs and maintains the appropriate control equipment designed to meet the emission limitation established in the regulations but is still unable to meet the limitation, the NO_x program allows the utility to apply for an alternative emission limitation (AEL) that corresponds to the level that the utility demonstrates is achievable.

Phase I of the program, which was delayed a year due to litigation, began on January 1, 1996, and affected two types of boilers, which were among those already targeted for Phase I SO_2 reductions: dry bottom wall-fired boilers and tangentially fired boilers. The regulations to govern the Phase II portion of the program, which began in 2000, were promulgated December 19, 1996. These regulations set lower emission limits for Group 1 boilers and established NO_x limitations for Group 2 boilers, which include boilers applying cell-burner technology, cyclone boilers, wet-bottom boilers, and other types of coal-fired boilers. The NO_x limitations and number of units affected in 2008 are provided in Table 8.4.

Title V: Permitting

The 1990 CAAA introduced a national permitting program to ensure compliance with all applicable requirements of the Clean Air Act and to enhance the EPA's ability to enforce the Act [18]. Sources are required to submit applications for Title V operating permits through the state agencies to EPA. Air pollution sources subject to the program must obtain an operating permit, states must develop and implement the program, and the EPA must issue permit program regulations, review each state's proposed program, and oversee the state's efforts to implement any approved program. The EPA must also develop and implement a federal permit program when a state fails to adopt and implement its own program.

Table 8.4 Number of NO$_x$ Affected Units by Boiler Type in 2008

Coal-Fired Boiler Type (all coverage for boilers larger than 25 MW unless otherwise noted)	Standard Emission Limit (lb/MM Btu)	Number of Units
Phase I Group 1 Tangentially fired	0.45	133
Phase I Group 1 Dry-bottom wall-fired	0.50	107
Phase II Group 1 Tangentially fired	0.40	300
Phase II Group 1 Dry-bottom wall-fired	0.46	294
Cell burners	0.68	37
Cyclones >155 MW	0.86	54
Wet-bottom >65 MW	0.84	20
Vertically fired	0.80	24
Total		**969**

The following sources are required to submit Title V permits:

- Major sources as determined under Title I:
    ≥100 short tons per year—and listed pollutant excluding CO
    ≥10 to100 short tons year—or sources in nonattainment areas depending on the classification of Marginal to Extreme
- All NSPS, PSD review sources, and NESHAPS sources
- Major sources as determined under Title III:
    ≥10 short tons per year any air toxic
    ≥25 short tons per year multiple air toxics
- Sources under Title IV
- Sources emitting ≥ 100 short tons per year of ozone depleting substances under Title VI
- Other sources required to have state or federal operating permits

8.2.5 Additional NO$_x$ Regulations and Trading Programs

Ozone Transport Commission NO$_x$ Budget Program (1992–2002)

Many urban areas do not meet the ozone standard and are classified as non-attainment areas. To address this, along with the fact that NO$_x$ can be transported great distances, more restrictive requirements for NO$_x$ emissions from electric power-generating plants and other large stationary boilers in 22 eastern states and the District of Columbia were established [24, 25]. The requirements set statewide NO$_x$ emissions budgets, including budget components for the electric power industry and certain industrial sources.

Title I of the 1990 CAAA includes provisions designed to address both the continued nonattainment of the existing ozone NAAQS and the transport of air pollutants across state boundaries. These provisions allow downwind states to petition for tighter controls on upwind states that contribute to their NAAQS non-attainment status. In general, Title I nitrogen oxide provisions require areas with an ozone nonattainment region to do the following:

- Require existing major stationary sources to apply reasonably available control technology (RACT), which is the lowest emission limitation that a particular source is capable of meeting by application of control technology that is reasonably available considering technological and economic feasibility
- Require new or modified stationary sources to offset their emissions and install controls representing the lowest achievable emission rate (LAER), which is the minimum emissions rate accepted by EPA for major new or modified sources in nonattainment areas
- Require each state with an ozone nonattainment region to develop a State Implementation Plan that may, in some cases, include reductions in stationary source NO_x emissions beyond those required by the RACT provisions of Title I

Section 184 of the Clean Air Act delineates a multistate ozone transport region (OTR) in the northeast and requires specific additional nitrogen oxide and volatile organic compound controls for all areas in this region. It also established the Ozone Transport Commission (OTC) for the purpose of assessing the degree of ozone transport in the OTR and recommending strategies to mitigate the interstate transport of pollution. The OTR consists of the states of Connecticut (CT), Delaware (DE), Maine (ME), Maryland (MD), Massachusetts (MA), New Hampshire (NH), New Jersey (NJ), New York (NY), Pennsylvania (PA), Rhode Island (RI), Vermont (VT), the northern counties of Virginia (VA), and the District of Columbia (DC). The OTR states confirmed that they would implement RACT on major stationary sources of NO_x (Phase I) and agreed to a phased approach for additional controls, beyond RACT, for power plants (larger than 25 MW) and other large fuel combustion sources (industrial boilers with a rated capacity larger than 250 million Btu per hour input; Phases II and III). This agreement, known as the OTC Memorandum of Understanding (MOU), was approved on September 27, 1994, with all OTR states, except Virginia, signing the MOU.

The MOU established an emission trading system to reduce the costs of compliance with the control requirements under Phase II, which began on May 1, 1999, and Phase II, which began on May 1, 2003. The OTC program target caps for the summer season (May 1 through September 30) NO_x emissions were approximately 219,000 short tons in 1999 and 143,000 short tons in 2003, which represented up to 65 and 75 percent reductions in NO_x, respectively, from the 1990 baseline emission level of 473,000 short tons. The program is summarized in Table 8.5. The OTC NO_x Budget Program sources successfully reduced the regional ozone season emissions in 2002 by nearly 280,000 short tons or about 60 percent, from 1990 baseline levels [24]. The program demonstrated a viable NO_x trading system and successful reduction in NO_x despite increases in NO_x emissions from mobile sources and continued transport of pollutants into the region from states downwind of the OTC states. However, ambient ozone conditions remained a serious problem in the region, and further reduction efforts both within and upwind of the OTC region were explored, as discussed in the following sections.

NO_x Budget Trading Program/NO_x SIP Call (2003–2008)

The EPA promulgated a rule on October 27, 1998, known as the NO_x SIP call, to address long-range transport of ozone. The purpose of this rule is to limit summer NO_x emissions in 22 northeastern states and the District of Columbia that EPA

Table 8.5 Summary of Nitrogen Oxide Reduction and Trading Programs

	Ozone Transport Commission (OTC) NO$_x$ Budget Program	NO$_x$ State Implementation Plan (SIP) Call	Section 126 Federal NO$_x$ Budget Trading Program	Acid Rain Program
Affected regions	DC and 12 states: CT, DE, MA, ME, MD, NH, NJ, NY, PA, RI, VA, VT	DC and 22 states: AL, CT, DE, GA, IL, IN, KY, MA, MD, MI, MO, NC, NJ, NY, OH, PA, RI, SC, TN, VA, WI, WV	DC and 22 states: AL, CT, DE, GA, IL, IN, KY, MA, MD, MI, MO, NC, NJ, NY, OH, PA, RI, SC, TN, VA, WI, WV	Entire nation
Compliance period	May 1–September 30 of each year	May 1–Sept. 30 of each year (in 2004 the compliance period begins May 31)	May 1–Sept. 30 of each year	Annual
Initial compliance year	1999	2003/2004 (Phase I) 2007 (Phase II)	2003	Phase I: Jan. 1, 1996 Phase II: Jan. 1, 2000
Emissions cap	219,000 short tons in 1999; 143,000 short tons in 2003		528,453 short tons in 2008	
NO$_x$ reductions	246,000 short tons in 1999; 322,000 short tons in 2003	1.2 million short tons in 2007	1.4 million short tons in 2008	340,000 short tons per year in Phase I; 2.06 million short tons per year in Phase II
Baseline year	1990	1995	1995	1980
Baseline emissions	473,000 million short tons NO$_x$			
Program owner	OTC: allowances set by OTC, program administered by EPA	States and EPA; States have the option of participating in the trading program and establishing unit allocations, program administered by EPA	EPA	EPA

Source: Modified from EPA (1998).

considers significant contributors to ozone nonattainment in downwind areas (see Table 8.5 for affected states). These states were required to amend their SIPs through a procedure established in Section 110 of the Clean Air Act to further reduce NO_x emissions by taking advantage of newer, cleaner control strategies. The EPA finalized a summer state NO_x budget and developed a state-implemented and federally enforced NO_x trading program to provide for emissions trading by certain electric and industrial stationary sources. Each affected state's NO_x budget is based on a population-wide 0.15 lb/MM Btu NO_x emission rate for large electric generating stations and a 60 percent reduction from uncontrolled emissions for large electric generating units. This effort was projected to reduce summer NO_x emissions by 1.2 million short tons in the affected 22 states and the District of Columbia by 2007 [27]. EPA began to administer the NO_x Budget Trading Program in 2003 [28]. In 2008, NO_x Budget Program sources emitted 481,420 tons of NO_x during the summer ozone season, a 62 percent decrease from 2000 levels and a 75 percent decrease from 1990 levels [28]. The NO_x Budget Program was replaced in 2008 by the Clean Air Interstate Rule.

8.2.6   Clean Air Interstate Rule

On March 10, 2005, EPA issued the Clean Air Interstate Rule (CAIR) to permanently cap emissions of SO_2 and NO_x in the eastern United States [29, 30]. The rule applies to 28 eastern states (Figure 8.2) that have trouble meeting air quality standards for

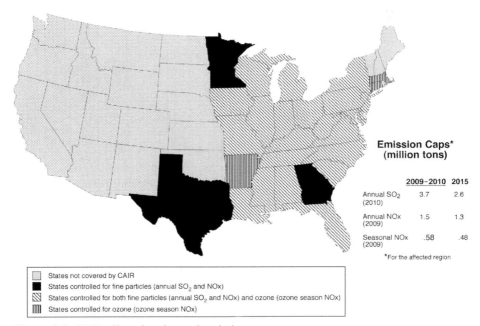

Emission Caps*
(million tons)

	2009–2010	2015
Annual SO₂ (2010)	3.7	2.6
Annual NOx (2009)	1.5	1.3
Seasonal NOx (2009)	.58	.48

*For the affected region

☐ States not covered by CAIR
■ States controlled for fine particles (annual SO₂ and NOx)
▨ States controlled for both fine particles (annual SO₂ and NOx) and ozone (ozone season NOx)
▥ States controlled for ozone (ozone season NOx)

Figure 8.2 CAIR affected region and emissions caps.
Source: From EPA (2005).

Table 8.6 CAIR Timeline and Emissions Caps [30]

	Emissions Caps (million short tons)	Year
Promulgate CAIR rule		2005
State Implementation Plans due		2006
Phase I annual cap in place for NO_x	1.5	2009
Phase I annual cap in place for SO_2	3.6	2010
Phase II annual cap in place NO_x	1.3	2015
Phase II annual cap in place for SO_2	2.5	2015
NO_x ozone season cap	0.58	2009
NO_x ozone season cap	0.48	2015

particulate and ozone. This rule provides an optional cap and trade program that is based on the Acid Rain and NO_x Budget Trading programs as a method to implement the necessary reductions. The EPA estimated that CAIR will reduce SO_2 emissions in these states by more than 70 percent and NO_x emissions by more than 60 percent from 2003 levels. CAIR's timeline and emissions caps are provided in Table 8.6.

On July 11, 2008, the U.S. Court of Appeals for the D.C. Circuit ruled on petitions for review of CAIR and CAIR federal Implementation Plans, including their provisions establishing the CAIR NO_x annual and ozone season and SO_2 trading programs. The Court determined that CAIR violated the Clean Air Act and vacated the rule. However, on December 23, 2008, the Court ruled that CAIR could remain in place until EPA issued a new rule to replace CAIR in accordance with the Clean Air Act [29, 30]. A replacement rule could take approximately two years because EPA must strictly follow the federal procedures before issuing a rule. This process generally takes a minimum of 18 months and involves gathering scientific and technical data, analyzing legal issues, proposing the rule, soliciting and responding to public comments, and issuing a final rule [31].

8.2.7 Clean Air Mercury Rule

The EPA prepared a Mercury Study Report, dated December 1997, which they submitted to Congress on February 24, 1998, as a requirement of the Section 112(n)(1)(B) of the 1990 CAAA [32]. The report provided an assessment of the magnitude of U.S. mercury emissions by source, the health and environmental implications of those emissions, and the availability and cost of control technologies. The report identified electric utilities as the largest remaining source of mercury emissions in the air as the EPA has regulated mercury emissions from municipal waste combustors, medical waste incinerators, and hazardous waste combustion.

The EPA also submitted the Utility Hazardous Air Pollutant Report to Congress on February 24, 1998, in which EPA examined 67 air toxics emitted from 52 fossil fuel–fired power plants and concluded that mercury is the air toxic of greatest concern [33]. Although not conclusive, the report finds evidence suggesting a link between utility emissions and the methylmercury found in soil, water, air, and fish from contaminated waters. The report identified the need for additional information on the amount of mercury in U.S. coals and mercury emissions from coal-fired power plants. Specifically, data identified by EPA included obtaining additional data on the quantity of mercury emitted from various types of generating units, the amount of mercury that is divalent versus elemental, and the effect of pollution control devices, fuel type, and plant configuration on emissions and speciation. To obtain these data, EPA issued a three-part Information Collection Request (ICR) for calendar year 1999 [34]. Part I collected information on the size and configuration of all coal-fired utility boilers greater than 25 MW and their pollution control devices. Part II obtained data quarterly on the origin, quantity, and analyses of coal shipments delivered to the generating units, which totaled more than 1,100, including a minimum of three analyses per month for mercury and chlorine contents, together with any other available analyses such as ash and sulfur contents and heating value. Part III required emission tests on 84 generating units selected at random from 36 categories representing different plant configurations and coal rank to measure total and speciated mercury concentrations in the flue gas before and after the final air pollution control device upstream of the stack.

On December 14, 2000, the EPA announced it will regulate mercury emissions from power plants [35]. On December 15, 2003, the EPA proposed a rule to permanently cap and reduce mercury emissions from power plants. The schedule required a final rule by December 2004, and implementation of controls by the end of 2007.

As noted earlier, in 2005 the United States became the first country in the world to regulate mercury emissions from coal-fired power plants when the EPA issued a federal rule to permanently cap and reduce these emissions [36]. The rule, called the Clean Air Mercury Rule (CAMR), created a market-based cap-and-trade program that will permanently cap utility mercury emissions in two phases. The first cap is 38 tons, and emissions will be reduced by taking advantage of cobenefit reductions—that is, mercury reductions achieved by reducing SO_2 and NO_x emissions under CAIR (see previous section). In the second phase, due in 2018, coal-fired power plants will be subject to a second cap, which will reduce emissions to 15 tons upon full implementation.

The CAMR, however, was challenged in the courts because EPA determined that the December 2000 Regulatory Finding on the Emissions of Hazardous Air Pollutants from Electric Utility Steam Generating Units lacked foundation and that recent information demonstrated that it was not appropriate or necessary to regulate coal- and oil-fired utility units under Section 112 of the Clean Air Act, and subsequently removed those utility units from the Section 112(c) list of source categories [36]. On February 8, 2008, the U.S. Court of Appeals for the D.C. Circuit Court vacated EPA's rule removing power plants from the Clean Air Act list of sources of hazardous pollutants and vacated the CAMR. Consequently, EPA has decided to develop emissions standards for power plants under the Clean Air Act (Section 112), consistent with the D.C. Circuit's opinion on the CAMR. This process is expected to require approximately two years until mercury legislation is implemented [31].

8.2.8 New Source Review

The New Source Review (NSR) program is one of many programs created by the Clean Air Act to reduce emissions of air pollutants—particularly "criteria" pollutants" that are emitted from a wide variety of sources and have an adverse impact on human health and the environment. The NSR program was established in parts C and D of Title I of the Clean Air Act to protect public health and welfare, as well as national parks and wilderness areas, as new sources of air pollution are built and when existing sources are modified in a way that significantly increases air pollutant emissions. Specifically, NSR's purpose is to ensure that when new sources are built or existing sources undergo major modifications, the air quality improves if the change occurs where the air currently does not meet federal air quality standards and air quality is not significantly degraded where the air currently meets federal standards [37].

The original intent of the NSR program was to ensure that major new facilities that are sources of emissions, or existing facilities that are modified and result in increased emissions, would install state-of-the-art controls. Subsequently, EPA provided interpretive guidance that complicated the review program and expanded it to include maintenance or improvement. While the determination of whether an activity is subject to the major NSR program is fairly straightforward for a newly constructed source, the determination of what should be classified as a modification subject to major NSR presents a more difficult issue. Consequently, installation of new technology, greater energy efficiency, and improved environmental performance at facilities was being inhibited. In addition, there was much controversy between the industry and the EPA over what "triggers" applicability of the major NSR program, which led to litigation between Wisconsin Electric Power Company (WEPCO) and the Environmental Protection Agency. In 1992, EPA promulgated revisions to the applicability regulations creating special rules for physical and operational changes at electric utility steam generating units.

On July 23, 1996, a number of changes were proposed to the existing major NSR requirements as part of a larger regulatory package [38]. This was followed by a Notice of Availability published by EPA in the federal Register on July 24, 1998, requesting comment on three of the proposed changes. After public comment, on December 31, 2002, EPA issued an NSR final rule to improve the New Source Review program and a proposed rule to provide a regulatory definition of routine maintenance, repair, and replacement. In summary, the final rule that became effective March 3, 2003:

- Reforms the emissions accounting system for determining when a change is triggered (now the trigger is based on actual emissions)
- Allows already-controlled "clean units" to make changes without triggering NSR
- Broadens the exclusion from NSR for projects intended for pollution control
- Sets new rules for establishing plant-wide applicability limits of specific pollutants under the program

EPA took action to improve the NSR program after performing a comprehensive review of the program and, in June 2002, issued a Report to the President on NSR [37]. The report concluded that the program, as was administered and as related to the energy sector, impeded or resulted in the cancellation of projects that would maintain or improve the reliability, efficiency, or safety of existing power plants. EPA issued the final rule improvements after the culmination of a ten-year process that included pilot studies and the engagement of state and local governments, environmental groups, private sector representatives, academia, and concerned citizens in an open and far-reaching public rulemaking process. In addition, the nation's governors and environmental commissioners, on a bipartisan basis, called for NSR reform.

The final rule implements the following major improvements to the New Source Review program [39]:

- *Plantwide Applicability Limits (PALs):* To provide facilities with greater flexibility to modernize their operations without increasing air pollution, facilities that agree to operate within strict site-wide emissions caps called PALs will be given flexibility to modify their operations without undergoing NSR as long as the modifications do not cause emissions to violate their plantwide cap.
- *Pollution Control and Prevention Projects:* To maximize investments in pollution prevention, companies that undertake certain specified environmentally beneficial activities will be free to do so upon submission to their permitting authority of a notice, rather than having to wait for adjudication of a permit application. EPA is also creating a simplified process for approving other environmentally beneficial projects.
- *Clean Unit Provision:* To encourage the installation of state-of-the-art air pollution controls, EPA will give plants that attain clean unit status flexibility in the future if they continue to operate within permitted limits. This flexibility is an incentive for plants to voluntarily install the best available pollution controls. Clean units must have an NSR permit or other regulatory limit that requires the use of the best air pollution technologies.
- *Emissions Calculation Test Methodology:* To provide facilities with a more accurate procedure for evaluating the effect of a project on future emissions, the final regulations improve how a facility calculates whether a particular change will result in a significant emissions increase and thereby trigger NSR permitting requirements.

The EPA's proposed rule would make improvements to the routine maintenance, repair, and replacement exclusion currently contained in their regulations. These proposed improvements will be subject to a full and open public rulemaking process. Since 1980, EPA regulations have excluded from NSR review all repairs and maintenance activities that are routine, but a complex analysis must be made to determine what activities meet the standard. This has deterred companies from conducting repairs and replacements that are necessary for the safe, efficient, and reliable operation of facilities.

After issuing the new NSR final and proposed rules, certain environmental groups and state and local governments petitioned the EPA to reconsider specific aspects of the final NSR reform rule. EPA announced on July 25, 2003, that it will reconsider parts of the NSR final rule [40]. The EPA's notice responds in part to the petitions by requesting comment on six limited areas. The EPA will take action on the remaining issues raised by the petitioners at a later date, and has not yet acted on the remaining issues but plans to make a decision within the time period it takes to reconsider the six items EPA has agreed to address.

The Environmental Protection Agency then solicited comments on the following six areas [40]:

- EPA's report, "Supplemental Analysis of the Environmental Impact of the 2002 Final NSR Improvement Rules," that concluded that the NSR improvement rule will likely result in greater environmental benefits than the prior program.
- The decision to allow certain sources of air emissions to maintain Clean Unit status after an area is redesignated from attainment to nonattainment for one of the six criteria air pollutants.
- EPA's inclusion of the reasonable possibility standard as it pertains to the need to maintain records and file certain reports when project actual emissions following a physical or operational change.
- The method for assessing air emissions from process units built after the 24-month baseline period used to establish PAL limits.
- The decision to allow a PAL to supersede existing emissions limits established for NSR applicability purposes. Compliance with plantwide applicability limits is then used to determine whether NSR requirements apply in the future.
- The method of measuring emission increases when the existing emission units are replaced.
- On October 27, 2003, EPA published its final NSR Equipment Replacement Rule [41]. This final version of the rule applies only to equipment replacement. It became effective December 26, 2003, and states will have up to three years to revise their state implementation plans to reflect these requirements. The regulation specifies that replacement components must be functionally equivalent to existing components in that there are no changes in the basic unit design or to pollutant emitting capacity. It also sets a 20 percent limit on replacement cost for equipment. If these restrictions are exceeded, the replacement work is subject to the NSR process.

8.2.9   Fine Particulate Matter

Epidemiological research over the past 10 to 15 years has revealed a consistent statistical correlation between levels of airborne fine particulate matter ($PM_{2.5}$) and adverse respiratory and cardiopulmonary effects in humans [42]. This has resulted in EPA's promulgation of NAAQS that limit the allowable mass concentrations of $PM_{2.5}$. Attainment of the $PM_{2.5}$ NAAQS requires an annual average mass concentration of less than 15 $\mu g/m^3$ and a daily maximum concentration of less than 35 $\mu g/m^3$. The daily maximum was reduced from 65 to 35 $\mu g/m^3$ on September 21, 2006, when EPA issued the strongest national air quality standards for particle pollution in U.S. history [43]. This was implemented because EPA identified 39 areas as not meeting the standards for $PM_{2.5}$ [44], and it is known that ambient $PM_{2.5}$ has also been found to contribute significantly to the impairment of long-range visibility (regional haze) in many areas of the United States [45]. EPA issued a Regional Haze Rule in 1999 that established goals for reducing regional haze in areas of the United States where long-range visibility has been determined to have exceptional value (Class I areas) and has outlined methods for achieving these goals [45] and followed it up with a fine particulate pollution standard [43].

It is generally recognized that coal-fired power plants can be important contributors to ambient $PM_{2.5}$ mass concentrations and regional haze. Consequently,

EPA required additional restrictions of coal power plant emissions in the 2006 time frame as they develop SIPs for achieving and/or maintaining compliance with the $PM_{2.5}$ NAAQS and Regional Haze Rule [43, 46].

8.2.10 Impending Legislation and Pollutants under Consideration for Regulation

Developing new, or modifying current, regulations of air pollutants emissions from coal-fired boilers is a continual process. This section summaries legislation that stalled in Congress and was not passed and impending legislation and pollutants under consideration for regulation, and it complements the previous discussion of NSR revisions and new regulations for SO_2, NO_x, and mercury (i.e., anticipated CAIR and CAMR rules). In addition, legislation to reduce greenhouse gas emissions is discussed, although this is a contentious topic, legislation language is constantly changing, and as of the fall of 2009, various forms of language were still being proposed.

Multipollutant Legislation

For nearly the last ten years, Congress has discussed and proposed, multipollutant legislation for more stringent control power plant emissions, including SO_2, NO_x, mercury, and CO_2. This includes CAMR and CAIR, discussed previously, which are currently undergoing revision. CAMR and CAIR, and now their revised versions, are the results from modifying recent proposals, including the Clear Skies Initiative, Clean Power Act, and Clean Air Planning Act, which were proposed in the 2002 to 2003 timeframe. Current greenhouse gas proposed legislation (2009) is also based on some of these preceding proposals.

The Clear Skies Initiative was the first of several multipollutant bills proposed. This legislation would create a mandatory program that would reduce power plant emissions of SO_2, NO_x, and mercury by setting a national cap on each pollutant. Clear Skies was proposed in response to a growing need for an emission reduction plan that will protect human health and the environment while providing regulatory certainty to the industry. The program was submitted as proposed legislation in July 2002. The program was reintroduced in the U.S. House of Representatives (H.R. 999) and the U.S. Senate (S. 485) on February 27, 2003 [47]. It has not passed to date, but EPA still lists Clear Skies as proposed legislation.

The emissions reductions from Clear Skies would help to alleviate air pollution–related health and environmental problems, including fine particles, ozone, mercury, acid rain, nitrogen deposition, and visibility impairment [48]. Specifically, Clear Skies would do the following:

- Reduce SO_2 emission by 73 percent, from year 2000 emissions of 11 million short tons to a cap of 4.5 million short tons in 2010 and to a cap of 3 million short tons in 2018.
- Reduce NO_x emissions by 67 percent, from year 2000 emissions of 5 million short tons to a cap of 2.1 million short tons in 2008 and to a cap of 1.7 million short tons in 2018.
- Reduce mercury emissions, through a first-ever national cap on mercury emissions, by 69 percent, from 1999 emissions of 48 short tons to a cap of 26 short tons in 2010 and to a cap of 15 short tons in 2018. The reduction in emission is illustrated in Figure 8.3.

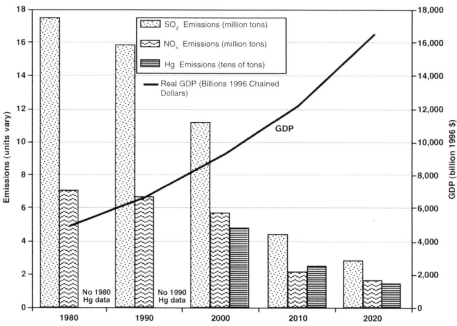

Figure 8.3 Emission levels of SO_2, NO_x, and mercury under the Clear Skies Act.
Source: EPA, Clear Skies Basic Information (2003).

Clear Skies—modeled on the cap-and-trade provisions of the 1990 CAAA's Acid Rain Program—would use an emission caps system that is predicted to do the following [48]:

- Protect against diseases by reducing smog and fine particles, which contribute to respiratory and cardiovascular problems
- Protect our wildlife, habitats, and ecosystem health by reducing acid rain, nitrogen, and mercury deposition
- Deliver a rapid reduction in emissions with certain improvements in air quality
- Enable power generators to continue to provide affordable electricity while quickly and cost effectively improving air quality and the environment
- Encourage the use of new and cleaner pollution control technologies that would further reduce compliance costs

Clear Skies is expected to provide significant benefits to public health and the environment at a reasonable cost [48]. The EPA projects that by 2020 the public health benefits alone would include more than 8,400 to 14,000 avoided premature deaths and $21 to $110 billion per year, depending on the methodology for calculating the health-related benefits. The annual cost of achieving the Clear Skies emission levels is project at $6.3 billion. Americans would also experience approximately 30,000 fewer visits to the hospital and emergency room, 23,000 fewer nonfatal heart attacks, 1.6 million fewer work loss days, and 200,000 fewer school absences each year. Benefits of improvements in visibility in national parks and wildernesses in 2020 are projected to be $3 billion annually. In addition, by 2020, an estimated 77 counties with 26 million people would meet the fine particulate NAAQS.

The future of the Clear Skies Act is still unknown. Congressional Republican support for Bush's planned legislation became increasingly uncertain [49]. Since July 2003, three bipartisan bills have been introduced in Congress that would reduce SO_2, NO_x, and mercury emissions to lower levels than the Clear Skies Initiative and on a faster timeline. In addition, the other bills have plans to limit CO_2 emissions. Hearings are planned to explore criticisms of the Clear Skies Initiative before Congress decides whether to move the legislation forward [49].

Two other proposals that received considerable attention were the Clean Power Act (S. 386) sponsored by Senator Jim Jeffords (Independent-Vermont) and the Clean Air Planning Act sponsored by Senators Tom Carper (Democrat-Delaware), Lincoln Chaffee (Republican-Rhode Island), and Judd Gregg (Republican-New Hampshire). These bills are currently being debated, and the key elements of each are summarized in Table 8.7. The Clear Skies Act is included in the table for comparison. In addition to the proposed legislation listed in Table 8.7, EPA announced CAIR in 2003. Of these bills, CAIR proceeded to the promulgation stage and is currently being rewritten as discussed previously.

Climate Change/Greenhouse Gas Emissions

Like the multipollutant proposals, climate change has been on Congress's agenda beginning with the 108th Congress and the Bush administration [51–54]. At that time, Congress considered several legislative initiatives with key issues being considered, including better coordination of government research, setting caps on greenhouse gas (GHG) emissions, and whether reporting systems should be voluntary or mandatory. Climate change debates occurred in connection with standalone climate change legislation, as well as during discussions of energy policy and the Clear Skies Initiative. There have been several bills proposed addressing climate change research and data management, managing risks of climate change, reporting of GHG emissions, and stabilization and caps on GHG emissions. This is a hotly debated topic and industry experts expect some form of legislation to be passed.

The second Bush administration also took steps in addressing GHG emissions. On June 25, 2003, U.S. Department of Energy secretary Spencer Abraham and energy ministers from around the world signed the first international framework for research and development on the capture and storage of CO_2 emissions. This initiative is called the Carbon Sequestration Leadership Forum and includes Australia, Brazil, Canada, China, Colombia, India, Italy, Japan, Mexico, Norway, Russian Federation, the United Kingdom, the United States, and the European Commission [52]. This was soon followed with the announcement on July 23, 2003, of the Bush administration's unveiling of a long-term strategic plan to study global change

Table 8.7 Overview of Multipollutant Bills Proposed in 2002–2003

	Clear Skies Act	Clean Power Act	Clean Air Planning Act
SO_2 cap	4.5 million short tons in 2008; 3.0 million short tons in 2018	2.25 million short tons by 2009 (0.28 million short tons in western region and 1.98 million short tons in eastern region)	4.5 million short tons by 2009 3.5 million short tons by 2013 2.25 million short tons by 2016
NO_x cap	2.1 million short tons in 2008; 1.7 million short tons in 2018	1.51 million short tons by 2009	1.87 million short tons by 2009 1.7 million short tons by 2013
CO_2 cap	None	2.05 billion short tons by 2009	In 2009, stabilize at 2006 levels ($\approx$2.57 billion short tons) plus flexibility measures In 2013, cap at 2001 levels ($\approx$2.47 billion short tons) plus flexibility measures
Mercury cap	26 short tons in 2010; 15 short tons in 2015	5 short tons by 2009	24 short tons by 2009 10 short tons by 2013 50 and 70% reduction required at each plant in 2009 and 2013, respectively
Emission trading	Trading allowed for SO_2, NO_x, and mercury	Trading allowed for SO_2, NO_x, and CO_2; no trading for mercury	Cap-and-trade for NO_x, SO_2, CO_2, and mercury, along with facility specific mercury requirements

Source: From Tatsutani (2003).

[53, 54]. Presented by Secretary of Commerce Don Evans and secretary of DOE Spencer Abraham, the 10-year plan set five goals, including the following:

- Identifying the natural variability in the earth's climate
- Understanding the forces that cause global warming
- Reducing the uncertainties in climate forecasting
- Improving the understanding of sensitivity and adaptability of ecosystems to climate change
- Developing more exact methods for calculating the risks of global warming

For the most part, climate change legislation stalled under the second Bush administration. It is anticipated that legislation will be implemented during the Obama administration because several forms of legislation have been proposed, and at least one bill is being prepared for introduction in October 2009. Two bills that have been proposed include the Lieberman-Warner Climate Security Act of 2008 (110th Congress) and the American Clean Energy and Security Act of 2009 (111th Congress) [55]. In the latter, emissions of CO_2 and other GHGs would be reduced 17 percent by 2020 from 2005 levels, 42 percent by 2030, and 83 percent by 2050. Electric utilities that capture and store GHG emissions could get up to $100 billion in bonus carbon pollution permits. Utilities would have to generate 15 percent of their electricity from renewable sources such as wind or solar power and show a 5 percent gain in energy efficiency by 2020. In addition to these bills, Senators Boxer and Kerry are also writing climate legislation that they plan to unveil in October 2009. Consequently, there is much activity in writing climate legislation, but there appears to be little consensus on the specifics, since multiple bills are being proposed. Since 2010 is a congressional midterm election year, there are some that think a climate change bill will not pass until 2011 [56].

In response to the Consolidated Appropriations Act (H.R. 2764; Public Law 110-161), the EPA has issued the Final Mandatory Reporting of Greenhouse Gases Rule, which was signed by the administrator on September 22, 2009 [57]. The rule requires reporting of greenhouse gas (GHG) emissions from large sources and suppliers in the United States and is intended to collect accurate and timely emissions data to inform future policy decisions. Under the rule, sources that emit 25,000 metric tons or more of GHG emissions are required to submit annual reports to EPA. The gases covered include carbon dioxide (CO_2), methane (CH_4), nitrous oxide (N_2O), hydrofluorocarbons (HFC), perfluorocarbons (PFC), sulfur hexafluoride (SF_6), and other fluorinated gases including nitrogen fluoride (NF_3), and hydrofluorinated ethers (HFE). It is the intent of the reporting system to identify where GHGs are coming from and guide development of the best possible policies and programs to reduce the emissions.

8.3 Emissions Legislation in Other Countries

This section examines regulatory requirements for sulfur dioxide, nitrogen oxides, particulate matter, trace elements, specifically mercury, and carbon dioxide for many countries. The emissions standards from other countries are compared to those from the United States.

8.3.1 Sulfur Dioxide

The United Nations Economic Commission for Europe's (UNECE) Convention on Long Range Transboundary Air Pollution (LRTAP) was the first legally binding instrument to address air pollution on a broad regional context [59]. The Convention was adopted in 1979, came into force in 1983, and now has been ratified by 48 countries, as listed in Table 8.8 [59, 60]. Under the Convention, the countries recognize the transboundary problems of air pollution and accept general responsibility to move toward a solution to these problems.

Following the LRTAP Protocol, the SO_2 Helinski Protocol—the "30% Club"—was signed in 1985 and came into force in 1987 [8.8]. Under this protocol, the signatories agreed to reduce their SO_2 emissions by 30 percent (based on 1980 values) by 1993 [58, 60, 61].

In June 1994, the Second Sulfur Protocol, the Protocol on Further Reductions of Sulfur Emissions, was signed by 27 European countries, the European Community

Table 8.8 Signatories to the UNECE LRTAP and Subsequent SO$_2$ Protocols

1979 LRTAP Convention	1985 First Sulfur Protocol	1994 Second Sulfur Protocol
Albania (Ac)	Albania (Ac)	
Armenia (Ac)		
Austria (R)	Austria (R)	Austria (R)
Azerbaijan (Ac)		
Belarus (R)	Belarus (At)	
Belgium (R)	Belgium (R)	Belgium (R)
Bosnia and Herzegovina (Sc)		
Bulgaria (R)	Bulgaria (Ap)	Bulgaria (R)
Canada (R)	Canada (R)	Canada (R)
Croatia (Sc)		Croatia (At)
Cyprus (Ac)		Cyprus (Ac)
Czech Republic (Sc)	Czech Republic (Sc)	Czech Republic (R)
Denmark (R)	Denmark (R)	Denmark (Ap)
Estonia (Ac)	Estonia (Ac)	
European Community (Ap)		European Community (Ap)
Finland (R)	Finland (R)	Finland (At)
France (Ap)	France (Ap)	France (Ap)
Georgia (Ac)		
Germany (R)	Germany (R)	Germany (R)
Greece (R)		Greece (R)
Holy See		
Hungary (R)	Hungary (R)	Hungary (R)
Iceland (R)		
Ireland (R)		Ireland (R)
Italy (R)	Italy (R)	Italy (R)
Kazakhstan (Ac)		
Kyrgyzstan (Ac)		
Latvia (Ac)		
Liechtenstein (R)	Liechtenstein (R)	Liechtenstein (At)
Lithuania (Ac)	Lithuania (Ac)	Lithuania (Ac)
Luxembourg (R)	Luxembourg (R)	Luxembourg (R)
Malta (Ac)		
Monaco (Ac)		Monaco (Ac)
Montenegro (Sc)		
Netherlands (At)	Netherlands (At)	Netherlands (At)
Norway (R)	Norway (R)	Norway (R)
Poland (R)		Poland
Portugal (R)		
Republic of Moldova (Ac)		
Romania (R)		
Russian Federation (R)	Russian Federation (At)	Russian Federation

Table 8.8 *Continued*

1979 LRTAP Convention	1985 First Sulfur Protocol	1994 Second Sulfur Protocol
San Marino		
Serbia (Sc)		
Slovakia (Sc)	Slovakia (Sc)	Slovakia (R)
Slovenia (Sc)		Slovenia (R)
Spain (R)		Spain (R)
Sweden (R)	Sweden (R)	Sweden (R)
Switzerland (R)	Switzerland (R)	Switzerland (R)
The Former Yugoslav Republic of Macedonia (Sc)		
Turkey (R)		
Ukraine (R)	Ukraine (At)	Ukraine
United Kingdom (R)		United Kingdom (R)
United States (At)		

Notes: R—ratification; Ac—accession; Ap—approval; At—acceptance; Sc—succession; *—committing to 30% reduction

Source: From Economic Commission for Europe (2007); Soud (2000) [59]; Wu (2002); and United States (2009).

(EC), and Canada. The protocol came into force in August 1998. All signatories, which are listed in Table 8.8, were allocated targets for 2000, while some countries agreed to additional targets for 2005 and 2010.

The UNECE Gothenburg Protocol to abate acidification, eutrophication, and ground-level ozone was signed by 27 countries in December 1999. As of September 2, 2009, it has 31 signatures and 25 ratifications [61]. The protocol sets emissions ceilings for SO_2, NO_x, volatile organic compounds, and ammonia for the year 2010. The Gothenburg Protocol is discussed in more detail later in this chapter.

On November 24, 1998, the EC adopted the Large Combustion Plants Directive (LCPD) and, with an amendment in December 1994, set targets and emission limits for air pollutants, including SO_2, for plants larger than 50 megawatts thermal (MW_t) for both exiting and new facilities. Revisions to the directive are proposing stricter SO_2 and NO_x emission ceilings to be achieved by the end of 2010, and regulations on smaller-sized units are being discussed [59].

In 1996, the European Union Environment Ministers adopted a Directive on Integrated Pollution Prevention and Control (IPPC) [59]. The Directive came into force in October 1999 and mandated implementation by October 1999 and applies to all new installations and those undergoing a substantial change. The main purpose of the IPPC Directive is to achieve integrated prevention and control of air pollution.

In most countries, the legislators target new and large facilities. More than 30 countries have adopted, or are in the processing of introducing, legislation limiting SO_2 emissions from their coal-fired power plants [59]. A range of national emissions standards for sulfur emissions, reported in milligrams SO_2 per cubic meter (mg SO_2/m^3), is given in Table 8.9 [59].

Table 8.9 Range of National Emission Standards for Sulfur Emissions (mg SO_2/m^3)[a]

Country	New Plants	Existing Plants	Coal Sulfur Limit (%)	Applicable Plant Size for Coal Sulfur Limit
Austria[b]	200–550	200–2,000	≤1 (brown coal)	Plants <10 MW_t
Belgium[b]	250–2,000	1,700	≤1	All plants
Bulgaria	610–3,335	1,875–3,500	—	—
Canada[c]	740	—	—	—
China				
Hong Kong	≈200	—	≤1	>550 MW_t
Mainland	1,200–2,000	Plant specific	—	—
Croatia	400–2,000[d]	2000[d]	<0.7 g/megajoule	<50 MW_t
Czech Republic	500–2,500	500–2,500	—	—
Denmark[b]	400–2,000	—[e]	0.9	Industrial plants
European Union	400–2,000[f]	—[g]	—	—
Finland[b]	380–620	620	≤390 g/gigajoule	Plants without FGD[h]
France[b]	400–2,000	400–2,000	—	—
Germany[b]	400–2,000	400–2,000	<1	Plants <1 MW_t
Greece[b]	400–2,000	—	—	—
Hungary	400–2,000	—	—	—
Indonesia	820	1,635[i]	—	—
Ireland[b]	400–2,000	—	—	—
Italy[b]	400–2,000	—	1	All plants
Japan	—[j]	—[j]	—	—
Korea, South	345	430–770	—	—
Luxembourg[b]	400–2,000	—	—	—
Netherlands[b]	200–700	400	1.2 (industrial)	Plant without FGD
Philippines	175–765	—	<1	All existing plants
Poland	540–1,755	675–2,890	—	—
Portugal[b]	100–2,000	—	—	—
Romania	400–2,000	—	—	—
Slovakia	400–2,000	500–2,500	—	—
Slovenia	400–2,000	2,000	—	—
Spain[b]	400–2,000	2,400–9,000	—	—
Sweden[b]	160–270	270–540	—	—
Switzerland	430–2,145	430–2,145	≤1	All plants
Taiwan	570–1,430	570–1,430	—	—
Thailand	180	290–390[k]	—	—
Turkey	430–2,500	430–2,500	—	—

Table 8.9 *Continued*

Country	New Plants	Existing Plants	Coal Sulfur Limit (%)	Applicable Plant Size for Coal Sulfur Limit
United Kingdom[b]	200–2,000	—[e]	—	—
United States	740–1,480	1,480[e]	—	—

[a]Based on dry flue gas at 6% O_2, STP (0°C (275K), 101.3kPa)
[b]European Union (EU) country
[c]National guidelines
[d]Proposed standards
[e]Based on annual quota totals
[f]EU proposed limits to come into operation after January 1, 2000. New plants 50–100 MW_t, 850 mg/m^3; >300 MW_t, 200 mg/m^3; 100–300 MW_t, sliding scale within upper and lower limits proposed
[g]European Parliament proposed limits for existing granted a license before January 1, 2000, to become mandatory from January 1, 2005. Existing plants 50–100 MW_t, 900 mg/m^3; >300 MW_t, 300 mg/m^3; 100–300 MW_t, sliding scale within upper and lower limits proposed
[h]Flue gas desulfurization
[i]Must meet new plant standards by January 1, 2000
[j]Set on a plant-by-plant basis according to nationally defined formula
[k]New and existing industrial plants
Source: From Soud (2000).

8.3.2   Nitrogen Oxides

NO_x emisson standards have been introduced or are becoming more stringent around the world with increased concerns about local, regional, and transboundary effects of NO_x emissions. Generally, international legislation has been an important factor in developing national regulations in many parts of the world. The recognition of the transboundary effect of air pollution has led to a number of international agreements [60, 61]. Those that pertain to NO_x emissions include the UNECE LRTAP (discussed in the previous section), EC directives, and World Bank environmental guidelines. In addition to SO_2 emissions discussed earlier, the UNECE Convention on LRTAP also addressed transboundary NO_x emissions. Table 8.10 lists the signatories to the UNECE Convention on LRTAP and the status of each country [61].

The Sofia NO_x protocol was signed in 1998 by 23 countries and come into force in 1991. The protocol requires that NO_x emissions be frozen at 1987 levels by the end of 1994 and that these levels be maintained in subsequent years. The protocol has 25 signatures and 33 ratifications, as listed in Table 8.10 [61].

Table 8.10 also contains the countries that signed the 1999 Protocol to Abate Acidification, Eutrophication, and Ground-Level Ozone. The protocol, signed in Gothenburg, Sweden, places NO_x emission limits of 400, 300, and 200 mg NO_x/m^3 for new installations with capacity of 50 to 100, 100 to 300, and larger than 300 MW_t, respectively [60]. Existing installations are limited to 650 mg NO_x/m^3 for solid fuels in general and 1,300 mg NO_x/m^3 for solid fuels with more than 10 percent volatile matter content.

The EC has adopted several directives and amendments, including the Directive on Controlling Emissions from Large Combustion Plants, the Directive on the

Table 8.10 Signatories to the UNECE Convention on LRTAP and Subsequent
NO_x Protocols as of September 2, 2009

1979 LRTAP Convention	1988 Sofia Protocol	1999 Gothenburg Protocol
Albania (Ac)	Albania (Ac)	
Armenia (Ac)		Armenia
Austria (R)	Austria (R)	Austria
Azerbaijan (Ac)		
Belarus (R)	Belarus (At)	
Belgium (R)	Belgium (R)*	Belgium (R)
Bosnia and Herzegovina (Sc)		
Bulgaria (R)	Bulgaria (R)	Bulgaria (R)
Canada (R)	Canada (R)	Canada
Croatia (Sc)	Croatia (Ac)	Croatia (R)
Cyprus (Ac)	Cyprus (Ac)	Cyprus (Ac)
Czech Republic (Sc)	Czech Republic (Sc)	Czech Republic (R)
Denmark (R)	Denmark (At)*	Denmark (Ap)
Estonia (Ac)	Estonia (Ac)	
European Community (Ap)	European Community (Ac)	
Finland (R)	Finland (R)*	Finland (Ac)
France (Ap)	France (Ap)*	France (Ap)
Georgia (Ac)		
Germany (R)	Germany (R)*	Germany (R)
Greece (R)	Greece (R)	Greece
Holy See		
Hungary (R)	Hungary (Ap)	Hungary (Ap)
Iceland (R)		
Ireland (R)	Ireland (R)	Ireland
Italy (R)	Italy (R)*	Italy
Kazakhstan (Ac)		
Kyrgyzstan (Ac)		
Latvia (Ac)		Latvia (Ac)
Liechtenstein (R)	Liechtenstein (R)*	Liechtenstein
Lithuania (Ac)	Lithuania (R)	Lithuania (Ac)
Luxembourg (R)	Luxembourg (R)	Luxembourg (R)
Malta (Ac)		
Monaco (At)		
Montenegro (Sc)		
Netherlands (At)	Netherlands (At)*	Netherlands (At)
Norway (R)	Norway (R)*	Norway (R)
Poland (R)	Poland (R)	Poland
Portugal (R)		Portugal (Ap)
Republic of Moldova (Ac)		
Romania (R)		Romania (R)
Russian Federation (R)	Russian Federation (At)	
San Marino		

Table 8.10 *Continued*

1979 LRTAP Convention	1988 Sofia Protocol	1999 Gothenburg Protocol
Serbia (Sc)		
Slovakia (Sc)	Slovakia (Sc)	Slovakia (R)
Slovenia (Sc)	Slovenia (Ac)	Slovenia (R)
Spain (R)	Spain (R)	Spain (R)
Sweden (R)	Sweden (R)•	Sweden (R)
Switzerland (R)	Switzerland (R)*	Switzerland (R)
The Former Yugoslav Republic of Macedonia (Sc)		
Turkey (R)		
Ukraine (R)	Ukraine (At)	
United Kingdom (R)	United Kingdom (R)	United Kingdom (R)
United States (At)	United States (At)	United States (Ac)

Notes: R—ratification; Ac—accession; Ap—approval; At—acceptance; Sc—succession; *—committing to 30% reduction

Source: From United Nations (2009).

Limitation of Emissions of Certain Pollutants into the Air from Large Combustion Plants, and the Directive on National Emission Ceilings for Certain Atmospheric Pollutants [60]. NO_x emission limits for solid fuel-fired boilers in the new EC directives are 600 and 500 mg NO_x/m^3 for existing installations with capacities of 50–500 and larger than 500 MW_t, respectively. These limits become stricter beginning January 1, 2016, for larger units, with the limit decreasing to 200 mg NO_x/m^3 for units larger than 500 MW_t. Emission limits for new installations with capacities of 50 to 100, 100 to 300, and more than 300 MW_t are, respectively, 400, 200 (300 for biomass-fired units), and 200 mg NO_x/m^3. National ceilings for NO_x emissions for 2010, under EC directive, are shown in Table 8.11 [60].

The World Bank has also developed environmental guidelines that must be followed in all the projects it funds, thereby covering a host of developing countries. The World Bank has determined that environmental standards of developed countries may not be appropriate for developing countries or economies in transition; therefore, their guidelines are flexible and try to maintain and improve environmental quality on an ongoing basis [60]. The World Bank's NO_x standards are 750 mg NO_x/m^3 for all coal-fired power plants except for those firing coal with less than 10 percent volatile matter content, where the NO_x emission limit is 1,300 mg NO_x/m^3.

National standards for NO_x emissions from coal-fired power plants have been adopted or are being introduced in more than 30 countries [60]. They vary widely between countries and are often determined by taking into account the technology available, the type of plant (new or existing), the size of the plant, and boiler configuration. A comparison of emission standards for various countries is shown in Table 8.12 [62].

Table 8.11 National NO_x Emissions Ceilings for 2010

	NO_x Emissions (metric kiloton)	
Country	1990[a]	2010
Austria	193	103
Belgium	339	176
Denmark	272	127
Finland	300	170
France	1,865	810
Germany	2,706	1,051
Greece	326	344
Ireland	118	65
Italy	1,935	990
Luxembourg	23	11
Netherlands	580	260
Portugal	317	250
Spain	1,156	847
Sweden	338	148
United Kingdom	2,756	1,167
Total	**13,227**	**6,519**

[a]The latest reported by each country to the LRTAP Convention
Source: From Wu (2002).

Table 8.12 NO_x Emission Standards Applicable to New and Existing Coal-Fired Power Plants of Thermal Capacity Larger than 300 MW_t

	NO_x Emissions (mg/m^3)[a]	
Country	New Plant	Existing Plant
Austria	200	200–300
Canada[b]	490–740	—
European Community	650–1,300	—
Denmark	200–650	—
Germany	200	200
Italy	200–650	200–650
Japan	410–515	410–515
Korea, South	720	720
Netherlands	200–400	650–1,100
Poland	405–460	610–1,335
Spain	650–1,300	—
United Kingdom	650–1,300	—
United States[b]	615–740	555–615

[a]Standards are given in mg/m^3 corrected to standard conditions (6% O_2, standard temperature and pressure (0°C (273 K), 101.3 kPa) on dry flue gas); when converting from lb/million Btu to mg/m^3, dry flue gas volume assumed to be 350 m^3/GJ (based on gross heat value); note that ranges exist because emission standards may vary according to plant type, size, location, construction/commissioning date, boiler configuration, and type of coal used.
[b]Federal standards only—state standards may be more stringent.
Source: From McConville (1997).

8.3.3   Particulate Matter

Standards for the control of particulate emissions were first introduced in the early 1900s in Japan, the United States, and Western Europe [63]. Over the decades they have become increasingly more stringent and widespread, with recent emphasis on the control of fine particulate matter.

Similarly to NO_x and SO_2 emissions, international agreements have been signed to reduce particulate emissions from coal-fired power plants. The EC set limits for particulate emissions from coal-fired power plants in the LCPD. These set particulate emissions limits of 100 and 50 mg/m^3 for new plants with capacities of 50 to 500 and larger than 500 MW$_t$, respectively [63]. Even stricter limits are under review to conserve the environment and human health.

Increasingly, more stringent national emission standards have been adopted in Japan, North America, and Western Europe. The growing importance of using coal in an environmentally acceptable manner for power generation, as well as in the industrial and residential sectors, has led to the introduction of particulate emission standards in other countries as well. Currently, 30 countries have emissions standards for particulate emissions from coal-fired power plants [63]. Examples of particulate emissions standards in some of these counties are provided in Table 8.13 [62].

Table 8.13 Particulate Emission Standards Applicable to New and Existing Coal-Fired Power Plants of Thermal Capacity Larger than 300 MW$_t$

Country	Particulate Emissions (mg/m^3)[a]	
	New Plant	Existing Plant
Austria	50	50
Canada[b]	145	—
European Community	50–100	—
Denmark	40	40–120
Germany	50	80–125
India	150–350	150–350
Italy	50	50
Japan	50–300	50–300
Korea, South	50–100	50–100
Netherlands	50	—
Poland	190–350	460–700
Spain	50–100	200–500
United Kingdom	50–100	—
United States[b]	40–125	40–125

[a]Standards are given in mg/m^3 corrected to standard conditions (6% O_2, standard temperature and pressure (0°C (273 K), 101.3 kPa) on dry flue gas); when converting from lb/million Btu to mg/m^3, dry flue gas volume assumed to be 350 m^3/GJ (based on gross heat value); note that ranges exist because emission standards may vary according to plant type, size, location, construction/commissioning date, boiler configuration, and type of coal used.
[b]Federal standards only—state standards may be more stringent.
Source: From McConville (1997).

8.3.4   Trace Elements/Mercury

Concern over environmental effects of trace elements emissions—specifically cadmium, chromium, copper, mercury, nickel, lead, selenium, and zinc—from human activities has lead to the introduction of legislation on emissions in many countries; however, this legislation sets limits for medical waste incinerators, municipal solid waste combustors, and hazardous waste incinerators [64]. Trace elements emissions from coal combustion are not currently regulated.

An overview of mercury regulation in the European Union, Japan, Australia, and United States reveals a variety of different approaches [65]. The 1998 United Nations Protocol on Heavy Metals set emissions limits for hazardous and municipal waste incinerators and directed signatories to the protocol to set limits for medical waste incinerators. The protocol had 36 signatures and 29 ratifications as of September 9, 2009, as shown in Table 8.14 [61].

Table 8.14 Signatories to the UNECE Convention of LRTAP and 1998 Heavy Metals Protocol

1979 LRTAP Convention	1998 Heavy Metals Protocol
Albania (Ac)	
Armenia (Ac)	
Austria (R)	Austria (R)
Azerbaijan (Ac)	
Belarus (R)	
Belgium (R)	Belgium (R)*
Bosnia and Herzegovina (Sc)	
Bulgaria (R)	Bulgaria (R)
Canada (R)	Canada (R)
Croatia (Sc)	Croatia (R)
Cyprus (Ac)	Cyprus (R)
Czech Republic (Sc)	Czech Republic (R)
Denmark (R)	Denmark (Ap)
Estonia (Ac)	Estonia (Ac)
European Community (Ap)	European Community (Ap)
Finland (R)	Finland (At)
France (Ap)	France (Ap)
Georgia (Ac)	
Germany (R)	Germany (R)
Greece (R)	
Holy See	
Hungary (R)	Hungary (R)
Iceland (R)	
Ireland (R)	
Italy (R)	
Kazakhstan (Ac)	

Table 8.14 *Continued*

1979 LRTAP Convention	1998 Heavy Metals Protocol
Kyrgyzstan (Ac)	
Latvia (Ac)	Latvia (R)
Liechtenstein (R)	Liechtenstein (At)
Lithuania (Ac)	Lithuania (R)
Luxembourg (R)	Luxembourg (R)
Malta (Ac)	
Monaco (At)	Monaco (Ac)
Montenegro (Sc)	
Netherlands (At)	Netherlands (At)
Norway (R)	Norway (R)
Poland (R)	
Portugal (R)	
Republic of Moldova (Ac)	Republic of Moldova (R)
Romania (R)	Romania (R)
Russian Federation (R)	
San Marino	
Serbia (Sc)	
Slovakia (Sc)	Slovakia (At)
Slovenia (Sc)	Slovenia (R)
Spain (R)	
Sweden (R)	Sweden (R)
Switzerland (R)	Switzerland (R)
The Former Yugoslav Republic of Macedonia (Sc)	
Turkey (R)	
Ukraine (R)	
United Kingdom (R)	United Kingdom (R)
United States (At)	United States (At)

*Notes:*R—ratification; Ac—accession; Ap—approval; At—acceptance; Sc—succession
Source: From United Nations (2009).

Separate from the United Nations protocol, the European Council issued a directive in 1996 ordering limit values and alert thresholds for a variety of air pollutants including mercury [66]. The directive resembles aspects of the U.S. federal regulation in that it assigns the responsibility to implement the limit values and all attainment programs to member states. In response to the directive, the EC proposed an ambient air quality standard of 0.05 mg/m^3 for elemental mercury. The proposed standard is rarely exceeded in Europe. The United States and Japan do not have such a standard.

The United States has regulated all significant sources of mercury emissions in a manner consistent with the United Nations protocol, with the exception of power plant emissions. The European Union also has not regulated mercury emissions from power plants but has seen mercury reductions due to cobenefits effects from existing SO_2 and NO_x regulations.

8.3.5   Carbon Dioxide

Fossil fuel consumption is projected to increase over the next two to three decades, with coal the leading energy source in some countries, especially certain developing countries. Consequently, carbon dioxide (CO_2) emissions are projected to increase. The increase in CO_2 emissions and the concern about global warming have received international attention.

The first major action was taken in New York on May 9, 1992, when the United Nations Framework Convention on Climate Change was adopted. The objective of the Convention is to achieve stabilization of greenhouse gas concentrations in the atmosphere at a level that would prevent dangerous interference with the climate system [67]. The Convention stated that such a level should be achieved naturally to climate change, to ensure that food production is not threatened, and to enable economic development to proceed in a sustainable manner. The Convention contains a legally binding framework that commits the world's governments to voluntary reductions of greenhouse gases or other actions such as enhancing greenhouse gas sinks, aimed at stabilizing atmospheric concentrations of greenhouse gases at 1990 levels by the year 2000 [68].

On June 12, 1992, at the Earth Summit in Rio de Janeiro, 154 nations, including the United States, signed the United Nations Framework Convention on Climate Change. In October 1992, the United States became the first industrialized nation to ratify the treaty, which came into force on March 21, 1994. The treaty was not legally binding, and because reducing emissions would likely cause great economic damage, many nations were not expected to meet the goal.

The Convention has become a cornerstone of global climate policy, representing a compromise between a wide range of different interests among member countries. The concept of a common goal but different responsibilities provided for different roles for industrialized and developing countries, notably in the obligations imposed on them in connection with climate protection policy [69]. This led to a grouping of the member states of the Convention into Annex I, Annex II, and non-Annex I countries, the last consisting of the developing countries with no commitments to reducing climate gases.

Annex I countries agreed to, among other issues, adopt national policies and take corresponding measures on the mitigation of climate change; periodically provide information on its policies and measures to mitigate climate change; and calculate emissions sources and removal through sinks. The developed countries in Annex II agreed to, along with additional provisions, provide new and additional financial resources to meet the agreed full costs incurred by developing countries in complying with their obligations.

Representatives from around the world met again in December 1997 at a conference in Kyoto to sign a revised agreement. The Clinton administration negotiators agreed to legally binding, internationally enforceable limits on the emissions of greenhouse gases as a key tenet of the treaty. The protocol called for a worldwide reduction of emissions of carbon-based gases by an average of 5.2 percent below 1990 levels by 2010. Different countries adopted different targets. Those countries that agreed to reduce specified amounts of climate gases within a specified time period are listed as Annex B countries, which is a subcategory of the Annex I countries. For example, the EU committed to a cut of 8 percent in climate gases, the United States to 7 percent, Japan to 6 percent, and Russia and the Ukraine agreed to stabilize at 1990 levels. Table 8.15 contains the listed Annex I, Annex II, and Annex B countries, along with the specified amounts of climate gases agreed upon by the Annex B countries [69, 70].

The conflict underlying the distribution of different obligations has become apparent since the Kyoto conference. In March 2001, the United States announced that it would not support the Kyoto Protocol [71]. The United States insists that the rules pertaining to the Annex B countries—the voluntary commitment to reducing climate gases—must be extended at least to the major developing countries and made this a precondition to ratifying the Kyoto Protocol.

In November 2001, the participating member countries of the United Nations' seventh Conference of Parties (COP-7) met in Marrakesh, Morocco, and reached final agreement for the procedures and institutions needed to make the Kyoto Protocol fully operational [71]. On March 4, 2002, the EU voted to ratify the protocol, committing its 15 member countries to reductions in greenhouse gas emissions as specified in the accord. No agreement has been reached among the EU member countries, however, with regard to the individual emission reductions that will be required. There is discontent that some countries feel they have been given a disproportionate share of the EU's total reduction burden [71]. The Kyoto Protocol enters into force 90 days after it has been ratified by at least 55 parties to the United Nations Framework Climate Change Convention, including a representation of Annex I countries accounting for at least 55 percent of the total 1990 CO_2 emissions from the Annex I group. Although the United States had the largest share of Annex I emissions in 1990 at 35 percent, even without U.S. participation, the Protocol entered into force in February 2005 for the other signatories because 184 Parties of the Convention ratified the Protocol [72].

8.4 Air Quality and Coal-Fired Emissions

The U.S. EPA evaluates the status and trends in the nation's air quality on a yearly basis and tracks air pollution by evaluating the air quality measured from more than 5,200 ambient air monitors located at more than 3,000 sites across the nation that are operated primarily by state, local, and tribal agencies [73]. In addition, the EPA tracks emissions from all sources for the last 30 years but does have data back to 1970. In the most recent report (for the year 2008) on the latest findings on air quality in the United States, the agency aggregate emissions of the six principal (i.e., criteria) air pollutants tracked nationally have been reduced by 60 percent since 1970 [74]. During this same period, the U.S. gross domestic product increased 209 percent, energy consumption increased 49 percent, the population increased 48 percent, and vehicle miles traveled increased 163 percent. This reduction

Table 8.15 Annex I, Annex II, and Annex B Countries of the UN's Framework Convention on Climate Change and the Kyoto Protcol

Annex I Countries	Annex II Countries	Annex B Countries (percent of base year or period)
Australia	Australia	Australia (108)
Austria	Austria	Austria (92)
Belarus[a]		
Belgium	Belgium	Belgium (92)
Bulgaria[a]		Bulgaria[a] (92)
Canada	Canada	Canada (94)
		Croatia (95)
Czechoslovakia[a]		Czechoslovakia[a] (92)
Denmark	Denmark	Denmark (92)
Estonia[a]		Estonia[a] (92)
European Economic Community	European Economic Community	European Economic Community (92)
Finland	Finland	Finland (92)
France	France	France (92)
Germany	Germany	Germany (92)
Greece	Greece	Greece (92)
Hungary[a]		Hungary[a] (94)
Iceland	Iceland	Iceland (110)
Ireland	Ireland	Ireland (92)
Italy	Italy	Italy (92)
Japan	Japan	Japan (94)
Latvia[a]		Latvia[a] (92)
		Liechtenstein (92)
Lithuania[a]		Lithuania[a] (92)
Luxembourg	Luxembourg	Luxembourg (92)
		Monaco (92)
Netherlands	Netherlands	Netherlands (92)
New Zealand	New Zealand	New Zealand (100)
Norway	Norway	Norway (101)
Poland[a]		Poland[a] (94)
Portugal	Portugal	Portugal (92)
Romania[a]		Romania[a] (92)
Russian Federation[a]		Russian Federation[a] (100)
		Slovakia[a] (92)
		Slovenia[a] (92)
Spain	Spain	Spain (92)
Sweden	Sweden	Sweden (92)
Switzerland	Switzerland	Switzerland (92)
Turkey	Turkey	
Ukraine[a]		Ukraine[a] (100)
United Kingdom	United Kingdom	United Kingdom (92)
United States	United States	United States (93)

[a]Countries that are undergoing the process of transition to a market economy.

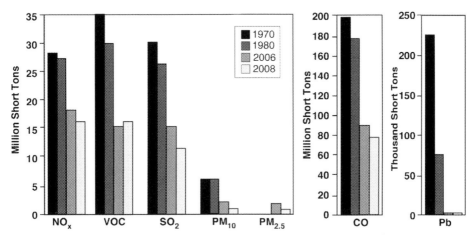

Figure 8.4 Comparison of 1970, 1980, 2006, and 2008 emissions of criteria air pollutants. *Source:* From EPA (2003) and EPA (2009).

in emissions in criteria air pollutants between 1970 and 2008 is illustrated in Figure 8.4. In addition, $PM_{2.5}$ emissions have increased by 57 percent since 1980, with 1 million tons emitted in 2008.

Despite this progress, about 123 million short tons of pollution are emitted into the air each year in the United States, and 158.5 million people live in counties where air monitored in 2007 was unhealthy because of high levels of at least one of the six criteria air pollutants [73, 75]. Most of the areas that experienced the unhealthy air did so because of particulate matter and/or ground-level ozone.

This section summarizes the air quality and emissions trends in the United States for the criteria pollutants: NO_2, ozone, SO_2, particulate matter, carbon monoxide, and lead, along with acid rain, trace elements, specifically mercury, and CO_2. Air quality is based on actual measurements of pollutant concentrations in the ambient air at monitoring sites. Trends are derived by averaging direct measurements from these monitoring stations on a yearly basis. Emissions of ambient pollutants and their precursors are estimated based on actual monitored readings or engineering calculations of the amounts and types of pollutants emitted by vehicles, factories, stationary combustion, and other sources.

8.4.1   Six Principal Pollutants

As previously discussed, under the Clean Air Act, the EPA established air quality standards to protect human health and public welfare. The agency has set national air quality standards for six principal or criteria air pollutants, which include nitrogen dioxide, ozone, sulfur dioxide, particulate matter, carbon monoxide, and lead. Four of these pollutants—NO_2, SO_2, CO, and lead—result primarily from direct emissions from a variety of sources. Particulate matter results from direct emissions but is also commonly formed when emissions of nitrogen oxides, sulfur oxides, ammonia, organic compounds, and other gases react in the atmosphere. Ozone is not directly emitted but is formed when nitrogen oxides and volatile organic compounds react in the presence of sunlight.

Nitrogen Dioxide

Nitrogen oxides (NO_x), the term used to describe the sum of NO, nitrogen dioxide (NO_2), and other oxides of nitrogen, contribute to the formation of ozone, particulate matter, haze, and acid rain. While the EPA traces national emissions of NO_x, the national monitoring network measures ambient concentrations of NO_2 for comparison to national air quality standards. The major sources of anthropogenic NO_x emissions are high-temperature combustion processes, such as those that occur in vehicles and power plants.

Over the period 1983–2003, monitored levels of NO_2 decreased 21 percent and from 2001 to 2007 decreased 20 percent [75]. All areas of the United States that once violated the NAAQS for NO_2 now meet the standard. While overall NO_x emissions are declining, emissions from some sources such as nonroad engines have actually increased since 1983.

Figures 8.5 through 8.7 illustrate trends in NO_2 air quality and NO_x emissions by sources [74, 75]. Of the approximately 10 million short tons of NO_x emitted from fuel combustion (Figure 8.7, which is the most recent comprehensive reporting), power plants contributed less than 4.5 million short tons [73]. All recorded concentrations in 2007 were well below the level of the annual standard of 0.053 ppm.

Ozone

Ground-level ozone, which is the primary constituent of smog, continues to be a pollution problem throughout many areas of the United States. Ozone is not emitted directly into the air but is formed by the reaction of volatile organic compounds

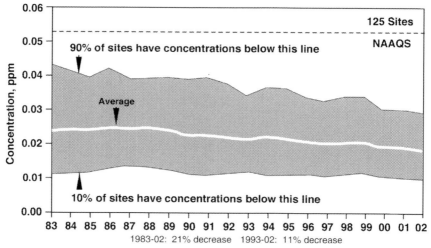

Figure 8.5 NO_2 air quality from 1983 to 2002.
Source: From EPA (2003).

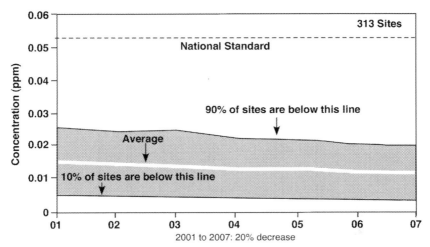

Figure 8.6 NO$_2$ air quality from 2001 to 2007.
Source: From EPA (2008).

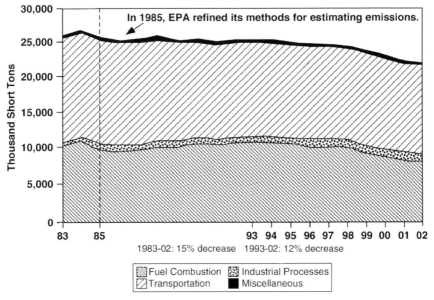

Figure 8.7 NO$_x$ emissions from 1983 to 2002.
Source: From EPA (2003).

(VOCs) and NO$_x$ in the presence of heat and sunlight. The trends of VOC emissions and their sources for the period 1983 to 2002 are shown in Figure 8.8 [73]. Fuel combustion contributes approximately 5 percent of the VOC emissions, with power stations comprising less than half of the 5 percent [4].

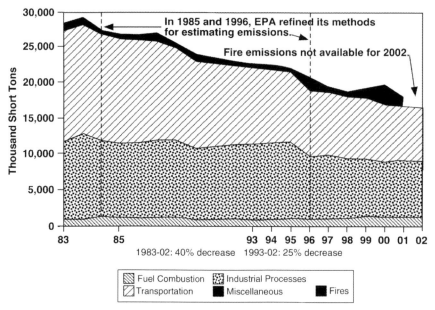

Figure 8.8 VOC Emissions from 1983 to 2002.
Source: From EPA (2003).

Over the period 1983 to 2002, national ambient ozone levels decreased 22 and 14 percent based on 1-hour and 8-hour data, respectively [73]. During this period, emissions of VOCs decreased 40 percent (excluding wildfires and prescribed burning). From 2001 to 2007, ozone concentrations declined 5 percent. Ozone air quality trends are illustrated in Figures 8.9 through 8.11 [73, 75]. Many areas measured concentrations above the 2008 national air quality standard for ozone (0.075 ppm).

Sulfur Dioxide

Nationally, average sulfur dioxide (SO_2) ambient concentrations decreased 54 percent from 1983 to 2002 and 39 percent from 1993 to 2002 (see Figure 8.12) [73]. SO_2 emissions decreased 33 percent from 1983 to 2002 and 31 percent from 1993 to 2002. Reductions in SO_2 concentrations and emissions since 1990 are the result of controls implemented under The EPA's Acid Rain Program beginning in 1995. As shown in Figure 8.13, fuel combustion, mainly coal and oil, accounts for most of the total SO_2 emissions. Coal combustion accounts for approximately 11 of the 15 million short tons of SO_2 emitted in 2002. Nationally, concentrations of SO_2 declined 24 percent between 2001 and 2007, as shown in Figure 8.14. All 2007 concentrations were below the annual standard of 0.03 ppm.

Particulate Matter

Between 1993 and 2002, PM_{10} concentrations decreased 13 percent, while PM_{10} emissions decreased 22 percent [73]. This can be seen in Figures 8.15 and 8.16,

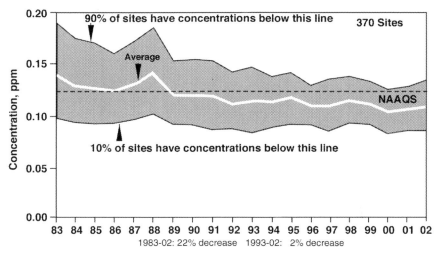

Figure 8.9 Ozone air quality from 1983 to 2002 based on 8-hour averages.
Source: From EPA (2003).

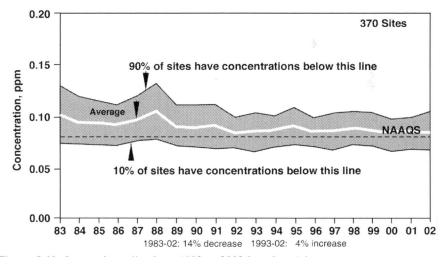

Figure 8.10 Ozone air quality from 1983 to 2002 based on 1-hour averages.
Source: From EPA (2003).

respectively, later in the chapter. Nationally, PM_{10} concentrations declined by 21 percent between 2001 and 2007, as shown in Figure 8.17 [75] (see page 359). Fuel combustion accounts for about one-third of total particulate emissions (see Figure 8.16), while electric utilities account for approximately 5 percent of the total particulate matter emitted [4].

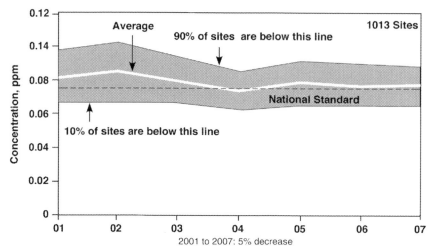

Figure 8.11 National 8-hour ozone air quality trend from 2001 to 2007.
Source: From EPA (2008).

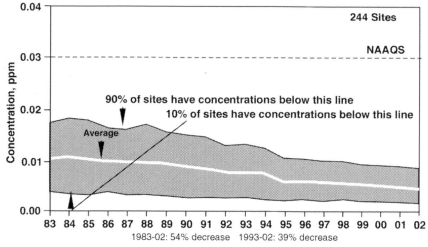

Figure 8.12 SO$_2$ air quality from 1983 to 2002.
Source: From EPA (2003).

Figure 8.18 shows that direct PM$_{2.5}$ emissions from anthropogenic sources decreased 10 percent nationally between 1992 and 2001 [73]. The figure tracks only directly emitted particles and does not account for secondary particles, which are primarily sulfates and nitrates, formed when emissions of NO$_x$, SO$_2$, ammonia, and other gases react in the atmosphere. Nationally, annual and 24-hour PM$_{2.5}$

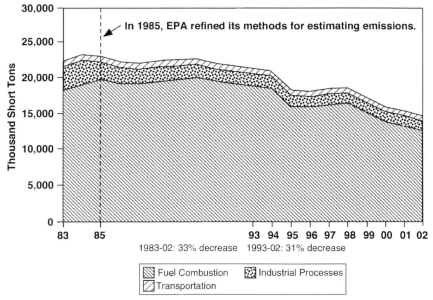

Figure 8.13 SO$_2$ emissions from 1983 to 2002.
Source: From EPA (2003).

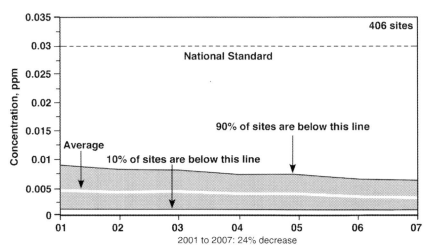

Figure 8.14 National SO$_2$ air quality trend from 2001 to 2007.
Source: From EPA (2008).

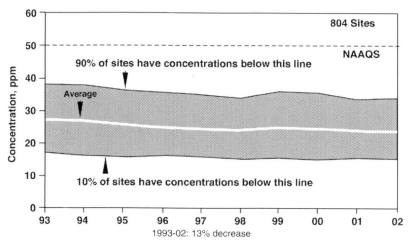

Figure 8.15 PM$_{10}$ air quality from 1993 to 2002.
Source: From EPA (2003).

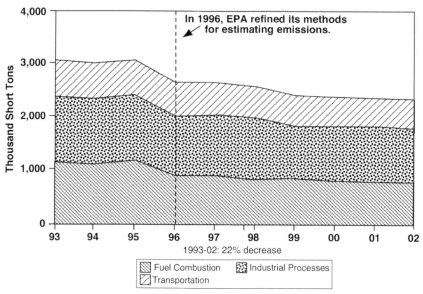

Figure 8.16 PM$_{10}$ emissions from 1993 to 2002.
Source: From EPA (2003).

concentrations declined by 9 and 10 percent, respectively, between 2001 and 2007, as shown in Figures 8.19 and 8.20, respectively [75].

Carbon Monoxide

Carbon monoxide (CO) is a component of motor vehicle exhaust, which contributes about 60 percent of all CO emissions nationwide. Other sources of CO emissions

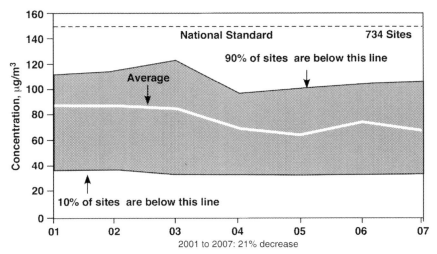

Figure 8.17 National PM$_{10}$ air quality trend from 2001 to 2007.
Source: From EPA (2008).

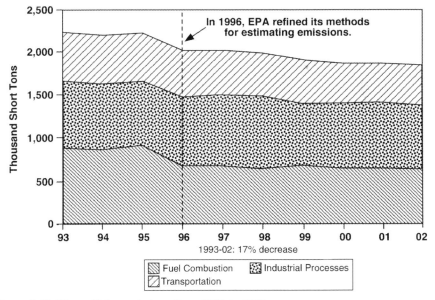

Figure 8.18 Direct PM$_{2.5}$ emissions from 1993 to 2002.
Source: From EPA (2003).

include industrial processes, nontransportation fuel combustion, and natural sources such as wildfires.

Nationally, the 2002 ambient average CO concentration was nearly 65 percent lower than that for 1983, which is illustrated in Figure 8.21 [73]. CO emissions decreased about 42 percent over the period 1993 to 2002 despite an approximately

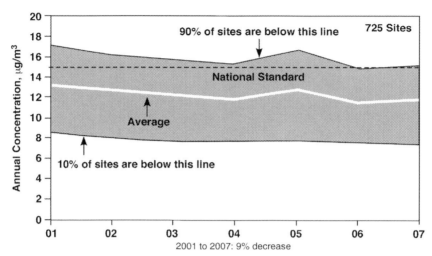

Figure 8.19 National $PM_{2.5}$ annual air quality trend from 2001 to 2007.
Source: From EPA (2008).

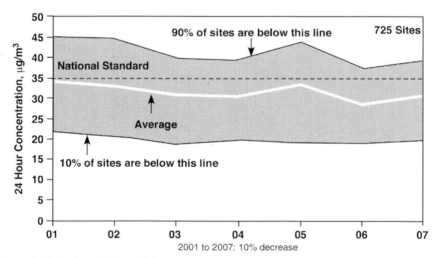

Figure 8.20 National $PM_{2.5}$ 24-hour air quality trend from 2001 to 2007.
Source: From EPA (2008).

23 percent increase in vehicle miles traveled. CO concentrations declined 39 percent between 2001 and 2007, as shown in Figure 8.22 [75].

Transportation sources are the largest contributors to CO emissions, with fuel combustion accounting for about 7 percent of the CO emissions. Electric utilities account for less than 0.5 percent of the total CO emissions [4]. The trend in CO emissions is shown in Figure 8.23 [73].

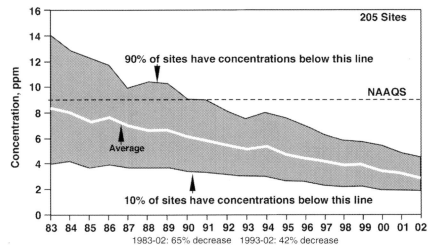

Figure 8.21 CO air quality from 1983 to 2002.
Source: From EPA (2003).

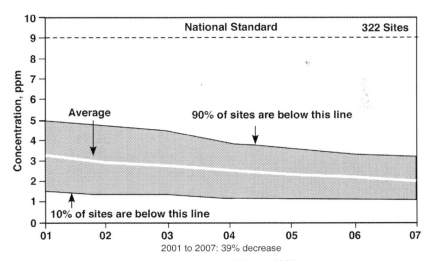

Figure 8.22 National CO air quality trend from 2001 to 2007.
Source: From EPA (2008).

Lead

In the past, automotive sources were the major contributor of lead (Pb) emissions to the atmosphere. The emissions of lead from the transportation sector have greatly declined over the last 20 years as leaded gasoline was phased out. Today, industrial processes, primarily metals processing, are the major sources of lead emissions to the atmosphere. As a result of the phaseout of leaded gasoline, lead concentrations

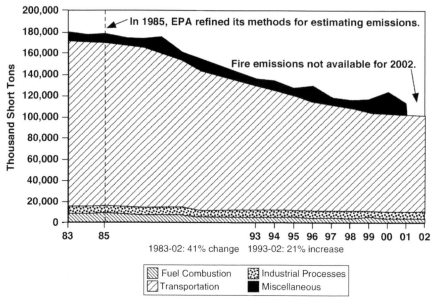

Figure 8.23 CO emissions from 1983 to 2002.
Source: From EPA (2003).

and emissions have decreased significantly as shown in Figures 8.24 through 8.26 [73, 75]. The 2002 average air quality concentration for lead is 94 percent lower than in 1982 and lead emissions decreased by 93 percent over the same period. Lead emissions from electric utilities are less than 10 percent of the total—that is, less than 500 short tons [4]—and the only violations of the lead NAAQS that occur today are near large industrial sources such as lead smelters and battery manufacturers [73].

8.4.2 Acid Rain

As discussed earlier, acid rain or acidic deposition occurs when emissions of sulfur dioxide and nitrogen oxides in the atmosphere react with water, oxygen, and oxidants to form acidic compounds. These compounds then fall to earth in either dry form (gas and particles) or wet form (rain, snow, and fog). In the United States, about 63 percent of annual SO_2 emissions and 22 percent of NO_x emissions are produced by electric utility plants that burn fossil fuels [73].

The EPA's Acid Rain Program will reduce annual SO_2 emissions by 10 million short tons from 1980 levels by 2010. The program sets a permanent cap of 8.95 million short tons on the total amount of SO_2 that may be emitted by power plants nationwide, which is about half of that emitted in 1980. Approximately 3,000 units are now affected by the Acid Rain Program. Figure 8.27 shows the SO_2 reductions

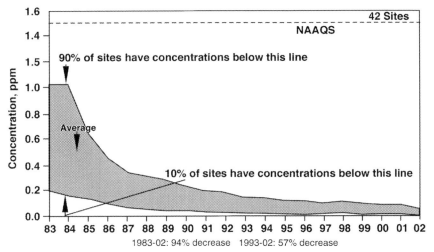

Figure 8.24 National lead air quality from 1983 to 2002.
Source: From EPA (2003).

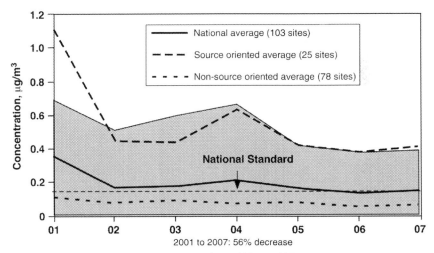

Figure 8.25 National lead air quality trend from 2001 to 2007.
Source: From EPA (2008).

achieved as of 2008, which illustrates that SO_2 emissions were reduced to about 7.6 million short tons in 2008 [76].

The NO_x component of the Acid Rain Program limits the emission rate for all affected utilities, resulting in a 2 million short ton NO_x reduction from 1980 levels by 2000. NO_x emissions, shown in Figure 8.28, have declined since 1990, with NO_x

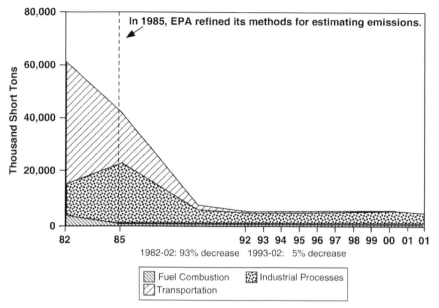

Figure 8.26 Lead emissions from 1982 to 2002.
Source: From EPA (2008).

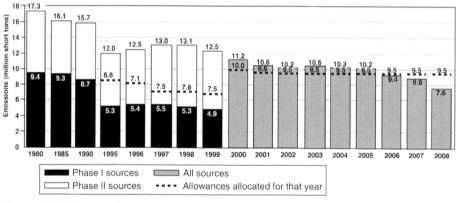

Figure 8.27 SO$_2$ emissions covered under the acid rain program.
Source: From EPA (2009).

emissions from about 1,000 affected sources totaling approximately 2.7 million short tons in 2008 [76].

Sulfate concentrations in the atmosphere, which is a major component of fine particles, especially in the eastern United States, have decreased since 1990, as shown in Figure 8.29 [77]. The pattern from 1990 to 2005 illustrates that significant reductions occurred in wet sulfate deposition, a major component of acid rain, in both the eastern United States and much of eastern Canada. Reductions in wet

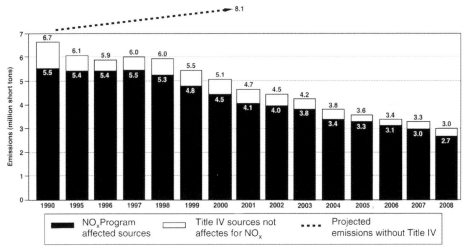

Figure 8.28 NO$_x$ emissions covered under the acid rain program.
Source: From EPA (2009).

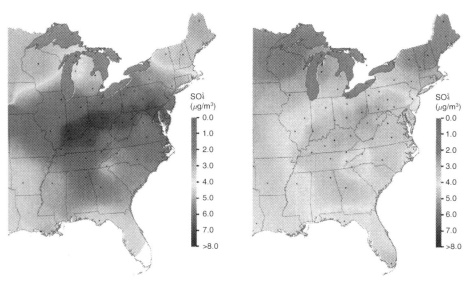

Figure 8.29 Wet sulfate deposition for 1990 and 2005.
Source: From EPA (2009).

nitrate deposition have generally been more modest than for wet sulfate deposition, as shown in Figure 8.30. Acid neutralizing capacity, a major indicator of recovery in acidified lakes and streams, is beginning to rise in streams in the Northeast, including the Adirondacks. This is an indicator that recovery from acidification is beginning in those areas.

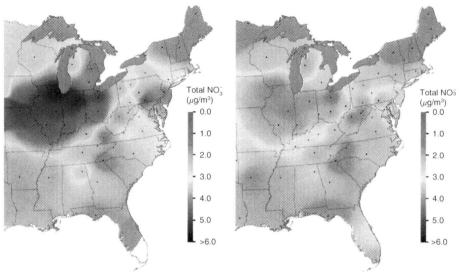

Figure 8.30 Wet nitrate deposition for 1990 and 2005.
Source: From EPA (2009).

8.4.3 Hazardous Air Pollutants

Currently, emissions of hazardous air pollutants (HAPs) are not regulated by the EPA, although mercury regulations are expected to be implemented during the 2011 to 2012 time frame. Emissions of HAPs from coal-fired power plants have been estimated by the EPA. The agency performed a study (i.e., Study of Hazardous Air Pollutant Emissions from Electric Utility Steam Generating Units) to determine the quantity of hazardous air pollutants being emitted from fossil fuel-fired power plants [11]. In this study, HAP emissions test data were gathered from 52 utility units (i.e., boilers), including a range of coal-, oil-, and natural gas-fired utility boilers. The emissions tests data, along with facility-specific information (e.g., boiler type, control devices, fuel usage) were used to estimate HAP emissions from all 684 utility plants in the United States. These utilities are fueled primarily by coal (59 percent), oil (12 percent), or natural gas (29 percent). Many plants have two or more units, and several plants burn more than one type of fuel (e.g., contain both coal- and oil-fired units). In 1990, there were 426 plants that burned coal as one of their fuels, 137 plants that burned oil, and 267 plants that burned natural gas. The overall summary of the study is presented in Table 8.16, which lists nationwide utility emissions estimates for 13 priority HAPs (EPA, February 1998). Table 8.17 contains estimated emissions for 9 priority HAPs from characteristic utility units.

In summary, the Utility Hazardous Air Pollutant Report to Congress analyzed 66 other air pollutants (other than mercury, which is discussed in the next section) from 684 power plants that are 25 MW or larger and burning coal, oil, or natural gas [11]. The report noted potential health concerns about utility emissions of dioxin, arsenic,

Table 8.16 Nationwide Utility (Coal-Fired) Emissions for 13 Priority HAPs[a]

HAP	Nationwide HAP Emission Estimates (tons per year)[b]		
	1990	1994	2010
Arsenic	61	56	71
Beryllium	7.1	7.9	8.2
Cadmium	3.3	3.2	3.8
Chromium	73	62	87
Lead	75	62	87
Manganese	164	168	219
Mercury	46	51	60
Nickel	58	52	69
Hydrogen chloride	143,000	134,000	155,000
Hydrogen fluoride	20,000	23,000	26,000
Acrolein	25	27	34
Dioxins[c]	0.000097	0.00012	0.00020
Formaldehyde	35	29	45

[a]Radionuclides are the one priority HAP not included on this table because radionuclide emissions are measured in different units (i.e., curies per year) and, therefore, would not provide a relevant comparison to the other HAPs shown.
[b]The emissions estimates in this table are derived from model projections based on a limited sample of specific boiler types and control scenarios. Therefore, there are uncertainties in these numbers.
[c]These emissions estimates were calculated using the toxic equivalency (TEQ) approach, which is based on the summation of the emissions of each congener after adjusting for toxicity relative to 2,3,7,8-tetrachlorodibenzo-p-dioxin (i.e., 2,3,7,8-TCDD).

Table 8.17 Estimated Emissions for 9 Priority HAPs from Characteristic Utility Units (1994; tons per year)[a]

Fuel Unit Size (MWe)	Coal 325
Arsenic	0.0050
Cadmium	0.0023
Chromium	0.11
Lead	0.021
Mercury	0.05
Hydrogen chloride	190
Hydrogen fluoride	14
Dioxins[b]	0.00000013
Nickel[c]	NC

[a]There are uncertainties in these numbers. Based on an uncertainty analysis, the EPA predicts that the emissions estimates are generally within a factor of roughly three of actual emissions.
[b]These emissions estimates were calculated using the toxic equivalency (TEQ) approach, which is based on the summation of the emissions of each congener after adjusting for toxicity relative to 2,3,7,8-tetrachlorodibenzo-p-dioxin (i.e., 2,3,7, 8-TCDD).
[c]Not calculated.

hydrogen chloride, hydrogen fluoride, and nickel, although uncertainties exit about the health data and emissions for these pollutants.

Mercury

The best estimate of annual anthropogenic U.S. emissions of mercury in 1999 is 115 short tons [78]. Approximately two-thirds of these emissions are from combustion sources, including waste and fossil fuel combustion. This is illustrated in Figure 8.31 [78]. Contemporary anthropogenic emissions are only one part of the mercury cycle. Releases from human activities today are adding to the mercury reservoirs that already exist in land, water, and air, both naturally and as a result of previous human activities. One estimate of the total annual global input to the atmosphere from all sources including natural, anthropogenic, and oceanic emissions is 1,230 to 2,890 metric tons [79, 80]. U.S. sources are estimated to have contributed about 3 percent of the 1999 total. Mercury emissions from U.S. coal-fired boilers are estimated to be 48 short tons per year.

In a report released by the United Nations Environment Program (UNEP), *Global Atmospheric Mercury Assessment: Sources, Emissions, and Transport,* coal-fired power plants were identified as the largest single anthropogenic source of mercury air emissions in most countries, although in Brazil, Indonesia, Colombia, and some other countries (in South American, Asia, and Africa in particular), artisan/small-scale gold mining is the largest single source. Mercury emissions, by source, are given in Table 8.18 for 2005 [79, 80]. Geographically, about two-thirds of global anthropogenic releases of mercury to the atmosphere appear to come from Asian sources, with China as the largest contributor worldwide. The United States and India are the second

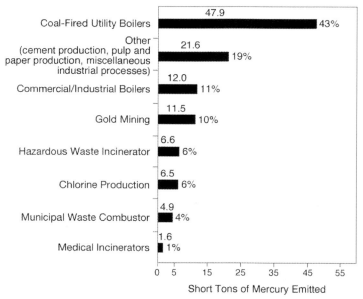

Figure 8.31 Annual mercury emissions in the United States.
Source: From EPA (2009).

Table 8.18 Global Anthropogenic Mercury Emissions (Metric Tons) by Sector in 2005

Sector	2005 Emissions (tons)	% of 2005 Emissions	Low-End Estimate	High-End Estimate
Fossil fuel combustion for power and heating	878	45.6	595	1160
Metal production (ferrous and nonferrous, excluding gold)	200	10.4	125	275
Large-scale gold production	111	5.8	65	155
Artisanal and small-scale gold production	350	18.2	225	475
Cement production	189	9.8	115	265
Chor-alkali industry	47	2.4	25	65
Waste incineration, waste, and other	125	6.5	50	475
Dental amalgam (cremation)	26	1.3	20	30
Total	**1,930**	**100**	**1,220**	**2,900**

Source: From UNEP (2008).

Table 8.19 Global Anthropogenic Mercury Emissions (Metric Tons) by Region in 2005

Continent	2005 Emissions (tons)	% of 2005 Emissions	Low-End Estimate	High-End Estimate
Africa	95	5.0	55	140
Asia	1281	66.5	835	1,760
Europe	150	7.8	90	310
North America	153	7.9	90	305
Oceania	39	2.0	25	50
Russia	74	3.9	45	130
South America	133	6.9	80	195
Total	**1,930**	**100.0**	**1,220**	**2,900**

Source: From UNEP (2008).

and third largest emitters, but their combined total emissions are only about one-third of China's emissions. Table 8.19 lists 2005 mercury emissions by region.

8.4.4 Carbon Dioxide

Carbon dioxide (CO_2) emissions from energy use are shown in Figure 8.32, which shows CO_2 emissions by sector and fuel for 1990 and 2001 and projections up to 2025 [81]. Petroleum products are the leading source of CO_2 emissions from energy use. In 2025, petroleum is projected to account for 971 million metric tons carbon equivalent, a 43 percent share of the projected total. Coal is the second leading source of CO_2 emissions, projected to produce 753 million metric tons carbon

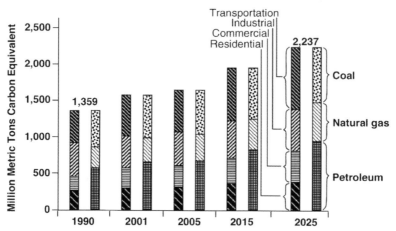

Figure 8.32 Current and projected carbon dioxide emissions by sector and fuel.
Source: From EIA (2003).

equivalent in 2025, or 34 percent of the total. In 2025, natural gas use is projected to produce a 23 percent share of the total CO_2 emissions, with 512 million metric tons carbon equivalent.

The use of fossil fuels in the electric power industry accounted for 39 percent of total energy-related CO_2 emissions in 2001, and the share is projected to be 38 percent in 2025, as shown in Figure 8.33 [81]. Coal is projected to account for 50 percent of the power industry's electricity generation in 2025 and to produce 81 percent of electricity-related CO_2 emissions. In 2025, natural gas is projected to account for 27 percent of electricity generation and 18 percent of electricity-related CO_2 emissions.

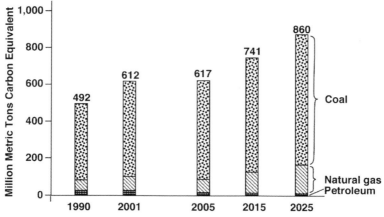

Figure 8.33 Current and projected carbon dioxide emissions from the electric power sector by fuel.
Source: From EIA (2003).

References

[1] B. Lomborg, The Skeptical Environmentalist, Measuring the Real State of the World, Cambridge University Press, 2001.

[2] U.C. Davis, University of California at Davis, *www-geology.ucdavis.edu/-GEL115/115CH11coal.html*, printed May 2003.

[3] C&EN (Chemical & Engineering News), Recognizing Pollution's Hazard's, C&E News 81 (15) (2003) 54–55.

[4] K.C.F. Wark, W.T. Warner, Davis, Air Pollution Its Origin and Control, third ed., Addison Wesley Longman, 1998.

[5] S. Dunn, King Coal's Weakening Grip on Power, World Watch (September/October) (1999) 10–19.

[6] EPA (U.S. Environmental Protection Agency), Clean Air Mercury Rule, *http://epa.gov/air/mercuryrule/*, March 27, 2009.

[7] EPA, Clean Air Interstate Rule, *http://epa.gove/air/interstateairquality/*, May 13, 2009.

[8] EPA, National Ambient Air Quality Standards (NAAQS), *www.epa.gov/air/criteria.html*, July 14, 2009.

[9] EPA, Subpart Da—Standards of Performance for Fossil-Fuel-Fired Steam Generators for Which Construction is Commenced After August 17, 1971, Fed Regist (1971).

[10] EPA, Standards of Performance for New Stationary Sources; Electric Utility Steam Generating Units, Decision in Response to Petitions for Reconsideration, Fed Regist 45 (26) (1980) 8210–8213.

[11] EPA, Revision of Standards of Performance for Nitrogen Oxide Emissions from New Fossil-Fuel Fired Steam Generating Units, Revisions to Reporting Requirements for Standards of Performance for New Fossil-Fuel Fired Steam Generating Units, Fed Regist 63 (179) (1998) 49442–49454.

[12] EPA, Fact Sheet—Revision of Standards of Performance for Nitrogen Oxides Emissions from Fossil-Fuel Fired Steam Generating Units, *www.epa.gov/ttn/oarpg*, August 7, 2001.

[13] EPA, Subpart Db—Standards of Performance for Industrial-Commercial-Institutional Steam Generating Units, Fed Register, November 25 (1986).

[14] EPA, Subpart Dc—Standards of Performance for Small Industrial-Commercial-Institutional Steam Generating Units, Fed Register, September 12 (1990).

[16] AP-42, Emission Factors, Chapter 1: External Combustion Sources, fifth ed., EPA Office of Air Quality Planning and Standards and Office of Air and Radiation, 1993 with latest revisions in 1998.

[17] EPA, Clean Air Act, *http://epa.gov/air/caa*, May 12, 2009.

[18] EPA, Overview: The Clean Air Act Amendments of 1990, *www.epa.gov/oar/caa/overview.txt*, November 15, 2002b.

[19] EPA, CAA: Original List of Hazardous Air Pollutants, *www.epa.gov/ttn/atw/orig189.html*, February 11, 2002.

[20] J. Makanski, Clean Air Act Amendments: The Engineering Response, Power 135 (6) (1991) 11–66.

[21] EPA, EPA Acid Rain Program 2001 Progress Report, Office of Air and Radiation, U.S. Government Printing Office, November 2002.

[22] M. Leone, Cleaning the Air the Market-Based Way, Power 129 (10) December (1990) 9–10.

[23] EPA, Acid Rain Program, Program Overview, *www.epa.gov/acidrain/overview.html*, April 1999.

[24] D.N. Smith, H.G. McIlvried, A.N. Mann, Understanding NO_x and How it Impacts Coal, Coal Age 105 (11) (2000) 35.

[25] EPA, Ozone Transport Commission NOx Budget Program, 1999–2002 Progress Report, Office of Air and Radiation, U.S. Government Printing Office, March 2003.

[26] EPA, NO_x Trading Programs, www.epa.gov/airmarkets/progregs/noxview.html, October 29, 2002.

[27] EPA, The Regional Transport of Ozone, Office of Air Quality Planning and Standards, U.S. Government Printing Office, September 1998.

[28] EPA, NO_x Budget Trading Program/NO_x SIP Call, 2003–2008, www.epa.gov/airmarkets/progsregs/nox/sip.html, May 28, 2009.

[29] EPA, Clean Air Interstate Rule, http://epa.gov/air/interstateairquality/, May 13, 2009.

[30] EPA, Clean Air Interstate Rule (CAIR)—Reducing Power Plant Emissions for Cleaner Air, Healthier People, and a Strong America, Office of Air and Radiation, U.S. Government Printing Office, March 2005.

[31] C&EN, Power Plant Pollution, vol. 87, Number 30, July 27, 2009, pp. 42–45.

[32] EPA, Mercury Study Report to Congress, Office of Air Quality Planning & Standards and Office of Research and Development, U.S. Government Printing Office, December 1997.

[33] EPA, Study of Hazardous Air Pollutant Emissions from Electric Utility Steam Generating Units—Final Report to Congress, Office of Air Quality Planning & Standards, U.S. Government Printing Office, February 1998.

[34] EPA, EPA ICR No. 1858: Information Collection Request for Electric Utility Steam Generating Unit Mercury Emissions Information Collection Effort, 1999.

[35] EPA, Regulatory Findings on the Emissions of Hazardous Air Pollutants From Electric Utility Steam Generating Units, Fed Regist 65 (245) (2000) 79825–79831.

[36] EPA, Clean Air Mercury Rule, http://epa.gov/air/mercuryrule.html, March 27, 2009.

[37] EPA, New Source Review: Report to the President, June 2002.

[38] EPA, Prevention of Significant Deterioration (PSD) and Nonattainment New Source Review (NSR): Final Rule and Proposed Rule, Fed Regist 67 (251) (2002) 80186–80289.

[39] EPA, EPA Announces Improvements to New Source Review Program, www.epa.gov/air/nsr-review/press_release.html, March 13, 2003.

[40] EPA, New Source Review, www.epa.gov/air/nsr-review/, July 25, 2003.

[41] CEP (Chemical Engineering Progress), EPA Finalizes New Source Review Rule, Chemical Engineering Progress 99 (12) (2003) 24.

[42] DOE (U.S. Department of Energy), Atmospheric Aerosol Source-Receptor Relationships: The Role of Coal-Fired Power Plants—Project Facts, Office of Fossil Energy, National Energy Technology Laboratory, January 2003.

[43] EPA, Regulatory Actions, http://epa.gov/air/particlepollution/actions.html, April 1, 2009.

[44] EPA, PM Standards, http://epa.gov/air/particlepollution/standards.html, October 14, 2008.

[45] EPA, Fact Sheet—Final Regional Haze Regulations for Protection of Visibility in National Parks and Wilderness Areas, www.epa.gov/oar/visibility/program.html, June 2, 1999.

[46] W.W. Aljoe, T.J. Grahame, The DOE-NETL Air Quality Research Program: Airborne Fine Particulate (PM2.5), in: Proceedings of the Conference on Air Quality III: Mercury, Trace Elements, and Particulate Matter, University of North Dakota, 2002.

[47] EPA, Clear Skies, www.epa.gov/clearskies/, July 17, 2003.

[48] EPA, Clear Skies Basic Information, www.epa.gov/clearskies/basic.html, July 10, 2003.

[49] C&EN, Bush Initiative Faces Skepticism in Congress, vol. 81, Number 29, July 21, 2003, p. 5.

[50] M. Tatsutani, Multi-Pollutant Proposals in the 108th Congress, Presented at the OTC Annual Meeting, Philadelphia, June 22, 2003.

[51] CEP, Congress Considers Climate Change, vol. 99, Number 6, June 2003, p. 23.

[52] C&EN, Global Initiative on CO$_2$ Storage, vol. 81, Number 26, June 30, 2003, p. 19.

[53] C&EN, Climate-Change Plan Released, vol. 81, Number 30, July 28, 2003, p. 39.

[54] CDT (Centre Daily Times), White House Seeks More Data on Global Climate Change, Thursday, July 24, 2003, p. A10.

[55] EPA, Climate Change Regulatory Initiatives, *www.epa.gov/climatechange/initiatives/index.html*, September 22, 2009.

[56] EPA, EPA Analyses of Climate Change Bills, *www.epa.gov/climatechange*.

[57] Power Engineering, Climate Bill Likely Won't Pass in 2009, *http://pepei.pennet.com*, September 21, 2009.

[58] Economic Commission for Europe, Strategies and Policies for Air Pollution Abatement, 2006 Review Prepared under the Convention on Long-Range Transboundary Air Pollution, United Nations, 2007.

[59] H.N. Soud, Developments in FGC, IEA Coal Research, March 2000.

[60] Z. Wu, NO$_x$ Control for Pulverized Coal-Fired Power Stations, IEA Coal Research, December 2002.

[61] United Nations, Status of the Convention on Long-Range Transboundary Air Pollution and its Related Protocols, as of September 2, 2009.

[62] A. McConville, An Overview of Air Pollution Emission Standards for Coal-Fired Plant Worldwide, Coal & Slurry Technology Association, 1997, pp. 1–12.

[63] H.N. Soud, Developments in Particulate Control for Coal Combustion, IEA Coal Research, April 1995.

[64] L.E. Clarke, L.L. Sloss, Trace elements—emissions from coal combustion and gasification, IEA Coal Research, 1992.

[65] L.L. Slossk, Economics of mercury control, IEA Coal Research, 2008.

[66] R. Lutter, E. Irwin, Mercury in the Environment, a Volatile Problem, Environment 44 (9) November (2002) 24–40.

[67] United Nations, United Nations Framework Convention on Climate Change, 1992.

[68] EPA, States Guidance Document Policy Planning to Reduce Greenhouse Gas Emissions, second ed., Office of Policy, Planning and Evaluation, May 1998.

[69] T. Jackson (Ed.), Mitigating Climate Change: Flexibility Mechanisms, Elsevier Science Ltd., 2001, p. 17.

[70] United Nations, Kyoto Protocol to the United Nations Framework Convention on Climate Change, 1997.

[71] EIA (Energy Information Administration), International Energy Outlook 2002, U.S. Department of Energy, March 2002.

[72] United Nations, Kyoto Protocol, *http://unfccc.int/kytoto_protocol/items*.

[73] EPA, Latest Findings on National Air Quality 2002 Status and Trends, Office of Air Quality Planning and Standards, U.S. Government Printing Office, August 2003.

[74] EPA, Air Trends, *http://www.epa.gov/airtrends*, June 4, 2009.

[75] EPA, National Air Quality Status and Trends through 2007, Office of Air Quality Planning and Standards, U.S. Government Printing Office, November 2008.

[76] EPA, Acid Rain Program, *http://www.epa.gov/airmarkets*, July 10, 2009.

[77] EPA, Acid Rain Challenge, *http://www.epa.gov/airmarkets/progsregs/usca/coop.html*, April 14, 2009.

[78] EPA, Controlling Power Plant Emissions: Emissions Progress, *http://www.epa.gov/mercury/control_emissions/emissions.htm*, August 3, 2009.

[79] EPA, Mercury Emissions: The Global Context, *http://www.epa.gov/mercury/control_emissions/global.htm*, August 3, 2009.

[80] United Nations Environmental Programme (UNEP), Global Atmospheric Mercury Assessment: Sources, Emissions and Transport, December 2008.

[81] EIA, Annual Energy Outlook 2003, U.S. Department of Energy, January 2003.

9 Emissions Control Strategies for Power Plants

For more than the last quarter century, power plant operators in the United States have been installing new pollution control technologies to meet ever-tightening regulatory standards for clean air. The Clean Air Act of 1970 (details of which can be found in Chapter 8, where the history of legislative action in the United States is discussed) established national standards to limit levels of air pollutants such as sulfur dioxide, nitrogen oxides, carbon monoxide, ozone, lead, and particulate matter. The Act, and its amendments in 1977, resulted in the development and installation of particulate matter and sulfur dioxide control technologies for coal-fired boilers. Particulate control devices, specifically electrostatic precipitators (ESPs) and fabric filter baghouses, have been installed in power plants. Efforts to develop new control technology, including flue gas desulfurization units, commonly called scrubbers, to remove sulfur from the flue gas, were initiated with units installed on many power generation facilities. In addition, technologies to reduce nitrogen oxides began to be developed.

The 1990 Clean Air Act Amendments contained major revisions to the Clean Air Act and required further reductions in power plant emissions, especially sulfur- and nitrogen-containing pollutants that contribute to acid rain. As a result, switching to low-sulfur fuels, installation of flue gas desulfurization units, and the development of a new market-based cap-and-trade system that requires power plants to either reduce their emissions or acquire allowances from other companies to achieve compliance are examples of sulfur dioxide control strategies that have been implemented. To meet the more stringent nitrogen oxide standards from the 1990 Clean Air Act Amendments, several technologies, specifically low-NO_x burners, selective catalytic and noncatalytic reduction, cofiring, and reburning, have been developed with varying levels of implementation. With recently enacted and impending legislation (e.g., Clean Air Interstate Rule or its successor), additional NO_x control is anticipated and is being planned for by the power-generating industry.

This chapter begins by summarizing the progress made over the last 45 years in reducing emissions from coal-fired power plants. Commercial control strategies for pollutants that are currently regulated, such as sulfur dioxide, nitrogen oxides, and particulate matter, are discussed. Control technologies that are under development for reducing mercury emissions, where regulations are anticipated to be promulgated during 2011 and 2012, are also discussed. The chapter concludes with a discussion of multipollutant control technologies. Carbon dioxide control options are discussed in Chapter 10, along with carbon sequestration technologies under development.

Clean Coal Engineering Technology. DOI: 10.1016/B978-1-85617-710-8.00009-1
Copyright © 2011 by Elsevier Inc. All rights of reproduction in any form reserved.

9.1 Currently Regulated Emissions

The pollutants that are of primary interest, and those currently regulated in the power-generation industry, include sulfur dioxide, nitrogen oxides, and particulate matter. Although pollutants such as carbon monoxide and volatile organic compounds (which lead to the production of ozone when reacted with nitrogen oxides) are important and are tracked nationally, there are no specific control technologies for these pollutants, since they are formed from incomplete combustion in the boiler. Power plants are minor contributors to nationwide carbon monoxide emissions. Similarly, power plants are insignificant contributors to nationwide hydrocarbon emissions. Carbon monoxide and hydrocarbon emissions are discussed in Chapter 8.

Emissions of sulfur dioxide, nitrogen oxides, and particulate matter in the United States have been dramatically reduced over the last 45 years due to legislative mandate and technological advances achieved through federal, state, and industrial efforts. Figures 9.1 and 9.2 show the decreases in overall, as well as individual, emissions while the use of coal for power generation is increasing. The steady decrease in emission rates is clearly shown in Figure 9.1, where the emission rates, given on a per billion kWh basis, are reported in five-year increments [1–3]. The decrease in the emission rates of these three pollutants is also shown in Figure 9.2, along with projected near-term emission rates [1–3]. The emission rates in Figure 9.2 are compared with the coal use for power generation for the same time periods, which illustrate how pollutant emissions per unit of coal burned have decreased significantly while at the same time coal use has increased. The technologies that are being used to achieve these reductions are discussed in the sections that follow.

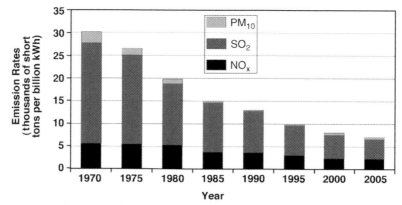

Figure 9.1 Emission rates of sulfur dioxide, nitrogen oxides, and particulate matter from coal-fired power plants for the period 1970 to the present.
Source: Data from DOE (2003) and EPA (2009).

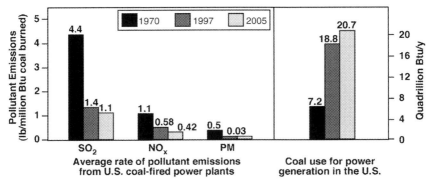

Figure 9.2 Past, present, and future emission rates of sulfur dioxide, nitrogen oxides, and particulate matter from coal-fired power plants.
Source: Data from DOE (2003) and EPA (2009).

9.1.1 Sulfur Dioxide

Sulfur dioxide (SO_2) is one of the most abundant air pollutants emitted in the United States, totaling about 11 million short tons in 2008, with fossil fuel combustion accounting for approximately 69 percent of the total anthropogenic emissions (i.e., 7.6 million short tons). Sulfur dioxide emissions have decreased 56 percent and 47 percent for the periods from 1980 to 2008 and from 1990 to 2008, respectively [4]. Reductions in SO_2 emissions and concentrations since 1990 are primarily due to controls implemented under the U.S. Environmental Protection Agency's (EPA) Acid Rain Program beginning in 1995. As of 2000, there were 248 coal-fired electric generators equipped with scrubbers, for a total generating capacity of nearly 102,000 MW [2].

The chemistry of sulfur dioxide formation is reviewed in this section, followed by technologies used to control SO_2 emissions. Control technologies will focus on commercially available and commercially used systems. Industry deployment of the SO_2 removal process worldwide is discussed as are the economics of flue gas desulfurization.

Chemistry of Sulfur Oxide Formation

Sulfur in coal occurs in three forms: as pyrite, organically bound to the coal, or as sulfates. The sulfates represent a very small fraction of the total sulfur, while pyritic and organically bound sulfur comprise the majority. The distribution between pyritic and organic sulfur is variable with up to approximately 40 percent of the sulfur being pyritic. During combustion, the pyritic and organically bound sulfur are oxidized to sulfur dioxide SO_2 with a small amount of sulfur trioxide (SO_3) being formed. The SO_2/SO_3 ratio is typically 40:1 to 80:1 [5].

The overall reaction for the formation of sulfur dioxide is

$$S + O_2 \rightarrow SO_2 \qquad \Delta H_f = -128,560 \text{ Btu/lb mole} \qquad (9.1)$$

and the overall reaction for the formation of sulfur trioxide is

$$SO_2 + \tfrac{1}{2}O_2 \leftrightarrow SO_3 \qquad \Delta H_f = -170,440 \text{ Btu/lb mole} \qquad (9.2)$$

It is proposed that sulfur monoxide, SO, is formed early in the reaction zone from sulfur containing molecules and is an important intermediate product [4]. The major SO_2 formation reactions are believed to be

$$SO + O_2 \rightarrow SO_2 + O \qquad (9.3)$$

and

$$SO + OH \rightarrow SO_2 + H \qquad (9.4)$$

with the highly reactive O and H atoms possibly entering the reaction scheme later.

The reactions involving SO_3 are reversible. The major formation reaction for SO_3 is the three-body process

$$SO_2 + O + M \rightarrow SO_3 + M \qquad (9.5)$$

where M is a third body that is an energy absorber [4]. The major steps for removal of SO_3 are thought to be the following:

$$SO_3 + O \rightarrow SO_2 + O_2 \qquad (9.6)$$

$$SO_3 + H \rightarrow SO_2 + OH \qquad (9.7)$$

$$SO_3 + M \rightarrow SO_2 + O + M \qquad (9.8)$$

Sulfur Dioxide Control

Methods to control sulfur dioxide emissions from coal-fired power plants include switching to a lower-sulfur fuel, cleaning the coal to remove the sulfur-bearing components such as pyrite, or installing flue gas desulfurization systems. In the past, building tall stacks to disperse the pollutants was a control method; however, this practice is no longer an alternative because tall stacks do not remove the pollutants but only dilute the concentrations to reduce the ground-level emissions to acceptable levels.

When fuel switching or coal cleaning is not an option, flue gas desulfurization (FGD) is selected to control sulfur dioxide emissions from coal-fired power plants (except for fluidized-bed combustion systems, which is discussed later in this chapter). FGD has been in commercial practice since the early 1970s. It has become the most widely used technique to control sulfur dioxide emissions next to the firing of low-sulfur coal. Many existing FGD systems are currently in use, and many others are under development. This section summarizes the worldwide application of FGD systems, with an emphasis on the United States. FGD processes are generally classified as wet scrubbers or dry scrubbers but can also be categorized as follows [6]:

- Wet scrubbers
- Spray dryers
- Dry (sorbent) injection processes
- Regenerable processes
- Circulating fluid-bed and moving-bed scrubbers
- Combined SO_2/NO_x removal systems

Based on the nature of the waste/by-product generated, a commercially available throwaway FGD technology may be categorized as wet or dry. A wet FGD process produces a slurry waste or a saleable slurry by-product. A dry FGD process application results in a solid waste, the transport and disposal of which is easier compared to the waste/by-product from wet FGD applications. Regenerable FGD processes produce a concentrated SO_2 by-product, usually sulfuric acid or elemental sulfur. Recently, there has been a focus on mercury removal in FGD systems, which is examined later in the chapter.

Worldwide Deployment of FGD Systems

Postcombustion control of sulfur dioxide emissions from pulverized coal combustion began in the early 1970s in the United States and Japan. Western Europe followed in the 1980s. In the 1990s, the application of FGD became more widespread, and countries in Central and Eastern Europe, Asia, and others installed FGD systems. Table 9.1 lists various control technologies and the amount of electricity generation that is being controlled in countries throughout the world [4]. According to Soud [4], as of 1999 there were 680 FGD systems installed in 27 countries, and 140 systems are currently under construction or planned in nine countries.

Worldwide, approximately 30,000 MW of generating capacity were controlled in 1980 compared to no controlled capacity in 1970. Controlled generating capacity has subsequently increased to approximately 130,000 MW in 1990 and approximately 230,000 MW in 2000. In the United States, controlled capacity has risen from zero in 1970 to 25,000 MW in 1980, to approximately 75,000 MW in 1990, and to approximately 100,000 MW in 2000.

Worldwide, FGD systems were installed (as of 1999) to control sulfur dioxide emissions from more than 229,000 MW of generating capacity. Of this, approximately 87 percent consists of wet FGD technology, 11 percent consists of dry FGD technology, and the balance consists of regenerable technology [7]. Of the worldwide capacity controlled with FGD technology, approximately 44 percent is in the United States alone, as shown in Table 9.2. In the United States, approximately 100,000 MW of capacity were equipped with FGD technology. Of these FGD systems, approximately 83, 14, and 3 percent consist of wet FGD, dry FGD, and regenerable technology, respectively. Worldwide, out of 668 units equipped with FGD, 522 were equipped with wet FGD, 124 with dry FGD, and 22 with regenerable FGD.

Table 9.1 Existing and Future FGD Systems

Country	Wet Lime/Limestone/Gypsum, MW	Wet Lime/Limestone/Other, MW	Spray Dry Scrubbers, MW	Sorbent Injection, MW	CFB and Moving-Bed Scrubbers, MW	Regenerable Systems, MW	Combined SO_2/NO_x Removal, MW
Existing							
Austria	825	–	835	–	260	–	–
Canada	1,495	–	–	600	–	–	–
China	2,485	–	510	300	–	–	325
Czech Republic	2,230	–	–	–	–	–	–
Denmark	2,260	–	1,200	250	–	–	305
Finland	1,875	–	530	600	–	–	–
France	1,800	–	–	600	–	–	–
Germany	43,670	445	2,245	25	295	885	400
Greece	300	–	–	–	–	–	–
Italy	5,420	–	–	75	–	–	–
India	–	500	–	–	185	–	–
Japan	21,725	365	–	350	–	–	–
Korea, Republic	7,500	–	–	–	–	–	–
Netherlands	3,915	–	–	–	–	–	–
Norway	–	30	–	–	–	–	–
Poland	7,390	50	315	1,720	–	–	–
Russian Federation	510	–	–	–	–	–	–
Slovakia	330	–	–	–	–	–	–
Slovenia	275	–	–	–	–	–	–
Spain	1,930	–	385	–	–	–	–
Sweden	–	–	360	130	–	–	–

Taiwan	7,100	–	–	–	–	–	–
Thailand	2,100	300	–	–	–	–	–
Turkey	3,355	–	–	–	–	–	–
Ukraine	150	–	–	–	–	–	–
United Kingdom	5,960	–	–	–	–	–	–
United States	21,680	62,475	12,000	970	80	2,870	1,845
Future							
China	720	–	–	200	–	–	–
Denmark	480	–	–	–	–	–	–
Germany	5,035	–	–	–	–	–	–
Israel	1,100	–	–	–	–	–	–
Japan	8,300	–	–	–	–	–	–
Korea Republic	2,000	–	–	–	–	–	–
Sri Lanka	–	300	–	–	–	–	–
Turkey	470	–	–	–	–	–	–
United States	801	1,780	–	520	–	–	–

Source: From EPA (2008).

Table 9.2 Worldwide Electrical Generating Capacity (in MW)
Equipped with FGD Technology

Technology	United States	Abroad	Total
Wet	82,859	116,374	199,233
Dry	14,386	11,008	25,394
Regenerable	2,798	2,059	4,857
Total FGD	**100,043**	**129,441**	**229,484**

Source: From Srivastava, Singer, and Jozewicz (2000).

Of the U.S. wet FGD technology population, 69 percent are limestone processes [7]. Abroad, limestone processes comprise as much as 93 percent of the total wet FGD technology installed. Of the worldwide capacity equipped with dry FGD technology, 74 percent use spray drying processes. This compares with 80 percent for spray drying processes in the United States.

A summary of the FGD systems in the United States, by process, is given in Table 9.3 for 1989 (actual capacity) and 2010 (projected capacity). The three primary processes are throwaway-product systems, including the two wet scrubbing systems using limestone and lime, where a synthetic gypsum ($CaSO_4$) is produced. A lack of commercial markets for the gypsum results in this material being disposed rather than utilized. Characteristics of these processes are provided in the next section.

A variety of FGD processes exist, and the selection of a system is dependent on site-specific consideration, economics, and other criteria. Elliot provides a ranking of various FGD processes used in the United States in Tables 9.4 and 9.5, where the cost, performance, and flexibility of application are assessed [9].

Wet limestone systems have been installed across the United States at plants of all sizes, firing all ranks of coal with sulfur contents varying from low to high. Wet lime systems have been installed at power plants of all sizes, firing both low- and high-sulfur coals at plants that are predominately in the Ohio River valley [9]. Some plants in the West use wet-lime systems, where the cost of lime delivered to the plant is less than limestone. Some plants firing high-sulfur coal in the Midwest have selected wet sodium-based dual-alkali systems. Dry scrubbing systems have typically been selected at power plants firing low-sulfur coals. Generally, dry scrubbing systems are considered more economical for power plants firing low-sulfur coal, while wet-based systems are selected for high-sulfur coal applications.

Techniques to Reduce Sulfur Dioxide Emissions

The primary methods used to control sulfur dioxide emissions from coal-fired power plants are to switch to a lower-sulfur fuel or install flue gas desulfurization systems and, to a lesser extent, clean coal to remove the sulfur-bearing components. These techniques are discussed in this section, with an emphasis on flue gas desulfurization technologies.

Table 9.3 Summary of FGD Processes in the United States (Percent of Total MW)

Process	Active Material (for scrubbing)	By-Product	% of MW 1989	% of MW 2010
Throwaway Product Wet Scrubbing				
Dual alkali	Na_2SO_3 solution regenerated by CaO or $CaCO_3$	$CaSO_3/CaSO_4$	3.4	2.3
Lime	$Ca(OH)_2$ slurry	$CaSO_3/CaSO_4$	16.3	13.5
Lime/alkaline fly ash	$Ca(OH)_2$/fly ash slurry	$CaSO_3/CaSO_4$	7.0	4.9
Limestone	$CaCO_3$ slurry	$CaSO_3/CaSO_4$	48.2	43.9
Limestone/alkaline fly ash	$CaCO_3$/fly ash slurry	$CaSO_3/CaSO_4$	2.4	1.6
Sodium carbonate	Na_2CO_3/Na_2SO_4 slurry	Na_2SO_4	4.0	3.3
Spray Drying				
Lime	Slaked $Ca(OH)_2$ slurry	$CaSO_3/CaSO_4$	8.8	7.9
Sodium carbonate	Na_2CO_3	Na_2SO_4	4.0	3.3
Reagent type not selected	Undecided	–	0.7	2.1
Dry Injection				
Lime	$Ca(OH)_2$ (dry)	$CaSO_3/CaSO_4$	0.2	0.1
Sodium carbonate	Na_2CO_3	Na_2SO_4	0	0.2
Reagent type not selected	Undecided	–	0	2.2
Process not selected	–	–	0	2.2
Saleable Product Wet Scrubbing				
Lime	$Ca(OH)_2$ slurry	$CaSO_4$	<0.1	<0.1
Limestone	$CaCO_3$ slurry	$CaSO_4$	4.1	4.6
Magnesium oxide	$Mg(OH)$ slurry	Sulfuric acid	1.4	1.0
Wellman Lord	Na_2SO_3 solution	Sulfuric acid	3.1	2.1
Spray Drying				
Lime	Slaked $Ca(OH)_2$ slurry	Dry scrubber waste	0	0.3
Process Undecided	–	–	0	7.8

Source: From Wark, Warner, and Davis (1998) [5] and Davis (2000) [8].

Using Low-Sulfur Fuels

One option for reducing sulfur dioxide emissions is to switch to fuels that contain less sulfur. Fuel switching includes using natural gas, liquefied natural gas, low-sulfur fuel oils, or low-sulfur coals in place of high-sulfur coals. In coal-fired boilers, switching from a high-sulfur coal to lower-sulfur noncoal fuels may make sense from both an economic and technological standpoint for smaller-sized industrial and utility boilers; however, the practice of switching power generation units from coal to natural gas is a questionable one. While this option may make good business sense (at least at the time), it is not good energy policy/energy security to take a

Table 9.4 Assessment Relative to Cost
and Performance

FGD Process	Operating Cost	Capital Cost	SO$_2$ Removal	Reliability	Commercial Use
Limestone					
Natural oxidation	M	M	M	M	H
Forced oxidation	M	M	M	M	H
MgO-lime	M	M	H	H	H
High-calcium lime	M	M	M	M	M
Dual-alkali					
Lime	M	M	H	H	M
Limestone	L	M	H	–	–
Dry scrubbing	H	M	L	H	H
Dry injection	M	L	L	–	–
Wellman-Lord	H	H	H	M	M
Regenerable MgO	M	H	H	M	L

Criterion header spans Operating Cost, Capital Cost, SO$_2$ Removal, Reliability, Commercial Use

[a]H = high; M = medium; L = low
Source: From Elliot (1989).

Table 9.5 Assessment with Respect to Flexibility
of Application

FGD Process	High Sulfur	Low Sulfur	Retrofit Ease	Waste Management	SO$_2$/NO$_x$ Removal
Limestone					
Natural oxidation	H	H	M	L	L
Forced oxidation	H	M	M	L	L
MgO-lime	H	L	M	L	M
High-calcium lime	H	M	M	L	L
Dual-alkali					
Lime	H	L	M	L	M
Limestone	H	L	M	L	L
Dry scrubbing	M	H	M	M	M
Dry injection	L	H	H	L	L
Wellman-Lord	H	L	L	M	M
Regenerable MgO	H	M	M	H	M

Criterion header spans High Sulfur, Low Sulfur, Retrofit Ease, Waste Management, SO$_2$/NO$_x$ Removal

[a]H = high; M = medium; L = low
Source: From Elliot (1989).

premium fuel and use it for power generation. This is discussed in more detail in Chapter 12.

Fuel switching to lower-sulfur coals is chosen by many power generators to achieve emissions compliance. In the United States, the replacement of high-sulfur eastern or midwestern bituminous coals with lower-sulfur Appalachian region bituminous coals or Powder River Basin coals is a control option that is widely exercised. This was illustrated in Chapter 1, where coal production by region is discussed. The option of using lower-sulfur coal has resulted in a large increase in western coal production and use. Table 9.6 illustrates the distribution of sulfur in U.S. coals by region as of January 1, 1997, the last year these data were reported in this manner [10]. Total recoverable reserves shown in the table differ than more recent values reported in Chapter 1 because the basis for determining the reserves periodically changes. Although Table 9.6 is dated, it provides a good relative distribution of the sulfur content in coal in the United States by region.

The relationship between sulfur content in the coal and pounds of sulfur per million Btu is provided in Table 9.7 for comparison. This listing, which was developed by the U.S. Department of Energy's Energy Information Agency (EIA), is used for approximate correlations with New Source Performance Standards (NSPS) and 1990 Clean Air Act Amendments criteria. With the exception of the low-sulfur coal, which meets NSPS requirements, the medium- and high-sulfur coals require control strategies. This includes emission reduction technologies or offsets through sulfur dioxide allowances.

Low-sulfur coals are also imported to the United States, specifically to coastal areas such as Florida or the eastern seaboard. Similar to the replacement of high-sulfur coals with noncoal fuels in power generation units, this practice of importing lower-sulfur coals to the United States for sulfur dioxide compliance also needs to be questioned from the standpoint of energy security.

Coal Cleaning

Coal preparation, or beneficiation, is a series of operations that remove mineral matter (i.e., ash) from coal. Preparation relies on different mechanical operations, which will not be discussed in detail, to perform the separation, such as size reduction, size classification, cleaning, dewatering and drying, waste disposal, and pollution control. Coal preparation processes, which are physical processes, are designed mainly to provide ash removal, energy enhancement, and product standardization [9]. Sulfur reduction is achieved because the ash material removed contains pyritic sulfur. Coal cleaning is used for moderate sulfur dioxide emissions control because physical coal cleaning is not effective in removing organically bound sulfur.

Chemical coal cleaning processes are being developed to remove the organic sulfur, but these are not used on a commercial scale. An added benefit of coal cleaning is that several trace elements, including antimony, arsenic, cobalt, mercury, and selenium, are generally associated with pyritic sulfur in raw coal, and they too are

Table 9.6 Estimated U.S. Recoverable Coal Reserves by Sulfur Range and Major Coal-Producing Region (Remaining as of January 1, 1997)

| Coal-Producing Region | Sulfur Content Categories (pounds of sulfur per million Btu) | | | | | | | | | | | |
| | Low Sulfur (≤0.60) | | Medium Sulfur (0.61–1.67) | | High Sulfur (≥1.68) | | Total | |
	Million Short Tons	Percent of Total	Million Short Tons	Percent of Total	Million Short Tons	Percent of Total	Million Short Tons	Percent of Total
Appalachia	11,675	11.6	20,337	24.0	23,283	25.9	55,295	20.1
Interior	769	0.8	10,041	11.8	57,966	64.4	68,776	25.0
Western	87,775	87.6	54,529	64.2	8,768	9.7	151,072	54.9
Total	100,219		84,907		90,017		275,143	

Source: From EIA (1999).

Table 9.7 Comparison of Sulfur Content in Coal with Pounds
of Sulfur per Million Btu

Qualitative Rating	Pounds of Sulfur per Million Btu	Approximate Range of Coal Sulfur Content (%)	
		High-Grade Bituminous Coal	High-Grade Lignite
Low sulfur	≤0.4 to 0.6	≤0.5 to 0.8	≤0.3 to 0.5
Medium sulfur	0.61 to 1.67	0.8 to 2.2	0.5 to 1.3
High sulfur	1.68 to >2.50	2.2 to >3.3	1.3 to >1.9

Source: EIA (1999).

reduced through the cleaning process. As the inert material is removed, the volatile matter content, fixed carbon content, and heating value increase, thereby producing a higher-quality coal. The moisture content, from residual water from the cleaning process, can also increase; this lowers the heating value, but it is usually minimal and has little impact on coal quality. Coal cleaning does add additional cost to the coal price; however, there are several benefits to reducing the ash content, including lower sulfur content, less ash to be disposed, lower transportation costs because more carbon and less ash is transported (since coal cleaning is usually done at the mine and not at the power plant), and increases in power plant peaking capacity, rated capacity, and availability [11]. Developing circumstances are making coal cleaning more economical and a potential sulfur control technology, and include the following [9]:

- Higher coal prices and transportation costs
- Diminishing coal quality because of less selective mining techniques
- The need to increase availability and capacity factors at existing boilers
- More stringent air quality standards
- Lower costs for improving fuel quality versus investing in extra pollution control equipment

Wet Flue Gas Desulfurization (Wet FGD)

Wet scrubbers are the most common FGD method currently in use (or under development) and include a variety of processes and the use of many sorbents, and they are manufactured by a large number of companies. The sorbents used by wet scrubbers include calcium-, magnesium-, potassium-, or sodium-based sorbents, ammonia, or seawater. Currently, there are no commercial potassium-based scrubbers in use and only a limited number of ammonia or seawater systems in use or being demonstrated. The calcium-based scrubbers are by far the most popular, and this technology is discussed in this section along with the use of sodium- and magnesium-based sorbents.

Limestone- and Lime-Based Scrubbers Wet scrubbing with limestone and lime are the most popular commercial FGD systems. The inherent simplicity, the availability of an inexpensive sorbent (limestone), production of a usable by-product (gypsum), reliability, and the high removal efficiencies obtained (which can be as high as 99 percent) are the main reasons for this popularity. Capital costs are typically higher than other technologies, such as sorbent injection systems; however, the technology is known for its low operating costs because the sorbent is widely available and the system is cost effective.

In a limestone/lime wet scrubber, the flue gas is scrubbed with a 5-15 percent (by weight) slurry of calcium sulfite/sulfate salts along with calcium hydroxide $(Ca(OH_2))$ or limestone $(CaCO_3)$. Calcium hydroxide is formed by slaking lime (CaO) in water according to the reaction

$$CaO(s) + H_2O(l) \rightarrow Ca(OH)_2(s) + heat \qquad (9.9)$$

In the limestone and lime wet scrubbers, the slurry containing the sulfite/sulfate salts and the newly added limestone or calcium hydroxide is pumped to a spray tower absorber and sprayed into it. The sulfur dioxide is absorbed into the droplets of slurry, and a series of reactions occur in the slurry. The reactions between the calcium and the absorbed sulfur dioxide create the compounds calcium sulfite hemihydrate $(CaSO_4 \cdot \frac{1}{2}H_2O)$ and calcium sulfate dihydrate $(CaSO_4 \cdot 2H_2O)$. Both of these compounds have low solubility in water and precipitate from the solution. This enhances the absorption of sulfur dioxide and further dissolution of the limestone or hydrated lime.

Reactions that occur in scrubbers are complex. Simplified overall reactions for limestone- and lime-based scrubbers are

$$SO_2(g) + CaCO_3(s) + \tfrac{1}{2}H_2O(l) \rightarrow CaSO_3 \cdot \tfrac{1}{2}H_2O(s) + CO_2(g) \qquad (9.10)$$

for a limestone scrubber and

$$SO_2(g) + Ca(OH)_2(s) + H_2O(l) \rightarrow CaSO_3 \cdot \tfrac{1}{2}H_2O(s) + \tfrac{3}{2}H_2O(l) \qquad (9.11)$$

for a lime scrubber.

Calcium sulfite hemihydrate can be converted to the calcium sulfate dihydrate with the addition of oxygen by the reaction

$$CaSO_3 \cdot \tfrac{1}{2}H_2O(s) + \tfrac{3}{2}H_2O(l) + \tfrac{1}{2}O_2(g) \leftrightarrow CaSO_4 \cdot 2H_2O(s) \qquad (9.12)$$

The actual reactions that occur, however, are much more complex and include a combination of gas-liquid, solid-liquid, and liquid–liquid ionic reactions. In the limestone scrubber, the following reactions describe the process [12]. In the gas-liquid contact zone of the absorber (see Figure 9.3 for a typical schematic diagram of a limestone scrubber system), sulfur dioxide dissolves into the aqueous state shown in Eq. (9.13).

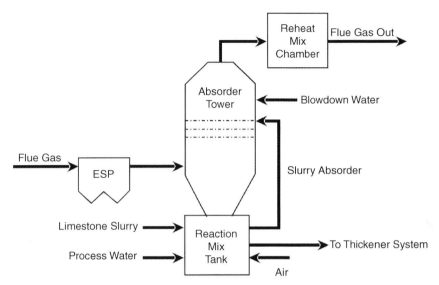

Figure 9.3 Limestone scrubber system with forced oxidation.

$$SO_2(g) \leftrightarrow SO_2(l) \tag{9.13}$$

and is hydrolyzed to form ions of hydrogen and bisulfate

$$SO_2(l) + H_2O(l) \leftrightarrow HSO_3^- + H^+ \tag{9.14}$$

The limestone dissolves in the absorber liquid and then forms ions of calcium and bicarbonate

$$CaCO_3(s) + H^+ \leftrightarrow Ca^{++} + HCO_3^- \tag{9.15}$$

which is followed by acid-base neutralization

$$HCO_3^- + H^+ \leftrightarrow CO_2(l) + H_2O(l) \tag{9.16}$$

stripping of the CO_2 from the slurry

$$CO_2(l) \leftrightarrow CO_2(g) \tag{9.17}$$

and dissolution of the calcium sulfite hemihydrates

$$CaSO_3 \cdot \tfrac{1}{2}H_2O(s) \leftrightarrow Ca^{++} + HSO_3^- + \tfrac{1}{2}H_2O(l) \tag{9.18}$$

In the reaction tank of a scrubber system, the solid limestone is dissolved into the aqueous state (Eq. (9.15)), acid-base neutralization occurs (Eq. (9.16)), the CO_2 is stripped out (Eq. (9.17)), and the calcium sulfite hemihydrate is precipitated by the reaction

$$Ca^{++} + HSO_3^- + \tfrac{1}{2}H_2O(l) \leftrightarrow CaSO_3 \cdot \tfrac{1}{2}H_2O(s) + H^+ \tag{9.19}$$

The dissolution of the calcium sulfite in the gas-liquid contact zone in the absorber is necessary in order to minimize scaling of the calcium sulfite hemihydrate in the absorber [5]. The equilibrium pH for calcium sulfite is about 6.3 at a CO_2 partial pressure of 0.12 atmospheres, which is the typical concentration of CO_2 in flue gas. Typically the pH is maintained below this level to keep the calcium sulfite hemihydrate from dissolving (i.e., to keep Eq. (9.18) from proceeding to the right).

The slurry returning from the absorber to the reaction tank can have a pH as low as 3.5, which is increased to 5.2 to 6.2 by the addition of freshly prepared limestone slurry into the tank [5]. The pH in the reaction tank must be maintained at a pH that is less than the equilibrium pH of calcium carbonate in water, which is 7.8 at 77°F.

The reaction equations for the lime scrubber are similar to those for the limestone scrubber, with the exception that the following reactions are substituted for Eqs. (9.15) and (9.16), respectively [12]:

$$Ca(OH)_2(s) + H^+ \leftrightarrow CaOH^+ + H_2O(l) \tag{9.20}$$

$$CaOH^+ + H^+ \leftrightarrow Ca^{++} + H_2O(l) \tag{9.21}$$

Limestone with Forced Oxidation Limestone scrubbing with forced oxidation (LSFO) is one of the most popular systems in the commercial market. A limestone slurry is used in an open spray tower with in situ oxidation to remove SO_2 and form a gypsum sludge. The major advantages of this process, relative to a conventional limestone FGD system (where the product is calcium sulfite rather than calcium sulfate (gypsum)), are easier dewatering of the sludge, more economical disposal of the scrubber product solids, and decreased scaling on the tower walls. LSFO is capable of greater than 90 percent SO_2 removal [13].

In the LSFO system, the hot flue gas exits the particulate control device, usually an ESP, and enters a spray tower, where it comes into contact with a sprayed dilute limestone slurry. The SO_2 in the flue gas reacts with the limestone in the slurry via the reactions listed earlier to form the calcium sulfite hemihydrate. Compressed air is bubbled through the slurry, which causes this sulfite to be naturally oxidized and hydrated to form calcium sulfate dihydrate. The calcium sulfate can be first dewatered using a thickener or hydrocyclones and then further dewatered using a rotary drum filter. The gypsum is then transported to a landfill for disposal. The formation of the calcium sulfate crystals in a recirculation tank slurry also helps to reduce the chance of scaling.

The absorbing reagent—limestone—is normally fed to the open spray tower in an aqueous slurry at a molar feed rate of 1.1 moles of $CaCO_3$/mole of SO_2 removed. This process is capable of removing more than 90 percent of the SO_2 present in the inlet flue gas. These are the advantages of LSFO systems [13]:

- Lower scaling potential on tower internal surfaces due to the presence of gypsum seed crystals and reduced calcium sulfate saturation levels. This in turn allows a greater reliability of the system.
- The gypsum product is filtered easier than the calcium sulfite ($CaSO_3$) produced with conventional limestone systems.
- A lower chemical oxygen demand in the final disposed product.
- The final product can be safely and easily disposed in a landfill.
- The forced oxidation allows the limestone utilization to be greater than conventional systems.
- The raw material (limestone) used as an absorbent is inexpensive.
- LSFO is an easier retrofit than natural oxidation systems, since the process uses smaller dewatering equipment.

A disadvantage of this system is the high energy demand due to the relatively higher liquid-to-gas ratio necessary to achieve the required SO_2 removal efficiencies.

Limestone with Forced Oxidation Producing a Wallboard Gypsum By-Product In the limestone/wallboard (LS/WB) gypsum FGD process, a limestone slurry is used in an open spray tower to remove SO_2 from the flue gas. The flue gas enters the spray tower where the SO_2 reacts with the $CaCO_3$ in the slurry to form calcium sulfite. The calcium sulfite is then oxidized to calcium sulfate in the absorber recirculation tank. The calcium sulfate produced with this process is of a higher quality, so it can be used in wallboard manufacture.

To achieve a higher-quality gypsum, this process is a little different. The LS/WB system uses horizontal belt filters to produce a drier product and provides sufficient cake washing to remove residual chlorides. Since the by-product is of higher quality, the use of the product handling system is replaced with a by-product conveying and temporary storage equipment. Sulfuric acid addition is used in systems with an external oxidation tank. The acid is used to control the pH of the slurry and neutralizes unreacted $CaCO_3$.

The limestone feed rate in this process is 1.05 moles $CaCO_3$/mole of SO_2 removed, which is slightly lower than the feed rate for the LSFO system [13]. Other advantages of this process are that the disposal area is kept to a minimum, since most of the by-product is reusable. The gypsum can be sold to cement plants and agricultural users. Also, SO_2 removal is slightly enhanced because of the high sulfite to sulfate conversion.

This process does have some disadvantages. A few full-scale operating systems that actually produce quality gypsum are in operation in the United States. To produce quality gypsum, specific process control and tight operator attention are constantly needed to ensure that chemical impurities do not lead to off-specification gypsum. Another disadvantage is the inability to use cooling tower blowdown as system makeup water due to chloride limits in the gypsum by-product.

Limestone with Inhibited Oxidation In the limestone with inhibited oxidation process, the hot flue gas exits the particulate control device and enters an open spray tower, where it comes into contact with a dilute $CaCO_3$ slurry. This slurry contains thiosulfate ($Na_2S_2O_3$), which inhibits natural oxidation of the calcium sulfite. The calcium sulfite is formed from the reaction with SO_2 in the flue gas and the $CaCO_3$ slurry. The slurry absorbs the SO_2 and then drains down to a recirculation tank below the tower. By inhibiting natural oxidation of the sulfite, gypsum scaling on process equipment is reduced along with gypsum relative saturation. The gypsum relative saturation is reduced below 1.0. Thiosulfate is either added directly as $Na_2S_2O_3$ to the feed tank, or is generated in situ by the addition of emulsified sulfur. In some cases, thiosulfate has the ability to increase the dissolution of the calcium carbonate and enlarge the size of the sulfite crystals to improve solids dewatering [13].

This process is capable of removing more than 90 percent of the SO_2 in the flue gas. The calcium sulfite slurry product is thickened, stabilized with fly ash and lime, and then sent to a landfill. The calcium carbonate feed rate is 1.10 moles Ca/mole of SO_2 removed. The effectiveness of thiosulfate is site specific, since the amount of thiosulfate required to inhibit oxidation strongly depends on the chemistry and operating conditions of each FGD system. Variables such as saturation temperature, dissolved magnesium, chlorides, flue gas inlet SO_2 and O_2 concentrations, and slurry pH affect the thiosulfate effectiveness [13].

Thiosulfate has been shown to increase limestone utilization when added to the system. This occurs because the thiosulfate reduces the gypsum relative saturation level, which in turn reduces the level of calcium dissolved in the liquor. The dissolution rate is increased by lowering the calcium concentration in the slurry. Thiosulfate also improves the dewatering characteristics of the sulfite product. By preventing the high concentrations of sulfate, the thiosulfate allows the calcium sulfite to form larger, single crystals. This increases the crystal's settling velocity and improves the filtering characteristics, which results in a higher solid content of dewatered product.

The process has a few disadvantages. The thiosulfate/sulfur reagent requires additional process equipment and storage facilities. Also, the reagent can cause corrosion of many stainless steels under scrubber conditions. Another disadvantage is that the thiosulfate is fairly temperature dependent, thereby requiring the system to operate within a particular temperature range.

Magnesium Enhanced Lime In the magnesium enhanced lime (MagLime) process, the hot flue gas exits the particulate control device and enters a spray tower, where it comes into contact with a magnesium sulfite/lime slurry. Magnesium lime, such as thiosorbic lime (which contains 4 to 8 percent MgO), is fed to the open spray tower in an aqueous slurry at a molar feed rate of 1.1 moles CaO/mole of SO_2 removed. The SO_2 is absorbed by the reaction with magnesium sulfite, forming magnesium bisulfite. This occurs through the following reactions [13]:

$$SO_2(g) + H_2O(l) \rightarrow H_2SO_3(aq) \rightarrow H^+ + HSO_3^- \qquad (9.22)$$

$$H^+ + MgSO_3(s) \rightarrow HSO_3^- + Mg^{++} \qquad (9.23)$$

The magnesium sulfite absorbs the H^+ ion and increases the HSO_3^- concentration in Eq. (9.23). This allows the scrubber liquor to absorb more of the SO_2. The absorbed SO_2 reacts with hydrated lime to form solid-phase calcium sulfite. The magnesium sulfite is reformed by the following reactions:

$$Ca(OH)_2(s) + 2HSO_3^- + Mg^{++} \rightarrow Ca^{++}SO_3^{--} + 2H_2O(l) + MgSO_3(s) \qquad (9.24)$$

$$Ca^{++}SO_3^{++} + \tfrac{1}{2}H_2O \rightarrow CaSO_3 \cdot \tfrac{1}{2}H_2O(s) \qquad (9.25)$$

Inside the absorber, some magnesium sulfite present in the solution is oxidized to sulfate. This sulfite reacts with the lime to form calcium sulfate solids. Calcium sulfite and sulfate solids are the main products of the MagLime process. The calcium sulfite sludge is dewatered using thickener and vacuum filter systems and then fixated using fly ash and lime prior to disposal in a lined landfill. The magnesium remains dissolved in the liquid phase.

The following are some of the advantages of the MagLime process compared to the LSFO process [13]:

- High SO_2 removal efficiency at low liquid-to-gas ratios
- Lower gas-side pressure drop due to lower liquid-to-gas ratios
- Reduced potential for scaling, which improves reliability of the system
- Lower power consumption due to a lower slurry recycle rate
- Lower capital investment due to smaller reagent handling equipment and no oxidation air compressor
- Reduction in fresh water use, since the process water may be recycled for the mist eliminator wash

The three major disadvantages of the process are the expense of the lime reagent compared to the limestone, the use of fresh water for lime slaking, and the difficult dewatering characteristics of the calcium sulfite/sulfate sludge. The sulfite can be oxidized to produce gypsum, but this requires extensive equipment and process control.

Limestone with Dibasic Acid The dibasic acid enhanced limestone process is very similar to the LSFO process. The hot flue gas exits the particulate control device and enters a spray tower, where it comes into contact with a diluted limestone slurry. The SO_2 in the flue gas reacts with the limestone and water to form hydrated calcium sulfite

$$SO_2(g) + CaCO_3(s) + \tfrac{1}{2}H_2O(l) \rightarrow CaSO_3 \cdot \tfrac{1}{2}H_2O(s) + CO_2(g) \qquad (9.10)$$

This equation is rate limited by the absorption of SO_2 into the scrubbing liquor

$$SO_2(l) + H_2O(l) \leftrightarrow HSO_3^- + H^+ \qquad (9.14)$$

The dissolved SO_2 ions then react with the calcium ions to form calcium sulfite. The hydrogen ions in solution are partly responsible for reforming SO_2.

After absorbing the SO_2, the slurry drains from the tower to a recirculation tank. Here the calcium sulfite is oxidized to calcium sulfate dihydrate using oxygen

$$CaSO_3 \cdot \tfrac{1}{2}H_2O(s) + \tfrac{3}{2}H_2O(l) + \tfrac{1}{2}O_2(g) \leftrightarrow CaSO_4 \cdot 2H_2O(s) \qquad (9.12)$$

Dibasic acid acts as a buffer by absorbing free hydrogen ions formed by Eq. (9.14). This then shifts the reaction to the right to form more sulfite ions, thus removing more SO_2. Alkaline limestone is added to replace the buffering capabilities of the acid, so there is no net consumption of the dibasic acid during SO_2 absorption.

The limestone dissolution rate is increased by increasing the SO_2 removal efficiency at a low slurry pH. This results in a lower reagent consumption due to an increase in calcium carbonate availability in the recirculation tank. The dibasic acid process offers some advantages compared to the LSFO process [13]:

- Increased SO_2 removal efficiency.
- Reduced liquid-to-gas ratio and the potential to decrease the reagent feed rate. This lowers capital and operating costs for the limestone grinding equipment, slurry handling, and landfill requirements.
- Reduced scaling because of the low pH and reduced gypsum relative saturation levels.
- Increased system reliability by reducing the maintenance requirements and increasing the flexibility of the system.

Disadvantages of the process include the following:

- More process capital is needed for the dibasic acid feed equipment.
- There is the potential for corrosion and erosion due to the low system pH.
- Odorous by-products are produced by the dibasic acid degradation. Although the SO_2 absorption reactions do not consume the dibasic acid, the acid does degrade by carboxylic oxidation into many short chain molecules. One of these molecules is valeric acid, which has a musty odor.
- Control problems may be caused by foaming in the recirculation and oxidation tanks due to the presence of the dibasic acid.

Sodium-Based Scrubbers Wet sodium-based systems have been in commercial operation since the 1970s. These systems can achieve high SO_2 removal efficiencies while burning coals with medium- to high-sulfur content. A disadvantage of these systems, however, is the production of a waste sludge that requires disposal.

Lime Dual Alkali In the lime dual alkali process, the hot flue gas exits the particulate control device and enters an open spray tower, where the gas comes into contact with a sodium sulfite (Na_2SO_3) solution that is sprayed into the tower [13]. An initial charge of sodium carbonate (Na_2CO_3) reacts directly with the SO_2 to form sodium sulfite and CO_2. The sulfite then reacts with more SO_2 and water to form sodium bisulfite ($NaHSO_3$). Some of the sodium sulfite is oxidized by excess oxygen in the flue gas to form sodium sulfate (Na_2SO_4). This does not react with SO_2 and cannot be reformed by the addition of lime to form calcium sulfate. The preceding process is described by the following reactions:

$$Na_2CO_3(s) + SO_2(g) \rightarrow Na_2SO_3(s) + CO_2(g) \qquad (9.26)$$

$$Na_2SO_3(s) + SO_2(g) + H_2O(l) \rightarrow 2NaHSO_3(s) \qquad (9.27)$$

$$Na_2SO_3(s) + \tfrac{1}{2}O_2(g) \rightarrow Na_2SO_4(s) \qquad (9.28)$$

along with the minor reaction:

$$2NaOH(aq) + SO_2(g) \rightarrow Na_2SO_3(s) + H_2O(l) \qquad (9.29)$$

The calcium sulfites and sulfates are reformed in a separate regeneration tank and are formed by mixing the soluble sodium salts (bisulfate and sulfate) with slaked lime. The calcium sulfites and sulfates precipitate from the solution in the regeneration tank. The scrubber liquor then has a pH of 6 to 7 and consists of sodium sulfite, sodium bisulfite, sodium sulfate, sodium hydroxide, sodium carbonate, and sodium bicarbonate [13].

The lime dual alkali process has several advantages over the LSFO process [13]:

- The system has a higher availability, since there is less potential for scaling and plugging of the soluble absorption reagents and reaction products.
- Corrosion and erosion is prevented with the use of a relatively high pH solution.
- Maintenance labor and materials are lower because of the high reliability of the system.
- The main recirculation pumps are smaller, since the absorber liquid/gas feed rate is less.
- Power consumption is lower due to the smaller pump requirements.
- There is no process blowdown water discharge stream.
- The highly reactive alkaline compounds in the absorbing solution allow for better turndown and load following capabilities.

There are two main disadvantages of the process compared to the LSFO system: The sodium carbonate reagent is more expensive than limestone, and the sludge must be disposed in a lined landfill because of sodium contamination of the calcium sulfite/sulfate sludge.

Regenerative Processes Regenerative FGD processes regenerate the alkaline reagent and convert the SO_2 to a usable chemical by-product. Two commercially accepted processes are discussed in this section. The Wellman-Lord process is the most highly demonstrated regenerative technology in the world, while the regenerative magnesia scrubbing process is in commercial service in the United States. Other processes have undergone demonstrations, are used on a limited basis, or are currently under development and include ammonia-based scrubbing, an aqueous carbonate process, and the Citrate process.

The Wellman-Lord Process The Wellman-Lord process uses sodium sulfite to absorb SO_2, which is then regenerated to release a concentrated stream of SO_2. Most of the sodium sulfite is converted to sodium bisulfite by reaction with SO_2, as in the dual alkali process. Some of the sodium sulfite is oxidized to sodium sulfate. Prescrubbing of the flue gases is necessary to saturate and cool the flue gas to about 130°F. This removes chlorides and any remaining fly ash and avoids excessive evaporation in the absorber. A schematic of the system is shown in Figure 9.4 [9].

The basic absorption reaction for the Wellman-Lord process is

$$SO_2(g) + Na_2SO_3(aq) + H_2O(l) \rightarrow 2NaHSO_3(aq) \tag{9.30}$$

The sodium sulfite is regenerated in an evaporator-crystallizer through the application of heat. A concentrated SO_2 stream (i.e., 90 percent) is produced at the same time. The overall regeneration reaction is

$$2NaHSO_3(aq) + heat \rightarrow Na_2SO_3(s) + H_2O(l) + SO_2(conc.) \tag{9.31}$$

The concentrated SO_2 stream that is produced may be compressed, liquefied, and oxidized to produce sulfuric acid or reduced to elemental sulfur. A small portion of collected SO_2 oxidizes to the sulfate form and is converted in a crystallizer to sodium sulfate solids that are marketed as a salt cake [9].

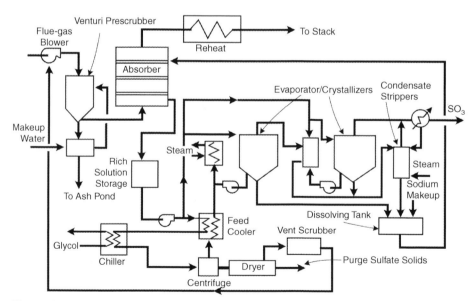

Figure 9.4 The Wellman-Lord process.
Source: From Elliot (1989).

The advantages of this process include minimal solid wastes production, low alkaline reagent consumption, and the use of a slurry rather than a solution, thereby preventing scaling and allowing the production of a marketable by-product. The disadvantage of the process is the high energy consumption and maintenance due to the complexity of the process and the large area required for the system. Another disadvantage is that the purge stream of about 15 percent of the scrubbing solution is required to prevent buildup of the sodium sulfate. Thiosulfate must be purged from the regenerated sodium sulfite.

Regenerative Magnesia Scrubbing In the magnesium oxide process, MgO in the slurry is used the same way limestone or lime is used in the lime scrubbing process. The primary difference between the processes is that the magnesium oxide process is regenerative, whereas lime scrubbing is generally a throwaway process.

The magnesium oxide process, shown in Figure 9.5, uses a slurry of slaked magnesium oxide ($Mg(OH)_2$) to remove SO_2 from the flue gas forming magnesium sulfite and sulfate via the basic reactions

$$Mg(OH)_2(s) + SO_2(g) \rightarrow MgSO_3(s) + H_2O(l) \tag{9.32}$$

$$MgSO_3(s) + \tfrac{1}{2}O_2(g) \rightarrow MgSO_4(s) \tag{9.33}$$

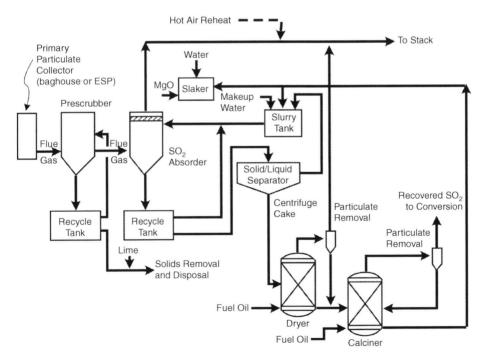

Figure 9.5 Regenerative magnesia scrubbing process.
Source: From Elliot (1989).

A bleed stream of scrubber slurry is centrifuged to form a wet cake containing 75 to 90 percent solids, which is then dried to form a dry, free-flowing mixture of magnesium sulfite and sulfate. This mixture is heated to decompose most of the magnesium sulfite/sulfate to SO_2 and MgO. A stream of 10 to 15 percent SO_2 is produced. Coke, or some other reducing agent, is added in the calcination step to reduce any sulfate present. The regenerated MgO is slaked and used in the absorber. The regeneration reactions are

$$MgSO_3(s) + heat \rightarrow MgO(s) + SO_2(g) \tag{9.34}$$

$$MgSO_4(s) + \tfrac{1}{2}C(s) + heat \rightarrow MgO(s) + SO_2(g) + \tfrac{1}{2}CO_2(g) \tag{9.35}$$

The SO_2 product gas is generally washed and quenched and fed to a contact sulfuric acid plant to produce concentrated sulfuric acid by-product. Sulfur production is possible but would be expensive, since the SO_2 stream is dilute.

Advantages of this process include high SO_2 removal efficiencies (up to 99 percent); minimum impact of fluctuations in inlet SO_2 levels on removal efficiency; low chemical scaling potential; the capability to regenerate the sulfate, which simplifies waste management; and more favorable economics compared to other available regenerative processes [9]. The main disadvantage of the process is its complexity and the need of a contact sulfuric acid plant to produce a salable by-product.

Dry Flue Gas Desulfurization Technology

Dry FGD technology includes lime or limestone spray drying; dry sorbent injection including furnace, economizer, duct, and hybrid methods; and circulating fluidized-bed scrubbers. These processes are characterized with dry waste products that are generally easier to dispose than waste products from wet scrubbers. All dry FGD processes are throwaway types.

Spray Dry Scrubbers Spray dry scrubbers are the second most widely used method to control SO_2 emissions in utility coal-fired power plants. Prior to 1980, the removal of SO_2 by absorption was usually performed using wet scrubbers. Wet scrubbing requires considerable equipment, and alternatives to wet scrubbing were developed, including spray dry scrubbers. Lime (CaO) is usually the sorbent used in the spray drying process, but hydrated lime ($Ca(OH)_2$) is also used. This technology is also known as semidry flue gas desulfurization and is generally used for sources that burn low- to medium-sulfur coal. In the United States, this process has been used in both retrofit applications and new installations on units burning low-sulfur coal [5, 7].

In this process, the hot flue gas exits the boiler air heater and enters a reactor vessel. A slurry consisting of lime and recycled solids is atomized/sprayed into the absorber. The slurry is formed by the reaction

$$CaO(s) + H_2O(l) \rightarrow Ca(OH)_2(s) + heat \tag{9.9}$$

The SO_2 in the flue gas is absorbed into the slurry and reacts with the lime and fly ash alkali to form calcium salts:

$$Ca(OH)_2(s) + SO_2(g) \rightarrow CaSO_3 \cdot \tfrac{1}{2}H_2O(s) + \tfrac{1}{2}H_2O(v) \tag{9.36}$$

$$Ca(OH)_2(s) + SO_3(g) + H_2O(v) \rightarrow CaSO_4 \cdot 2H_2O(s) \tag{9.37}$$

Hydrogen chloride (HCl) present in the flue gas is also absorbed into the slurry and reacts with the slaked lime. The water that enters with the slurry is evaporated, which lowers the temperature and raises the moisture content of the scrubbed gas. The scrubbed gas then passes through a particulate control device downstream of the spray drier. Some of the collected reaction product, which contains some unreacted lime, and fly ash is recycled to the slurry feed system, while the rest is sent to a landfill for disposal. Factors affecting the absorption chemistry include the flue gas temperature, SO_2 concentration in the flue gas, and the size of the atomized slurry droplets. The residence time in the reactor vessel is typically about 10 to 12 seconds.

The lime spray dryer process offers a few advantages over the LSFO process [13]. Only a small alkaline stream of scrubbing slurry must be pumped into the spray dryer. This stream contacts the gas entering the dryer instead of the walls of the system. This prevents corrosion of the walls and pipes in the absorber system. The pH of the slurry and dry solids is high allowing for the use of mild steel materials rather than expensive alloys. The product from the spray dryer is a dry solid that is handled by conventional dry fly ash particulate removal and handling systems, which eliminates the need for dewatering solids handling equipment and reduces associated maintenance and operating requirements. Overall power requirements are decreased, since less pumping power is required. The gas exiting the absorber is not saturated and does not require reheat, thereby capital costs and steam consumption is reduced. Chloride concentration increases the SO_2 removal efficiencies (whereas in wet scrubbers increasing chloride concentration decreases efficiency), which allows the use of cooling tower blowdown for slurry dilution after completing the slaking of the lime reagent. The absorption system is less complex, so operating, laboratory, and maintenance manpower requirements are lower than that required for a wet scrubbing system.

The lime spray dryer has some disadvantages compared to the LSFO system, and these, along with the advantages, must be evaluated for specific applications [13]. A major product of the lime spray dryer process is calcium sulfite, since only 25 percent or less oxidizes to calcium sulfate. The solids handling equipment for the particulate removal device has to have a greater capacity than conventional fly ash removal applications. Fresh water is required in the lime slaking process, which can represent approximately half of the system's water requirement. This differs from the wet scrubbers, where cooling tower water can be used for limestone grinding circuits and most other makeup water applications. The lime spray dryer process requires a higher reagent feed ratio than conventional systems to achieve high removal efficiencies. Approximately 1.5 moles CaO/mole of SO_2 removed are needed for

90 percent removal efficiency. Lime is also more expensive than limestone, so the operating cost is increased. This cost can be reduced if higher coal chloride levels and/or calcium chloride spiking are used, since chlorides improve removal efficiency and reduce reagent consumption. A higher inlet flue gas temperature is needed when a higher sulfur coal is used, which in turn reduces the overall boiler efficiency.

Combining spray dry scrubbing with other FGD systems such as furnace or duct sorbent injection and particulate control technology such as a pulse-jet baghouse allows the use of limestone as the sorbent instead of the more costly lime [6]. Sulfur dioxide removal efficiencies can exceed 99 percent with such a combination.

Sorbent Injection Processes A number of dry injection processes have been developed to provide moderate SO_2 removal that are easily retrofitted to existing facilities and are low capital cost. Of the five basic processes, two are associated with the furnace–furnace sorbent injection and convective pass (economizer) injection, and three are associated with injection into the ductwork downstream of the air heater–in-duct injection, in-duct spray drying, and hybrid systems. Combinations of these processes are also available. Sorbents include calcium- and sodium-based compounds, but the use of calcium-based sorbents is more prevalent. Furnace injection has been used in some small plants using low-sulfur coals. Hybrid systems may combine furnace and duct sorbent injection or introduce a humidification step to improve removal efficiency. These systems can achieve as high as 70 percent removal and are commercially available [6]. Process schematics for dry injection SO_2 control technologies are illustrated in Figure 9.6.

Figure 9.7 is a representation of the level of SO_2 removal that the dry calcium-based sorbent injection processes achieve and the temperature regimes in which they operate [14]. The peak at approximately 2,200°F represents furnace sorbent injection, the peak at about 1,000°F is convective pass/economizer injection, and the peak at the low temperature represents all of the processes downstream of the air heater.

Another dry limestone injection technique, LIMB (limestone injection into a multistage burner), was developed in the 1960s to the 1980s but has not been adopted on a commercial scale for utility applications and is not discussed in detail. In this low capital cost process, which is used in some industrial-scale applications where low SO_2 removal is required, limestone is added to the coal stream and fed with the coal directly to the burner. This process gives poor SO_2 removal (typically about 15 but in rare cases as much as 50 percent), experiences dead-burning (i.e., sintering or melting of the sorbent thereby reducing surface area and lowering sulfur capture), is difficult to introduce in a uniform manner, and can cause operational problems such as tube fouling and impairment of ESP performance because of excessive sorbent addition [9].

In addition, sorbents under development are not discussed in this section, but rather this section focuses mainly on commercial applications. Calcium organic salts (e.g., calcium acetate, calcium magnesium acetate, and calcium benzoate), pyrolysis liquor, and other sorbents are under development for use in injection processes.

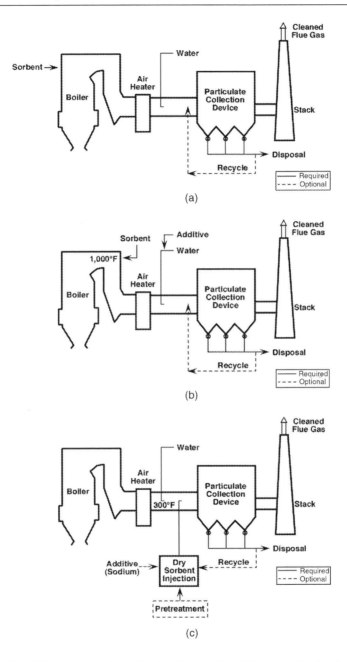

Figure 9.6 Simplified process schematics for dry injection SO$_2$ control technologies: (a) furnace sorbent injection; (b) economizer injection; (c) duct sorbent injection: dry sorbent injection;

(Continued)

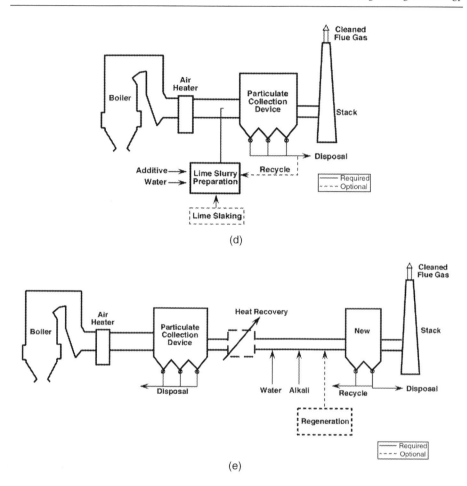

Figure 9.6—Simplified process schematics for dry injection SO$_2$ control technologies:
(d) duct sorbent injection: duct spray drying; and (e) a hybrid system.
Source: Modified from Rhudy, McElroy, and Offen (1986).

Furnace Sorbent Injection With the exception of limestone injection into a multistage
burner, furnace, sorbent injection (FSI) is the simplest dry sorbent process. In this
process, illustrated in Figure 9.6a, pulverized sorbents, most often calcium hydroxide and sometimes limestone, are injected into the upper part of the furnace to react
with the SO$_2$ in the flue gas. The sorbents are distributed over the entire cross section of the upper furnace, where the temperature is in the range 1,400 to 2,400°F and
the residence time for the reactions is 1 or 2 seconds. The sorbents decompose and
become porous solids with high surface area. At temperatures higher than about
2,300°F, dead-burning or sintering is experienced.

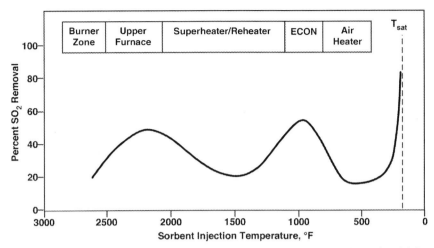

Figure 9.7 SO_2 capture regimes for hydrated calcitic lime at a Ca/S molar ratio of 2.0.
Source: From Rhudy, McElroy, and Offen (1986).

When limestone is used as the sorbent, it is rapidly calcined to quicklime when it enters the furnace

$$CaCO_3(s) + heat \rightarrow CaO(s) + CO_2(g) \tag{9.38}$$

Sulfur dioxide diffuses to the particle surface and heterogeneously reacts with the CaO to form calcium sulfate

$$CaO(s) + SO_2(g) + \tfrac{1}{2}O_2(g) \rightarrow CaSO_4(s) \tag{9.39}$$

Sulfur trioxide, although present at a significantly lower concentration than SO_2, is also captured using calcium-based sorbents

$$CaO(s) + SO_3(g) \rightarrow CaSO_4(s) \tag{9.40}$$

Approximately 15 to 40 percent SO_2 removal can be achieved using a Ca/S in the flue gas molar ratio of 2.0. The optimum temperature for injecting limestone is approximately 1,900 to 2,100°F.

The calcium sulfate formed travels through the rest of the boiler flue gas system and is ultimately collected in the existing particulate control device with the fly ash and unreacted sorbent. Some concerns exist regarding increased tube deposits as a result of injecting solids into the boiler, and the extent of calcium deposition is influenced by overall ash chemistry, ash loading, and boiler system design.

The following overall reactions occur when using hydrated lime as the sorbent:

$$Ca(OH)_2(s) + heat \rightarrow CaO(s) + H_2O(v) \tag{9.41}$$

$$CaO(s) + SO_2(g) + \tfrac{1}{2}O_2(g) \rightarrow CaSO_4(s) \tag{9.39}$$

$$CaO(s) + SO_3(g) \rightarrow CaSO_4(s) \tag{9.40}$$

Approximately 50 to 80 percent SO_2 removal can be achieved using hydrated lime at a Ca/S molar ratio of 2.0. The hydrate is injected at very nearly the same temperature window as limestone, and the optimum range is 2,100 to 2,300°F.

The FSI process can be applied to boilers burning low- to high-sulfur coals. The factors that affect the efficiency of the FSI system are flue gas humidification (to condition the flue gas to counter degradation that may occur in ESP performance from the addition of significant quantities of fine, high-resistivity sorbent particles), type of sorbent, efficiency of ESP, and temperature and location of the sorbent injection. The process is better suited for large furnaces with lower heat release rates [13]. Systems that use hydrated calcium salts sometimes have problems with scaling; however, this can be prevented by keeping the approach to adiabatic saturation temperature above a minimum threshold.

The FSI system has several advantages [13]. One advantage is simplicity of the process. The dry reagent is injected directly into the flow path of the flue gas in the furnace, and a separate absorption vessel is not required. The injection of lime in a dry form allows for a less complex reagent handling system. This in turn lowers operating labor and maintenance costs and eliminates the problems of plugging, scaling, and corrosion found in slurry handling. Power requirements are lower, since less equipment is needed. Steam is not required for reheat, whereas most LSFO systems require some form of reheat to prevent corrosion of downstream equipment. The sludge dewatering system is eliminated, since the FSI process produces a dry solid, which can be removed by conventional fly ash removal systems.

The FSI process has a few disadvantages when compared to the LSFO process [13]. One major disadvantage is that the process removes only up to 40 percent and 80 percent SO_2 when using limestone and hydrated lime at Ca/S molar ratios of 2.0, respectively, whereas the LSFO process can remove greater than 90 percent SO_2 using 1.05 to 1.1 moles CaO/mole SO_2 removed. This is further illustrated in Figure 9.8, which shows calcium utilization (defined as the percent SO_2 removed divided by the Ca/S ratio) of hydrated lime and limestone at various injection temperatures (modified from [14]). Hence, more sorbent is needed in the FSI process, and lime, which works better than limestone, is more expensive than limestone. There is a potential for solids deposition and boiler convective pass fouling, which occurs during the humidification step through the impact of solid droplets onto surfaces. Also, there is a potential for corrosion at the point of humidification, the ESP, the downstream ductwork, and the stack. The corrosion at the point of humidification is caused by operating below the acid dew point, whereas downstream corrosion is caused by the humidified gas temperature being close to the water saturation temperature.

Plugging can also occur, thereby affecting system pressures. The efficiency of an ESP can be reduced by increased particulate loading and changes in the ash resistivity. This can, in turn, lead to the installation of additional particulate collection devices. Sintering of the sorbent is a concern if it is injected at too high of a

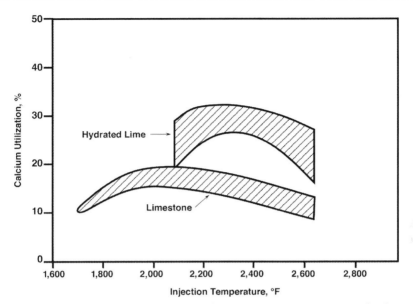

Figure 9.8 Calcium utilization as a function of sorbent injection temperature for furnace sorbent injection.
Source: Modified from Rhudy, McElroy, and Offen (1986).

temperature (e.g., higher than 2,300°F for hydrated lime). Multiple injection ports in the furnace wall may be needed to ensure proper mixing and follow boiler load swings and thus shifting temperature zones. Hydration of the free lime in the product may be required. Lime is very reactive when exposed to water and can pose a safety hazard for disposal areas.

Economizer Injection In an economizer injection process (which is shown earlier in Figure 9.6b), hydrated lime is injected into the flue gas stream near the economizer inlet where the temperature is between 950 and 1,050°F. This process is not commercially used at this time but was extensively studied because it was found that the reaction rate and extent of sulfur capture (refer to Figure 9.7) are comparable to FSI. However, the economizer temperatures are too low for dehydration of the hydrated lime (only about 10 percent of the hydrated lime forms quicklime), and the hydrate reacts directly with the SO_2 to form calcium sulfite

$$Ca(OH)_2(s) + SO_2(g) \rightarrow CaSO_3(s) + H_2O(v) \tag{9.42}$$

This process is best suited for older units in need of a retrofit process and can be used for low- to high-sulfur coals. The advantages and disadvantages of this system are similar to the FSI process (but will not be discussed in detail, since this process is not presently being used in the power industry), with the notable exception that there is no reactive CaO is contained in the waste.

Duct Sorbent Injection—Duct Spray Drying Spray dry scrubbers are the second most widely used method to control SO_2 emissions in utility coal-fired power plants. Lime is usually the sorbent used in this technology, but sodium carbonate is also used, specifically in the western United States. Spray dryer FGD systems have been installed on more than 12,000 MW of total FGD capacity, as shown earlier in Table 9.1, as well as numerous industrial boilers.

The first commercial dry scrubbing system installed on a coal-fired boiler in the United States was in mid-1981 at the Coyote station (jointly owned by Montana-Dakota Utilities, Northern Municipal Power Agency, Northwestern Public Service Company, and Ottertail Power Company) near Beulah, North Dakota. The 425 MW unit burns lignite from a mine-mouth plant and initially used soda ash (Na_2CO_3) as the sulfur removal reagent. The spray dryer was modified about ten years later, and the unit currently uses lime as the reagent. The second dry scrubbing system that was installed on a coal-fired utility boiler was on two 440 MW units that became operational in 1982 and 1983 at the Basin Electric Power Cooperative's Antelope Valley station, also located near Beulah, North Dakota. These units fire mine-mouth lignite and use a slaked lime slurry to remove SO_2 in the spray dryer.

A slaked lime slurry is sprayed directly into the ductwork to remove SO_2 (see Figure 9.6c). The reaction products and fly ash are captured downstream in the particulate removal device. A portion of these solids is recycled and reinjected with the fresh sorbent. Dry spray drying (DSD) is a relatively simple retrofit process capable of 50 percent SO_2 removal at a Ca/S ratio of 1.5. The concept is the same as conventional spray drying except that the existing ductwork provides the residence time for drying instead of a reaction vessel. The main difference is that the residence time in the duct is much shorter—1 or 2 seconds—compared to 10 to 12 seconds in a spray drying vessel.

The slaked lime is produced by hydrating raw lime to form calcium hydroxide. This slaked lime is atomized and absorbs the SO_2 in the flue gas. The SO_2 reacts with the slurry droplets as they dry to form equimolar amounts of calcium sulfite and calcium sulfate. The water in the lime slurry improves SO_2 absorption by humidifying the gas. The reaction products, unreacted sorbent, and fly ash are collected in the particulate control device located downstream. Some of the unreacted sorbent may react with a portion of the CO_2 in the flue gas to form calcium carbonate. Also, a little more SO_2 removal is achieved in the particulate control device. The reactions occurring in the process are

$$CaO(s) + H_2O(g) \rightarrow Ca(OH)_2(s) + heat \tag{9.9}$$

$$Ca(OH)_2(s) + SO_2(g) \rightarrow CaSO_3 \cdot \tfrac{1}{2}H_2O(s) + \tfrac{1}{2}H_2O(v) \tag{9.36}$$

$$Ca(OH)_2(s) + SO_2(g) + \tfrac{1}{2}O_2(g) + H_2O(v) \rightarrow CaSO_4 \cdot 2H_2O(s) \tag{9.43}$$

$$Ca(OH)_2(s) + CO_2(g) \rightarrow CaCO_3(s) + H_2O(v) \tag{9.44}$$

There are two different methods for atomizing the slurry. One method is the use of rotary atomizers, with the ductwork providing the short gas residence time of 1 or

2 seconds. When using this atomizer, the ductwork must be sufficiently long to allow for drying of the slurry droplets. There must also be no obstructions in the duct.

The second method for atomizing the slurry is the use of dual-fluid atomizers, where compressed air and water are used to atomize the slurry. This is called the Confined Zone Dispersion (CZD) process. The dual-fluid atomizer has been shown to be more controllable due to the adjustable water flow rate. They are also relatively inexpensive and have a long and reliable operating life with little maintenance. The spray is confined in the duct, which allows better mixing with the flue gas rather than impinging on the walls.

The DSD process has several advantages over the wet processes. The DSD process is less complex, since the reagent is injected directly into the flow path of the flue gas and a separate absorption vessel is not needed. Less equipment is needed and thereby power requirements are lower. The waste from this process does not contain reactive lime like the FSI process and therefore does not require special handling.

Some of the problems encountered by the DSD system are also common to other dry processes. A main disadvantage of the system includes limited SO_2 removal efficiency (i.e., ≈ 50 percent) and low calcium utilization compared to wet processes. Quicklime is more expensive than limestone. If an ESP is used, there is the potential for reduced efficiency due to changes in fly ash resistivity and the increased dust loading in the flue gas. Additional collection devices may be needed as well as humidification to improve ESP collection efficiency. There must be sufficient length—that is, residence time of the ductwork to ensure complete droplet vaporization prior to the particulate collection device. This is necessary for good sulfur capture and to avoid plugging and deposition, which in turn results in an increased pressure drop that the induced draft fans must overcome.

Duct Sorbent Injection—Dry Sorbent Injection Dry sorbent injection (DSI), also referred to as in-duct dry injection, was illustrated in Figure 9.6d. Hydrated lime is the sorbent typically used in this process, especially for power generation facilities. However, sodium-based sorbents have been tested extensively, including full-scale utility demonstrations, and are used in industrial systems, such as municipal and medical waste incinerators for acid gas control.

When using hydrated lime in this process, it is injected either upstream or downstream of a flue gas humidification zone. In this zone, the flue gas is humidified to within $20°F$ of the adiabatic saturation temperature by injecting water into the duct downstream of the air preheater [13]. The SO_2 in the flue gas reacts with the calcium hydroxide to form calcium sulfate and calcium sulfite:

$$Ca(OH)_2(s) + SO_2(g) + \tfrac{1}{2}O_2(g) + H_2O(v) \rightarrow CaSO_4 \cdot 2H_2O(s) \qquad (9.43)$$

$$Ca(OH)_2(s) + SO_2(g) \rightarrow CaSO_3 \cdot \tfrac{1}{2}H_2O(s) + \tfrac{1}{2}H_2O(v) \qquad (9.36)$$

The water droplets are vaporized before they strike the surface of the wall or enter the particulate control device. The unused sorbent, along with the products and fly ash, is collected in the particulate control device. About half of the collected material is shipped to a landfill, while the other half is recycled for injection with the fresh sorbent into the ducts [13].

The DSI system offers many of the same advantages and disadvantages that other dry systems offer [13]. The process is less complex (i.e., no slurry recycle and handling, no dewatering system, less pumps, and no reactor vessel) than a wet system, specifically LSFO. The humidification water and hydrated lime are injected directly into the existing flue gas path. No separate SO_2 absorption vessel is necessary. The handling of the reagent is simpler than in wet systems. The costs for DSI systems are less than in wet systems because there is less equipment to install, and because there is less equipment, operating and maintenance costs are reduced. The waste product is free of reactive lime, so no special handling is required.

Some of the problems encountered by the DSI system and its disadvantages compared to the LSFO system are common to other dry processes. Sulfur dioxide removal efficiencies are lower (as is calcium utilization) than wet systems and range from 30 to 70 percent for a Ca/S ratio of 2.0. Quicklime is more expensive than limestone. When an ESP is used for particulate control, there is the potential for reduced efficiency due to increased fly ash resisitivity and dust loading in the flue gas. Additional collection devices may be required. A sufficient length of ductwork is necessary to ensure a residence time of 1 or 2 seconds in a straight, unrestricted path. Plugging of the duct can occur if the residence time is insufficient for droplet vaporization leading to increased system pressure drop.

In the dry sodium desulfurization process, a variety of sodium-containing crystalline compounds may be injected directly into the flue gas. The main compounds of interest include the following [15]:

- Sodium carbonate (Na_2CO_3), a refined product of $\approx$98 percent purity
- Sodium bicarbonate ($NaHCO_3$), a refined product of $\approx$98 percent purity
- Nacholite ($NaHCO_3$), a natural material of $\approx$76 percent purity containing high levels of insolubles
- Sodium sesquicarbonate ($NaHCO_3 \cdot Na_2CO_3 \cdot 2H_2O$), a refined product of $\approx$98 percent purity
- Trona ($NaHCO_3 \cdot Na_2CO_3 \cdot 2H_2O$), a natural material of $\approx$88 percent purity containing high levels of insolubles

Sodium bicarbonate and sodium sesquicarbonate have been the most extensively tested in pilot, demonstration, and full-scale utility applications due to proven success and commercial availability. In addition, sodium bicarbonate is extensively used in industrial applications for acid gas control.

Of the preceding compounds, sodium bicarbonate has demonstrated to have the best sulfur capture in coal-fired boiler applications. This is illustrated in Figure 9.9, where SO_2 removal as a function of normalized stoichiometric ratio (NSR) for sodium bicarbonate injection into a coal-fired pilot-scale test facility and industrial boiler equipped with fabric filter baghouses is shown. Note that the NSR represents the molar ratio between the injected sodium compound and the initial SO_2 concentration in the flue gas considering that it takes two moles of sodium to react with only one mole of SO_2.

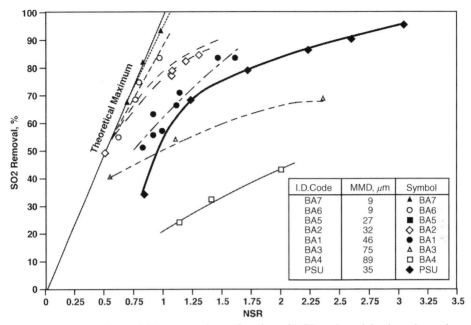

Figure 9.9 Comparison of SO_2 removal as a function of NSR and particle size when using sodium bicarbonate. Tests identified by the BA code were performed in a coal-fired pilot-scale facility, while the tests coded PSU were performed in a coal-fired industrial boiler system. Both facilities were equipped with fabric filter baghouses. The Penn State tests were performed with flue gas temperature of about 380°F, while the pilot-scale tests were performed at about 300°F. All tests were performed with sorbent from the same vendor. *Source:* From Bland and Martin (1990) and Miller et al. (2000) [16].

The flue gas stream must be above 240°F for rapid decomposition of the sodium bicarbonate when it is injected, or little SO_2 capture will occur. While SO_2 will react directly with the sodium bicarbonate, in the presence of nitric oxide (NO) this reaction is inhibited and does not result in significant sulfur capture. Therefore, for acceptable SO_2 capture to progress rapidly and attain acceptable levels of utilization, the bicarbonate component must begin to decompose. As flue gas temperatures increase into the optimum range of 240 to 320°F, the carbonate is decomposed and the subsequent sulfation reaction occurs [15]. When the bicarbonate component decomposes, carbon dioxide and water vapor are evolved from the particle interior, creating a network of void spaces. Sulfur dioxide and NO can diffuse to the fresh sorbent surfaces where the heterogeneous reactions to capture SO_2 (and to a lesser extent NO) take place. The decomposition and sulfation reactions are

$$2NaHCO_3(s) + heat \rightarrow Na_2CO_3(s) + H_2O(v) + CO_2(g) \tag{9.44}$$

$$SO_2(g) + Na_2CO_3(s) + \tfrac{1}{2}O_2(g) \rightarrow Na_2SO_4(s) + CO_2(g) \tag{9.45}$$

Lower NO_x emissions also result from injection of dry sodium compounds [15]. The mechanism is not well understood, but reductions of up to 30 percent have been demonstrated; this is a function of SO_2 concentration and the NSR ratio. There is a side effect of this reduction, though. Nitric oxide is oxidized to NO_2 (a reddish brown gas), and not all of the NO_2 is reacted with the sorbent. As the NO_2 concentration increases in the stack, an undesirable coloration in the plume can be created.

Hybrid Systems Hybrid sorbent injection processes are typically a combination of the FSI and the DSI systems with the goal of achieving greater SO_2 removal and sorbent utilization [6]. Various types of configurations have been tested, including injecting secondary sorbents, such as sodium compounds, into the ductwork or humidifying the flue gas in a specially designed vessel. Humidification reactivates the unreacted CaO and can increase the SO_2 removal efficiency. Advantages of hybrid processes include high SO_2 removal; low capital and operating costs; less space necessary, thereby lending to easy retrofit; easy operation and maintenance; and no wastewater treatment [6].

In some hybrid systems, a new baghouse is installed downstream of an existing particulate removal device (generally an ESP). The existing ESP continues to remove the ash, which can be either sold or disposed. Sulfur dioxide removal is accomplished in a manner similar to in-duct injection, with the sorbent injection upstream of the new baghouse [14].

The potential advantages of this system include the potential for toxic substances control, since a baghouse is the last control device (this is further discussed later in this chapter), easier waste disposal, the potential for sorbent regeneration, separate ash and product streams, and more efficient recycle without ash present [14]. The major issue is the high capital cost of adding a baghouse, although the concept of adding one with a high air-to-cloth ratio (3–5 acfm/ft^2) can minimize this cost. Hybrid systems are discussed in more detail later in this chapter.

Circulating Fluidized-Bed Scrubbers CFB scrubbers are the least used commercial option with few or no systems planned in the future (see Table 9.1). CFB scrubbers include dry and semidry systems [6]. Commercial application of the dry CFB is more widespread of the two, and this process can achieve SO_2 removal efficiencies of 93 to 97 percent at a Ca/S molar ratio of 1.2 to 1.5. In this system, hydrated lime is injected directly into a CFB reactor along with water to obtain operation close to the adiabatic saturation temperature. The main features of this process include [6] simplicity and reliability with proven high availability; mild steel construction, which does not require lining; no moving parts or slurry nozzles; all solids that require handling are dry; water injection is independent of reagent feed; moderate space requirements; and greater flexibility to handle varying SO_2 and SO_3 concentrations. The CFB scrubber uses hydrated lime rather than the less expensive limestone commonly used in wet FGD technology processes. Additionally, due to a higher particulate matter concentration downstream of the scrubber, improvements to the particulate removal device, specifically an ESP, may be needed to meet the required particulate emission levels.

Fluidized-Bed Combustion

Fluidized-bed combustion is not an SO_2 control technology per se. However, this combustion technology does offer the capability to control SO_2 emissions during the combustion process rather than postcombustion where FGD systems need to be installed. Fluidized-bed combustion and the role of sorbents in controlling SO_2 are discussed in Chapters 5 and 7. The sorbents used in a fluidized-bed combustor are usually limestones, but sometimes dolomites (a double carbonate of calcium and magnesium) are used. The calcination and sulfation chemistry is discussed in Chapters 5 and 7.

Economics of Flue Gas Desulfurization

The costs of an FGD system are site specific and are made up of capital and operating costs. The capital costs of an FGD system depend on many factors, including the following [17]:

- Market conditions
- Geographical location
- Preparatory site work required
- Volume of flue gas to be scrubbed
- Concentration of SO_2 in the flue gas
- Extent of SO_2 removal required
- Quality of the products produced
- Process and wastewater treatment
- The need for flue gas reheat
- The degree of reliability and redundancy required
- The life of the system

The capital costs of wet FGD systems have been decreasing in the United States over the last 30 years. The prototype of flue gas desulfurization systems of the 1970s cost \$400/kW and experienced many problems [18]. With standardization, better chemistry, and improved materials, the cost of FGD systems in the 1980s dropped to \$275/kW. Additional developments, such as reduced redundancy, fewer modules, increased competition from foreign vendors, and use of well-engineered packages, have reduced capital cost to about \$100/kW today.

Operating costs are divided into variable and fixed costs [17]. Variable costs include the costs of the sorbents/reagents, costs associated with disposal or utilization of the by-products, and steam, power, and water costs. Fixed costs include costs of operating labor, maintenance, and administration. The operating and maintenance costs of an FGD system can be significant. According to Soud [6], operating and maintenance costs for the various subsystems in a pulverized coal-fired power generating facility with state-of-the-art environmental protection are 78 percent for the boiler/turbine/generator, 10 percent for the flue gas desulfurization system, 6 percent for a selective catalytic reduction system, 4 percent for wastewater treatment, and 2 percent for an ESP.

In the following sections, the costs associated with different FGD systems are briefly discussed. It must be noted, however, that it is difficult to compare costs (i.e., whether it is capital or operating) between systems because costs have many site-specific factors, are dependent on the age of the system, are influenced by economies of scale, and are higher for retrofit applications than for new installations.

Wet Processes

In a compilation of costs for various processes, Wu [17] reports that the capital and operating costs (for new installations) for a limestone wet FGD process are approximately $100/kW and $100/kWh, respectively, while the capital and operating costs are approximately $50/kW and $125/kWh, respectively, for magnesium-based systems, and $50/kW and 200/kWh, respectively, for sodium-based scrubbers. The variability in operating costs, though, is evident in data reported by Blythe and colleagues [19] and summarized in Figure 9.10, which shows the total operating costs for various systems for low-, medium-, and high-sulfur coals. Similarly, capital costs for recent LSFO retrofits varied from $180 to 348/kW [17].

Spray Dry Processes

The spray dry process generally has lower capital cost but higher and more expensive sorbent use, which is typically lime, than wet processes. This process is used mostly for small-to-medium plants firing low- to medium-sulfur coals and is preferable for retrofits [17]. For example, Blythe and colleagues [19] reported total

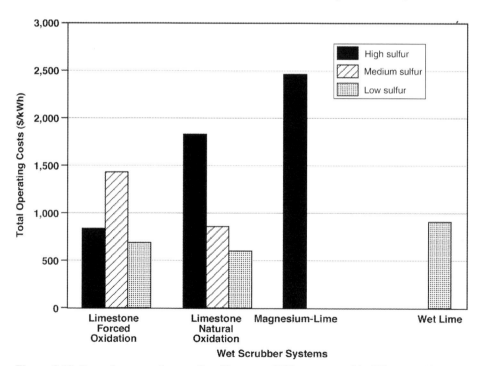

Figure 9.10 Operating costs for wet lime/limestone FGD systems with different sulfur content coals.

operating costs for lime spray dry processes of $836/kWh compared to $695/kWh for LSFO systems (see Figure 9.10) for low-sulfur coal applications.

Sorbent Injection Processes
The sorbent injection process, with a moderate SO_2 removal efficiency, has a relatively low capital cost. The capital cost for furnace sorbent injection is approximately 70-120$/kW [17].

Circulating Fluidized-Bed Processes
The circulating fluidized-bed processes have a relatively low capital cost, similar to that for the spray dry process. The process also has a low to moderate fixed operating cost, but its variable operating cost is relatively high [17].

Regenerative Processes
Regenerative processes generally have a high capital cost and power consumption [17]. The net variable operating cost is moderate because the processes produce saleable by-products. However, the fixed operating cost is substantially higher than other processes.

9.1.2 *Nitrogen Oxides*

Approximately 3 million short tons of nitrogen oxide (NO_x) were emitted from power plants in 2008 [4]. All sources affected by the EPA's Acid Rain Program (ARP) NO_x requirements reduced their combined NO_x emissions by 51 percent over the period 1995 to 2008 [4]. These reductions have been achieved, while the amount of fuel burned to produce electricity, as measured by heat input, increased 40 percent since 1990 [4]. In 2008, all 969 units that are subjected to the ARP NO_x program achieved compliance.

NO_x formation mechanisms are reviewed in this section, followed by technologies used to control NO_x emissions. Similar to the discussion on SO_2 control technologies, NO_x control technologies will focus on commercially available, commercially used systems with the focus on pulverized coal-fired boiler systems.

NOx Formation Mechanisms

NO_x formation from coal combustion was introduced in Chapter 4, and the discussion will be expanded in this section. NO_x formation during pulverized coal-combustion will be discussed, since this is the most widely used process for power generation.

The majority of nitrogen oxides emitted from power plants are in the form of nitric oxide (NO), with only a small fraction as nitrogen dioxide (NO_2) and nitrous oxide (N_2O). Collectively they are referred to as NO_x. NO originates from the coal-bound nitrogen and nitrogen in the air used in the combustion process and is produced through three mechanisms: thermal NO, prompt NO, and fuel NO. Fuel-bound nitrogen accounts for 75 to 95 percent of the total NO generated, while thermal and prompt NO accounting for the balance, with prompt NO being no more than 5 percent of the total NO [20].

The factors that influence NO_x emissions in pulverized coal-fired boilers can be generally categorized as boiler design, boiler operation, and coal properties [20]. However, NO_x formation is complex, and many parameters influence its production [21, 22]. Boiler design factors include boiler type, capacity, burner type, number and capacity of the burners, burner zone heat release, residence times, and presence of overfire air ports. Similarly, boiler operation factors include load, mills in operation, excess air level, burner tilt, and burner operation. Coal properties that influence NO_x production include volatiles release and nitrogen partitioning, ratio of combustibles-to-volatile matter, heating value, rank, and nitrogen content.

Thermal NO

Thermal NO formation involves the high temperature ($>2,370°F$) reaction of oxygen and nitrogen from the combustion air [20]. The principal reaction governing the formation of NO is the reaction of oxygen atoms formed from the dissociation of O_2 with nitrogen. These reactions, referred to as the Zeldovich mechanism, are

$$N_2 + O\bullet \leftrightarrow NO + N\bullet \tag{9.46}$$

$$N\bullet + O_2 \leftrightarrow NO + O\bullet \tag{9.47}$$

These reactions are sensitive to temperature, local stoichiometry, and residence time. High temperature is required for the dissociation of oxygen and to overcome the high activation energy for breaking the triple bond of the nitrogen molecule. These reactions dominate in fuel-lean high-temperature conditions. Under fuel-rich conditions, there are increased hydroxyl and hydrogen radical concentrations, which initiate the oxidation of the nitrogen radicals and at least one additional step should be included in this mechanism:

$$N\bullet + OH\bullet \leftrightarrow NO + H\bullet \tag{9.48}$$

Equations (9.46) through (9.48) are usually referred to as the extended Zeldovich mechanism. In addition, the following reactions can occur in fuel-rich conditions [20]:

$$H\bullet + N_2 \leftrightarrow N_2H \tag{9.49}$$

$$N_2H + O\bullet \leftrightarrow NO + NH\bullet \tag{9.50}$$

Thermal NO is of greater significance in the postflame region than within the flame. Consequently, several technologies have been developed for reducing thermal NO by lowering the peak temperature in the flame, minimizing the residence time in the region of the highest temperature, and controlling the excess air levels.

Prompt NO

Prompt NO is the fixation of atmospheric (molecular) nitrogen by hydrocarbon fragments in the reducing atmosphere in the flame zone [20]. The proposed mechanism is [22]

$$CH \bullet + N_2 \leftrightarrow HCN + N \bullet \tag{9.51}$$

$$HCN + O\bullet \leftrightarrow NH \bullet + CO \tag{9.52}$$

$$NH \bullet + O\bullet \leftrightarrow NO \bullet + H\bullet \tag{9.53}$$

The main reaction product of hydrocarbon radicals with N_2 is HCN with the amount of NO formed governed by the reactions of the nitrogen atoms with available radical species. In fuel-rich environments, therefore, the formation of N_2 is favored due to the reduced concentrations of hydroxyl and oxygen radical concentrations [20].

Fuel NO

Nitrogen in the coal, which typically ranges from 0.5 to 2.0 wt.%, occurs mainly as organically bound heteroatoms in aromatic rings or clusters [20]. Pyrrolic (5-membered ring) nitrogen is the most abundant form and contributes 50 to 60 percent of the total nitrogen. Pyridinic (6-membered ring) nitrogen comprises about 20 to 40 percent of the total nitrogen. The remaining 0 to 20 percent nitrogen is thought to be amine or quaternary nitrogen forms.

Coal nitrogen is first released during volatilization in the coal flame as an element in aromatic compounds referred to as tar. The tar undergoes pyrolysis to convert most of the nitrogen to HCN as well as some NH_3 and NH. Some nitrogen is expelled from the char as HCN and occasionally NH_3, but this occurs at a much slower rate than evolution from the volatiles. The partitioning of nitrogen between volatiles and char is important in NO_x formation.

NO formation proceeds along two paths [20]. The nitrogen from the char reacts with oxygen to form NO. The NH_3 and NH released from the volatile matter, and to a lesser extent the coal, react with oxygen atoms, forming NO. HCN is converted to NO via a pathway of hydrogen abstraction to form ammonia species and subsequently NO. Volatile nitrogen species can also be converted to nitrogen atoms through a series of fuel-rich pyrolysis reactions. Also, reactions between NO and volatile nitrogen species and carbon particles can result in the formation of nitrogen molecules:

$$C + NO \leftrightarrow \tfrac{1}{2}N_2 + CO \tag{9.54}$$

$$CH \bullet + NO \leftrightarrow HCN + O\bullet \tag{9.55}$$

In a fuel-rich environment, the main product of the reaction of NO with hydrocarbon radicals is HCN, which is then converted to N_2 in an oxygen-deficient environment. This is the basis for reburning, discussed later in this chapter, where a secondary hydrocarbon fuel is injected into combustion products containing NO.

Approximately 15 to 40 percent of the fuel nitrogen is converted to NO, and the formation of NO is influenced by stoichiometry, flame temperature, coal nitrogen content, and coal volatile matter content [20]. Approximately 25 percent of the char nitrogen is converted to NO. The reason for the fuel NO dominance (i.e., 75 to 95 percent of total NO_x production) is because the N-H and N-C bonds, common in fuel-bound nitrogen, are weaker than the triple bond in molecular nitrogen, which must be dissociated to produce thermal NO.

Nitrogen Dioxide and Nitrous Oxide

Small amounts of nitrogen dioxide (NO_2) and nitrous oxide (N_2O) are formed during coal combustion, but they comprise less than 5 percent of the total NO_x production. The oxygen levels are too low and the residence times are too short in high-temperature coal flames for much of the NO to be oxidized to NO_2. Nitrous oxide, however, can be formed in the early part of fuel-lean flames by gas phase reaction by the reactions [20]

$$O \bullet + N_2 \leftrightarrow N_2O \tag{9.56}$$

$$NH \bullet + NO \leftrightarrow N_2O + H \bullet \tag{9.57}$$

$$NCO \bullet + NO \leftrightarrow N_2O + CO \tag{9.58}$$

NO_x Control in Pulverized Coal Combustion

Technologies for control of NO_x emissions from pulverized coal-fired power plants can be divided into two groups:

1. Combustion modifications, where the NO_x production is reduced during the combustion process
2. Flue gas treatment, which removes the NO_x from flue gas following its formation

Sometimes the practice of injecting reducing agents to reduce NO_x to molecular nitrogen (N_2) is classified separately, but in this section it is included as a flue gas treatment.

Table 9.8 lists various NO_x control technologies with a summary of their attributes [23]. The abatement or emission control principal for these various control methods include reducing peak flame temperatures, reducing the residence time at peak flame temperatures, chemically reducing NO_x, oxidizing NO_x with subsequent absorption, removing nitrogen, using a sorbent, or a combination of these methods.

Reducing combustion temperature is accomplished by operating at non-stoichiometric conditions to dilute the available heat with an excess of fuel, air, flue gas, or steam [23]. The combustion temperature is reduced by using fuel-rich mixtures to limit the availability of oxygen; using fuel-lean mixtures to dilute energy input; injecting cooled oxygen-depleted flue gas into the combustion air to dilute energy; injecting cooled flue gas with the fuel; or injecting water or steam.

Table 9.8 NO_x Control Technologies

Technique	Description	Advantages	Disadvantages	Impacts	Applicability
Less excess air (LEA)	Reduces oxygen availability	Easy modification	Low NO_x reduction	High CO Flame length Flame stability	All fuels
Off stoichiometric **a.** Burners Out of Service (BOOS) **b.** Overfire Air (OFA)	Staged combustion	Low cost No capital cost for BOOS	**a.** Higher air flow for CO reduction **b.** High capital cost	Flame length Fan capacity Header pressure	All fuels Multiple burners required for BOOS
Low NO_x burner	Internal staged combustion	Low operating cost Compatible with FGR	Moderately high capital cost	Flame length Fan capacity Turndown capability	All fuels
Flue Gas Recirculation (FGR)	<30% flue gas recirculated with air, decreasing temperature	High NO_x reduction potential for low nitrogen fuels	Moderately high capital and operating costs Affects heat transfer and system pressures	Fan capacity Furnace pressure Burner pressure drop Turndown stability	All fuels
Water/steam injection	Reduces flame temperature	Moderate capital cost NO_x reduction similar to FGR	Efficiency penalty Fan power higher	Flame stability Efficiency penalty	All fuels
Reduced air preheat	Air not preheated, reduces flame temperature	High NO_x reduction potential	Significant efficiency loss (1%/40°F)	Fan capacity Efficiency penalty	All fuels

Continued

Table 9.8 NO_x Control Technologies—Cont'd

Technique	Description	Advantages	Disadvantages	Impacts	Applicability
Selective Catalytic Reduction (SCR)	Catalyst located in air flow, promotes reaction between ammonia and NO_x	High NO_x removal	Very high capital cost High operating cost Catalyst siting Increased pressure drop Possible water wash required	Space requirements Ammonia slip Hazardous materials Disposal	All fuels
Selective Non-Catalytic Reduction (SNCR) **a.** urea **b.** ammonia	Inject reagent to react with NO_x	**a.** Low capital cost Moderate NO_x removal Nontoxic chemical **b.** Low operating cost Moderate NO_x removal	**a.** Temperature dependent NO_x reduction less at lower loads **b.** Moderately high capital cost Ammonia storage, handling, injection system	**a.** Furnace geometry Temperature profile **b.** Furnace geometry Temperature profile	All fuels
Fuel reburning	Inject fuel to react with NO_x	Moderate cost Moderate NO_x removal	Extends residence time	Furnace temperature profile	All fuels (pulverized solid)
Combustion optimization	Change efficiency of primary combustion	Minimal cost	Extends residence time	Furnace temperature profile	All fuels

Inject oxidant	Chemical oxidant injected into flow	Moderate cost	Nitric acid removal	Add-on	All fuels
Oxygen instead of air	Uses oxygen as oxidizer	Moderate to high cost Intense combustion Eliminate thermal NO_x	Eliminate prompt NO_x Furnace alteration	Equipment to handle oxygen	All fuels
Ultra-low nitrogen fuel	Uses low-nitrogen fuel	Eliminates fuel NO_x No capital cost	Possible rise in operating cost	Minimal change	All ultra-low nitrogen fuels
Sorbent injection a. Combustion B. Duct to baghouse C. Duct to esp	Uses a chemical to absorb NO_x, or an adsorber to capture it, or to reduce it	Can control other pollutants as well as NO_x Moderate operating cost	Cost of sorbent Space for the sorbent storage and handling	Add-on	All fuels
Air staging	Admit air in separated stages	Reduce peak combustion temperature	Extend combustion to a longer residence time at lower temperature	Add ducts and dampers to control air Furnace modification	All fuels
Fuel staging	Admits fuel in separated stages	Reduce peak combustion temperature	Extend combustion to a longer residence time at lower temperature	Adds fuel injectors to other locations Furnace modification	All fuels

Source: From EPA (1999).

Reducing residence time at high combustion temperature is accomplished by restricting the flame to a short region to keep the nitrogen from becoming ionized. Fuel, steam, more combustion air, or recirculating flue gas is then injected immediately after this region. Chemically reducing NO_x removes oxygen from the nitrogen oxides. This is accomplished by reducing the valence level of nitrogen to zero after the valence has become higher. Oxidizing NO_x intentionally raises the valence of the nitrogen ion to allow water to absorb to it. This is accomplished by using a catalyst, injecting hydrogen peroxide, creating ozone within the air flow, or injecting ozone into the air flow.

Removing nitrogen from combustion is accomplished by removing nitrogen as a reactant either by using low nitrogen content fuels or using oxygen instead of air. The ability to vary coal nitrogen contents, however, is limited. Treatment of flue gas by injection sorbents such as ammonia, limestone, aluminum oxide, or carbon can remove NO_x and other pollutants. This type of treatment has been applied in the combustion chamber, flue gas, and particulate control device. Many of these methods can be combined to achieve a lower NO_x concentration than can be achieved alone by any one method. In some cases, technologies that are used to control other pollutants, such as SO_2, can also reduce NO_x.

Combustion Modifications

Primary NO_x control technologies involve modifying the combustion process. Several technologies have been developed and applied commercially, including the following:

- Low-NO_x burners
- Furnace air staging
- Flue gas recirculation
- Fuel staging (i.e., reburn)
- Process optimization

Options to control NO_x during combustion and their effects are different for new and existing boilers. For new boilers, combustion modifications are easily made during construction, whereas for existing boilers, viable alternatives are more limited. Modifications can be complicated and unforeseen problems may arise. When combustion modifications are made, it is important to avoid adverse impacts on boiler operation and also the formation of other pollutants such as N_2O or CO. Issues pertaining to low-NO_x operation include the following:

- Safe operation (e.g., stable ignition over the desired load range)
- Reliable operation to prevent corrosion, erosion, deposition, and uniform heating of the tubes
- Complete combustion to limit formation of other pollutants such as CO, polyorganic matter, or N_2O
- Minimal adverse impact on the flue gas cleaning equipment
- Low maintenance costs

Combustion modification technologies redistribute the fuel and air to slow mixing, reduce the availability of oxygen in the critical NO_x formation zones, and decrease the amount of fuel burned at peak flame temperatures. In addition, reburning chemically destroys the NO_x formed by hydrocarbon radicals during the combustion process. The commercially applied technologies are discussed in detail in the following sections. One technology listed in Table 9.8, low excess air (LEA), is the simplest of the combustion control strategies, but it is not discussed in detail because it has limited success in coal-fired applications (i.e., 1 to 15 percent). In this technique, excess air levels are reduced until there are adverse impacts on CO formation and flame length and stability. Similarly, a technique called burners out of service (BOOS) has limited success with coal and is not discussed in detail. In this technique, the fuel flow to selected burner is stopped, but airflow is maintained to create staged combustion in the furnace. The remaining burners operate fuel-rich, which limits oxygen availability, lowering peak flame temperatures, and reducing NO_x formation. The unreacted products combine with the air form the burners out of service to complete burnout before exiting the furnace.

Low-NO_x Burners Prior to concern with NO_x emissions in the early 1970s, coal burners were designed to provide highly turbulent mixing and combustion at peak flame temperature to ensure high combustion efficiency, a condition that is ideal for NO_x formation [24]. In 1971, the industry began developing low-NO_x burners for coal-fired boilers with the promulgation of New Source Performance Standards. By the mid-1970s, low-NO_x burners were being demonstrated and commercial operation started in the late 1970s [25]. They have undergone considerable improvements in design spurred by the 1990 Clean Air Act Amendments Title IV, Phase II Acid Rain regulations, and Title I Ozone regulations [26, 27]. The technology is well proven for NO_x control in both wall- and tangentially fired boilers and is commercially available, with a significant number of them installed worldwide.

Low-NO_x burners work under the principle of staging the combustion air within the burner to reduce NO_x formation. Rapid devolatilization of the coal particles occurs near the burner in a fuel-rich, oxygen-starved environment to produce NO. NO_x formation is suppressed because oxygen molecules are not available to react with the nitrogen released from the coal and present in the air, and the flame temperature is reduced. Hydrocarbon radicals that are generated under the substoichiometric conditions then reduce the NO that is formed to N_2. The air required to complete the burnout of the coal is added after the primary combustion zone where the temperature is sufficiently low so that additional NO_x formation is minimized.

Larger and more branched flames are produced by staging the air [20]. This flame structure limits coal and air mixing during the initial devolatilization stage while maximizing the release of volatiles from the coal. The more volatile nitrogen that is released with the volatiles and the longer the residence time in the fuel-rich zone, the lower the amount of fuel NO that is produced. An oxygen rich layer is produced around the flame that aids in carbon burnout.

An example of this concept is shown in Figure 9.11, which is a schematic of a low-NO_x burner (i.e., Ahlstom Power's Radially Stratified Fuel Core burner)

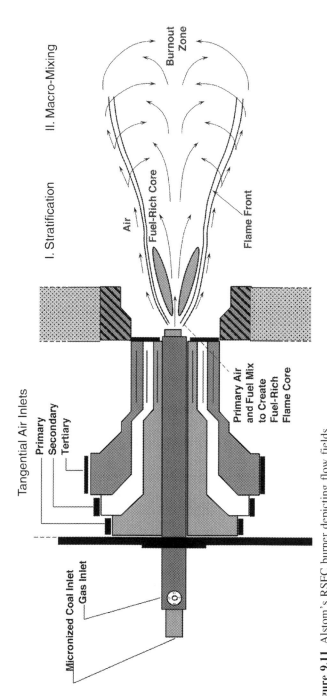

Figure 9.11 Alstom's RSFC burner depicting flow fields.
Source: From Patel et al. (1996).

Secondary Air Inlet   **Tertiary Air Inlet**

Dampers for Controlling Amount of Air into Each Zone   **Primary Air Inlet**   **Dampers to Control Tertiary Air Swirl Number**

Figure 9.12 Photograph of the RSFC burner showing internal components.

illustrating a typical flow field emanating from it [28]. A photograph of the burner, which is a 20 million Btu/h prototype used for developmental work at Penn State prior to its commercialization and worldwide deployment, is shown in Figure 9.12. The photograph shows the various dampers and air scoops for channeling and controlling the quantity and degree of swirl of the various air streams [29].

Low-NO$_x$ burners are designed to accomplish the following [30]:

- Maximize the rate of volatiles evolution and total volatile yield from the fuel with the fuel nitrogen evolving in the reducing part of the flame.
- Provide an oxygen-deficient zone where the fuel nitrogen is evolved to minimize its conversion to NO$_x$ but ensuring there is sufficient oxygen to maintain a stable flame.
- Optimize the residence time and temperature in the reducing zone to minimize the fuel nitrogen conversion to NO$_x$.
- Maximize the char residence time under fuel rich conditions to reduce the potential for NO$_x$ formation from the nitrogen remaining in the char after devolatilization.
- Add sufficient air to complete combustion.

All low-NO$_x$ burners employ the air-staging principle but the designs vary widely between manufacturers. All of the major boiler manufacturers have one or more versions of low-NO$_x$ burners employed in boilers throughout the world. Mitchell [20] reported that there were more than 370 units worldwide fitted with low-NO$_x$ burners with a total generating capacity of more than 125 GW prior to 1998. The number of units installing low-NO$_x$ burners has increased significantly as DOE reports that low-NO_x burners are currently found on more than 75 percent of United States coal-fired power capacity [31]. This is significant as there are 1,470 coal-fired steam-electric generators with nameplate capacity of 336 GW producing more than 1,994 billion kWh of electricity (as of 2008) [2, 32].

Low-NO_x burners, based on air-staging alone, are capable of achieving 30 to 60 percent NO_x reduction. In addition, they should perform in such ways that the following occurs [30]:

• The overall combustion efficiency is not significantly reduced.
• Flame stability and turndown limits are not impaired.
• The flame has an oxidizing envelope to minimize the potential for high temperature corrosion at the furnace walls.
• Flame length is compatible with furnace dimensions.
• The performance should be acceptable for a wide range of coals.

The major concern with low-NO_x burners is the potential for reducing combustion efficiency and thereby increasing the unburned carbon level in the fly ash. An increase in the unburned carbon level will lower the fly ash resistivity, which can reduce the efficiency of an ESP. In addition, it may also affect the sale of the ash. Some operating parameters that can be adjusted to mitigate the impact of the unburned carbon are the following [30]:

• Fire coal with high reactivity and high volatile matter content
• Reduce the size of the coal particles
• Balance coal distribution to the burners
• Use advanced combustion control systems

Furnace Air Staging A technique to stage combustion is done by installing secondary and even tertiary overfire air (OFA) ports above the main combustion zone. This is a well-proven, commercially available technology for NO_x reduction at coal-fired power plants and is applicable to both wall and tangentially fired boilers [30].

When OFA is employed, 70 to 90 percent of the combustion air is supplied to the burners with the coal (i.e., primary air), with the balance introduced to the furnace above the burners (i.e., overfire air). The primary air and coal produce a relatively low-temperature, oxygen-deficient, fuel-rich environment near the burner, which reduces the formation of fuel-NO_x. The overfire air is injected above the primary combustion zone, producing a relatively low-temperature secondary combustion zone that limits the formation of thermal-NO_x.

Overfire air in combustion with low-NO_x burners can reduce NO_x emissions by 30 to 70 percent. Advanced OFA systems such as separated overfire air (SOFA), where the overfire air is introduced some distance above the burners, and close-coupled overfire air (CCOFA), where the ovefire air nozzles are immediately above the burners, can achieve higher NO_x reduction efficiency [20]. Mitchell [20] reports that furnace air staging is used in about 300 pulverized coal-fired units, with a total generating capacity of more than 100 GW. There are a number of advanced overfire air systems commercially available with designs varying between suppliers [30].

Furnace air staging can increase unburned carbon levels in the ash by 35 to 50 percent, with the degree of increase being dependent on the reactivity of the coal used [30]. In addition, operational problems can be experienced including waterwall corrosion, changes in slagging and fouling patterns, and a loss in steam temperature.

Flue Gas Recirculation Flue gas recirculation (FGR) involves recirculating part of the flue gas back into the furnace or the burners to modify conditions in the combustion zone by lowering the peak flame temperature and reducing the oxygen concentration, thereby reducing thermal NO_x formation. FGR has been used commercially for many years at coal-fired units. However, unlike gas- and oil-fired boilers, which can achieve high-NO_x reduction, coal-fired boilers typically realize less than 20 percent NO_x reduction due to a relatively low contribution of thermal NO_x to total NO_x.

In conventional FGR applications, 20 to 30 percent of the flue gas is extracted from the boiler outlet duct upstream of the air heater (at $\approx$570–750°F) and is mixed with the combustion air. This process reduces thermal NO_x formation without any significant effect on fuel NO_x.

A major consideration of FGR is the impact on boiler thermal performance [30]. The reduced flame temperature lowers heat transfer, potentially limiting the maximum heating capacity of the unit, which results in a reduction in steam generating capacity.

Fuel Staging (Reburn) Reburn is a comparatively new technology that combines the principles of air and fuel staging. In this technology, a reburn fuel, which can be coal, oil, gas, orimulsion, biomass, coal-water mixtures, and so on, is used as a reducing agent to convert NO_x to N_2. The process does not require modifications to the existing main combustion system and can be used on wall-, tangential-, and cyclone-fired boilers.

Reburn is a combustion hardware modification in which the NO_x produced in the main combustion zone is reduced downstream in a second combustion zone (i.e., the reburn zone). This, in turn, is followed by a zone where overfire air is introduced to complete burnout. This is illustrated in Figure 9.13 [33].

In the primary combustion zone, the burners are operated at a reduced firing rate with low excess air (stoichiometry of 0.9 to 1.1) to produce lower fuel and thermal NO_x levels. The reburn fuel, which can be 10 to 30 percent of the total fuel input, on a heat input basis, is injected above the main combustion zone to create a fuel-rich zone (stoichiometry of 0.85–0.95) [33]. In this zone, most of the NO_x reduction occurs with hydrocarbon radicals formed in the reburn zone reacting with the NO_x forming N_2 and water vapor. The temperature in this zone must be greater than 1,800°F. The remaining combustion air is injected above the reburn zone to produce a fuel-lean burnout zone.

While reburn technology is considered relatively new and numerous pilot-scale tests and full-scale demonstrations have been conducted, the concept was proposed in the late 1960s [34]. The concept was based on the principle of Myerson and colleagues [35] that CH fragments can react with NO.

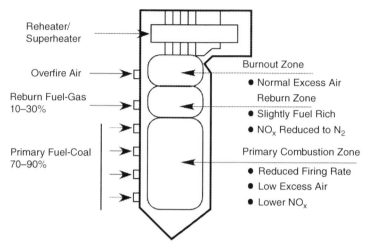

Figure 9.13 The reburn process.
Source: From EPA (1996).

The major chemical reactions for the reburn process are [33]

$$\text{Hydrocarbon fuel} \xrightarrow[\text{Heat\&O_2 deficiency}]{} \bullet CH_2 \tag{9.59}$$

where hydrocarbon radicals are produced due to the pyrolysis of the fuel in the oxygen-deficient, high-temperature reburn zone. The hydrocarbon radicals then mix with the combustion gases from the primary combustion zone:

$$\bullet CH_3 + NO \rightarrow HCN + H_2O \tag{9.60}$$

$$N_2 + \bullet CH_2 \rightarrow \bullet NH_2 + HCN \tag{9.61}$$

$$\bullet H + HCN \rightarrow \bullet CN + H_2 \tag{9.62}$$

The radicals then react with the NO to form molecular nitrogen:

$$NO + \bullet NH_2 \rightarrow N_2 + H_2O \tag{9.63}$$

$$NO + \bullet CN \rightarrow N_2 + CO \tag{9.64}$$

$$2NO + 2CO \rightarrow N_2 + 2CO_2 \tag{9.65}$$

An oxygen-deficient atmosphere is critical for Eqs. (9.60) through (9.62) to occur. If oxygen levels are high, the NO_x reduction reactions will not occur, and the following will predominate:

$$CN + O_2 \rightarrow CO + NO \tag{9.66}$$

$$NH_2 + O_2 \rightarrow H_2O + NO \tag{9.67}$$

To complete the combustion process, air is introduced above the reburn zone. Some NO_x is formed from conversion of HCN and ammonia compounds; however, the net effect is to significantly reduce the total quantity of NO_x emitted from the boiler. The reactions with HCN and ammonia are

$$HCN + \tfrac{5}{4}O_2 \rightarrow NO + CO + \tfrac{1}{2}H_2O \tag{9.68}$$

$$NH_3 + \tfrac{5}{4}O_2 \rightarrow NO + \tfrac{3}{2}H_2O \tag{9.69}$$

$$HCN + \tfrac{3}{4}O_2 \rightarrow \tfrac{1}{2}N_2 + CO + \tfrac{1}{2}H_2O \tag{9.70}$$

$$NH_3 + \tfrac{3}{4}O_2 \rightarrow \tfrac{1}{2}N_2 + \tfrac{3}{2}H_2O \tag{9.71}$$

Reburn offers the advantages of being able to operate over a wide range of NO_x reduction using a variety of reburn fuels. A reburn system can be varied from relatively low levels of reduction, 25 to 30 percent, using an overfire air system without any reburn fuel, to about 70 percent reduction when reburn fuel is added [20, 30]. This allows for fine-tuning to meet emissions limits.

Concerns regarding the use of reburn technology are similar to those for other combustion modification processes. This includes concerns about incomplete combustion (i.e., CO and hydrocarbon production, and unburned carbon in the fly ash), changes in slagging and fouling characteristics, different ash characteristics and fly ash loadings, corrosion of boiler tubes in reducing atmospheres, higher fan power consumption, and pulverizer constraints (if pulverized coal is used as the reburn fuel).

Cofiring Cofiring is the practice of firing a supplementary fuel, such as coal-water slurry fuel (CWSF) or biomass, with a primary fuel (i.e., coal) in the same burner or separately but into the main combustion zone. This technology was originally developed to utilize opportunity fuels; however, various levels of NO_x reduction were achieved and provide an option for NO_x reduction without investing in a postcombustion system when the emissions are near the regulatory requirements. This technique is not currently used as a commercial means for NO_x reduction, but it is briefly discussed in this section because several demonstrations of this technology have been conducted, with a few still ongoing. In addition, it is considered a viable option for NO_x trimming, especially if used in conjunction with legislation that mandates a percentage of electricity be generated from renewable/sustainable sources. Such legislation has been seriously discussed in the United States and has been included in congressional bills although they have not passed to date.

CWSF technology was originally developed as a fuel oil replacement with considerable research and development from the late 1970s to the late 1980s. During the late 1980s and early 1990s, coal suppliers and coal-fired utilities began to evaluate the production of CWSF using bituminous coal fines from coal cleaning circuits in an effort to reduce dewatering/drying costs and/or to recover and utilize low-cost impounded coal fines [36, 37]. This marked a philosophical change in the driving force behind utilizing CWSF in the United States as well as the coal-water slurry fuel characteristics of these two fuel types, since cofire CWSFs are quite different from fuel oil replacement CWSFs; that is, they have low solids content (50 percent) and no additive package to wet the coal, provide stability, and modify rheology compared to high solids content ($\approx$70 percent) with an expensive additive package, respectively.

Several companies and universities performed extensive testing, culminating with waste impoundment characterizations and several utility demonstrations in pulverized coal-fired boilers (including both wall- and tangentially fired units) and cyclone-fired boilers. Funding for these demonstrations was provided by industry, U.S. DOE, Electric Power Research Institute, and state agencies. Penn State provided fuel support in all but one of these demonstrations, which is summarized in a CWSF preparation and operation manual prepared by Morrison and colleagues [38], where the CWSFs were being developed to provide coal preparation plants a means for utilizing hard to dewater fines, cleaning up waste coal impoundments thereby reducing coal mine liability, and supplying utilities with a low-cost fuel that also serves as a low-cost NO_x reduction technology. NO_x reductions varying from about 11 percent in cyclone-fired boilers [39] to about 30 percent in wall-fired boilers [40, 41] to about 35 percent in tangentially fired boilers [42] were achieved. Several mechanisms were responsible for the NO_x reduction including lower flame temperature from the addition of the water, staged combustion from cofiring in low-NO_x burners, and the CWSF acting as a reburn fuel when injected in upper level burners.

Biomass cofiring has been demonstrated and deployed at a number of power plants in the United States and Western Europe using a variety of materials including sawdust, urban wood waste, switchgrass, straw and other similar materials [43]. Biomass fuels have been cofired with all ranks of coal: bituminous and subbituminous coals and lignites. The benefits of biomass cofiring include reduced NO_x, fossil CO_2, SO_2, and mercury emissions.

Cofiring biomass, particularly sawdust and urban wood waste, but also switchgrass to a lesser extent, in large-scale pulverized coal- and cyclone-fired units has been demonstrated at several utilities, with seven commercial installations in the United States [44, 45]. Many of the demonstrations were conducted to achieve NO_x reduction, which can vary significantly but can be as high as almost 35 percent. Tillman [43] noted that the dominant mechanism for NO_x reduction is to support deeper staging of combustion when staging has not been particularly extensive. When biomass can introduce or accentuate staging by the early release of volatile matter, then NO_x reduction can be significant [43, 46]. A secondary mechanism for NO_x reduction is the influence of cofiring on furnace exit gas temperature (FEGT). Data indicate that cofiring has a minimal impact on flame temperatures but can have a pronounced impact on FEGT, thereby reducing NO_x emissions. A third influence is the reduction in fuel nitrogen content when a low-nitrogen fuel such as sawdust is used.

Process Optimization Several software packages have been developed, or are under development, that apply optimization procedures to the distributed control system of the boiler to provide tighter control of plant operation parameters [30]. The combustion process is optimized, resulting in lower NO_x emissions and improved boiler efficiency, while maintaining safe, reliable, and consistent unit operation. Also, combustion optimization approaches have been developed where advanced computational and experimental approaches are used to make design and operational modifications to the process equipment and boiler as a whole [47].

Many software packages are under development and are in use, with some of the main ones including the ULTRAMAX Method, Generic NO_x Control Intelligent System (GNOCIS/GNOCIS Plus), Boiler OP, QuickStudy, and Smart Burn [30, 47]. The use of these packages has resulted in NO_x reductions of 10 to 40 percent, reduced unburned carbon levels by 25 to 50 percent, increased boiler efficiencies by 1 to 3 percent, and increased heat rates by 0.5 to 5 percent.

Flue Gas Treatments

Flue gas treatment technologies are postcombustion processes to convert NO_x to molecular nitrogen or nitrates. The two primary strategies that have been developed for postcombustion control and are commercially available are selective catalytic reduction (SCR) and selective non-catalytic reduction (SNCR). There are additional concepts under development, including combining SCR and SNCR technologies (known as hybrid SCR/SNCR) and rich reagent injection; however, these are not extensively used at this time. Of these technologies, SCR is being identified by utilities as the strategy to meet stringent NO_x requirements. These technologies are discussed in the following sections, with an emphasis on SCR.

Selective Catalytic Reduction Selective catalytic reduction of NO_x using ammonia (NH_3) as the reducing gas was patented in the United States by Englehard Corporation in 1957 [48]. This technology can achieve NO_x reductions in excess of 90 percent and is widely used in commercial applications in Western Europe and Japan, which have stringent NO_x regulations, and is becoming the postcombustion technology of choice in the United States and China. Stringent NO_x regulations in Western Europe essentially mandate the installation of secondary NO_x reduction systems, the majority of which is SCR, with only a few boilers using the selective noncatalytic reduction process [49].

Germany was the first country to install SCRs and has been installing them since 1986. Similarly, SCR technology was introduced into commercial service in Japan in 1980 and has been applied to coal-fired generating since. United States utilities initially deployed SCR for coal-fired units for new and retrofit applications in 1991 and 1993, respectively [50]. As of early 2009, more than 200 SCR units have been installed on fossil fuel-fired power generation facilities with overall capacity greater than 100 GW. China has a very large power plant construction program underway [51].

It is expected that by 2020 more than 900,000 MW of coal-fired boilers will be in operation, and more than 60 percent will be fitted with SCR. This 540,000 MW of SCR compares to a worldwide total today of 300,000 MW. The forecast for SCR at coal plants in the United States in 2020 is 290,000 MW. Germany will have less than 20 percent as much SCR as China, and no other country will have even 10 percent as much. In 2010, China is expected to retrofit 11,000 MW of existing plants with SCR and will start up 30,000 MW of new coal units with SCR for an increase of 41,000 MW. This is more SCR capacity than exists in any country except the United States and Germany.

The SCR process uses a catalyst at approximately 570°F to 750°F to facilitate a heterogeneous reaction between NO_x and an injected reagent, vaporized ammonia, to produce nitrogen and water vapor. Ammonia chemisorbs onto the active sites on the catalyst. The NO_x in the flue gas reacts with the adsorbed ammonia to produce nitrogen and water vapor. The principal reactions are as follows [52]:

$$4NO + 4NH_3 + O_2 \rightarrow 4N_2 + 6H_2O \tag{9.72}$$

$$2NO_2 + 4NH_3 + O_2 \rightarrow 3N_2 + 6H_2O \tag{9.73}$$

A small fraction of the sulfur dioxide is oxidized to sulfur trioxide over the SCR catalyst. In addition, side reactions may produce the undesirable by-products ammonium sulfate $((NH_4)_2SO_4)$ and ammonium bisulfate (NH_4HSO_4), which can cause plugging and corrosion of downstream equipment. These side reactions are as follows [48]:

$$SO_2 + \tfrac{1}{2}O_2 \rightarrow SO_3 \tag{9.2}$$

$$2NH_3 + SO_3 + H_2O \rightarrow (NH_4)_2SO_4 \tag{9.74}$$

$$NH_3 + SO_3 + H_2O \rightarrow NH_4HSO_4 \tag{9.75}$$

The three SCR system configurations for coal-fired boilers are known as high-dust, low-dust, and tail-end systems. These are shown schematically in Figure 9.14 [52]. In a high-dust configuration, the SCR reactor is placed upstream of the particulate removal device between the economizer and the air preheater. This configuration (also referred to as hot side, high dust) is the most commonly used, particularly with dry-bottom boilers [32] and is the principle type planned for the U.S. installations [49]. In this configuration, the catalyst is exposed to the fly ash and chemical compounds present in the flue gas that have the potential to degrade the catalyst by ash erosion and chemical reactions (i.e., poisoning). However, these can be addressed by proper design, as evidenced by the extensive use of this configuration.

In a low-dust installation, the SCR reactor is located downstream of the particulate removal device. This configuration (also referred to as hot side, low dust) reduces the degradation of the catalyst by fly ash erosion. However, this configuration requires a costly hot-side ESP or a flue gas reheating system to maintain the optimum operating temperature.

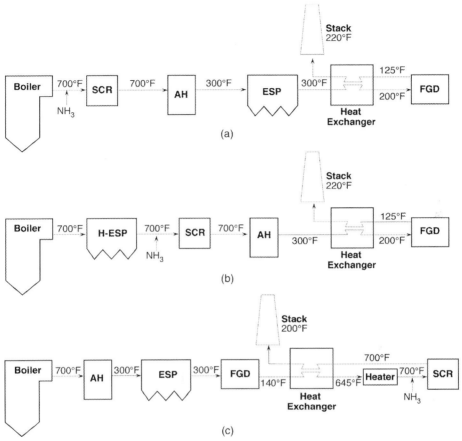

Figure 9.14 SCR configurations with typical system temperatures: (a) high-dust system, (b) low-dust system, and (c) tail-end system.
Source: From EPA (1997).

In tail-end systems (also referred to as cold side, low dust), the SCR reactor is installed downstream of the FGD unit. It may be used mainly in wet-bottom boilers and also on retrofit installations with space limitations [30]. However, this configuration is typically more expensive than the high-dust configuration due to flue gas reheating requirements. This configuration does have the advantage of longer catalyst life and the use of more active catalyst formulations to reduce overall catalyst cost.

Several issues must be considered in the design and operation of SCR systems, including coal characteristics, catalyst and reagent selections, process conditions, ammonia injection, catalyst cleaning and regeneration, low-load operation, and process optimization [30]. Coals with high sulfur in combination with significant quantities of alkali or alkaline earth elements, arsenic, or phosphorus in the ash can severely deactivate a catalyst and reduce its service life. In addition, the SO_3 can react with residual ammonia resulting in ammonium sulfate deposition in the air preheater and loss of performance.

The two leading geometries of SCR catalysts are honeycomb and plate [48]. The honeycomb form usually is an extruded ceramic with the catalyst either incorporated throughout the structure (homogenous) or coated on the substrate. In the plate geometry, the support material is generally coated with catalyst. The catalyst commonly consists of a vanadium pentoxide active material on a titanium dioxide substrate.

For optimum SCR performance, the reagent must be well mixed with the flue gas and in direct proportion to the amount of NO_x reaching the catalyst. Anhydrous ammonia has been commonly used as reagent, accounting for more than 90 percent of current world selective catalytic reduction applications [30]. It dominates planned installations in the United States, although numerous aqueous systems will be installed. Recently, urea-based processes are being developed to address utilizing anhydrous ammonia, which is a hazardous and toxic chemical. When using urea $CO(NH_2)_2$, it produces ammonia, which is the active reducing agent, by the following reactions:

$$NH_2 - CO - NH_2 \rightarrow NH_3 + HNCO \tag{9.76}$$

$$HNCO + H_2O \rightarrow NH_3 + CO_2 \tag{9.77}$$

During the operation of the SCR, the catalyst is deactivated by fly ash plugging, catalyst poisoning, and/or the formation of binding layers. The most common method of catalyst cleaning has been the installation of steam soot blowers, although acoustic cleaners have been successfully tested. Once the catalyst has been severely deactivated, it is conventional practice to add additional catalyst or replace it; however, several regeneration techniques have evolved over the last few years that provide extended service life for catalysts [30].

Low-load boiler operation can be problematic with SCR operation, specifically with high-sulfur coals. There is a minimum temperature below which the SCR should not be operated; therefore, system modifications such as economizer bypass to raise the SCR temperature during low-load operation may be required [30].

Selective Non-Catalytic Reduction SNCR is a proven, commercially available technology that has been applied since 1974 with more than 300 systems installed worldwide on various combustion sources, including coal-fired utility applications [30]. The SNCR process involves injecting nitrogen-containing chemicals into the upper furnace or convective pass of a boiler within a specific temperature window without the use of an expensive catalyst. Several different chemicals can be used to selectively react with NO in the presence of oxygen to form molecular nitrogen and water, but the two most common are ammonia and urea. Other chemicals that have been tested in research include amines, amides, amine salts, and cyanuric acid. In recent years, urea-based reagents such as dry urea, molten urea, or urea solution have been increasingly used, replacing ammonia at many plants because anhydrous ammonia is the most toxic and requires strict transportation, storage, and handling procedures [30]. The main reactions when using ammonia or urea are, respectively,

$$4NO + 4NH_3 + O_2 \rightarrow 4N_2 + 6H_2O \qquad (9.72)$$

$$4NO + 2CO(NH_2)_2 + O_2 \rightarrow 4N_2 + 2CO_2 + 6H_2O \qquad (9.78)$$

A critical issue is finding an injection location with the proper temperature window for all operating conditions and boiler loads. The chemicals then need to be adequately mixed with the flue gases to ensure maximum NO_x reduction without producing too much ammonia. Ammonia slip from an SNCR can affect downstream equipment by forming ammonium sulfates.

The temperature window varies for most of the reducing chemicals used, but generally is between 1,650 and 2,100°F. Ammonia can be formed below the temperature window, and the reducing chemicals can actually form more NO_x above the temperature window. Ammonia has a lower operating temperature than urea: 1,560 to 1,920°F compared to 1,830 to 2,100°F, respectively. Enhancers such as hydrogen, carbon monoxide, hydrogen peroxide (H_2O_2), ethane (C_2H_6), light alkanes, and alcohols have been used in combination with urea to reduce the temperature window [53]. Several processes use proprietary additives with urea in order to reduce NO_x emissions [54].

The efficiency of reagent utilization is significantly less with SNCR than with SCR. In commercial SNCR systems, the utilization is typically between 20 and 60 percent; consequently, usually three to four times as much reagent is required with SNCR to achieve NO_x reduction similar to that of SCR. SNCR processes typically achieve 20 to 50 percent NO_x reduction, with stoichiometric ratios of 1.0 to 2.0.

The major operational impacts of SNCR include air preheater fouling, ash contamination, N_2O emissions, and minor increases in heat rate. A major plant impact of SNCR is on the air preheater, where residual ammonia reacts with the SO_3 in the flue gas to form ammonium sulphate and bisulphate (see Eqs. (9.2), (9.74), and (9.75)), causing plugging and downstream corrosion. High levels of ammonia slip can contaminate the fly ash and reduce its sale or disposal. Significant quantities of N_2O can be formed when the reagent is injected into areas of the boiler that are below the SNCR optimum operating temperature range. Urea injection tends to produce a higher level of N_2O compared to ammonia. The unit heat rate is increased slightly due to the latent heat losses from vaporization of injected liquids and/or increased power requirements for high-energy injection systems. The overall efficiency and power losses normally range from 0.3 to 0.8 percent [30].

Hybrid SNCR/SCR SCR generally represents a relatively high capital requirement, whereas SNCR has a high reagent cost. A hybrid SNCR/SCR system balances these costs over the life cycle for a specific NO_x reduction level, provides improvements in reagent utilization, and increases overall NO_x reduction [30]. However, there is limited experience with these hybrid systems, since full-scale power plant operation to date has only been in demonstrations. They are discussed here because they have demonstrated NO_x reductions as high as 60 to 70 percent.

In a hybrid SNCR/SCR system, the SNCR operates at lower temperatures than stand-alone SNCRs, resulting in greater NO_x reduction but also higher ammonia slip. The residual ammonia feeds a smaller-sized SCR reactor, which removes the ammonia slip and decreases NO_x emissions further. The SCR component may achieve only 10 to 30 percent NO_x reduction, with reagent utilization as high as 60 to 80 percent [30]. Hybrid SNCR/SCR systems can be installed in different configurations, including the following [30]:

- SNCR with conventional reactor-housed SCR
- SNCR with in-duct SCR, which uses catalysts in existing or expanded flue gas ductwork
- SNCR with catalyzed air preheater, where catalytically active heat transfer elements are used
- SNCR with a combination of in-duct SCR and catalyzed air heater

Rich Reagent Injection Cyclone burners, with their turbulent and high-temperature environment, are conducive for NO_x production. Lower-cost methods than installing SCRs to reduce NO_x production in cyclone-fired boilers have been tested, such as CWSF or biomass cofiring, while others are under development. One such process currently under development is the rich reagent injection (RRI) process. It involves injection of amine reagents in the fuel-rich zone above the main combustion zone at temperatures of 2,370 to 3,100°F. NO_x in the flue gas is converted to molecular nitrogen, and reductions of 30 percent have been achieved. The capital costs for an RRI system are consistent with those of SNCR, but operating costs are expected to be two to three times that of SNCR due to increased reagent usage.

NO_x Control in Fluidized-Bed Combustion

The fluidized-bed combustion (FBC) process, which is described in Chapter 5, inherently produces lower NO_x emissions due to its lower operating temperature (i.e., bed temperature of about 1,600°F). Also, the bed is a reducing region where available oxygen is consumed by carbon, thereby reducing ionization of nitrogen. Additional combustion modifications or flue gas treatment for NO_x control, discussed previously in this chapter, can also be employed. Techniques currently used for FBC include reducing the peak temperature by flue gas recirculation (FGR), natural gas reburning (NGR), overfire air (OFA), low excess air (LEA), and reduced air preheat [23]. Postcombustion control is also used, including SCR, SNCR, and fuel reburning, which achieve 35 to 90 percent NO_x reduction. Also, low-nitrogen fuel can be used (e.g., sawdust), thereby reducing the amount of fuel nitrogen available. Injecting sorbents into the combustion chamber or in the ducts can reduce NO_x by 60 to 90 percent [23].

NO_x Control in Stoker-Fired Boilers

NO_x control in stokers, specifically traveling grate and spreader stokers, include abatement methods to reduce the peak temperature, reduce the residence time at peak temperature, chemically reduce the NO_x, use low-nitrogen fuels, and inject a sorbent [23]. In traveling grate stokers, the peak temperature can be reduced by FGR, NGR, combustion optimization, OFA, LEA, water or steam injection, and reduced air preheat, thereby achieving 35 to 50 percent NO_x reduction. Air or fuel staging, which reduces the residence time at peak temperature, can achieve 50 to 70 percent NO_x reduction while using SCR, SNCR, or fuel reburning technologies can achieve 55 to 80 percent NO_x reduction. Sorbent injection, which can achieve 60 to 90 percent NO_x reduction, and using fuels with low nitrogen content are technologies also employed. NO_x technologies used for spreader stokers are similar to traveling grate stokers with slightly different results. FGR, natural gas reburning, low-NO_x burners, combustion optimization, OFA, LEA, water or steam injection, and reduced air preheat temperature are control options to reduce peak temperatures and can achieve 50 to 65 percent NO_x reduction. Air or fuel staging or steam injection, which reduces the residence time at peak temperature, can achieve 50 to 65 percent NO_x reduction, while using SCR, SNCR, or fuel reburning technologies achieve 35 to 80 percent NO_x reduction. Additional NO_x reduction technologies include sorbent injection, which can achieve 60 to 90 percent reduction, and using lower nitrogen fuels.

Economics of NO_x Reduction/Removal

The costs for NO_x reduction/removal techniques are site and performance specific, thereby making it difficult to compare generalized system costs. These techniques depend on several factors, including degree of retrofit difficulty, unit size, uncontrolled NO_x levels, and required NO_x reduction [30]. This section will summarize costs for the various systems using published data.

Low-NO_x Burners
Wu [30] reports that the capital cost for a low-NO_x burner is in the range of $650 to $8,300/MM Btu. The operating cost can range from $340 to $1,500/MM Btu. The levelized cost can vary from $240 to $4,300/ton of NO_x removed, with the average cost closer to the lower end of the range [55].

Furnace Air Staging
The costs for furnace air staging are similar to those for low-NO_x burners [30]. The capital cost is about $8 to $23/kW, and the levelized cost ranges from $110 to $210 per short ton of NO_x removed. If furnace air staging is combined with low-NO_x burners, the capital cost will increase to $15 to $30/kW, while the levelized cost remains relatively unchanged. Retrofits of furnace air staging in tangentially fired boilers are generally more expensive than those in wall-fired boilers, $5 to $11/kW and $11 to $23/kW, respectively.

Flue Gas Recirculation
The capital costs for conventional fuel gas recirculation (FGR) is similar to that for low-NO_x burners and OFA: $8 to $35/kW [55]. However, capital costs of induced FGR, a design derivative of conventional forced flue gas desulfurization, has been reduced to $1 to $3/kW.

Fuel Staging (Reburn)

The capital cost for reburn technology depends on the size of the unit, ease of retrofit, control system upgrade requirements, and, for natural gas reburn, availability of natural gas at the plant [30]. The retrofit costs are typically about $15 to $20/kW for natural gas, coal, or oil reburn, excluding the cost of any natural gas pipeline. The operating cost for a reburn retrofit is mainly due to the differential cost of the reburn fuel over the main fuel. For coal reburn this is zero, while reburn fuels like natural gas or oil are usually more expensive than the main fuel. This differential, however, can be offset by reductions in SO_2 emissions, ash remediation and disposal, and pulverizer power. The levelized cost for reburn is about $110 to $210/short ton of NO_x removed [17].

Cofiring

Cofiring of CWSF is not commercially used at this time. Biomass cofiring, on the other hand, is currently being demonstrated at several plants, with commercial operations being performed at seven utilities. The capital costs for biomass cofiring ranges from $175 to $250/kW [44].

Process Optimization

The total turnkey installation cost for an advanced combustion control system ranges from $150,000 to $500,000 [30]. It is possible to achieve moderate cost reductions on a per unit basis for similar units at the same power plant site. The size of the unit typically has little impact on the cost of a system.

Selective Catalytic Reduction

The capital costs for an SCR system depend on the level of NO_x removal and other site-specific conditions, such as inlet NO_x concentration and unit size and ease of retrofit, and they range from $80 to $160/kW [56]. The capital costs of an SCR system include the following [30]:

* Catalyst and reactor system
* Flow control skid and valving system
* Ammonia injection grid
* Ammonia storage
* Piping
* Ducts, expansion joints, and dampers
* Fan upgrades/booster fans
* Air preheater changes
* Foundations, structural steel, and electricals
* Installation

The operating costs can vary from $1,500 to $5,800/MM Btu, and the levelized cost can range from $1,800 to $10,900/MM Btu [30]. The operating costs include the following:

* Ammonia usage
* Pressure drop changes
* Excess air change
* Unburned carbon change

- Ash disposal
- Catalyst replacement
- Vaporization/injection energy requirements
- Other auxiliary power usage

Selective Non-Catalytic Reduction

SNCR is less capital-intensive than SCR. The cost of an SNCR retrofit is $10 to $20/kW, whereas incorporating SNCR into a new boiler is $5 to $10/kW [30]. The difference is due to the cost associated with modifying the existing boiler to install the reagent injection ports. The operating costs associated with the reagent, auxiliary power, and potential adverse plant impacts are on the order of 1 to 2 mills/kWh. The levelized costs average approximately $1,000/short ton of NO_x removed. A new, single-level approach approach to SNCR, SNCR trim, offering 20 to 30 percent NO_x reduction at about half the cost of conventional SNCR, is being tested by the Electric Power Research Institute [57]. SCNR trim has low operating costs, equivalent to only about $850/short ton of NO_x removed.

Other Flue Gas Treatment Processes

Little data are available for hybrid SNCR/SCR systems, since it is still in the demonstration phase. A levelized cost estimate for a 500 MW boiler with 50 percent NO_x reduction is about $5,800/short ton of NO_x removed [30]. Similarly, cost data on the rich reagent injection process, which is under development, are not available.

Hybrid Flue Gas Treatment and Combustion Modifications

A combination of flue gas treatment with combustion modification is being increasingly used. This technology provides higher overall NO_x reduction and can be more cost effective than stand-alone technology for the same level of NO_x control [30]. The costs of SCR can be reduced when it is used in combination with combustion modifications such as low-NO_x burners and overfire air [30]. Capital costs are lowered because combustion modifications lower the inlet NO_x concentration, which reduces the catalyst volume, support systems, and installation cost of SCR. In addition, operating costs are lowered due to reductions in catalyst replacement and reagent consumption.

SNCR can be combined with low-NO_x burners or gas reburn. SNCR and gas reburn have comparable economics at the same level of NO_x reduction; however, combining the two technologies considerably lowers cost while achieving a slightly higher NO_x reduction. An example of annual costs, reported by Wu [30], are about $1,140, $1,120, and $730 per short ton NO_x removed, respectively, for urea SNCR, gas reburn, and urea SNCR/gas reburn.

9.1.3 Particulate Matter

Emissions standards for particulate matter (PM) were first introduced in Japan, the United States, and Western European nations in the early to mid-1900s. The following decades found many countries also setting standards for particulate emissions, including those in Asia, Eastern Europe, Australia, and India. The importance of utilizing coal in an environmentally friendly manner for power generation has led to the introduction or proposal of particulate emissions standards in more than 40 nations [58]. In recent years, there has been increasing concern for control of fine particulate matter, and existing particulate emissions standards have progressively become more stringent over the years. The most stringent measures are associated with wealthy countries such as Japan and those in North America and Western Europe [58].

Particulate matter (PM) emissions from coal-fired electric utility boilers in the United States have decreased significantly since the implementation of the 1970 Clean Air Act Amendments. In 2008, approximately 12 million short tons of dry particulate matter, reported as PM_{10} (i.e., particles with an aerodynamic diameter less than or equal to 10 microns), were emitted from inventoried point and area sources, of which approximately 122,000 short tons (or ≈ 1.0 percent of the total) were emitted by coal-fired electric utility boilers [59]. This is a substantial decrease from a total of approximately 1.7 million short tons of PM_{10} being emitted from coal-fired power plants in 1970, especially since coal consumption for electricity generation has increased more than 150 percent over this period, and the reduction is due to the application of particulate control technologies. Similarly, annual emissions of dry particulates smaller than 2.5 μm (i.e., $PM_{2.5}$) in 2008, which is a subset of PM_{10}, were 10,000 short tons, or less than 0.4 percent of the total primary $PM_{2.5}$ emitted from all sources.

The application of control technologies to combustion sources is illustrated in Figures 9.1 and 9.2, which show the improvements in emissions rates from coal-fired power plants since 1970 as well as near-term projected emissions rates. As of 2005 (the most recent data as of January 2009), there were 1,216 coal-fired electric generators equipped with particulate collectors with a total of more than 356,000 MW generating capacity [2].

Several particulate control technologies for coal-fired power plants are available, including electrostatic precipitators (ESPs), fabric filters (baghouses), wet particulate scrubbers, mechanical collectors (cyclones), and hot-gas particulate filtration [60]. Of these, ESPs and fabric filters are currently the technologies of choice, since they can meet current and pending legislation particulate matter levels while cleaning large volumes of flue gas, achieve very high collection efficiencies, and remove fine particles. When operating properly, ESPs and baghouses can achieve overall collection efficiencies of 99.9 percent of primary particulates (> 99 percent control of PM_{10} and 95 percent control of $PM_{2.5}$), thereby achieving the 1978 New Source Performance Standards required limit of 0.03 lb PM/million Btu [61]. The primary particulate matter collection devices used in the power generation industry—ESPs and fabric filters (baghouses)—are discussed in this section. In addition, hybrid systems that are under development, where ESPs and fabric filters are combined in a single overall system, are presented.

Electrostatic Precipitators

Particulate and aerosol collection by electrostatic precipitation is based on the mutual attraction between particles of one electrical charge and a collection electrode of opposite polarity. This concept was pioneered by F. G. Cottrell in 1910

[5]. The advantages of this technology include the ability to handle large gas volumes (ESPs have been built for volumetric flow rates up to 4,000,000 ft^3/min), achieve high collection efficiencies (which vary from 99 to 99.9 percent), maintain low pressure drops (0.1–0.5 inches of water column), collect fine particles (0.5–200 μm), and operate at high gas temperatures (gas temperatures up to 1,200°F can be accommodated). In addition, the energy expended in separating particles from the gas stream acts solely on the particles and not on the gas stream.

ESPs have been used in the control of particulate emissions from coal-fired boilers used for steam generation for about 60 years [8]. Initially, all ESPs were installed downstream of the air preheaters at temperatures of 270 to 350°F, and are referred to as cold-side ESPs. ESPs have been installed upstream of air preheaters where the temperature is in the range of 600°F to 750°F (i.e., hot-side ESPs) as a result of using low-sulfur fuels with lower fly ash resistivity. In the early 1970s, ESPs were the preferred choice for a high-efficiency particulate control device [8]. Nearly 90 percent of U.S. coal-based electric utilities use ESPs to collect fine particles [62].

Operating Principles

Several basic geometries are used in the design of ESPs, but the common design used in the power generation industry is the plate and wire configuration. In this design, shown schematically in Figure 9.15, the ESP consists of a large hopper-bottomed box containing rows of plates forming passages through which the flue gas flows. Centrally located in each passage are electrodes energized with high-voltage

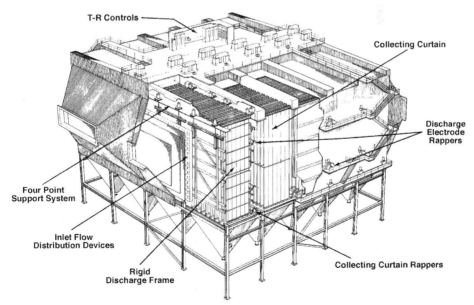

Figure 9.15 Electrostatic precipitator.
Source: From B&W (1982) [63].

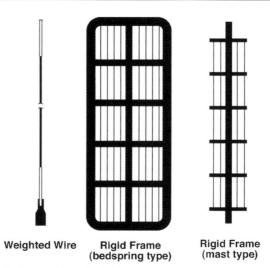

Weighted Wire   **Rigid Frame**   **Rigid Frame**
                        (bedspring type)   **(mast type)**

Figure 9.16 Rigid discharge electrode designs.
Source: From Elliot (1989).

(45–70 kV), negative-polarity, direct current (dc) provided by a transformer-rectifier set [9]. Examples of various designs of rigid discharge electrodes are shown schematically in Figure 9.16 [9]. The most commonly used discharge electrode in the United States is the weighted wire electrode, while the rigid frame electrode is commonly used in Europe [9]. The flow is usually horizontal, and the passageways are typically 8 to 10 inches wide. The height of a plate varies from 18 to 40 feet with a length of 25 to 30 feet. The ESP is designed to reduce the flue gas flow from 50 to 60 feet/sec to less than 10 feet/sec as it enters the ESP so the particles can be effectively collected.

Electrostatic precipitation consists of three steps:

1. Charging the particles to be collected via a high-voltage electric discharge
2. Collecting the particles on the surface of an oppositely charged collection surface
3. Cleaning the collection surface

These are illustrated in Figure 9.17 [60].

The electrodes discharge electrons into the flue gas stream, ionizing the gas molecules. These gas molecules, with electrons attached, form negative ions. The gas is heavily ionized in the vicinity of the electrodes, resulting in a visible blue corona effect. The fine particles are then charged through collisions with the negatively charged gas ions, resulting in the particles becoming negatively charged. Under the large electrostatic force, the negatively charged ash particles migrate out of the gas stream toward the grounded plates, where they collect and form an ash layer. These plates are periodically cleaned by a rapping system to release the layer into the ash hoppers as an agglomerated mass.

The speed at which the migration of the ash particles takes place is known as the migration or drift velocity. It depends on the electrical force on the charged particle as well as the drag force developed as the particle attempts to move perpendicularly

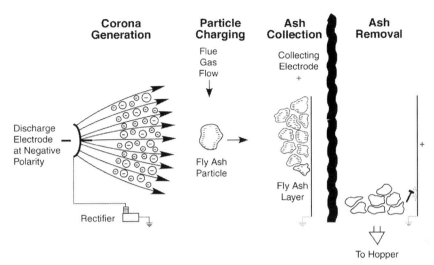

Figure 9.17 Basic concept of charging and collecting particles in an ESP.
Source: Modified from Soud and Mitchell (1997).

to the main gas flow toward the collecting electrode [5]. The drift velocity, w, is defined as

$$w = \frac{2.95 x 10^{-12} p E_c E_p d_p}{\mu_g} K_c \tag{9.79}$$

where w is in meters per second, p is the dielectric constant for the particles (which typically lies between 1.50 and 2.40), E_c is the strength of the charging field (v/m), E_p is the collecting field strength (v/m), d_p is the particle diameter (μm), K_C is Cunningham correction factor for particles with a diameter less than roughly 5 μm (dimensionless), and μ_g is the gas viscosity (kg/m sec).

The Cunningham correction factor in Equation (9.79) is defined as [5]

$$K_C = 1 + \frac{2\lambda}{d_p}\left[1.257 + 0.400e^{\left(\frac{0.55d_p}{\lambda}\right)}\right] \tag{9.80}$$

where λ is the mean free path of the molecules in the gas phase. This quantity is given by

$$\lambda = \frac{\mu_g}{0.499\rho_g u_m} \tag{9.81}$$

where u_m is the mean molecular speed (m/sec) and ρ_g is the gas density (kg/m³). From the kinetic theory of gases, u_m is given by

$$u_m = \left[\frac{8R_u T}{\pi M}\right]^{\frac{1}{2}} \tag{9.82}$$

where M is the molecular weight of the gas, T is temperature (°K), and R_u is the universal gas constant (8.31×10^3 m^2/sec^2 mole °K).

The drift velocity is used to determine collection efficiency using the Deutsch-Anderson equation

$$\eta = 1 - e^{\left(-\frac{wA}{Q}\right)} \tag{9.83}$$

where w is the drift velocity, A is the area of collection electrodes, and Q is the volumetric flow rate. The units of w, A, and Q must be consistent because the factor wA/Q is dimensionless.

The ratio, A/Q, is often referred to as the specific collection area (SCA) and is the most fundamental ESP size descriptor [9]. Collection efficiency increases as SCA and w increase. The value of w increases rapidly as the voltage applied to the emitting voltage is increased; however, the voltage cannot be increased above that level at which an electric short circuit, or arc, is formed between the electrode and ground.

The collecting plates are periodically cleaned to release the layer into the ash hoppers as an agglomerated mass by a mechanical (rapping) system in a dry ESP or by water washing in the case of a WESP. The hopper system must be adequately designed to minimize ash reentrainment into the gas stream until the hopper is emptied. The strength of the electric field and ash bonding on the plates, mass gas flow, and the striking energy must be matched to ensure that ash is not reentrained into the gas stream [64]. The ideal situation is where the electric field holding the ash layer that is directly adjacent to the plate be of such strength that the strike energy just breaks this bond and gravity dislodges the particulate matter into the ash hopper.

Factors that Affect ESP Performance
Several factors affect electrostatic precipitators performance. Of these, fly ash resistivity is the most important.

Fly Ash Resistivity Fly ash resistivity plays a key role in dust-layer breakdown and the ESP performance. Resistivity is dependent on the flue gas temperature and chemistry, and the chemical composition of the ash itself. Electrostatic precipitation is most effective in collecting dust in the resistivity range of 10^4 to 10^{10} ohm-cm [5]. In general, resistivities above 10^{11} ohm-cm are considered to be a problem because the maximum operating field strength is limited by the fly ash resistivity. Back corona, the migration of positive ions generated in the fly ash layer toward the emitting electrodes, which neutralize the negatively charged particles, will result if the ash resistivity is greater than 10^{12} ohm-cm. If the fly ash resistivity is below 2×10^{10} ohm-cm, it is not considered to be a problem because the maximum operating field strength is limited by other factors other than resistivity.

Examples of low- and high-resistvity fly ashes are shown in Figure 9.18, where resistivity is plotted as a function of temperature for two U.S. lignite (from North

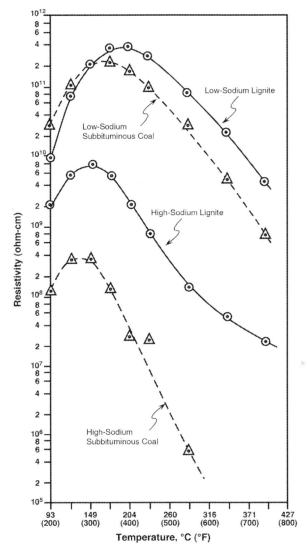

Figure 9.18 The effect of ash composition on fly ash resistivity for coals from the same geographical location.
Source: From Miller (1986).

Dakota) and two subbituminous coal samples (from the Powder River Basin) [65]. The differences in fly ash resistivity are due to variations in ash composition. The low-resistivity fly ashes were produced from coals that contained higher levels of sodium in the coal ash. Higher sodium levels result in lower resistivity. Similarly, higher concentrations of iron lower resistivity. Higher levels of calcium and magnesium have the opposite effect on resistivity. This is illustrated in Figure 9.19,

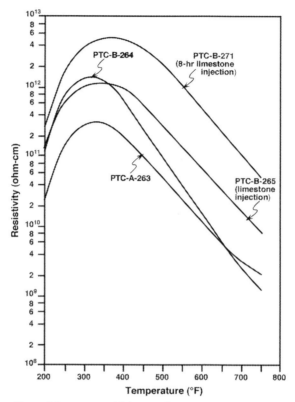

Figure 9.19 The effect of limestone addition on fly ash resistivity.
Source: Miller et al. (1984).

where the fly ash resistivities of two Texas lignites are shown along with the fly ash resistivities from the same two coals when injecting limestone for SO_2 control [66]. The addition of calcium through sorbent injection resulted in increasing the fly ash resistivities.

Flue gas properties also affect fly ash resistivity. The two properties that have the most influence on ash resistivity are temperature and humidity. The effect of temperature can be observed in Figures 9.18 and 9.19. Similarly, as moisture content in the flue gas is increased, the fly ash resistivity decreases. The dome-shaped curves shown in those figures are typical of fly ashes. The shape of the curves is due to a change in the mechanism of conduction through the bulk layer of particles as the temperature is varied [5]. The predominant mechanism below 300°F is surface conduction where the electric charges are carried in a surface film adsorbed on the particle. As the temperature is increased above 300°F, the phenomenon of adsorption becomes less effective and the predominant mechanism is volume or intrinsic conduction. Volume conduction involves passage of electric charge through the particles.

Other Factors The three primary mechanical deficiencies in operating units are gas sneakage, fly ash reentrainment, and flue gas distribution [9]. Flue gas sneakage—flue gas that is bypassing the effective region of the electrostatic precipitators—increases the outlet dust loading. Reentrainment occurs when individual dust particles are not collected in the hoppers but are caught up in the gas stream, increasing the dust loading to the ESP and resulting in higher outlet dust loadings. Nonuniform flue gas distribution throughout the entire cross section of the ESP decreases the collection ability of the unit.

Many other factors can affect the performance of an ESP, including the quality and the type of fuel. Changes in coal and ash composition, grindability, and the burner/boiler system are all important. Fly ash resistivity increases with decreasing sulfur content, an issue that must be considered when switching to lower-sulfur coals. Moisture content and ash composition affect resistivity as discussed earlier. Changes in coal grindability can affect pulverizer performance by altering particle size distribution, which in turn can impact combustion performance and ESP performance. Modifications to the boiler system can affect temperatures or combustion performance and thereby impact ESP performance.

Methods to Enhance ESP Performance

Difficulties in collecting high-resistivity fly ash and fine particulates have led to very large units being specified, unacceptable increases in ESP power consumption, and, in extreme cases, the use of fabric filters in lieu of electrostatic precipitators [9]. As a result, concepts have been developed to overcome technical limitations and maintain competitiveness with fabric filters. This includes [9] pulse energization, where a high-voltage pulse is superimposed on the base voltage to enhance ESP performance during operation under high-resistivity conditions; intermittent energization, where the voltage to the ESP is turned off during selected periods, allowing for a longer period between each energization cycle, which limits the potential for back corona; and wide plate spacing to reduce capital and maintenance costs and allow for thicker discharge electrodes and increased current density.

Another approach to achieving electrical resistivities in the desired range is the addition of conditioning agents to the flue gas stream. This technique is applied commercially to both hot-side and cold-side ESPs. Conditioning modifies the electrical resistivity of the fly ash and/or its physical characteristic by changing the surface electrical conductivity of the dust layer deposited on the collecting plates, increasing the space charge on the gas between the electrodes, and/or increasing dust cohesiveness to enlarge particles and reduce rapping reentrainment losses [9]. Over 200 utility boilers are equipped with some form of conditioning in the United States [9].

The most common conditioning agents are sulfur trioxide (SO_3), ammonia (NH_3), compounds related to them, and sodium compounds. Sulfur trioxide is most widely applied for cold-side ESPs, while sodium compounds are used for hot-side ESPs [9]. While results vary between coal and system, the injection of 10 to 20 ppm of SO_3 can reduce the resisitivity to a value that will permit good collection efficiencies. In select cases, SO_3 injection of 30 to 40 ppm has resulted in reductions of fly ash resisitivity of 2 to 3 orders of magnitude (e.g., from 10^{11} to about 10^8 ohm-cm) [5]. Disadvantages of SO_3 injection systems include the possibility of plume color degradation. Disadvantages for sodium compounds are the potential problems with increased deposition and interference from certain fuel constituents, which affects the economics of the injection [5]. Combined SO_3-NH_3 conditioning is used with the SO_3, adjusting the resistivity downward, while the NH_3 modifies the space-charge effect, improves agglomeration, and reduces rapping reentrainment losses [5].

Wet ESPs

Dry ESPs, which have been discussed up to this point, have been successfully used for many years in utility applications for coarse and fine particulate removal. Dry ESPs can achieve 99+ percent collection efficiency for particles 1 to 10 μm in size. However, dry ESPs cannot remove toxic gases and vapors that are in a vapor state at 400°F, cannot efficiently collect very small fly ash particles, cannot handle moist or sticky particulate that would stick to the collection surface, require a lot of space for multiple fields due to reentrainment of particles, and rely on mechanical collection methods to clean the plates, which require maintenance and periodic shutdowns [67].

Wet electrostatic precipitators (WESPs) address these issues and are a viable technology to collect finer particulate than existing technology, while also collecting aerosols. WESPs have been commercially available since their first introduction by F. G. Cottrell in 1907 [68]. However, most of their use has been in small, industrial-type settings as opposed to utility power plants. WESPs have been in service for almost 100 years in the metallurgical industry and in many other applications. They are used to control acid mists, submicron particulate (as small as 0.01 μm with 99.9 percent removal), mercury, metals, and dioxins/furans as the final polishing device within a multipollutant control system [68]. When integrated with upstream air pollution control equipment, such as an SCR, dry ESP, and wet scrubber, multiple pollutants can be removed, with the WESP acting as the final polishing device.

WESPs operate in the same three-step process as dry ESPs: charging, collecting, and cleaning the particles from the collecting electrode [69]. However, cleaning the collecting electrode is performed by washing the collection surface with liquid, rather than by mechanically rapping the collection plates.

Wet electrostatic precipitators operate in a wet environment in order to wash the collection surface; therefore, they can handle a wider variety of pollutants and gas conditions than dry ESPs [69]. WESPs find their greatest use in the following situations:

- The gas in question has a high moisture content.
- The gas stream includes sticky particulate.
- The collection of submicron particulate is required.
- The gas stream has acid droplets of mist.
- The temperature of the gas stream is below the moisture dew point.

WESPs continually wet the collection surface and create a dilute slurry that flows down the collecting wall to a recycle tank, never allowing a layer of particulate cake to build up [69]. As a result, captured particulate is never reentrained. Also, when firing low-sulfur coal, which produces a high-resistivity dust, the electrical field does not deteriorate, and power levels in a wet electrostatic precipitator can be dramatically higher than in a dry ESP: 2,000 watts/1,000 scfm versus 100–500 watts/ 1,000 scfm, respectively.

Similar to a dry ESP, WESPs can be configured either as tubular precipitators (i.e., the charging electrode is located down the center of a tube) with vertical gas flow or as plate precipitators with horizontal gas flow [70]. For a utility application, tubular WESPs are appropriate as a mist eliminator above a flue gas desulfurization scrubber, while the plate type can be employed at the back end of dry ESP train for final polishing of the gas.

Fabric Filters

Historically, ESPs have been the principle control technology for fly ash emissions in the electric power industry. Small, relatively inexpensive electrostatic precipitators could be installed to meet early federal and state regulations. However, as particulate control regulations have become more stringent, electrostatic precipitators have become larger and more expensive. Also, increased use of low-sulfur coal has resulted in the formation of fly ash with higher electric resistivity that is more difficult to collect. Consequently, ESP size and cost have increased to maintain high collection efficiency [71]. As a result, interest in baghouses has increased. Baghouses are a technology with extremely high collection efficiency (i.e., 99.9 to 99.99+ percent), are capable of filtering large volumes of flue gas, and with size and efficiency that are relatively independent of the type of coal burned [71]. Baghouses are essentially huge vacuum cleaners consisting of a large number of long, tubular filter bags arranged in parallel flow paths. As the ash-laden flue gas passes through these filters, the particulate is removed.

Advantages of fabric filters include high collection efficiency over a broad range of particles sizes; flexibility in design provided by the availability of various cleaning methods and filter media; a wide range of volumetric capacities in a single installation, which may range from 100 to 5 million ft^3/min; reasonable operating pressure drops and power requirements; and the ability to handle a variety of solid materials [5]. Disadvantages of baghouses include large footprints, so space factors may prohibit consideration of baghouses, possibility of an explosion or fires if sparks are present in the vicinity of a baghouse, and hydroscopic materials usually cannot be handled, owing to cloth cleaning problems.

The first utility baghouse in the United States was installed on a coal-fired boiler in 1973 by the Pennsylvania Power and Light Company at its Sunbury Station [71]. This baghouse, as well as the next several baghouses installed, were small, and it was not until 1978 that the first large baghouse was installed on a utility boiler. This baghouse serviced a 350 MW pulverized coal-fired boiler at the Harrington Station of Southwestern Public Service Company. Starting in 1978, there has been a steady increase in the installation of utility commitments to baghouse technology, and now more than 110 baghouses are in operation on utility boilers in the United States, servicing more than 26,000 MW of generating capacity [71, 72].

Filtration Mechanisms

Filtration occurs when the particulate-laden flue gas is forced through a porous, solid medium, which captures the particles. In a baghouse, this solid medium is the filter bag and/or the residual dust cake on the bag. The important filtering mechanisms are three aerodynamic capture mechanisms: direct interception, inertial impaction, and diffusion. Electrostatic attraction may also play a role with certain types of dusts/fiber combinations [5].

Direct interception occurs if the gas streamlines carrying the particles are close to the filter elements for contact. Inertial impaction occurs when the particles have sufficient momentum and cannot follow the gas stream when the stream is diverted by the filter element and the particles strike the filter. Diffusion results when the particle mass is very low and Brownian diffusion superimposes random motion on the streamline trajectory, thereby increasing the particle's probability of contacting and being captured by the filter [71]. Particles may be attracted to or repulsed by filters due to a variety of Coulombic and polarization forces. Particles larger than 1 μm are removed by impaction and direct interception, whereas particles from 0.001 to 1 μm are removed mainly be diffusion and electrostatic separation [5].

The effectiveness of a filter in capturing particles is reported in terms of collection efficiency or particle penetration. Particle penetration, P, is defined as the ratio of the particle concentration (mass or number of particles per unit volume of gas), also referred to as dust loading, on the outlet of the filter (i.e., cleaned flue gas stream) to that on the inlet side of the filter (i.e., dirty flue gas stream). Collection efficiency, η, is defined as

$$\eta = 1 - P \tag{9.84}$$

Typically, both penetration and collection efficiency are multiplied by 100 and reported as a percent.

The filtration process can be divided into three distinct time regimes [71]:

1. Filtration by a clean fabric, which occurs only once in the life of a bag
2. Establishment of a residual dust cake, which occurs after many filtering and cleaning cycles
3. Steady-state operation in which the quantity of particulate matter removed during the cleaning cycles equals the amount collected during each filter cycle

In general, the initial collection efficiency of new filters is quite low (<99 percent and can be on the order of 75–90 percent), whereas a conditioned bag (i.e., a bag that has retained residual particles in the fibers of the filter that cannot be removed by cleaning) may have a collection efficiency of 99.99+ percent. A dust cake will form on the filters where the adhesive and cohesive forces acting between the particles and filter elements, and among the particles, respectively, are sufficiently strong to allow particulate agglomerates to bridge the filter pores. The accumulated dust cake forms a secondary filter of much higher efficiency than the clean fabric. On a seasoned bag, residual dust cakes generally weigh 10 to 20 times as much as the ash deposited during an average cleaning cycle [71].

Operating Principles

Baghouses remove particles from the flue gas in compartments arranged in parallel flow paths, with each compartment containing several hundred large, tube-shaped filter bags. Figure 9.20 is a cutaway view of a typical ten-compartment baghouse [71]. A baghouse on a 500 MW coal-fired unit may be required to handle in excess of 2 million ft^3/min of flue gas at temperatures of 250 to 350°F. From an inlet manifold, the dirty flue gas, with typical dust loadings from 0.1 to 10 grains/ft^3 of gas (0.23 to 23 grams/m^3), enters hopper inlet ducts that route it into individual compartment hoppers. From each hopper, the gas flows upward through the bags, where the fly ash is deposited. The clean gas is drawn into an outlet manifold, which carries it out of the baghouse to an outlet duct. Periodic operation requires shutdown of portions of the baghouse at regular intervals for cleaning. Cleaning is accomplished in a variety of ways, including mechanical vibration or shaking, pulse jets of air, and reverse air flow.

The two fundamental parameters in sizing and operating baghouses are the air-to-cloth (A/C) ratio and pressure drop across the filters. Other important factors that affect the performance of the fabric filter include the flue gas temperature, dew point, and moisture content, and particle size distribution and composition of the fly ash [73].

The air-to-cloth ratio, which is a fundamental fabric filter descriptor denoting the ratio of the volumetric flue gas flow (ft^3/min) to the amount of filtering surface area (ft^2), is reported in units of ft/min [9]. For fabric filters, it has been generally observed that the overall collection efficiency is enhanced as the A/C ratio—that is, superficial filtration velocity—decreases. Factors to be considered with the A/C

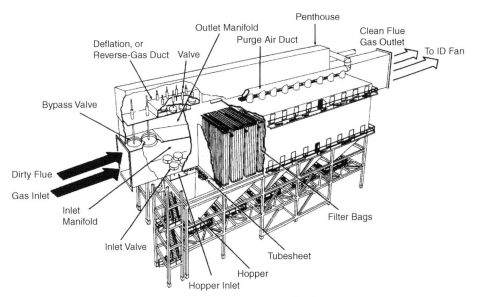

Figure 9.20 Cutaway view of a typical 10-compartment baghouse.
Source: From Bustard et al. (1988).

ratio include type of filter fabric, type of coal and firing method, fly ash properties, duty cycle of the boiler, inlet fly ash loading, and cleaning method [73]. The A/C ratio determines the size of the baghouse and thus the capital cost.

Pressure drop is a measure of the energy required to move the flue gas through the baghouse. Factors affecting pressure drop are boiler type (which influence the fly ash particle size), filtration media, fly ash properties, and flue gas composition [73]. The pressure drop is an important parameter because it determines the capital cost and energy requirements of the fans.

As the filter cake accumulates on the supporting fabric, the removal efficiency typically increases, but the resistance to flow increases as well. For a clean filter cloth, the pressure drop is about 0.5 inches water column (W.C.), and the removal efficiency is low. After sufficient filter cake buildup, the pressure drop can increase to 2 to 3 inches W.C., with the removal efficiency 99+ percent [5]. When the pressure drop reaches 5 to 6 inches W.C., it is usually necessary to clean the filters.

The pressure drop for both the cleaned filter and the dust cake, ΔP_T, may be represented by Darcy's equation [5]

$$\Delta P_T = \Delta P_R + \Delta P_C = \frac{\mu_g x_R V}{K_R} + \frac{\mu_g x_C V}{K_C} \tag{9.85}$$

where ΔP_R is the conditioned residual pressure drop, ΔP_C is the dust cake pressure drop, K_R and K_C are the filter and dust cake permeabilities, respectively, V is the superficial velocity, μ_g is the gas viscosity, and x_R and x_C are the filter and dust cake thicknesses, respectively. The permeabilities K_R and K_C are difficult quantities to predict with direct measurements, since they are functions of the properties of the filter and dust such as porosity, pore size distribution, and particle size distribution. Therefore, in practice ΔP_R is usually measured after the bags are cleaned, and ΔP_C is determined using the equation

$$\Delta P_C = K_2 C_i V^2 t \tag{9.86}$$

where C_i is the dust loading and, along with V, is assumed constant during the filtration cycle; t is the filtration time; and K_2, which is the dust resistance coefficient, is estimated from

$$K_2 = \frac{0.00304}{\left(d_{g,mass}\right)^{1.1}} \left(\frac{\mu_g}{\mu_{g,70°F}}\right) \left(\frac{2600}{\rho_p}\right) \left(\frac{V}{0.0152}\right)^{0.6} \tag{9.87}$$

where d_g is the geometric mass median diameter (m), μ_g is the gas viscosity (kg/m-s), ρ_p is the particle density (kg/m^3), and V is the superficial velocity (m/s).

Basic Types of Fabric Filters The three basic types of baghouses are reverse-gas, shake-deflate, and pulse-jet. They are distinguished by the cleaning mechanisms and by their A/C ratio. Air-to-cloth ratios for fabric filters range from a low of 1.0

to 12.0 ft/min, depending on the type of cleaning mechanism used and characteristics of the fly ash [5]. The two most common baghouse designs are the reverse-gas and pulse-jet types.

Ash that accumulates on the bags in excess of the desired residual dust cake must be removed by periodic bag cleaning to reduce the gas flow resistance, and thus induced draft fan power requirements, and to reduce bag weight. In U.S. utility baghouses, cleaning is done off-line by isolating individual compartments for cleaning.

Reverse-Gas Reverse-gas fabric filters are generally the most conservative design of the fabric filter types. It typically operates at low A/C ratio ranging from 1.5 to 3.5 ft/min [5, 73]. Fly ash collection is on the inside of the bags, since the flue gas flow is from the inside of the bags to the outside, as illustrated in Figure 9.21 [71]. Reverse-gas baghouses use off-line cleaning where compartments are isolated and cleaning air is passed from the outside of the bags into the inside, causing the bags to partially collapse to release the collected ash. The dislodged ash falls into the hopper. A simplified schematic showing the cleaning cycle is shown in Figure 9.22 [60]. A variation of the reverse-gas cleaning method is the use of sonic energy for bag cleaning. With this method, low-frequency (from <250–300 Hz), high sound pressure (0.3–0.6 inches WC) pneumatic horns are sounded simultaneously with the normal reverse-gas flow to add energy to the cleaning process. Reverse-gas fabric filters are widely used in the United States, with approximately 90 percent of the utility baghouse employing this reverse-gas cleaning [71].

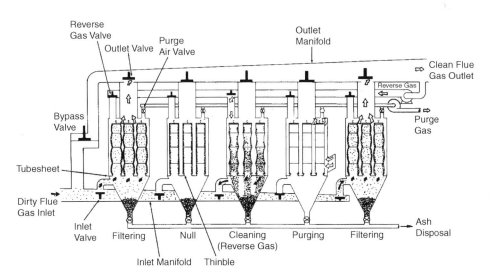

Figure 9.21 The compartments in a reverse-gas baghouse showing the flue gas and cleaning air flows during the various cycles of operation.
Source: From Bustard et al. (1988).

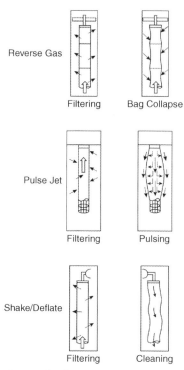

Figure 9.22 Baghouse cleaning mechanisms.
Source: Modified from Soud and Mitchell (1997).

Shake-Deflate Shake-deflate baghouses are another low A/C type system (2 to 4 ft/ min), and they collect dust on the inside of the bags similar to the reverse-gas systems [5]. With shake-deflate cleaning, a small quantity of filtered gas is forced backward through the compartment being cleaned, which is done off-line. The reversed filtered gas relaxes the bags but does not completely collapse them. As the gas is flowing, or immediately after it is shut off, the tops of the bags are mechanically shaken for 5 to 20 seconds at frequencies ranging from 1 to 4 Hz and at amplitudes of 0.75 to 2 inches [71]. The operating cycles of a shake-deflate baghouse are illustrated in Figure 9.23 [71], with a simplified cleaning cycle shown in Figure 9.22. Operating experience with shake-deflate baghouses in utility service has been good [9].

Pulse-Jet In pulse-jet fabric filters, the flue gas flow is from the outside of the bag inward; this is illustrated in Figure 9.24 [71]. The air-to-cloth ratio is higher than reverse-air units and is typically 3 to 4 ft/min allowing for a more compact installation, but the ratio can vary from 2 to 5 ft/min [5]. Cleaning is performed with a high-pressure burst of air into the open end of the bag, as shown in the simplified schematic of Figure 9.22. Pulse-jet systems required metal cages on the inside of the bags to prevent bag collapse. Bag cleaning can be performed on-line by pulsing

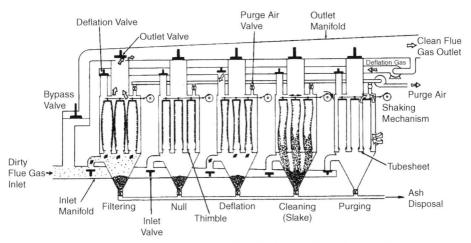

Figure 9.23 The compartments in a shake-deflate baghouse illustrating the flue gas and cleaning air flows during the various cycles of operation.
Source: From Bustard et al. (1988).

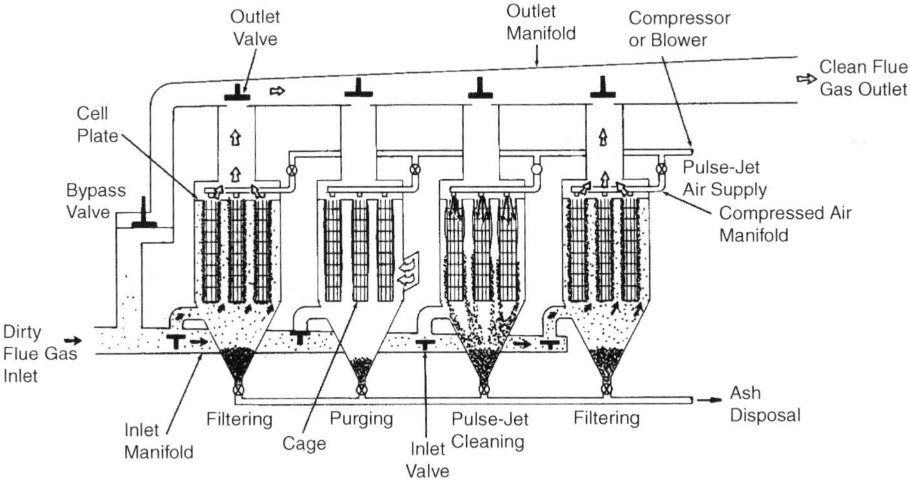

Figure 9.24 The compartments in a pulse-jet baghouse illustrating the flue gas and cleaning air flows during the various cycles of operation.
Source: From Bustard et al. (1988).

selected bags, while the remaining bags continue to filter the flue gas. Three cleaning methods have evolved for the pulse-jet systems [73]:

1. High-pressure (40–100 psig), low-volume pulse
2. Intermediate-pressure (15–30 psig) and intermediate-volume pulse
3. Low-pressure (7.5–10 psig), high-volume pulse

The first method is used mainly in the United States, while the latter two methods are used mainly in larger boilers in Australia, Canada, and Western Europe [73].

Pulse-jet cleaning results in lower resistance to gas flow than the other two baghouse types, thereby allowing smaller baghouses to filter the same volume of flue gas. Despite this, pulse-jet cleaning is not the preferred choice in the United States for utility boilers because the more rigorous cleaning method may result in lower particulate collection efficiency and shorter bag life. Pulse-jet baghouses are used in the United States, as well as Japan, for industrial boilers [73]. In Canada and Europe, pulse-jet systems are used in industrial plants and some large-sized utility plants. Much work has been done on improving fabrics for the filters, and the pulse-jet technology is becoming more attractive to utilities.

Fabric Filter Characteristics

Fabric filters are made from woven, felted, and knitted materials with filter weights that generally range from as low as 5 ounces/yd^2 to as high as 25 ounces/yd^2 [5]. Filtration media are selected depending on the type of baghouse, their efficiency in capturing particles, system operating temperature, physical and chemical nature of the fly ash and flue gas, durability for a long bag life, and the cost of the fabric. There is a tendency toward using needle felts or polytetrafluorethylene (PTFE) membranes on woven glass, due to their ability to withstand higher temperatures (during system upsets that result in temperature excursions) and improve bag performance [60]. To protect bags against chemical attack, the fabrics are usually coated with other materials such as Teflon, silicone, graphite, and GORE-TEX® [71].

The most common bag material in coal-fired utility units with pulse-jet fabric filters is polyphenylene sulfide (PPS) needled felt. In addition to PPS, fiberglass, acrylic, polyester, polypropylene, Nomex®, P84®, special high-temperature fiberglass media, membrane covered media, and ceramic are used in various applications. Table 9.9 provides a summary of selected filter media characteristics [74]. Typical bag size is 127 or 152 mm (5 or 6 inches) diameter round or oval with a length of 3 to 8 m (10 to 26 ft).

Factors that Affect Baghouse Performance

Key factors in proper baghouse design and operation are flue gas flow and properties, fly ash characteristics, and coal composition [9]. The baghouse must minimize pressure drop, maintain appropriate temperature and velocity profiles, and distribute the ash-laden flue gas evenly to the individual compartments and bags.

Particle size distribution of the fly ash and loading of the flue gas vary with the type of combustion system [73]. Stoker-fired units produce ash with high carbon content, moderate loading, and large particle size distribution (compared to other combustion systems). Pulverized coal-fired systems produce ash with low carbon content, high loading, and fine particle size distribution. Cyclone-fired units produce ash with low carbon content, moderate loadings, and very fine particle size distribution. Fluidized-bed systems generally produce ash with high carbon content, high loading, and fine particle size distribution [9].

Table 9.9 Fabric Filter Media Characteristics

Glass
Glass fabrics offer outstanding performance in high-heat applications. In general, by using a proprietary finish they become resistant to acids, except by hydrofluoric and hot phosphoric acids in their most concentrated forms. They are attached by strong alkalis at room temperature and weak alkalis at higher temperatures. Glass is vulnerable to damage caused by abrasion and flex. However, the proprietary finishes can lubricate the fibers and reduce the internal abrasion caused by flexing. Maximum operating temperature is 260°C (500°F).

Polyphenylene Sulfide
Polyphenylene sulfide (PPS) fibers offer excellent resistance to acids, good-to-excellent resistance to alkalis, have excellent stability and flexibility, and provides excellent filtration efficiency. Maximum operating temperature is 190°C (375°F).

Acrylic
The resistance of homopolymer acrylic fibers is excellent in organic solvents, good in oxidizing agents and mineral and organic acids, and fair in alkalis. They dissolve in sulfuric acid concentrations. Maximum operating temperature is 127°C (260°F).

Polyester
Polyester fabrics offer good resistance to most acids, oxidizing agents, and organic solvents. Concentrated sulfuric and nitric acids are the exception. Polyesters are dissolved by alkalis at high concentrations. Maximum operating temperature is 132°C (270°F).

Polypropylene
Polypropylene fabrics offer good tensile strength and abrasion resistance. They perform well in organic and mineral acids, solvents, and alkalis. Polypropylene is attacked by nitric and chlorosulfonic acids, and sodium and potassium hydroxide at high temperatures and concentrations. Maximum operating temperature is 93°C (200°F).

Nomex®
Nomex fabrics resist attack by mild acids, mild alkalis, and most hydrocarbons. Resistance to sulfur oxides above the acid dew point at temperatures above 150°F is better than polyester. Flex resistance of Nomex is excellent. Maximum continuous operating temperature is 204°C (400°F).

P-84®
P-84 fabrics resist common organic solvents and avoid high pH levels. They provide good acid resistance. P-84 offers superior collection efficiency due to irregular fiber structure. Maximum operating temperature is 260°C (500°F).

Source: Modified from Kitto and Stultz (2005).

Sulfur content of the coal has been correlated to fabric filter operation. The cohesiveness of ash produced from high-sulfur coals is greater than from Western low-sulfur coals [9]. Also, maintaining the baghouse above the acid dew point is critical in high-sulfur coal applications.

The fly ash properties are important since they affect the adhesion and cohesion characteristics of the dust cake. This in turn affects the properties of the residual dust cake, collection efficiency, and cleanability of the bags.

Methods to Enhance Filter Performance

The most recognized method to enhance fabric filter performance is the application of sonic energy, which was discussed previously. Virtually all reverse-gas baghouses have included sonic horns [9].

Gas conditioning has been explored for improving filter performance, although this is not done commercially [9]. Low concentrations of ammonia and/or sulfur trioxide have been added in test programs to control fine particulate emissions and reduce pressure drop when firing low-rank fuels.

Hybrid Systems

Although the discussions of technologies in this chapter have mainly focused on commercial systems, this section will briefly discuss two concepts that are under development for improving particulate capture. They are discussed because these technologies are expected to become commercial in the very near future, especially as particulate emissions become more stringent.

Hybrid systems have been under development for more than 10 years, since utilities are required to meet increasingly tighter emissions regulations for particulate matter as well as sulfur dioxide. Fly ash resisitivity and dust loadings are affected by switching to low-sulfur coals or injecting sorbents for sulfur dioxide control, which in turn can reduce ESP efficiency. The desire to reduce fine particulate emissions is also leading to innovative technologies to reduce particulate emissions. Two such systems, the Compact Hybrid Particulate Collector (COHPAC) and the Advanced Hybrid Particulate Collector (AHPC), have been developed to address these issues.

Compact Hybrid Particulate Collector

The COHPAC, developed by the Electric Power Research Institute (EPRI), involves the installation of a pulse-jet baghouse downstream of the ESP or retrofitted into the last field of an ESP [75, 76]. Since the pulse-jet collector is operating as a polisher for achieving lower particulate emissions, the low-dust loading to the baghouse allows the filter to be operated at higher A/C ratios (8–20 ft/min) without increasing the pressure drop. This system allows for the ability to retrofit existing units and achieve high efficiencies at relatively low cost. The COHPAC technology has been demonstrated at the utility scale, including full-scale operation at Alabama Power's E.C. Gaston Station (272 MW) and TU Electric's Big Brown Plants (2 units, each 575 MW) [75, 77]. Results from COHPAC operation have been positive. For example, at E.C. Gaston Station the COHPAC has been operated with both on-line cleaning and long filter bags (i.e., 23 feet) at filtration rates of 8.5 ft/min while providing low outlet emissions levels (<0.01 lb/MM Btu) and reduced pressure drops, even with occasional high-inlet dust loadings. COHPAC is a promising technology for polishing particulate emissions and is expected to help utilities in meeting more stringent particulate emissions standards.

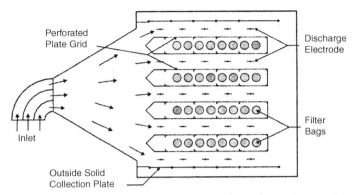

Figure 9.25 Perforated Plate Grid, Discharge Electrode, Inlet, Filter Bags, Outside Solid Collection Plate

Figure 9.25 A graphical representation of the Advanced Hybrid Particulate Collector. *Source:* From DOE (2001).

Advanced Hybrid Particulate Collector Technology

Another hybrid system under development is the Advanced Hybrid Particulate Collector (AHPC). This technology, which was developed by the University of North Dakota Energy and Environmental Research Center (EERC) and is being demonstrated by EERC, DOE, W. L. Gore & Associates, and Otter Tail Power Company, is unique because instead of placing the ESP and fabric filter in series, the filter bags are placed directly between ESP collection plates [78]. A schematic diagram of the AHPC is shown in Figure 9.25 [79]. The collection plates are perforated with 45 percent open area to allow dust to reach the bags; however, because the particles become charged before they pass through the plates, more than 90 percent of the particulate mass is collected on the plates before it ever reaches the bags [80]. The low-dust loading to the bags allows them to be operated at high-filtration velocity (i.e., smaller device as 65 to 75 percent fewer bags are needed) and be cleaned without the normal concern for dust reentrainment [81]. When pulses of air are used to clean the filter bag, the dislodged particles are injected into the ESP fields, where they have another opportunity to be collected on the plates. Because these bags will not need to be cleaned as often as in typical baghouse operations, they are expected to have excellent performance over a long operating life, thereby leading to lower operating costs.

Particulate capture efficiencies of greater than 99.99 percent have been achieved in a 2.5 MW slipstream demonstration [82]. The AHPC technology is expected to increase fine particulate ($PM_{2.5}$) collection efficiency by one or two orders of magnitude (i.e., 99.99–99.999 percent) [82]. A 450 MW demonstration is currently being conducted in the Big Stone cyclone-fired power plant, operated by Otter Tail Power Company and co-owned by Montana-Dakota Utilities, Northwestern Public Service, and Otter Tail Power Company burning coal from Wyoming's Powder River Basin.

Economics of Particulate Matter Control

As with other pollution control technology costs, the costs for particulate control systems are site specific and vary from country to country. They are influenced by the required emission limit and type of coal. This section summarizes the costs for ESPs, fabric filters, and hybrid systems using published data.

ESPs

The capital cost for a new electrostatic precipitator is between $40 to $60 kW, with the higher cost associated with higher collection efficiencies [17]. Since most coal-fired power plants are already fitted with ESPs, much of the published data relate to costs for upgrading existing ESPs. ESP rebuilds are less costly today due to greater market competition, the emergence of new construction techniques, and the use of wide plate spacing, requiring less collecting plates. Wide plate spacing is one of the most economic and effective approaches to replacing internals. The cost benefits result from the need for less internal elements, materials, and erection savings due to reuse of part of the original casing, and the weight savings effects on the existing support structures and foundations [17]. Costs for upgrading ESPs has been estimated at about $12/kW per field for a 500 MW unit, with the increased operating costs estimated to be $100,000 per year [17].

Flue gas conditioning has proved to be more cost effective than adding new fields. With difficult to collect fly ashes, conditioning allows operation without adding new fields. The reduction in ESP size with conditioning also lowers the operating cost because fewer fields and hoppers are used, decreasing the number of heaters, and consequently the power consumption [17]. A native SO_3 conditioning system for a 500 MW power plant requires a capital cost of $4.50/kW. Adding an anhydrous ammonia conditioning system to an existing SO_3 system would cost about $1/kW for a 50 MW unit, with the operating cost increasing by $50,000/year [17].

Staehle and colleagues [68] performed an economic analysis for using WESPs for SO_3 control at three different levels of control: 50, 80, and 95 percent. The capital costs for the three levels of control were $10, 15, and 20/kW, respectively. The total operating costs, based on 8,000 hours of operation per year, were $120,000, $160,000, and $200,000 per year, respectively.

Fabric Filters

Fabric filters are reported to cost between $50 and $70/kW [17]. Reverse-gas baghouses have higher capital and operating costs than pulse-jet baghouses because reverse-gas baghouses operate at a lower A/C ratio. Fabric filters are generally more expensive than ESPs for collection efficiencies up to 99.5 percent; however, baghouses become more cost effective for higher collection efficiencies. In addition, high-resistivity fly ashes need to be upgraded to achieve high collection efficiencies, and baghouses have economic advantages over ESPs for fly ash resistivity greater than 10^{13} to 10^{14} ohm-cm. Operating costs for baghouses are also higher than ESPs due to bag replacement and auxiliary power requirements.

Hybrid Systems

A cost analysis that was performed for a COHPAC that uses a pulse-jet bag filter following an ESP estimated that the capital cost varied from $57 to $70/kW and an operating cost of $320,000 to $570,000 per year, both depending on the unit size [17]. The analysis was performed for upgrading ESPs at a few coal-fired units ranging in size from 150 to 300 MW.

Less information is available for the AHPC, since it is in the early stages of commercialization. According to Gebert and colleagues [78], retrofit or ESP conversion jobs have been quoted in North America and Europe comparing the AHPC to a COHPAC design and, in those cases where the ESP was old and in need of significant upgrades for the hybrid filter system to function well, the AHPC has the economic advantage. For example, the 450 MW Big Stone power plant conversion has a project cost for the overall filter system of $25/kW. Gebert and colleagues [78] anticipate that these costs will decline further as more systems are built and the design is further refined and optimized.

9.2 Pollutants with Pending Compliance Regulation

As discussed in Chapter 8, on March 15, 2005, the EPA issued the first-ever federal rule to permanently cap and reduce mercury emissions from coal-fired power plants, making the United States the first country in the world to regulate mercury emissions from coal-fired power plants. The rule, called the Clean Air Mercury Rule (CAMR), created a market-based cap-and-trade program that was to permanently cap utility mercury emissions, with initial compliance required by the end of 2007. The CAMR, however, was challenged in the courts because the EPA determined that the December 2000 Regulatory Finding on the Emissions of Hazardous Air Pollutants from Electric Utility Steam Generating Units lacked foundation and that recent information demonstrated that it was not appropriate or necessary to regulate coal- and oil-fired utility units under Section 112 of the Clean Air Act, and subsequently it removed those utility units from the Section 112(c) list of source categories. On February 8, 2008, the U.S. Court of Appeals for the D.C. Circuit vacated EPA's rule removing power plants from the Clean Air Act list of sources of hazardous pollutants and vacated the CAMR. Consequently, EPA has decided to develop emissions standards for power plants under the Clean Air Act (Section 112), consistent with the D.C. Circuit's opinion on the CAMR. This process is expected to require approximately 2 years until mercury legislation is implemented.

In the period leading up to the CAMR and subsequent court hearings, substantial research and development activities were underway to control mercury emissions. This section discusses some of the leading options for mercury control. It is not inclusive, since many technologies are being investigated at the bench- and pilot-scale; however, it does discuss several of the options closest to commercialization. Although CAMR was vacated, federal legislation is expected in the 2011 to 2012 timeframe, and mercury control options will be required. In addition, several states are in the process of imposing mercury legislation at the state level.

9.2.1 Mercury

Mercury exists in trace amounts in fossil fuels, vegetation, crustal material, and waste products [83]. Mercury vapor can be released to the atmosphere through combustion or natural processes, where it can drift for a year or more, spreading over the globe. An estimated 5,500 short tons of mercury were emitted globally in 1995 from both natural and anthropogenic sources, with coal-fired power plants in the United States contributing about 48 short tons (less than 1 percent) of the total [83].

The complexity of the mercury control issue is illustrated in a simple example from DOE. If one imagined the Houston (Texas) Astrodome filled with Ping-Pong balls that represent the quantity of flue gas emitted from coal-fired power plants in the United States each year, there would be 30 billion Ping-Pong balls present. Mercury emissions would be represented by 30 intermixed colored balls. The challenge of the industry is to remove 21 of the 30 colored balls (for 70 percent compliance) or 27 of the 30 colored balls (for 90 percent compliance) from the 30 billion balls. There are several technologies under development or being demonstrated that involve removal of mercury from the flue gas. They include sorbent injection, particulate collection systems, catalysts, or chemical additives to promote the oxidation of elemental mercury and facilitate its capture in particulate and sulfur dioxide control systems, and fixed structures in flue gas ducts that adsorb mercury.

Mercury in U.S. Coal

Over 40,000 fuel samples were analyzed as part of the Information Collection Request (ICR), and a summary of the ICR coal data, by point of origin for six regions and corresponding coal rank, is provided in Table 9.10 [84]. Appalachian bituminous coal and Western subbituminous coal accounted for about 75 percent of U.S. coal production in 1999 and more than 80 percent of the mercury entering coal-fired power plants. The composition of these coals is quite different, which can affect their mercury emissions. Appalachian coals typically have high mercury, chlorine, and sulfur contents and low calcium content, resulting in a high percentage of oxidized mercury (i.e., Hg^{2+}), whereas Western subbituminous coals typically have low concentrations of mercury, chlorine, and sulfur contents and high calcium content, resulting in a high percentage of elemental mercury (i.e., $Hg°$).

Emissions from Existing Control Technologies from Coal-Fired Power Plants

Estimates for mercury emissions from coal-fired power plants with various control technologies, based on the 1999 ICR data, are given in Table 9.11 [84]. These data show that mercury emissions are estimated at about 49 short tons in 1999. This estimate is based on 84 units tested in the third phase of the ICR (out of more than 1,100 units in the United States), and because questions about bias based on the number of samples from Eastern versus Western coal-fired boilers arise, various estimates of mercury emissions range from 40 to 52 short tons/year [84]. The ICR data indicate that the speciation of mercury exiting the stack of the boilers is primarily gas-phase oxidized (i.e., 43 percent) or elemental (i.e., 54 percent) mercury, with some particulate-bound (i.e., 3 percent) mercury present [83].

Table 9.11 provides information on the influence of various existing air pollution control devices (APCDs) on mercury removal; however, mercury capture across the APCDs can vary significantly based on coal properties; fly ash properties, including unburned carbon; specific APCD configurations; and other factors [83]. Mercury

Table 9.10 Summary of ICR Data on Mercury in Coal

Coal Region	Bituminous			Subbituminous	Lignite		Totals
	Appalachian	Interior	Western	Western	Fort Union	Gulf Coast	
Number of samples	19,530	3,763	1,471	7,989	424	623	33,800
Average ICR Coal Analysis (dry basis)							
Hg, ppm	0.126	0.09	0.049	0.068	0.09	0.119	
Cl, ppm	948	1,348	215	124	139	221	
S, %	1.7	2.5	0.6	0.5	1.2	1.4	
Ash, %	11.7	10.4	10.5	7.9	13.4	23.6	
Btu/lb	13,275	13,001	12,614	11,971	10,585	9,646	
Other Coal-Related Factors							
Ca, ppm, dry basis	2,700	6,100	7,000	14,000	32,000	33,000	
Fe, ppm, dry basis	16,000	23,000	4,200	10,000	12,000	20,000	
Moisture, % as received	2.5	6.6	4.2	19.4	37.3	34.5	
Typical heat rate, Btu/kWh	10,002	10,067	10,047	10,276	10,805	10,769	
Regional Coal Production for Utility Use							
Million short tons, as rec'd.	342	67	75	336	23	57	900
Million short tons, dry coal	333	63	72	271	14	37	790
Mercury in Coal Used by Utilities							
Short tons of Hg	42.1	5.4	3.5	18.4	1.3	4.5	75.1
Pounds of Hg/10¹² Btu	9.5	6.6	3.9	5.7	8.3	12.5	
Pounds of Hg/GWh	0.0951	0.07	0.039	0.0584	0.09	0.134	

Source: From Pavlish et al. (2003).

Table 9.11 Estimated Mercury Removal by Various Control Technologies

Control Technology	Short Tons of Mercury Entering	Number of U.S. Power Plants	Number of ICR Part III Test Sites	Estimated Mercury Removals (%)[a]	Hg Emission Calculation, EPRI ICR (short tons)
ESP cold	39.4	674	18	27	28.8
ESP cold + FGD wet	16.8	117	11	49	8.6
ESP hot	5.5	120	9	4	5.3
Fabric filter	2.9	58	9	58	1.2
Venturi particulate scrubber	2.2	32	9	18	1.8
Spray dryer + fabric filter	1.6	47	10	38	1
ESP hot + FGD wet	1.6	20	6	26	1.2
Fabric filter + FGD wet	1.5	14	2	88	0.2
Spray dryer + ESP cold	0.3	5	3	18	0.2
FBC + fabric filter	3.4	39	5	86	0.5
Integrated Gasification Combined Cycle	0.07	2	2	4	0.1
FBC + ESP cold	0.02	1	1	–	0.1
Totals	**75.3**	**1,128**	**84**		**48.8**

[a]Removals as percentage of mercury in coal calculated by EPRI
Source: From Pavlish et al. (2003).

removals across cold-side ESPs averaged 27 percent, compared to 4 percent for hot-side ESPs [84]. Removals for fabric filters were higher, averaging 58 percent, owing to additional gas-solid contact time for oxidation. Both wet and dry FGD systems removed 80 to 90 percent of the gaseous oxidized mercury, but elemental mercury was not affected. High mercury removals (i.e., 86 percent) in fluidized-bed combustors with fabric filters were attributed to mercury capture on high carbon content fly ash.

Differences in mercury emissions as a function of coal rank are one of the most significant findings of the ICR and subsequent DOE testing, as can be observed in Table 9.12. The table lists the ranges (note that values less than zero in the ICR data, due to mercury measurement limitations, have been changed to zero for averaging purposes) and average cobenefit mercury capture by coal rank and APCD configuration [85]. Specifically, it is the fact that units burning subbituminous coal and lignite frequently demonstrate worse mercury capture than similarly equipped bituminous coal-fired plants. An exception to this is the case where a subbituminous coal is fired in a boiler containing an SCR and cold-side ESP. The SCR is oxidizing the elemental mercury, which is enhancing the ESP's effectiveness in capturing the mercury.

These data also demonstrate the improved mercury capture effectiveness of wet FGD systems when an upstream SCR system is in service. For example, average mercury capture for bituminous coal-fired plants equipped with a cold-side ESP

Table 9.12 Average Cobenefit Mercury Capture by Coal Rank and APCD Configurations

APCD Configuration	Average Percentage Mercury Capture (range of mercury capture)				
	Bituminous Coal		Subbituminous Coal		Lignite
	w/o SCR	w/ SCR	w/o SCR	w/ SCR	w/o SCR
CS-ESP	28 (0–92)	8 (0–18)	13 (0–61)	69 (58–79)	8 (0-18)
CS-ESP+Wet FGD	69 (41–91)	85 (70–97)	29 (2–60)	N/A	44 (21–56)
HS-ESP	15 (0–43)	N/A	7 (0–27)	N/A	N/A
HS-ESP+Wet FGD	49 (38–59)	N/A	29 (0–49)	N/A	N/A
FF	90 (84–93)	N/A	72 (53–87)	N/A	N/A
FF+Wet FGD	98 (97–99)	N/A	N/A	N/A	N/A
SDA+FF	98 (97–99)	95 (89.99)	19 (0–47)	N/A	4 (0–8)
SDA+CS-ESP	N/A	N/A	38 (0–63)	N/A	N/A
PS	N/A	N/A	9 (5–14)	N/A	N/A
PS+Wet FGD	32 (7–58)	91 (88–93)	10 (0–74)	N/A	33 (9–51)

Note: CS-ESP, cold-side ESP; HS-ESP, hot-side ESP; PS, particulate scrubber; SDA, spray dryer absorber; FF, fabric filter; SCR, selective catalytic reactor; FGD, flue gas desulfurization; w/, with; w/o, without
Source: From Feeley et al. (2003).

and wet FGD increased from 69 to 86 percent with the addition of an SCR. FGD systems are more efficient at mercury removal with upstream SCR systems, but FGD systems also remove significant mercury quantities without an SCR system. The data indicate that for pulverized coal-fired units, the greatest cobenefit for mercury control is obtained for bituminous coal-fired units equipped with a fabric filter for particulate matter control and either a WFGD or spray dryer absorber for sulfur dioxide control. In these cases, average mercury capture of 98 percent was observed for both cases. The worst-performing pulverized bituminous coal-fired units were those equipped only with a hot-side ESP [83].

The rank-dependency on mercury removal is due to the speciation of the mercury in the flue gas, which can vary significantly between power plants, depending on the coal properties. Power plants that burn bituminous coal typically have higher levels of oxidized mercury than power plants that burn subbituminous coal or lignite, which is attributed to the higher chlorine and sulfur content of the bituminous coal. The oxidized mercury, as well as the particulate mercury, can be effectively captured in some conventional control devices such as an ESP, fabric filter, or FGD system, while elemental mercury is not as readily captured. The oxidized mercury can be more readily adsorbed onto fly ash particles and collected with the ash in either an ESP or a fabric filter. Also, because the most likely form of oxidized mercury present in the flue gas, mercuric chloride ($HgCl_2$), is water soluble, it is more readily absorbed in the scrubbing slurry of plants equipped with wet FGD systems compared to elemental mercury, which is not water soluble [83]. The use of SCRs to significantly increase oxidation and improve removal of mercury is actively being pursued.

Technologies for Mercury Control

Many research organizations, federal agencies, technology vendors, and industrial companies are actively in the process of identifying, developing, and demonstrating cost-effective mercury control technologies for the electric utility industry. There are many technology options at various levels of testing, demonstration, and commercialization but based on the current state of development, coal treatment/combustion modifications, sorbent injection, and FGD enhancement/oxidation represent the best potential for reducing mercury emissions and meeting future mercury regulations [86].

Sorbent Injection

Although existing APCD can capture some mercury, innovative control technologies will be needed to comply with the CAMR Phase II mercury emissions cap. To date, activated carbon injection (ACI) has shown the most promise as a near-term mercury control technology, although continuous long-term operation is required to determine the effect on plant operations. In a typical configuration, Powdered activated carbon (PAC) is injected downstream of the power plant's air heater and upstream of the particulate control device—either an ESP or a fabric filter. The PAC adsorbs the mercury from the combustion flue gas and is subsequently captured along with the fly ash in the particulate control device. A variation of this concept is the TOXECON™ process, where a separate baghouse is installed after the primary particulate collector (especially when it is a hot-side ESP) and an air heater and PAC is injected prior to the TOXECON unit (i.e., TOXECON I™). This concept allows for separate treatment or disposal of fly ash collected in the primary particulate control device. A variation of the process is the injection of PAC into a downstream ESP collection field to eliminate the requirement of a retrofit fabric filter and allow for potential sorbent recycling (i.e., TOXECON II™ configuration).

Overview of Powdered Activated Carbon Injection for Mercury Control The performance of PAC in capturing mercury is influenced by the flue gas characteristics, which is determined by factors such as coal type, APCD configuration, and additions to the flue gas, including SO_3 for flue gas conditioning [87]. Research has shown that HCl and sulfur species (i.e., SO_2 and SO_3) in the flue gas significantly impact the adsorption capacity of fly ash and activated carbon for mercury. Specifically, the following results have been found [86]:

- HCl and H_2SO_4 accumulate on the surface on the carbon.
- HCl increases the mercury removal effectiveness of activated carbon and fly ash for mercury, particularly as the flue gas concentration increases from 1 to 10 ppm. The relative enhancement in mercury removal performance is not as great above 10 ppm HCl. Other strong Brønsted acids, such as the hydrogen halides HCl, HBr, or HI should have a similar effect. Halogens such as Cl_2 and Br_2 should also be effective at enhancing mercury removal effectiveness, but this may be the result of the halogens reacting directly with mercury rather than the halides thereby promoting the effectiveness of the activated carbon.

- SO$_2$ and SO$_3$ reduce the equilibrium capacity of activated carbon and fly ash for mercury. Activated carbon catalyzes SO$_2$ to H$_2$SO$_4$ in the flue gas. Because the concentration of SO$_2$ is much higher than mercury in the flue gas, the overall adsorption capacity of mercury is likely dependent on the SO$_2$ and SO$_3$ concentrations in the gas, since these form H$_2$SO$_4$ on the surface of the carbon.

Figure 9.26 shows some results from conventional PAC injection tests performed through numerous U.S. DOE test programs [86]. Conventional PAC injection was the focus of initial field testing (in 2001–2002) and serves as the benchmark for all field PAC injection tests. This work showed that a maximum of approximately 65 percent mercury capture could be achieved when firing a subbituminous coal in a power plant using an electrostatic precipitator. Also, when using conventional PAC costs of 50¢/lb, it was found that the cost to remove 70 percent mercury from a bituminous coal-fired plant with an ESP would cost about $70,000/lb of mercury removed, which is higher than the targets of $30,000 to $45,000/lb of mercury removed (in 1999 dollars) [88].

The conventional powdered activated carbon testing was followed by work with chemically treated PACs, which were developed for low-rank coal applications due the low mercury capture results in the initial testing. Some results from the chemically treated PAC testing are shown in Figure 9.27 [86]. With PAC injection at 1 lb/MMacf (million actual cubic feet of flue gas), mercury removal ranges from 70 to 95 percent for low-rank coals. For 90 percent mercury removal, costs are estimated at about $6,000/lb mercury removed for a subbituminous coal/SDA-FF combination to more than $20,000/lb mercury removed for a subbituminous coal/ESP combination assuming chemically treated PAC costs of $0.75 to $1.00/lb [86].

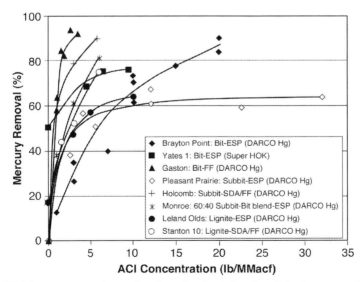

Figure 9.26 Mercury removal as a function of activated carbon injection concentration. *Source:* From Feeley (2006).

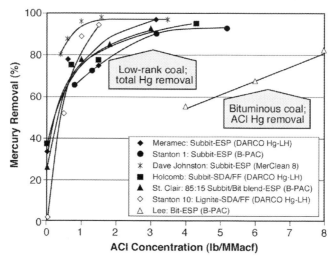

Figure 9.27 Mercury removal as a function of chemically treated activated carbon injection concentration.
Source: From Feeley (2006).

Table 9.13 Demonstrated Mercury Removal at 2 or 5 lb/MMacf

	ESP (5 lb/MMacf)	ESP with SO$_3$[a] (5 lb/MMacf)	TOXECON (2 lb/MMacf)	SDA+FF (2 lb/MMacf)
Low S, very low Cl (PRB or North Dakota lignite)	78–95% with brominated PAC	40–91% with brominated PAC	70–90%	90–95% with brominated PAC
Low S, >50 ppm Cl (some Texas lignites)	78–95%	40–91%	70–90%	90–95% with brominated PAC
Low S, bituminous coal	55–75%	40–91%	ND[b]	ND
Low S, bituminous coal with SCR	15–70%	NA[c]	ND	ND
Low S, bituminous coal	<15%	NA	NA	NA

[a]SO$_3$ from SO$_3$ injection
[b]Not available
[c]Not applicable or configuration unlikely
Source: Modified from Sjostrom et al. (2007).

Results from testing using the TOXECON technology are given in Table 9.13 (fabric filter configuration) and Figure 9.28 (ESP configuration) [87]. In the fabric filter configuration, mercury removals of 70 to 90 percent have been achieved using low-rank coals, while removal efficiencies of 50 to 90 percent are expected when using low-sulfur bituminous coals [87]. From Figure 9.28, 50 to 80 percent mercury

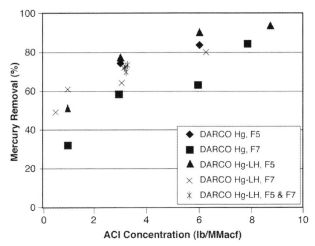

Figure 9.28 Mercury removal as a function of activated carbon injection concentration using the TOXECON II configuration at the Independence Station.
Source: From Feeley (2006).

reduction was achieved when injecting chemically treated PAC at 4–5 lb/MMacf into the next-to-last ESP field (i.e., F5) and last ESP field (i.e., F7) [86].

The improved mercury capture efficiency of the advanced chemically treated sorbent injection systems has given U.S. coal-fired power plant operators the confidence to begin deploying the technology. As of April 2008, nearly 90 full-scale ACI systems have been ordered by U.S. coal-fired power generators [89]. These contracts include both new and retrofit installations. The ACI systems have the potential to remove more than 90% of the mercury in many applications.

Balance-of-Plant Issues Long-term tests are necessary to evaluate the effect of PAC injection on plant performance, and projects addressing this have recently started. To date, through short-term field tests with plants using ESPs, no impact of PAC injection upstream of the air preheater has been observed. In general, no balance-of-plant issues, including increases in particulate emissions and changes in ESP operation, have been identified as a result of PAC injection in the ductwork upstream of the APCDs. One exception was an increase in the arc rate at low boiler load conditions as compared to baseline arcing during PAC injection [87]. Particulate measurements collected at the outlet of the ESP when injecting PAC in the TOXECON II configuration indicated that the PAC injection did not impact particulate emissions [87]. However, stack opacity and ESP outlet particulate spikes associated with PAC injection following fourth field raps were observed.

Two significant balance-of-plant issues were discovered during TOXEON I testing at Presque Isle [87]. The first was that PAC can self-ignite in the hoppers of the baghouse when it is allowed to accumulate and when exposed to external heating from hopper heaters. The second was that the ash/PAC mixture becomes stickier than ash alone and harder to remove from the hoppers and transport with the ash removal system when it is heated. The first issue is being addressed by frequently evacuating the hoppers and controlling the maximum temperature of the heating elements to less than 300°F. The second issue can be addressed through equipment design.

No balance-of-plant problems have been noted at sites using spray dryer absorbers during PAC injection. This includes opacity or changes in spray dryer absorber or fabric filter operation.

Wet Flue Gas Desulfurization

Although PAC injection has shown the most promise as a near-term mercury control technology, testing is underway to enhance mercury capture for plants equipped with wet FGD systems. These FGD-related technologies include the following:

- Coal and flue gas chemical additives with fixed-bed catalysts to increase levels of oxidized mercury in the flue gas
- Wet FGD chemical additives to promote mercury capture and prevent reemissions of previously captured mercury from the FGD absorber vessel

There is much interest in these activities, since the use of FGD systems at coal-fired power plants will likely increase significantly over the next 15 years due to the implementation of CAIR, or a modified version of the rule, that establishes a market-based allowance cap-and-trade program to permanently cap emissions of SO_2 and NO_x in 28 eastern U.S. states and the District of Columbia. In the initial version of CAIR, the SO_2 emission caps were based on percent reductions from the total number of Title V, Phase II allowances currently allocated to sources in the affected states—a 50 percent reduction for 2010 and 65 percent reduction for 2015. When fully implemented, CAIR was to reduce SO_2 emissions by more than 70 percent from 2003 levels. It is anticipated that the target levels in the new rule will be similar to the original with later dates for implementation. Table 9.14 provides a summary of the power generation capacity equipped with FGD controls in 2004, categorized by coal rank and FGD type [85]. Total FGD capacity is projected to increase from 100 GW in 2004 to about 150 GW in 2010 to about 180 GW in 2015 to 231 GW in 2020 [90]. It is anticipated that most of this will be wet-based systems with dry FGD increasing from 10 GW in 2004 to 21 GW by 2015.

Table 9.14 2004 U.S. Coal-Fired Generation Capacity with FGD Controls, GW

	Bituminous Coal	Subbituminous Coal	Lignite	Total
Wet FGD	57	24	9	90
Dry FGD	4	5	1	10
Total FGD	61	29	10	100
Total coal-fired	**213**	**96**	**15**	**324**

Source: From Feeley et al. (2003).

Wet FGD systems, especially those associated with bituminous coal-fired power plants equipped with SCR systems, appear to be good candidates for capturing mercury. With the projected increase in wet FGD systems for bituminous coal-fired power plants, the cobenefit of capturing of mercury with SO_2 can be realized. Research is currently underway to evaluate technologies that facilitate mercury oxidation and to ensure that captured mercury is not reemitted from FGD systems. At this time, research is encouraging.

Coal Cleaning

Coal cleaning is an option for removing mercury from the coal prior to utilization. Of the more than 1 billion short tons of coal mined each year in the United States, about 600 to 650 million short tons are processed to some degree [91]. Coal cleaning removes pyritic sulfur and ash. Mercury tends to have a strong inorganic association (i.e., it is associated with the pyrite), especially for Eastern bituminous coals, but mercury removal efficiencies reported for physical coal cleaning vary considerably. Physical coal cleaning is effective in reducing the concentration of many trace elements, especially if they are present in the coal in relatively high concentrations. The degree of reduction achieved is coal specific, relating in part to the degree of mineral association of the specific trace element and the degree of liberation of the trace element–bearing mineral. High levels of mercury removal (up to ≈80 percent) have been demonstrated with advanced cleaning techniques such as column flotation and selective agglomeration [92], while conventional cleaning methods, such as heavy media cyclone, combined water-only cyclone/spiral concentrators, and froth flotation have been shown to remove up to 62 percent of the mercury [93]. In both the conventional and advanced cleaning techniques, the results varied widely and were coal dependent.

Cost Estimates to Control Mercury Emissions

Cost estimates to control mercury emissions are quite variable. They represent "snapshots" in time based on many assumptions and operating conditions. As a consequence, the economics are plant- and condition-specific and are based on relatively small data sets [94]. This section will present cost estimate data primarily to illustrate comparative information between different control strategies and current data on the leading control technology at this time—that is, PAC injection.

Coal Cleaning

The costs of cleaning Eastern bituminous coals for mercury removal range from no additional cost (for coals already washed for sulfur removal) to a cost of $33,000 per pound of mercury removed [95]. The costs for cleaning Powder River Basin subbituminous coals are higher and approach $58,000 per pound of mercury removed [84]. However, mercury reductions from washing methods currently being applied are already built into the ICR mercury data for delivered coal; consequently, to realize a benefit from coal cleaning, higher levels must be employed. Advanced cleaning methods can remove additional mercury, but they are generally not economical.

Table 9.15 Estimates of Current and Projected Annualized Operating Costs
for Mercury Emissions Control Technology

Coal		Existing Controls	Retrofit Control[a]	Current Cost (mills/kWh)	Projected Cost (mills/kWh)
Type[b]	S, %				
Bit	3	ESP cold + FGD	PAC	0.727–1.197	0.436–0.718
Bit	3	Fabric filter + FGD	PAC	0.305–0.502	0.183–0.301
Bit	3	ESP hot + FGD	PAC + PFF	1.501–NA[c]	0.901–NA
Bit	0.6	ESP cold	SC + PAC	1.017–1.793	0.610–1.076
Bit	0.6	Fabric filter	SC + PAC	0.427–0.753	0.256–0.452
Bit	0.6	ESP hot	SC + PAC + PFF	1.817–3.783	1.090–2.270
Subbit	0.5	ESP cold	SC + PAC	1.150–1.915	0.690–1.149
Subbit	0.5	Fabric filter	SC + PAC	0.423–1.120	0.254–0.672
Subbit	0.5	ESP hot	Sc + PAC + PFF	1.419–2.723	0.851–1.634

[a]PAC = powdered activated carbon; SC = spray cooling; PFF = polishing fabric filter
[b]Bit = bituminous coal; subbit = subbituminous coal
[c]NA = not available
Source: From Kilgroe and Srivastava (2000).

General Economic Comparisons

As previously mentioned, cost estimates for mercury compliance are very approximate and vary from $5,000 to $70,000 per pound of mercury removed, from 0.03 to 0.8 cents/kWh, and from $1.7 to $7 billion annually for the total national cost depending on technical advances [84]. A breakdown of costs by various technology options is provided in Table 9.15 [96]. The costs are dated (from 2000), but they are provided to compare costs between technologies.

Activated Carbon Injection

Activated carbon injection (ACI) is the leading technology under development in the United States for mercury control in power plants, since there has been a significant amount of research, development, and demonstrations performed. The work started at the fundamental level and pilot-scale with full-scale tests being a priority at this time. Since about 2001, the U.S. Department of Energy, National Energy Technology Laboratory (NETL), has managed full-scale tests of mercury control technologies at nearly 50 U.S. coal-fired power generation facilities [97]. NETL has observed an improvement in both the cost and performance of mercury control during full-scale tests with chemically treated (or brominated) ACI. The improved mercury capture efficiency of these sorbent injection systems has give coal-fired power plant operators the confidence to begin deploying this technology.

Evaluating the cost of mercury control with sorbents such as PAC can be done with three scenarios [98]:

1. Using only activated carbon
2. Using activated carbon with an additive—for example, brominated PAC
3. Installation of a baghouse to handle the additional particulate load of the PAC

The third scenario is the most costly, but if a baghouse is installed for the carbon capture, the quality of the fly ash does not change significantly and has a minimal effect on fly ash sales if the fly ash was being sold.

The capital costs for installing a PAC injection system depend on the scale of the power plant. Estimates are reported as $3 to $4/kW for power plants greater than 360 MW$_e$, $4 to $6/kW for power plants ranging from 200 to 360 MW$_e$, and $6 to $9/kW for power plants smaller than 200 MW$_e$ [98]. Installing a new baghouse is $55 to $70/kW regardless of power plant size. Operating and maintenance costs vary significantly from plant to plant. Costs as high as $22 million/year have been estimated [98].

NETL has published economic data on mercury control at the various field-testing sites on which they have performed testing [97]. These data provide plant-specific cost estimates and were performed so NETL could assess its ability to achieve a target of reducing the baseline (1999) mercury control costs estimate of $60,000/ lb mercury removed by 25 to 50 percent. Figure 9.29 shows the 20-year levelized cost estimates for the incremental cost of mercury control for 90 percent ACI mercury removal (using chemically treated ACI) at seven of NETL's field-testing sites [97]. NETL finds the results encouraging, especially for the lower-rank Powder River Basin (PRB) and lignite coals. Figure 9.29 also shows an assessment of the potential for ACI to negatively impact the sale and disposal of the fly ash and therefore impact the overall cost of mercury control. Although it was demonstrated that ACI systems have the potential to remove more than 90 percent of the mercury

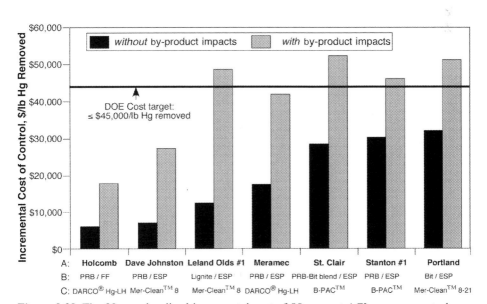

Figure 9.29 The 20-year levelized incremental cost of 90 percent ACI mercury control. A: Power plant name; B: Coal rank/particulate control device; C: Sorbent name. *Source:* From Feeley and Jones (2008).

in many applications at a cost estimate below $10,000/lb mercury removed, NETL points out that only through experience gained during long-term continuous operation of these technologies in a range of commercial applications will their actual costs and performance be determined.

9.3   Multipollutant Control

The concept of controlling/removing more than one pollutant from a single control device has been of interest to the coal-fired industry for many years. Initially, technologies to simultaneously control NO_x and SO_2, particulate matter, and NO_x, or all three pollutants, were developed and tested at various scales in DOE's Clean Coal Technology program starting in the late 1980s (see Chapter 11 for details of this program). Simultaneous NO_x and SO_2 control was demonstrated using the SNOX™, SNRB™, and Integrated Dry NO_x/SO_2 Emissions Control systems approaches [99]. Each of these demonstration projects involved a unique combination of control technologies to achieve reduction of NO_x and SO_2 emissions. The SNOX process uses an SCR, catalytic SO_2 converter (to SO_3), and a wet-gas sulfuric acid tower for removing 93 to 94 percent and 95 percent of the NO_x and SO_2, respectively.

The SNRB process combines the removal of SO_2, NO_x, and particulates in one unit, a high-temperature baghouse (fitted with a catalyst in the bag cages for NO_x reduction) located between the economizer and the combustion air preheater with a calcium- or sodium-based sorbent injected for SO_2 control, and achieves 80 percent, 50 to 95 percent, and 99+ percent control of SO_2, NO_x, and particulate matter, respectively. The Integrated Dry NO_x/SO_2 Emissions Control System is comprised of low-NO_x burners, overfire air, SNCR, and duct sorbent injection and achieves up to 80 and 70 percent reductions in NO_x and SO_2, respectively. These three technologies are potentially applicable to flue gas cleaning for all types of conventional coal-fired units, including stoker, cyclone, and pulverized coal-fired boilers. Capital costs for these three systems are estimated at $305/kW, $253/kW, and $190/kW, respectively, for SNOX, SNRB, and the Integrated Dry NO_x/SO_2 Emissions Control System. The operating costs are estimated at 12.1 mills/kWh for the SNRB process while the SNOX process generates a 6.1 mills/kWh credit [99]. No data are available for the integrated process.

Similarly, technologies to simultaneously remove NO_x and particulate matter were also developed and tested. These technologies, which consisted of coating substrates such as ceramic membrane filters with SCR catalysts, were initially being developed for high-pressure, high-temperature atmospheres found in gasification systems. This work was applied under conventional pulverized coal-fired conditions to investigate simultaneous NO_x and particulate removal [16, 100, 101]. The technology removed NO_x and particulate matter (particulate matter removal efficiencies greater than high efficiency P-84 polyimide bags were observed), and it had the added benefit of removing a significant amount of mercury as 79 percent of the mercury was removed across a 6,000 acfm ceramic membrane filter system [101]. However, unacceptable pressure drops across the ceramic filters were not solved (i.e., fine particles embedded into the ceramic substrate), hindering the commercialization of this process [102].

Interest in multipollutant control, or integrated emissions control, has intensified in the United States over the last several years primarily as a result of recently passed multipollutant legislation such as the Clean Air Interstate Rule and impending legislation discussed in Chapter 8. The pollutants of interest include SO_2, NO_x, mercury, SO_3 and other acid gases, fine particulate matter, and, more recently, CO_2. The list of multipollutant technologies is long, continues to grow, and has proceeded through pilot- and demonstration-scale testing [103], and in some cases, these technologies are commercially available. Similarly, Canada is pursuing multipollutant control for reducing SO_x, NO_x, particulate, and mercury emissions levels comparable to a natural gas combined cycle plant with SCR [104].

Several promising technologies are currently being tested, and a few of these are reviewed. Several of these evaluations are cofunded by various DOE programs, state agencies, and industry. Note that several of the technologies discussed in previous sections are also considered multipollutant control options.

9.3.1   ECO Process

Powerspan Corporation has developed an integrated air pollution control technology that has achieved major reductions in emissions of NO_x, SO_2, fine particulate matter, and mercury from coal-fired power plants. In initial tests using a 1 MW_e slipstream at FirstEnergy Corporation's R.E. Burger Plant in Ohio, the following reductions were achieved: NO_x 90 percent, SO_2 98 percent, fine particulate matter 95 percent, and mercury 80 to 90 percent [105]. The success of this work has lead to 50 MW_e commercial application, again at the R.E. Burger Plant, which has been in operation since 2004 [106]. The patented technology, named Electro-Catalytic Oxidation (ECO), also reduces emissions of air toxic compounds such as arsenic and lead as well as acid gases such as hydrochloric acid (HCl). Recently, Powerspan Corporation has modified the system to include CO_2 capture as well. A 120 MWe demonstration of this technology is planned at Basin Electric Power Cooperative's Antelope Valley Power Station in North Dakota [106].

In commercial operation, the ECO process is to be installed downstream of a power plant's existing ESP or baghouse [105, 106]. It treats the flue gas in three steps to achieve multipollutant removal. In the first process step, a barrier discharge reactor oxidizes gaseous pollutants to higher oxides—in other words, nitric oxide to nitrogen dioxide, a portion of the sulfur dioxide to sulfuric acid, and mercury to mercuric oxide. Following the barrier discharge reactor is the ammonia scrubber, which removes unconverted sulfur dioxide and nitrogen dioxide produced in the barrier discharge. A WESP follows the scrubber, and it, along with the scrubber, captures acid aerosols produced by the discharge reactor, fine particulate matter, and oxidized mercury. The WESP also captures aerosols generated in the ammonia scrubber. Liquid effluent from the ammonia scrubber contains dissolved sulfate and nitrate salts, along with mercury and captured particulate matter. It is sent to a by-product recovery system, which includes filtration to remove ash and activated carbon adsorption for mercury removal. The treated by-product stream, free of mercury and ash, can be processed to form ammonia sulfate/nitrate fertilizer. When the CO_2 module is used, an ammonia-based solution is used to scrub the CO_2 from the flue gas [106].

Powerspan's capital cost estimate, in 2003, is about \$200/kW, including balance of plant modifications [105]. The levelized operating and maintenance costs are estimated to be 2.0 to 2.5 mills/kWh. This is for the Electro-Catalytic Oxidation process without the CO_2 module.

9.3.2 Airborne Process

Airborne Pollution Control, in cooperation with LG&E Energy Corporation, The Babcock and Wilcox Company, and USFilter HPD Systems, developed an emerging multipollutant, postcombustion control system [107]. The technology combines the use of dry sodium bicarbonate injection, coupled with enhanced wet sodium carbonate scrubbing to provide SO_x, NO_x, mercury, and other heavy metal reductions. Although sodium bicarbonate scrubbing is well known as an effective flue gas cleanup process, commercial application has been prevented by the high cost of sodium bicarbonate, the limited economic value of the scrubber product (i.e., sodium sulfate), and the economic and environmental issues associated with sodium sulfate disposal [103]. Airborne has developed a recycling process that will regenerate sodium sulfate back into sodium bicarbonate and a sulfate-based fertilizer product that may eliminate the financial and disposal barriers. Currently, the process is being marketed by Airborne Clean Energy [108]. Testing has reached the 5 MW_e level, and China has expressed interest in the technology.

9.3.3 Multipollutant Control for Smaller Coal-Fired Boilers

In a U.S. Department of Energy Power Plant Improvement Initiative (PPII; the PPII is discussed in more detail in Chapter 11) project that was completed in October 2008 at the AES Greenidge 100 MW_e power plant, cost-effective removal of multiple pollutants from older power plants (i.e., retrofits) was successfully demonstrated [109]. The demonstration focused on older power plants because more than 400 smaller coal-fired plants with capacities between 50 and 300 MW_e exist in the United States. These power plants produce about 60 GW_e, which is about 20 percent of the country's coal-based capacity.

The multipollutant control equipment included a hybrid SNCR/SCR system for NO_x control, a circulating fluidized-bed scrubbing system for SO_2, mercury, acid gas, and particulate matter control, and an activated carbon injection system to reduce mercury. The system demonstrated average pollutant reductions of 98 percent for SO_2, 95 percent for SO_3, 97 percent for HCl, and 98 percent for mercury.

9.3.4 Nalco Mobotec Systems

Nalco Mobotec, formerly Mobotec USA, Inc., has developed a multipollutant control system that consists of furnace sorbent injection using limestone or trona in combination with ROFA (rotary opposed fire air) and ROTAMIX (a second-generation SNCR and sorbent injection process) for SO_2 and mercury removal. Mixing in the ROFA and ROTAMIX systems create optimal conditions for achieving multipollutant reduction by providing ample turbulence and residence time within a specific temperature window [103, 110]. Tests with trona and limestone have achieved reductions in SO_2 (69 and 64 percent), SO_3 (90 and 90 percent), HCl (0 and 75 percent), mercury (89 and 67 percent), NO_x (4 and 11 percent), and particulate matter (18 and 80 percent), respectively [111].

9.3.5 Others

Combining sorbent injection for mercury control with other technologies for NO_x and/or SO_x removal represents another multipollutant control option. Many companies are exploring sorbent and chemical injection techniques that can remove mercury at reasonable costs, including Sorbent Technologies Corporation, ADA-Environmental Solutions, URS Corporation, EPRI, and Alstom Power, as well as Universities such as the University of North Dakota Energy and Environmental Research Center and Penn State University, and Federal agencies such as DOE's National Energy Technology Laboratory.

References

[1] DOE (U.S. Department of Energy), National Energy Technology Laboratory Accomplishments FY 2002, Office of Fossil Energy, August 2003.
[2] EIA (U.S. Energy Information Agency), Electric Power Annual 2007, U.S. Department of Energy, Office of Coal, Nuclear, Electric and Alternate Fuels, U.S. Government Printing Office, January 21, 2008.
[3] EPA (U.S. Environmental Protection Agency), Latest Findings on National Air Quality Status and Trends through 2006, Office of Air Quality Planning and Standards, U.S. Government Printing Office, 2009.
[4] EPA, Acid Rain and Related Programs: 2008 Emission, Compliance, and Market Analysis, Office of Air Quality Planning and Standards, U.S. Government Printing Office, January 2008.
[5] K. Wark, C.F. Warner, W.T. Davis, Air Pollution Its Origin and Control, third ed., Addison Welsey Longman, 1998.
[6] H.N. Soud, Developments in FGD, IEA Coal Research, 2000.
[7] R.K. Srivastava, C. Singer, W. Jozewicz, SO_2 Scrubbing Technolgoies: A Review, in: Proceedings of the AWMA 2000 Annual Conference and Exhibition, 2000.
[8] W.T. Davis (Ed.), Air Pollution Engineering Manual, second ed., John Wiley & Sons, 2000.
[9] T.C. Elliot (Ed.), Standard Handbook of Powerplant Engineering, McGraw-Hill, 1989.
[10] EIA, U.S. Coal Reserves: 1997 Update, U.S. Department of Energy, Office of Coal, Nuclear, Electric and Alternate Fuels, U.S. Government Printing Office, February 1999.
[11] C.D. Harrison, Fuel Options to Mitigate Emissions Reduction Costs, in: Proceedings of the 28th International Technical Conference on Coal Utilization & Fuel Systems, Coal & Slurry Technology Association, 2003.

[12] S.C. Stultz, J.B. Kitto (Eds.), Steam: Its Generation and Use, fortieth ed., The Babcock and Wilcox Company, 1992.

[13] P.T. Radcliffe, Economic Evaluation of Flue Gas Desulfurization Systems, Electric Power Research Institute, 1991.

[14] R. Rhudy, M. McElroy, G. Offen, Status of Calcium-Based Dry Sorbent Injection SO$_2$ Control, in: Proceedings of the Tenth Symposium on Flue Gas Desulfurization, November 17–21, 1986, pp. 9.69–9.84.

[15] V.V. Bland, C.E. Martin, Full-Scale Demonstration of Additives for NO$_2$ Reduction with Dry Sodium Desulfurization, Electric Power Research Institute, EPRI GS-6852, June 1990.

[16] B.G. Miller, A.L. Boehman, P. Hatcher, H. Knicker, A. Krishnan, J. McConnie, et al., The Development of Coal-Based Technologies for Department of Defense Facilities Phase II Final Report, Prepared for the U.S. Department of Energy Federal Energy Technology Center, Pittsburgh, July 31, 2000, DE-FC22-92PC92162, 784 pages.

[17] Z. Wu, Air Pollution Control Costs for Coal-Fired Power Stations, IEA Coal Research, 2001.

[18] D.J. Smith, Cost of SO$_2$ Scrubbers Down to $100/kW, Power Engineering (September 2001) 63–68.

[19] G. Blythe, B. Horton, R. Rhudy, EPRI FGD Operating and Maintenance Cost Survey, in: Proceedings of the EPRI-DOE-EPA Combined Utility Air Pollution Control Symposium: The MEGA Symposium: Volume I: SO$_2$ Controls, 1999, pp. 1–21 to 1–34.

[20] S.C. Mitchell, NO$_x$ in Pulverized Coal Combustion, IEA Coal Research, 1998.

[21] R.M. Davidson, How Coal Properties Influence Emissions, IEA Coal Research, 2000.

[22] R. Moreea-Taha, NO$_x$ Modelling and Prediction, IEA Coal Research, 2000.

[23] EPA, Technical Bulletin, Nitrogen Oxides (NO$_x$), Why and How They Are Controlled, Office of Air Quality Planning and Standards, U.S. Government Printing Office, November 1999.

[24] C.J. Lawn (Ed.), Principles of Combustion Engineering for Boilers, Academic Press, 1987.

[25] C. Tsiou, H. Lin, S. Laux, J. Grusha, Operating Results from Foster Wheeler's New Vortex Series Low-NO$_x$ Burners, in: Proceedings of Power Gen, 2000.

[26] T.H. Steitz, R.W. Cole, Field Experience in Over 30,000 MW of Wall Fired Low NO$_x$ Installations, in: Proceedings of Power Gen, 1996.

[27] T.H. Steitz, J. Grusha, R. Cole, Wall Fired Low NO$_x$ Burner Evolution for Global NO$_x$ Compliance, in: Proceedings of the 23rd International Technical Conference on Coal Utilization & Fuel Systems, Coal & Slurry Technology Association, 1998.

[28] R.L. Patel, D.E. Thornock, R.W. Borio, B.G. Miller, A.W. Scaroni, Firing Micronized Coal with a Low NO$_x$ RSFC Burner in an Industrial Boiler Designed for Oil and Gas, in: Proceedings of the Thirteenth Annual International Pittsburgh Coal Conference, 1996.

[29] R.W. Borio, R.L. Patel, D.E. Thornock, B.G. Miller, A.W. Scaroni, J.G. McGowan, Task 5—Final Report: One Thousand Hour Demonstration Test in the Penn State Boiler, Prepared for U.S. Department of Energy, Federal Energy Technology Center, No. DE-AC22-91PC91160, March 1998.

[30] Z. Wu, NO$_x$ Control for Pulverized Coal Fired Power Stations, IEA Coal Research, 2002.

[31] DOE, Innovation for Existing Plants, Office of Fossil Energy, October 2007.

[32] EIA, Electric Power Monthly, U.S. Department of Energy, Office of Coal, Nuclear, Electric and Alternate Fuels, U.S. Government Printing Office, September 11, 2009.

[33] EPA, Control of NO_x Emissions by Reburning, Office of Research and Development, U.S. Government Printing Office, February 1996.
[34] J.O.L. Wendt, C.V. Sternling, M.A. Matovich, Reduction of Sulfur Trioxide and Nitrogen Oxides by Secondary Fuel Injection, in: Proceedings of the 14th Symposium (International) on Combustion, Combustion Institute, 1973, pp. 897–904.
[35] A.L. Myerson, F.R. Taylor, B.G. Faunce, Ignition Limits and Products of the Multistage Flames of Propane-Nitrogen Dioxide Mixtures, in: 6th Symposium (International) on Combustion, The Combustion Institute, 1957, pp. 154–163.
[36] R.D. Stoesssner, E. Zawadzki, Coal Water Slurry Dual Firing Project for Homer City Station—Phase I Test Results, 16th International Technical Conference on Coal Utilization & Fuel Systems, Coal & Slurry Technology Association, 1991, pp. 599–608.
[37] S. Falcone Miller, B.G. Miller, A.W. Scaroni, S.A. Britton, D. Clark, W.P. Kinneman, et al., Coal-Water Slurry Fuel Combustion Program, Pennsylvania Electric Power Company, 1993, 98 pages.
[38] J.L. Morrison, B.G. Miller, A.W. Scaroni, Determining Coal Slurryability: A UCIG/Penn State Initiative, Electric Power Research Institute, WO3852-06, January 1998.
[39] R.A. Ashworth, T.M. Sommer, Economical Use of Coal Water Slurry Fuels Produced from Impounded Coal Fines, in: Proceedings of Effects of Coal Quality on Power Plants, Electric Power Institute, 1997.
[40] S. Falcone Miller, J.L. Morrison, A.W. Scaroni, The Effect of Cofiring Coal-Water Slurry Fuel Formulated from Waste Coal Fines with Pulverized Coal on NO_x Emissions, in: Proceedings of the 21st International Technical Conference on Coal Utilization & Fuel Systems, Coal & Slurry Technology Association, 1996.
[41] B.G. Miller, S. Falcone Miller, J.L. Morrison, A.W. Scaroni, Cofiring Coal-Water Slurry Fuel with Pulverized Coal as a NO_x Reduction Strategy, in: Proceedings of the Fourteenth International Pittsburgh Coal Conference, 1997.
[42] J.J. Battista, personal communication, 1997.
[43] D.A. Tillman, NO_x Reduction Achieved through Biomass Cofiring, in: Proceedings of the 20th Annual International Pittsburgh Coal Conference, September 2003.
[44] D.A. Tillman, N.A. Foster Wheeler, personal communication, November 2003.
[45] D.A. Tillman, N.S. Harding, Fuels of Opportunity: Characteristics and Uses in Combustion Systems, Elsevier, 2004.
[46] D.A. Tillman, B.G. Miller, D. Johnson, Analyzing Opportunity Fuels for Firing in Coal-Fired Boilers, in: Proceedings of the 20th Annual International Pittsburgh Coal Conference, September 2003.
[47] E.R. Vasquez, H. Gadalla, K. McQuistan, F. Iman, R.E. Sears, NO_x Control in Coal-Fired Cyclone Boilers using SmartBurn Combustion Technology, in: Proceedings of the EPRI-DOE-EPA Combined Power Plant Air Pollution Control MEGA Symposium, 2003.
[48] DOE, Clean Coal Technology, Control of Nitrogen Oxide Emissions: Selective Catalytic Reduction (SCR), Topical Report Number 9, U.S. Department of Energy, July 1997.
[49] R.W. McIlvaine, H. Weiler, W. Ellison, SCR Operating Experience of German Powerplant Owners as Applied to Challenging, U.S., High-Sulfur Service, in: Proceedings of the EPRI-DOE-EPA Combined Power Plant Air Pollution Control MEGA Symposium, 2003.
[50] J.E. Cichanowicz, L.L. Smith, L.J. Muzio, J. Marchetti, 100 GW of SCR: Installation Status and Implications of Operating Performance on Compliance Strategies, in: Proceedings of the EPRI-DOE-EPA Combined Power Plant Air Pollution Control MEGA Symposium, 2003.

[51] McIlvaine Company, China to Dominate the NO_x Market as a Purchaser and a Supplier, www.mcilvainecompany.com/, August 2009.

[52] EPA, Performance of Selective Catalytic Reduction on Coal-Fired Steam Generating Units, Office of Air and Radiation, U.S. Government Printing Office, June 25, 1997.

[53] P. Lodder, J.B. Lefers, Effect of Natural Gas, C_2H_6, and CO on the Homogenous Gas Phase Reduction of NO_x by NH_3, Chemical Engineering Journal 30 (3) (1985) 161.

[54] V. Ciarlante, M.A. Zoccola, Conectiv Energy Successfully Using SNCR for NO_x Control, Power Engineering 105 (6) (2001) 61–62.

[55] N. Frederick, R.K. Agrawai, S.C. Wood, NO_x Control on a Budget: Induced Flue Gas Recirculation, Power Engineering 107 (7) (2003) 28–32.

[56] B. Hoskins, Uniqueness of SCR Retrofits Translates into Broad Cost Variations, Power Engineering 107 (5) May (2003) 25–30.

[57] EPRI (Electric Power Research Institute), EPRI 2002 Annual Report, Electric Power Research Institute, 2003, pp. 11–12.

[58] Q. Zhu, Developments in Particulate Control, IEA Clean Coal Centre, 2003.

[59] EPA, Technology Transfer Network Clearing House for Inventories and Emissions, National Emissions Trends, www.epa.gov/ttn/chief/index.htmlm, 2009.

[60] H.N. Soud, S.C. Mitchell, Particulate Control Handbook for Coal-Fired Plants, IEA Coal Research, 1997.

[61] DOE, Description—PM Emissions Control, www.netl.doe.gov/coalpower/environment/ pm/description.html, last updated December 2, 2003.

[62] DOE, Controlling Air Toxics with Electrostatic Precipitators Fact Sheet, Office of Fossil Energy, 1997.

[63] B&W, Electrostatic Precipitator Product Sheet, PS151 2M A 12/82, The Babcok & Wilcox Company, December 1982.

[64] B.G. Miller, D.A. Tillman (Eds.), Combustion Engineering Issues for Solid Fuel Systems, Academic Press, 2008.

[65] B.G. Miller, unpublished data, 1986.

[66] B.G. Miller, S.J. Miller, G.P. Lamb, J.A. Luppens, Sulfur Capture by Limestone Injection during Combustion of Pulverized Panola County Texas Lignite, in: Proceedings of Gulf Coast Lignite Conference, 1984.

[67] W. Buckley, I. Ray, Application of Wet Electrostatic Precipitation Technology in the Utility Industry for PM2.5 Control, in: Proceedings of the EPRI-DOE-EPA Combined Power Plant Air Pollution Control MEGA Symposium, 2003.

[68] R.C. Staehle, R.J. Triscori, G. Ross, K.S. Kumar, E. Pasternak, The Past, Present and Future of Wet Electrostatic Precipitators in Power Plant Applications, in: Proceedings of the EPRI-DOE-EPA Combined Power Plant Air Pollution Control MEGA Symposium, 2003.

[69] R. Altman, G. Offen, W. Buckley, I. Ray, Wet Electrostatic Precipitation Demonstrating Promise for Fine Particulate Control—Part I, Power Engineering 105 (1) (2001) 37–39.

[70] R. Altman, W. Buckley, I. Ray, Wet Electrostatic Precipitation Demonstrating Promise for Fine Particulate Control—Part II, Power Engineering 105 (2) (2001) 42–44.

[71] C.J. Bustard, K.M. Cushing, D.H. Pontius, W.B. Smith, R.C. Carr, Fabric Filters for the Electric Utility Industry, Volume 1 General Concepts, Electric Power Research Institute, 1988.

[72] DOE, National Energy Technology Laboratory, NETL Coal Power Database 2000, www.netl.doe.gov/energy-analyses/pubs, 2002.

[73] H.N. Soud, Developments in Particulate Control for Coal Combustion, IEA Coal Research, 1995.

[74] J. Kitto, S. Stultz (Eds.), Steam, Its Generation and Use, fortyfirst ed., The Babcock & Wilcox Company, 2005.

[75] R.L. Miller, W.A. Harrison, D.B. Prater, R. Chang, Alabama Power Company E.C. Gaston 272 MW Electric Steam Plant—Unit No. 3 Enhanced COHPAC I Installation, in: Proceedings of the EPRI-DOE-EPA Combined Utility Air Pollution Control Symposium: The MEGA Symposium: Volume III: Particulates and Air Toxics, 1997.

[76] C.J. Bustard, S.M. Sjostrom, R. Chang, Predicting COHPAC Performance, in: Proceedings of the EPRI-DOE-EPA Combined Utility Air Pollution Control Symposium: The MEGA Symposium: Volume III: Particulates and Air Toxics, 1997.

[77] K.M. Cushing, W.A. Harrison, R.L. Chang, Performance Response of COHPAC I Baghouse during Operation with Normal and Artificial Changes in Inlet Fly Ash Concentration and during Injection of Sorbents for Control of Air Toxics, in: Proceedings of the EPRI-DOE-EPA Combined Utility Air Pollution Control Symposium: The MEGA Symposium: Volume III: Particulates and Air Toxics, 1997.

[78] R. Gebert, C. Rinschler, D. Davis, U. Leibacher, P. Studer, W. Eckert, et al., Commercialization of the Advanced Hybrid Filter Technology, Conference on Air Quality III: Mercury, Trace Elements, and Particulate Matter, University of North Dakota, 2002.

[79] DOE, Advanced Hybrid Particulate Collector Fact Sheet, Office of Fossil Energy, June 2001.

[80] DOE, Control Technology Advanced Hybrid Particulate Collector, www.netl.doe.gov/coalpower/environment/pm/con_tech/hybrid.html, last updated December 2, 2003.

[81] S. Blankinship, Hybrid Filter Technology Weds ESPs with Bag Filters, Power Engineering 106 (2) (2002) 9.

[82] DOE, Demonstration of a Full-Scale Retrofit of the Advanced Hybrid Particulate Collector (AHPC) Collector Fact Sheet, Office of Fossil Energy, February 2003.

[83] T.J. Feeley, J. Murphy, J. Hoffman, S.A. Renninger, A Review of DOE/NETL's Mercury Control Technology R&D Program for Coal-Fired Power Plants, DOE/NETL Hg R&D Review, www.netl.doe.gov/, April 2003.

[84] J. Pavlish, J.E.A. Sondreal, M.D. Mann, E.S. Olson, K.C. Galbreath, D.L. Laudal, et al., Status Review of Mercury Control Options for Coal-Fired Power Plants, Fuel Processing Technology 82 (2003) 89–165.

[85] T.J. Feeley, J. Murphy, J. Hoffman, S.A. Renninger, A Review of DOE/NETL's Mercury Control Technology R&D Program for Coal-Fired Power Plants, DOE/NETL Hg R&D Review, www.netl.doe.gov/, April 2003.

[86] T.J. Feeley, U.S. DOE's Hg Control Technology RD&D Program—Significant Progress, But More Work to be Done!, Mercury 2006, Conference on Mercury as a Global Pollutant, Madison, August 6–11, www.netl.doe.gov/technologies/coalpower/ewr/index.html, 2006.

[87] S. Sjostrom, T. Campbell, J. Bustard, R. Stewart, Activated Carbon Injection for Mercury Control: Overview, The 32nd International Technical Conference on Coal Utilization & Fuel Systems, Clearwater, June 10–16, 2007.

[88] A.P. Jones, J.W. Hoffmann, D.N. Smith, T.J. Feeley III, J.T. Murphy, DOE/NETl's Phase II Mercury Control Technology Field Testing Program, Updated Economic Analysis of Activated Carbon Injection, May www.netl.doe.gov/technologies/coalpower/ewr/index.html, 2007.

[89] T. Feeley, A. Jones, An Update on DOE/NETL's Mercury Control Technology Field Testing Program, DOE White Paper, www.netl.doe.gov/technologies/coalpower/ewr/mercury/index.html, U.S. Department of Energy White Paper, 2008.

[90] Multipollutant Regulatory Analysis: CAIR/CAMR/CAVR, U.S. Environmental Protection Agency, *www.epa.gov/airmarkets/mp/cair_camr_cavr.pdf*, October 2005.

[91] National Research Council, Coal Waste Impoundments: Risks, Responses, and Alternatives, National Academy Press, 2002.

[92] M.C. Jha, F.J. Smit, G.L. Shields, N. Moro, Engineering Development of Advanced Physical Fine Coal Cleaning for Premium Fuel Applications Project Final Report, DOE Contract No. DE-AC22-92PC92208, September 1997.

[93] D.J. Akers, C.E. Raleigh, The Mechanism of Trace Element Removal during Coal Cleaning, Coal Preparation 19 (3) (1998) 257–269.

[94] A.P. Jones, J.W. Hoffman, D.N. Smith, T.J. Feeley, J.T. Murphy, DOE/NETL's Phase II Mercury Control Technology Field Testing Program: Preliminary Economic Analysis of Activated Carbon Injection, Environ. Sci. Technol. 41 (4) (2007) 1365–1371.

[95] D.J. Akers, B. Toole-O'Neil, Coal Cleaning for HAP Control: Cost and Performance, in: Proceedings of the 23rd International Technical Conference on Coal Utilization & Fuel Systems, Coal & Slurry Technology Association, 1998.

[96] J.D. Kilgroe, R.K. Srivastava, Technical Memorandum: Control of Mercury Emissions from Coal-Fired Electric Utility Boilers, U.S. Environmental Protection Agency, September 2000.

[97] T.J. Feeley, A.P. Jones, An Update on DOE/NETL's Mercury Control Technology Field Test Program, U.S. Department of Energy White Paper Office of Fossil Energy, July 2008.

[98] L.L. Sloss, Economics of Mercury Control, IEA Coal Research, 2008.

[99] DOE, Clean Coal Technology, Technologies for the Combined Control of Sulfur Dioxide and Nitrogen Oxides Emissions from Coal-Fired Boilers, Topical Report Number 13, U.S. Department of Energy, May 1999.

[100] B.G. Miller, A.L. Boehman, P. Hatcher, H. Knicker, A. Krishnan, J. McConnie, et al., The Development of Coal-Based Technologies for Department of Defense Facilities Phase II Final Report, Prepared for the U.S. Department of Energy Federal Energy Technology Center, Pittsburgh, July 31, 2000, DE-FC22-92PC92162, 784 pages.

[101] B.G. Miller, S. Falcone Miller, R.T. Wincek, A.W. Scaroni, A Demonstration of Fine Particulate and Mercury Removal in a Coal-Fired Industrial Boiler Using Ceramic Membrane Filters and Conventional Fabric Filters, in: Proceedings of the EPRI-DOE-EPA Combined Utility Air Pollutant Symposium "The Mega Symposium," 1999.

[102] B.G. Miller, S. Falcone Miller, R.T. Wincek, A.W. Scaroni, A Preliminary Evaluation of Ceramic Filters Including the Use of Nondestructive X-Ray Computerized Tomography, in: Proceedings of the 26th International Technical Conference on Coal Utilization & Fuel Systems, 2001.

[103] B.K. Schimmoller, Lack of Environmental Certainty Renews Emphasis on Low-Cost Emissions Control, Power Engineering 107 (9) (2003) 32–38.

[104] D.H. Cameron, C.E. Martin, W.A. Campbell, R.A. Stobbs, The Future of Multipollutant Control for Coal-Fired Boilers: A Canadian Perspective, in: Proceedings of the EPRI-DOE-EPA Combined Power Plant Air Pollution Control MEGA Symposium, 2003.

[105] C.R. McLarnon, D. Steen, Combined SO_2, NO_x, PM, and Hg Removal from Coal Fired Boilers, in: Proceedings of the EPRI-DOE-EPA Combined Power Plant Air Pollution Control MEGA Symposium, 2003.

[106] Powerspan ECO Process, *www.powerspan.com*, 2009.

[107] M.E. Mortson, F.C. Owens II, Multi Pollutant Control with the Airborne Process, in: Proceedings of the EPRI-DOE-EPA Combined Power Plant Air Pollution Control MEGA Symposium, 2003.

[108] Airborne Clean Energy, *www.airbornecleanenergy.com*, 2009.

[109] U.S. Department of Energy, Clean Coal Demonstrations, Power Plant Improvement Initiatives, *www.netl.doe.gov/technologies/coalpower/cctc/PPII*, 2009.

[110] Nalco Mobotec Technologies, *www.nalcomobotec.com*, 2009.

[111] E. Haddad, J. Ralson, G. Green, S. Castagnero, Full-Scale Evaluation of a Multipollutant Reduction Technology: SO_2, Hg, and NO_x, in: Proceedings of the EPRI-DOE-EPA Combined Power Plant Air Pollution Control MEGA Symposium, 2003.

10 CO_2 Capture and Storage

As discussed in previous chapters, coal-fired power plants have continued to make significant progress in reducing emissions of sulfur and nitrogen oxides, particulate matter, and mercury since the passage of the 1970 Clean Air Act. However, with the CO_2 concentration in the atmosphere increasing from fossil fuel combustion causing concern about global warming, regulations on CO_2 emissions are expected in the near future. At present, however, there are no cost-effective CO_2 control technologies available for coal-fired plants, and no capture technologies have been demonstrated at scale [1]. Recognizing this, the U.S. Department of Energy (DOE) National Energy Technology Laboratory (NETL) has initiated a research and development program that is directed specifically at postcombustion and oxy-combustion CO_2 capture technologies that can be retrofitted to existing coal-fired power plants, as well as designed into new plants. The goal of the program is to develop advanced CO_2 capture and compression technologies for both existing and new coal-fired power plants that, when combined, can achieve 90 percent CO_2 capture at less than a 35 percent increase in cost of electricity (COE) at a commercial scale by 2020 [1]. The timeline is shown in Figure 10.1.

Coal-fired power plants generate approximately 50 percent of the electricity in the United States, and it is projected that the more than 300 gigawatts (GW) of coal-fired generating capacity currently in operation will increase to more than 400 GW by 2030 [3]. The existing fleet of coal-fired power plants emits about 2 billion metric tons of CO_2 annually, as illustrated in Figure 10.2, accounting for about two-thirds of the total emissions in the U.S. power sector [1, 4]. Moreover, as shown in Figure 10.3, more than 90 percent of the coal-fired CO_2 emissions projected to be released from 2007 through 2030 will originate from today's existing coal-fired power plants, since less than 4 GW of capacity is projected to retire during that period [3].

Carbon capture and storage (CCS) is one way to address the rising CO_2 concentration in the power industry. It is not the use of fossil fuels per se that needs to be reduced but rather the emission of CO_2 that is produced in their use. This chapter contains an overview of the technologies being investigated for CO_2 capture and sequestration in anticipation of future CO_2 emissions legislation. The status and types of technologies under development are continually updated by DOE on their Carbon Sequestration Program website *www.netl.doe.gov/technologies/carbon_seq/index.html*.

For a given energy content, coal, being primarily carbon, produces the most CO_2, oil will produce less, and natural gas, which derives a significant amount of its energy content from the hydrogen component of methane, will produce the least. The EPA has published the following CO_2 emissions factors for fossil fuel combustion: 207 lb/MM Btu for coal, 168 lb/MM Btu for oil, and 117 lb/MM

Clean Coal Engineering Technology. DOI: 10.1016/B978-1-85617-710-8.00010-8
Copyright © 2011 by Elsevier Inc. All rights of reproduction in any form reserved.

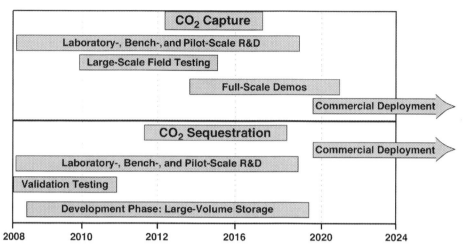

Figure 10.1 Research, development, and deployment timeline to commercial deployment of CO_2 capture technologies.
Source: From Ciferno (2009) [2].

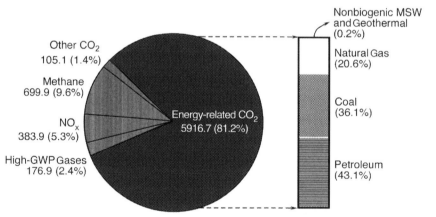

Figure 10.2 The 2007 U.S. greenhouse gas emissions (in million metric tons) by gas and U.S. CO_2 emissions from the energy sector.
Source: From DOE (2006).

Btu for natural gas [5]. Also, lower-rank coals, such as lignite and the low-sulfur subbituminous coal, which is commonly used to replace high-sulfur bituminous coal, produce more carbon dioxide per unit of heat than the higher-rank bituminous coals. Consequently, average yearly emissions of CO_2 from electric utilities in the United States have been steadily increasing (e.g., from 206.7 to 208.2 lb of CO_2/MM Btu from 1980 to 1997, respectively) due to switching from high-rank to low-rank coals [5].

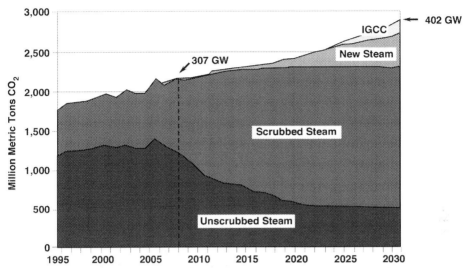

Figure 10.3 U.S. coal-fired electric power generation CO$_2$ emissions projection.
Source: From Ciferno (2009).

Switching from a high-carbon fuel to a low-carbon fuel, such as from coal to natural gas, for electricity generation would greatly reduce CO$_2$ emissions. This strategy, however, is dependent on abundant, affordable natural gas, which the United States does not have. Natural gas availability and affordability has been extremely volatile, as is evident each fall and winter since fall 1999. Switching the United States' electricity generating capacity from coal to natural gas is not sound energy policy; thus, options for capturing and sequestering CO$_2$ from coal-fired generators are being developed and are reviewed in the following sections.

Improving energy efficiency is considered a third method for reducing CO$_2$ emissions. A measure of efficiency of electricity generation is the heat rate, or the Btu consumed per kilowatt-hour (kWh) generated. A generating plant operating at 33 percent efficiency would have a heat rate of 10,400 Btu/kWh. The electric power industry in the United States has been on a trend toward increased efficiency since 1949, but that trend has been very slight in the past two decades [5]. Improving power plant efficiency is a major goal of not only DOE but also industry.

10.1   CO$_2$ Capture Technologies

Capture of CO$_2$ from electric power generation can be accomplished by three general methods: precombustion CO$_2$ capture, where carbon is removed from the fuel prior to combustion; oxy-fuel combustion, where coal is combusted in an oxygen and CO$_2$-enriched environment; and postcombustion CO$_2$ capture, where coal is combusted normally in a boiler and CO$_2$ is removed from the flue gas. Figures 10.4 through 10.6 are generalized flow diagrams of the three processes, respectively,

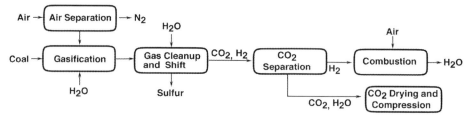

Figure 10.4 Flow diagram of precombustion CO_2 capture.

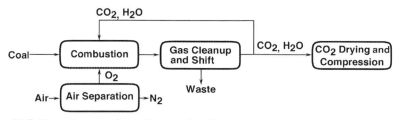

Figure 10.5 Flow diagram of oxy-fuel combustion.

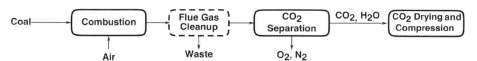

Figure 10.6 Flow diagram of postcombustion CO_2 capture.

showing the various components for comparison. Precombustion and oxy-fuel combustion are summarized in the following sections because they were discussed in detail in Chapter 7.

10.1.1 Precombustion (IGCC) CO_2 Capture

Precombustion CO_2 capture is characterized by removing the carbon from the coal prior to utilization in a gas turbine (see Chapter 7 for a detailed discussion of this concept). In this concept, coal is gasified through partial oxidation, using air or oxygen, to produce synthesis gas (syngas), which is composed of hydrogen (H_2), carbon monoxide (CO), and minor amounts of other constituents including methane (CH_4). The syngas then passes through gas cleanup stages and a shift reactor to convert the CO to CO_2 and increases the CO_2 and H_2 molar concentrations to approximately 40 percent and 55 percent, respectively. The CO_2 then has a high partial pressure and high chemical potential, which improves the driving force for various types of separation and capture technologies. After CO_2 removal, the hydrogen-rich syngas can be fired in a combustion turbine to produce electricity. Additional electricity can be generated by extracting energy from the combustion turbine flue gas using a heat

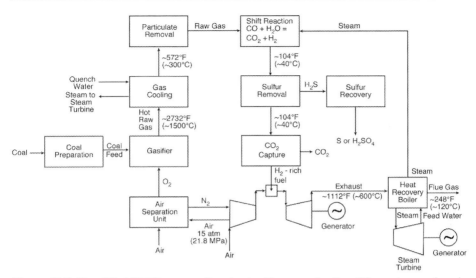

Figure 10.7 Simplified IGCC process flowsheet with precombustion CO₂ capture and coal gas cleanup.
Source: Modified from Kather et al. (2008) and Henderson (2003).

recovery steam generator—that is, in an integrated gasification combined cycle (IGCC) power plant. The CO_2 that is removed is dried and compressed and can be sequestered. For IGCC systems, the preferred separation method is physical absorption using a solvent, since this technology is commercially available. An IGCC process flowsheet with precombustion CO_2 capture and coal gas cleanup is shown in Figure 10.7 ([6, 7]).

10.1.2 Oxy-Fuel Combustion

A second option for CO_2 capture is the combustion of coal in an atmosphere of oxygen and CO_2, rather than in air (see Chapter 7 for details on this technology). Under these conditions, the primary products of combustion are CO_2 and H_2O, so CO_2 separation is not necessary. In this scenario, CO_2 can be captured from the flue gas by condensing the water followed by compression and storage. A schematic diagram of the most common oxy-fuel process, which involves the combustion of pulverized coal in pure oxygen (>95 percent) mixed with recycled flue gas is shown in Figure 10.8 ([6, 8]).

10.1.3 Postcombustion CO₂ Capture

Technologies for capturing CO_2 from emissions streams have been used for many years to produce a pure stream of CO_2 from natural gas or industrial processing for use in the food processing and chemical industries. The methods listed on the next page are currently used for CO_2 separation:

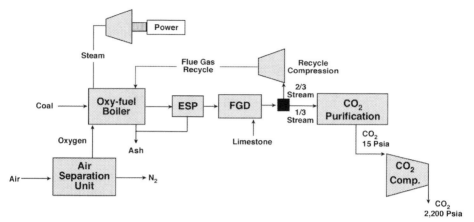

Figure 10.8 A pulverized coal oxy-fuel combustion system.
Source: Modified from Kather (2008) and Figueroa et al. (2008).

- Physical and chemical solvents, particularly monoethanolamine (MEA)
- Various types of membranes
- Adsorption onto solids
- Cryogenic separation

These methods can be used on a range of industrial processes; however, their use for removing CO_2 from high-volume, low-CO_2 concentration flue gases, such as those produced by coal-fired power plants, is more problematic. The high capital costs for installing postcombustion separation systems to process the large volume of flue gas is a major impediment to postcombustion capture of CO_2. In addition, a large amount of energy is required to release the CO_2 from solvents or solid adsorbents after separation. Several major technical and cost challenges must be overcome before retrofit of exiting power plants with postcombustion capture systems becomes an effective mitigation option. Capture technologies will likely improve in the coming years, thereby improving economics. This section discusses the status of technologies that are currently available for CO_2 capture, as well as some key ones that are under research and development.

Postcombustion CO_2 capture technologies, including state-of-the-art amine-based scrubbing systems, are discussed in this section along with emerging technologies such as scrubbing with carbonates and aqueous ammonia, the use of adsorbents, membranes, metal organic frameworks, and ionic liquids. Table 10.1 lists a number of possibilities to perform postcombustion CO_2 capture [6]. Note, however, that some of the technologies that are listed in Table 10.1 are more suited for precombustion applications (e.g., physical absorption such as Rectisol and Selexol) due to concentrated CO_2 streams and were discussed in Chapter 7 and will only be summarized here. Similarly, oxy-combustion, which in itself is not truly a capture technology, but rather an approach to modify the combustion process so the flue gas has a high concentration of CO_2, making it easier to remove the CO_2, is only

Table 10.1 Technology Options for Postcombustion Capture

Chemical Absorption		Physical Absorption	Adsorption	Alternative Approaches		
Amines	**Alternative Solvents**			**Membranes**	**Cryogenic**	**Others**
Primary	Ammonia	Rectisol	PSA	Gas	Distillation	Ionic
Secondary	Alkali-	Selexol	TSA	absorption	Frosting	liquid
Tertiary	compound	Pressurized	Solid	membrane		
Sterically	Aminosalt	water	sorbents			
hindered						

summarized in this chapter because it was discussed in detail in Chapter 7. Biomass cofiring will be discussed, since this is a retrofit option for CO₂ reduction in which coal and biomass are cofired, thereby reducing the carbon footprint of the power plant. This chapter concludes with a discussion of the economics of CO₂ capture and storage.

Postcombustion CO₂ capture mainly applies to coal-fired power plants but may also be applied to gas-fired combustion turbines. In a typical coal-fired power plant (Figure 10.9), fuel is burned with air in a boiler to produce steam, which drives a turbine to generate electricity. The boiler exhaust gas (i.e., flue gas) consists of mostly nitrogen, CO₂, O₂, moisture, and trace impurities. Separating the CO₂ from this gas stream is challenging because the CO₂ is present in dilute concentrations (i.e., 13–15 volume percent in coal-fired systems and 3–4 volume percent in natural gas-fired systems) and at low pressure (15–25 psia). In addition, trace impurities

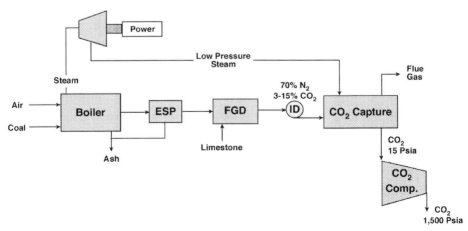

Figure 10.9 Block diagram illustrating a post combustion system.
Source: Modified from Figueroa et al. (2008).

(particulate matter, sulfur dioxide, and nitrogen oxides) in the flue gas can degrade sorbents and reduce the effectiveness of some of the CO_2 capture processes. Also, compressing the captured or separated CO_2 from atmospheric pressure to pipeline pressure (about 2,000 psia) represents a large auxiliary power load on the overall power plant system.

In spite of these difficulties, postcombustion capture has the greatest near-term potential for reducing CO_2 emissions because it can be retrofitted to existing units that generate about 67 percent of the CO_2 emissions in the power sector [8]. Retrofitting will primarily be of interest in power plants with advanced efficiencies to compensate for the large efficiency loss from postcombustion capture. For this reason, capture-ready fossil fuel plants that are retrofitted with a postcombustion capture process some time after the start of initial operation could play an important role in the near future. While postcombustion capture is still being developed, new plants with increased efficiencies will continue to be constructed.

During the lifetime of these plants, it is realistic to expect that postcombustion capture will become sufficiently mature. Some of the options for postcombustion CO_2 capture are discussed on the following pages. These range from state-of-the art amine-based systems to emerging technologies under development. Of the technologies presented in this section, chemical absorption technologies using amine or alternative solvents scrubbing are the nearest to commercial use in utility boilers. The others presented in this section are emerging technologies in various stages of development. A general ordering of the technologies from nearest to commercialization in utility boilers to longer lead times until commercialization are amine solvents, aqueous ammonia, chilled ammonia < solid sorbents, membranes < ionic liquids, metal organic frameworks.

Amine-Based Liquid Solvent Systems

Since the partial pressure of CO_2 in the flue gas of fossil fuel–fired power plants is low, technologies that are driven by high-CO_2 partial pressure differentials, such as physical solvents or membranes, are not efficiently applicable for postcombustion capture [6]. At partial pressures typical of coal-fired power plants, the possible loading (in weight percent CO_2 per unit solvent) is very limited for physical solvents (e.g., about 1 compared to about 9 for chemical solvents), resulting in large solvent circulation rates and ultimately in large amounts of heat required for solvent regeneration. Only chemical solvents show an absorption capacity large enough to be applicable for CO_2 capture for CO_2 concentrations typical of coal-fired power plants.

A system for postcombustion capture of CO_2 by chemical absorption is shown in Figure 10.10. The flue gas is usually cooled before entering the absorber column at the bottom. As the flue gas rises in the column, the CO_2 is absorbed by a solvent in countercurrent flow. The CO_2-free gas is vented to the atmosphere. The absorbed CO_2 is stripped from the rich solution, which occurs in a separate regeneration column. At the bottom of the absorber, the CO_2-rich solvent is collected and pumped through a heat exchanger, where it is preheated before entering the regenerator. In the regenerator, the CO_2-rich solvent flows downward. While flowing downward,

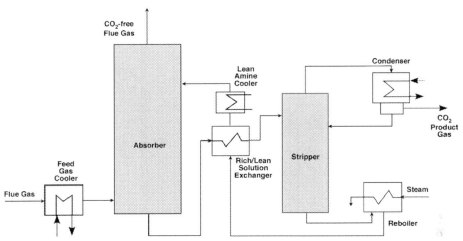

Figure 10.10 A postcombustion capture system using chemical absorption.

the temperature increases, thereby releasing CO_2, which rises to the top of the column and is removed.

The solvent most commonly used for CO_2 capture, monoethanolamine (MEA), is amine based. Amines are organic compounds with a functional group that contains nitrogen as the key atom. Structurally amines resemble ammonia, where one or more hydrogen atoms are replaced by an organic substituent. Primary amines arise when one of the three hydrogen atoms in ammonia is replaced by an organic substituent; the secondary amines have two hydrogen atoms replaced and tertiary amines have all three hydrogen atoms substituted. The following are the types of amine solvents [9]:

- Simple alkanolamines
- Primary: monoethanolamine (MEA)—$(C_2H_4OH)NH_2$
- Secondary: methylmonoethanolamine (MMEA), diethanolamine (DEA)—$(C_2H_4OH)_2NH$
- Tertiary: dimethylmonoethanolamine (DMMEA), methyldiethanolamine (MDEA)
- Hindered amines
- Mildly hindered primary: alamine (ALA)
- Moderately hindered: aminomethylpropoanol (AMP)
- Cyclic diamines: piperazine

CO_2 solvent extraction is based on the reaction of weak alkanolamine base with CO_2, which is a weak acid for producing a water-soluble salt. Amines react with CO_2 to form carbamate and bicarbonate via Eqs. 10.1 and 10.2, respectively.

$$2RNH_2 + CO_2 \leftrightarrow RNH_3^+ + RNHCOO^- \tag{10.1}$$

$$RNH_2 + H_2O + CO_2 \leftrightarrow RNH_3^+ + HCO_3^- \tag{10.2}$$

Chemical absorption with amines has been used for CO_2 separation in several industrial applications and in slipstream tests on coal-fired boilers; however, no amine-based CO_2 absorption system has been fully adapted to the specific conditions for complete flue gas treatment in a large coal-fired utility boiler that would produce around 700 metric tons CO_2 per hour for a state-of-the-art 1,000 MW_e plant. Amine, specifically MEA, based systems are the leading technology for retrofit applications, since they are applicable for low-CO_2 partial pressure streams and provide recovery rates of up to 98 percent and product purity of more than 99 vol%. The disadvantages of MEA systems include high-process energy consumption; degradation of the solvent due to SO_x and NO_x forming nonregenerable, heat-stable salts; degradation of the solvent in the presence of oxygen; and high corrosion potential of the solvent [6]. However, improvements to amine-based systems for postcombustion CO_2 capture are being pursued by a number of developers, including Fluor, Mitsubishi Heavy Industries, and Cansolv Technologies [8].

Aqueous Ammonia Process

Ammonia-based wet scrubbing is similar in operation to amine systems. The process scheme is similar to the configuration shown in Figure 10.10. The main difference with that of the amine system includes the precipitation of crystalline solid salts at temperatures below 80°C and the relatively high vapor pressure of ammonia, which requires regeneration of the solvent at very high pressures to minimize ammonia slip.

In the absorber column, CO_2 from the flue gas is absorbed by reacting CO_2 with ammonium carbonate to form ammonium bicarbonate:

$$(NH_4)_2CO_3 + CO_2 + H_2O \leftrightarrow 2NH_4HCO_3 \tag{10.3}$$

Ammonium bicarbonate partly precipitates in the absorber as a crystalline product [6]. In doing so, the absorption process is promoted, and the CO_2 capacity increased. The CO_2-rich slurry, consisting mainly of ammonium bicarbonate, is pumped through a heat exchanger to the high-pressure regenerator.

If the absorption/desorption cycle is controlled in such a way that it is primarily cycling between ammonium bicarbonate and ammonium carbonate, the regeneration energy can be reduced to approximately 30 percent of the necessary energy of a MEA system. In addition, Eq. (10.3) has a significantly lower heat of reaction than amine-based systems, resulting in energy savings. Ammonia-based absorption has a number of other advantages over amine-based systems, such as the potential for high-CO_2 capacity, lack of solvent degradation during absorption/regeneration, tolerance to oxygen in the flue gas, and potential for regeneration at high pressure. It also offers multipollutant control by reacting with sulfur and nitrogen oxides to form fertilizer (ammonium sulfate and ammonium nitrate) as a salable by-product.

Disadvantages of the process include ammonia's higher volatility than MEA, which often results in ammonia slip into the exit gas and the requirement of the flue gas being cooled to the 60 to 80°F range to enhance CO_2 absorptivity of the ammonia compounds. Also, ammonia is consumed through the irreversible formation of ammonium sulfates and nitrates, as well as the removal of HCl and HF.

A variation of the aqueous ammonia process is under development by Alstom and is called the chilled ammonia process. In this process, the absorber is kept at a significantly lower temperature—0 to 10°C. This process uses the same ammonium carbonate/ammonium bicarbonate absorption chemistry as the aqueous system, but it is different in that no fertilizer is produced, and a slurry of aqueous ammonium carbonate and ammonium bicarbonate and solid ammonium bicarbonate is circulated to capture CO$_2$ [8]. The main advantages of this process configuration are water condensation, smaller volume flow, less power duty for the blower, and less ammonia slip in the absorber [6]. Technical hurdles include cooling the flue gas and absorber to maintain operating temperatures below 50°F, mitigating the ammonia slip during absorption and regeneration, achieving 90 percent removal efficiencies in a single stage, and avoiding fouling of heat transfer and other equipment by ammonium bicarbonate deposition [8].

Alkali Carbonate-Based Systems

Carbonate systems are based on the ability of a soluble carbonate to react with CO$_2$ to form a bicarbonate, which when heated releases CO$_2$ and reverts back to a carbonate. A major advantage of carbonates over amine-based systems is the significantly lower energy required for regeneration.

Several processes (e.g., Benfield^{TM} and Catacarb^{TM}) are available for the removal of CO$_2$ from high-pressure gases that utilize aqueous potassium carbonate (K$_2$CO$_3$), and these are in industries that produce natural gas and hydrogen for ammonia synthesis [6]. These processes are attractive because they are not expensive, use nonvolatile compounds, and are nontoxic. Currently, however, no plants use potassium carbonate for CO$_2$ removal from flue gas, but activities are underway to explore this option. In fact, work is underway using sodium carbonate solutions as well, since sodium carbonate solutions are nonhazardous, nonvolatile, and have low corrosion rates. During absorption, the CO$_2$ becomes chemically bound according to the reaction

$$M_2CO_3 + H_2O + CO_2 \leftrightarrow 2MHCO_3 \qquad (10.4)$$

where M is potassium (K) or sodium (Na).

Solid Sorbents

Solid particles can be used to capture CO$_2$ from flue gas through chemical absorption, physical adsorption, or a combination of the two. Amines and other chemicals, such as sodium carbonate, can be immobilized on the surface of solid supports to create a sorbent that reacts with CO$_2$. Solid sorbents that physiosorb the CO$_2$ onto the surface include activated carbon, carbon nanotubes, and zeolites. The following are potential advantages of solid sorbents [10]:

- Ease of material handling because coal-fired plants are experienced with solids handling
- Safe for local environment
- High CO$_2$ capacity

- Lower regeneration capacity
- Multipollutant control capabilities

Possible configurations for contacting the flue gas with the solid particles include fixed, moving, and fluidized beds. A number of solids can be used to react with CO_2 to form stable compounds at one set of operating conditions and then be regenerated to liberate the adsorbed CO_2 at another set of conditions. In the adsorption phase, the CO_2 is adsorbed by a molecular sieve, activated carbon, or zeolites. In the regeneration phase, the CO_2 is released in a second process stage using pressure swing adsorption (PSA) or temperature swing adsorption (TSA), although less attention is being given to TSA due to the longer cycle times needed for heating during regeneration.

Chemical sorbents that react with CO_2 in the flue gas include a high surface area support with an immobilized amine or other reactant on the surface. Examples of commonly used supports are alumina or silica, while common reactants include chemicals such as sodium carbonate (Na_2CO_3) or amines such as polyethylenimine.

Research is underway to investigate a dry, inexpensive, regenerable, supported sorbent, sodium carbonate (Na_2CO_3), which reacts with CO_2 and water to form sodium bicarbonate ($NaHCO_3$) [8] via the equation

$$Na_2CO_3 + H_2O + CO_2 \leftrightarrow 2NaHCO_3 \tag{10.5}$$

A temperature swing is used to regenerate the sorbent and produce a pure CO_2/water stream that can be separated during cooling and compression. The pure CO_2 stream can then be sequestered.

An example of a polyethylenimine (PEI) sorbent for CO_2 capture is the novel nanoporous-supported polymer sorbent, known as a molecular basket sorbent (MBS), which is under development by Penn State [11–15]. In this concept, a CO_2-philic polymer sorbent, PEI, is loaded onto the surface of a porous material, such as as MCM-41, SBA-15, or others, to increase the accessible sorption sites per weight/volume unit of sorbent and to improve the mass transfer rate in the sorption/desorption process by increasing the gas-PEI interface. This has the following advantages:

- High sorption capacity and selectivity for CO_2
- Higher sorption-desorption rate
- Good regenerability and stability in the sorption-desorption cycles
- Low energy consumption
- Special functionality
- Promoting effect of moisture in the gas on sorption capacity
- No or reduced corrosion

Membranes

Although still in the development stage, membranes have the potential to selectively separate CO_2 from the flue gas. Membrane-based CO_2 capture uses permeable or semipermeable materials that allow for the selective transport and separation of CO_2 from flue gas. Key technical challenges to the use of membrane systems are processing large flue gas volumes, relatively low CO_2 concentration, low flue gas pressure, flue gas contaminants, and the need for a high-membrane surface area [16]. Current membranes are either not selective enough toward CO_2 (i.e., the separated gas contains unacceptable concentrations of other flue gas constituents) or not permeable enough (i.e., not separating enough of the CO_2). Researchers may be able to improve membrane performance until it can be an economically feasible carbon capture technology [10].

Membranes for postcombustion capture are better applicable when used as gas-liquid contactors and are referred to as gas absorption membranes. In the absorber, the gas phase remains separated from the liquid absorbent, and CO_2 diffuses from the gas stream through the membrane and is then absorbed by the solvent. This type of system has a number of advantages over conventional absorption systems [6]:

- Separated gas and liquid flow, which prevents the problems encountered in packed/tray columns such as flooding, foaming, channeling, and entrainment
- No need for wash sections after the absorber to recover absorption liquid that is carried over
- Operation at thermodynamically optimal conditions not dictated by the hydrodynamic conditions of the contacting equipment

Metal Organic Frameworks

Metal organic frameworks (MOFs) are a new class of hybrid material built from metal ions with well-defined coordination geometry and organic bridging liquids [8]. They are extended structures with carefully sized cavities that can adsorb CO_2. High storage capacity is possible, and the heat required for recovery of the adsorbed CO_2 is low.

Cryogenics

The use of cryogenics for removing CO_2 has been considered for nearly 20 years [6]. Three general methods are being investigated: compressing the flue gas to about 1,100 psia with water used for cooling/condensing and removing the CO_2; compressing the flue gas to 250 to 350 psia at 10°F to 70°F, dehydrating the feed stream with activated alumina or a silica gel dryer, and distilling the condensate in a stripping column; and cooling the flue gas stream to condense CO_2 at atmospheric pressure. The last method is also known as frosting or antisublimation (i.e., phase change of a gas to a solid) on a cold surface. For removal of 90 percent of the CO_2 from a coal-fired flue gas, the sublimation temperature lies in the range of −103° to −122°C.

Ionic Liquid Systems

Ionic liquids are a broad category of salts, typically containing an organic cation and either an inorganic or organic anion. They are generally liquid at room temperature, nonvolatile, thermally stable, and nonflammable. Ionic liquids can dissolve gaseous CO_2 and are stable at temperatures up to several hundred degrees Centigrade. Their good temperature stability offers the possibility of recovering CO_2 from the flue gas without having to cool it first. Also, since ionic liquids are physical solvents, little heat is required for regeneration. Due to their nonvolatility, solvent slip is basically eliminated. At this time, however, ionic liquids most suited for CO_2 separation have only been synthesized in small quantities in laboratories, and major developmental work is underway.

10.1.4 Biomass Cofiring

Cofiring is a near term, low-cost option (when it is compared to other capture technologies) for efficiently, and cleanly, converting biomass to electricity by feeding biomass as a partial substitute fuel in coal-fired boilers. This concept has been demonstrated successfully in more than 150 installations worldwide and has resulted in some commercially operating units, using a variety of feedstocks in all boiler types commonly used by electric utilities, including pulverized coal, cyclone, stoker, and FBC boilers. Common feedstocks include woody and herbaceous materials, energy crops, and agricultural and construction residues [17, 18]. Extensive demonstrations and tests have confirmed that biomass energy can provide as much as 15 to 20 percent of the total energy input with only a fuel feed system and burner modifications.

Cofiring biomass with coal offers several environmental benefits, including reductions in SO_2, NO_x, and greenhouse gas emissions. Carbon dioxide emissions from coal-fired power plants are essentially reduced by the percentage of biomass cofired on a heat input basis (Figure 10.11) [19] because the biomass is considered carbon neutral—that is, the CO_2 emitted from burning the biomass was removed from the atmosphere by the biomass-based material. When cofiring biomass at rates of 5 percent and 15 percent by heat input, greenhouse gas emissions have been reduced by 5.4 percent and 18.2 percent, on a CO_2-equivalent basis, respectively [20]. In studies comparing carbon emissions when producing biofuels and electricity from biomass, it has been shown that carbon displacement is significantly less when producing liquid fuels [21]. Thus, from the standpoint of reducing carbon emissions, it is better to use biomass to produce electricity. This would especially be the case if carbon were captured and sequestered from the biomass.

In addition to the environmental benefits, there are other benefits as well. For example, it has been shown that total energy consumption is lowered when cofiring biomass in coal-fired boilers by as much as 3.5 percent and 12.4 percent when cofiring 5 percent and 15 percent biomass, by heat input, respectively [20].

Although biomass cofiring is considered a proven technology, certain technical issues must be considered when cofiring biomass with coal. Specifically, this includes fuel handling, storage, and preparation, NO_x formation, carbon conversion, and ash effects such as deposition, corrosion, impacts on selective catalytic reduction and other downstream processes, and loss of sales in selling fly ash for beneficial uses. These issues are dependent on fuel characteristics, operating conditions, and boiler design. Proper combinations of coal and biomass and operating conditions can minimize or eliminate most impacts for most fuels.

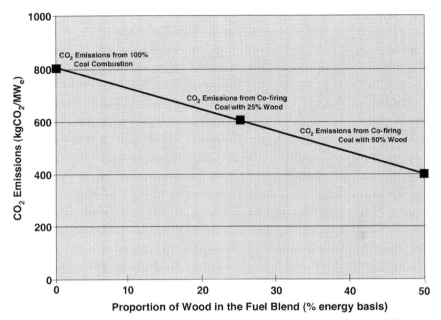

Figure 10.11 Carbon dioxide emissions reduction when cofiring wood in a coal-fired boiler. *Source:* From Patumsawad (2007).

10.2 Transport of CO₂

It is anticipated that most early CCS projects will be located near a geologic storage site (CO_2 storage is discussed in the following section). However, as the scale of CCS activities grows, it will become necessary to transport large volumes of CO_2 from the point of capture to the storage site. The primary method for CO_2 transport will be pipelines. Pipelines currently operate as a mature technology and are the most common method for transporting CO_2. Gaseous CO_2 is typically compressed to pressures above 8 MPa, so it remains a supercritical fluid (i.e., avoids two-phase flow and increases the density), thereby making it easier and less costly to transport [22]. The energy required for compression corresponds to a loss in power plant efficiency by about 2 to 4 percentage points.

The first long-distance CO_2 pipeline came into operation in the early 1970s [22]. In the United States, 45 million metric tons (Mt) per year are transported more than 3,500 miles of pipe for enhanced oil recovery (EOR) [23]. The existing infrastructure for CO_2 transport is shown in Figure 10.12 [23]. For comparison, the existing U.S. natural gas pipeline network transports 455 Mt per year of natural gas more than 300,000 miles of pipe.

CO_2 can also be transported as a liquid in ships, over the road, and by rail in tankers that carry CO_2 in insulated tanks at a temperature well below ambient and at much lower pressures [22]. In some situations, transport of CO_2 by ship may be more economical if the CO_2 has to be moved over large distances. CO_2 can be transported by

Figure 10.12 The existing U.S. CO_2 pipeline infrastructure.
Source: From CCSReg Project (2009).

ship much in the same way liquefied petroleum gases currently are (typically at 0.7 MPa pressure). Road and rail tankers are also technically feasible options. These systems transport CO_2 at a temperature of $-20°C$ and at 2 MPa pressure. Compared to pipelines and ships, road and rail tankers are uneconomical except on a very small scale and therefore are unlikely to be relevant to large-scale CCS [22].

10.3 CO_2 Storage

CO_2 (carbon) storage is defined as the placement of CO_2 into a repository where it will likely remain stored or sequestered permanently. This is the final step in carbon management. At present, the world is releasing about 8.5 Gt/year of carbon, or about 30 Gt/year of CO_2, into the environment. Storage volumes could amount to several thousand Gt of CO_2 over the century; consequently, many different technologies are being considered for long-term CO_2 storage. These include storing the CO_2 in geologic formations, in the ocean, terrestrially, and through mineral sequestration. These various options are discussed in this section.

10.3.1   Geologic Storage

Underground storage of CO_2 in geological formations is the most developed disposal option for CO_2 because it has the advantages of existing experience and low cost. Geologic sequestration involves the injection of CO_2 into underground reservoirs that have the ability to securely contain it over long periods of time. The geologic formations considered for CO_2 storage are deep saline or depleted

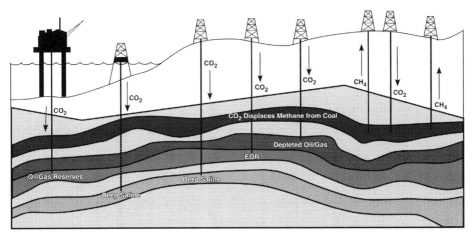

Figure 10.13 Primary geologic storage options.

oil and gas reservoirs, which are layers of porous rock capped by a layer or multiple layers of nonporous rock above them, and unmineable coal beds. These are depicted in Figure 10.13. The DOE has performed a high-level estimate of theoretical storage capacity for North America and estimated that there is geologic capacity to sequester between 919 and 3,389 Gt of CO$_2$ on- or near-shore [23].

Oil and gas reservoirs have shown to be impermeable over millions of years. This, coupled with the extensive exploration and removal of oil and gas during which the composition of the strata have become precisely known, makes depleted oil and gas reservoirs prime candidates for CO$_2$ storage. Depleted oil and gas fields provide an estimated sequestration capacity of 700 to 1,800 Gt of CO$_2$. In addition, the injection of CO$_2$ into active reservoirs for EOR is an applicable method for CO$_2$ storage. EOR capacity has been estimated at 70 to 200 Gt of CO$_2$ [24]. However, the biggest problem for safely storing CO$_2$ in these structures is posed by old, abandoned drill holes, which in some cases may be present in oil and gas fields in large numbers. These drill holes need to be filled to ensure that CO$_2$ is not readmitted into the atmosphere.

Similar to enhanced oil recovery, high-pressure CO$_2$ can also be used for enhanced methane recovery for unmineable coalbeds (i.e., too deep or too thin to be economically mined), provided the permeability is sufficient [22, 25]. The CO$_2$ is immobilized by either physical or chemical absorption onto the coal's surface. While CO$_2$ injection is known to displace methane, a greater understanding of the displacement mechanism is needed to optimize CO$_2$ storage and to understand the problems of coal swelling and permeability [26]. Coal has been shown to swell when it absorbs CO$_2$, and in an underground formation, swelling can cause a sharp drop in permeability, which restricts the flow of CO$_2$ into the formation and impedes the recovery of coalbed methane [25].

Saline aquifers are highly porous sedimentary rocks that are saturated with a strong saline solution (i.e., brine). Compared to coal seams or oil and gas reservoirs, saline formations are more common and offer the added benefits of greater proximity, higher CO_2 storage capacity, and fewer existing well penetrations [25, 26]. However, much less is known about saline formations because they lack the characterization experience that the industry has acquired through resource recovery from oil and gas reservoirs and coal seams. Therefore, there is more uncertainty regarding the suitability of saline formations for CO_2 storage. Storage capacities for CO_2 in saline formations, as estimated by DOE, range from a low of 3,300 Gt to a high of 12,600 Gt [25].

Two geologic formations under study that are in their infancy, but might prove to be optimal storage sites for stranded emissions sources, are organic-rich shales and basalt formations [25, 26]. Shale is a sedimentary rock and is characterized by thin, horizontal layers with very low permeability in the vertical direction. Many shales contain 1 to 5 percent organic material, which provides an adsorption substrate for CO_2 storage, similar to CO_2 storage in coal seams. Basalt formations, which are formations of solidified lava, have a unique chemical makeup that could potentially convert all of the injected CO_2 to a solid mineral form.

The uncertainty in geologic storage lies in the long-term stability of reservoirs. Generally, the risk of technical plants (e.g., separation equipment, compressors, pipelines) is considered to be low or manageable with the usual technical means and controls. The primary issue is long-term safety and long-term leakage rate, which ultimately will define the capacity of these technologies. Storage times under discussion vary from 1,000 to 10,000 years. The following are some important processes that could compromise the safety and permanence CO_2 storage:

- Geochemical processes such as the dissolution of carbonate rocks through acidic CO_2-water mixtures
- Pressure-induced processes such as the expansion of existing small fissures in the seal rock through the overpressure of CO_2 injection
- Leakage through existing drill holes
- Leakage via undiscovered migration paths in the seal rock
- Lateral expansion of the formation of water, which is displaced by the injected CO_2

10.3.2   Ocean Storage

A potential CO_2 storage option is to inject captured CO_2 directly into the deep ocean where most of it would be isolated from the atmosphere for centuries. Estimates for ocean storage capacity in excess of 100,000 billion GtC (i.e., 360,000 Gt CO_2) have been given; however, such numbers are only feasible if alkalinity (e.g., NaOH) were added to the ocean to neutralize the carbonic acid produced because an addition of 1,000 Gt of carbon is projected to change ocean chemistry [24].

A number of CO_2 injection methods have been investigated, but deep-ocean (at depths greater than 1,000 meters) injection is the option that would offer long-term (e.g., centuries) sequestration. This can be achieved by transporting CO_2 via pipelines or ships to an ocean storage site where it is injected into the water column of the ocean or at the sea floor (Figure 10.14). Below 2,700 meters, the density of compressed CO_2 is higher than that of seawater, and therefore, CO_2 sinks to the bottom. In addition,

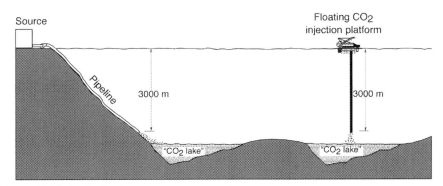

Figure 10.14 Methods of ocean storage.
Source: Modified from Metz et al. (2005) and Sarv (1999) [27].

CO$_2$ will react with seawater to form a solid clathrate, which is an icelike cage structure with approximately six water molecules per CO$_2$ molecule [24].

Ocean storage research has been performed since 1977; however, the technology has not yet been deployed or demonstrated at the pilot scale [22]. There is much resistance to this option due to environmental concerns related to long-term chronic issues of altering the ocean chemistry, as well as on the local effects of low pH (which can be as low as 4) and its effect on marine organisms. The ocean is an open system, and it is difficult, if not impossible, to monitor the distribution of stored carbon to confirm residence times of CO$_2$. It is most unlikely that ocean storage will be a CO$_2$ storage option in the foreseeable future.

10.3.3 Terrestrial Storage

Terrestrial carbon sequestration is defined as the removal of CO$_2$ from the atmosphere by soil and plants, both on land and in aquatic environments such as wetlands and tidal marshes, and/or the prevention of CO$_2$ net emissions from terrestrial ecosystems into the atmosphere [25, 26]. This is considered a low-cost option and can offer other benefits such as habitat and/or water quality improvements. Terrestrial sequestration efforts include tree plantings, no-till farming, wetlands restoration, land management on grasslands and grazing lands, fire management efforts, and forest preservation. It has been reported that terrestrial uptake of CO$_2$ offsets approximately one-third of global anthropogenic CO$_2$ emissions. This uptake is expected to decrease 13 percent over the next 20 years as Northeast forests mature. Opportunities for enhanced terrestrial uptake include 1.5 million acres of land damaged by past mining practices, 32 million acres of CRP (Conservation Reserve Program) farmland, and 120 acres of pastureland [28].

10.3.4 Mineral Carbonation

Mineral carbonation is the fixation of CO$_2$ by the use of alkaline and alkaline-earth oxides, such as magnesium oxide (MgO) and calcium oxide (CaO), that are present as naturally occurring silicate rocks such as serpentine and olivine [22].

The chemical reactions between these materials and dissolved CO_2 produce magnesium carbonate ($MgCO_3$) and calcium carbonate ($CaCO_3$). The process involves leaching out the Mg or Ca from the rock into the aqueous phase and reacting them with dissolved CO_2 to form the carbonate. Mineral carbonation produces silica and carbonates that are stable over long time periods and therefore can be disposed in areas such as silicate mines or reused in construction processes. The main challenges of this storage method are the slow dissolution kinetics and the large energy requirements that are associated with mineral processing. However, researchers' latest advancements have shown great potential to reduce the cost of mineral carbonation [24].

A commercial process would require mining, crushing, and milling of the mineral-bearing ores and transporting them to a processing plant that receives a CO_2 stream from a power plant. The carbonation energy would be 30 to 50 percent of the power plant output. Factoring in the additional energy required to capture the CO_2 at the power plant, a CCS system with mineral carbonation would require 60 to 180 percent more energy input per kilowatt-hour than a reference electricity producing plant without capture or mineral carbonation [22].

10.4   Economics of CO_2 Sequestration

The costs associated with CO_2 sequestration have been the focus of many studies, and a wide range in costs has been reported. The total cost for applying carbon capture and storage to a coal-fired power plant contains three main components: capture, transportation, and storage. This section summarizes the cost estimates for CO_2 capture (i.e., the cost of physically capturing the CO_2 at the plant and compressing it to transportation pressure), transportation (the cost of transporting the CO_2 from the plant to the storage site), and storage (the cost of injecting and storing the CO_2, including insurance, monitoring, and verification costs [37]). In general, the capture costs dominate the CCS. Costs reported in this section are in U.S. dollars.

10.4.1   Capture Costs

Technical issues for retrofitting coal-fired power plants for carbon capture were presented earlier in the chapter. In this section, the economics of retrofitting plants for carbon capture are addressed. Capturing CO_2 from power plants is expensive, and one can find a wide range of cost estimates for both retrofit situations and new plant construction. Increased costs are due to parasitic energy loss and an increase in capital costs, thereby resulting in an overall cost of electricity. Numbers vary from 5 to 30 percent parasitic energy losses, 35 to 110 percent increase in capital cost, and 30 to 80 percent increases in the cost of electricity [29]. The primary cost issues and their effect on the cost of electricity are summarized in this section.

Postcombustion CO₂ Capture Economics

The U.S. DOE performed a study to evaluate the technical and economic feasibility of various levels of CO_2 capture (i.e., 90 percent, 70 percent, 50 percent, and 30 percent) for retrofitting an existing pulverized coal-fired power plant (basing it on the Conesville #5 unit with a net plant out of 433.8 MW_e) using advanced amine-based capture technology—that is, with a solvent regeneration level of 1,550 Btu/lb CO_2 [30]. Impacts on plant output, efficiency, and CO_2 emissions resulting from the addition of CO_2 capture systems on an existing coal-fired power plant were considered. Cost estimates were developed for the systems required to produce, extract, clean, and compress CO_2, which would then be available for sequestration or other uses.

Adding CO_2 capture technology impacts net plant output and thermal efficiency, as illustrated in Figure 10.15. The efficiency decrease is essentially a linear function of CO_2 recovery level. Similarly, the specific investment cost is also a nearly linear function of CO_2 recovery level, with specific investment costs increasing from $540 to $1,319/$kW_e$ as CO_2 capture level increases from 30 percent to 90 percent (Figure 10.16). The specific investment cost is a nearly linear function of CO_2 recovery level. In all cases studied, significant increases to the levelized cost of electricity (LCOE) are incurred as a result of CO_2 capture, which is illustrated in Figure 10.17. For the ranges studied, the incremental LCOE is most impacted by the following parameters in given order: CO_2 by-product selling price, CO_2 capture level, solvent regeneration energy, capacity factor, investment cost, and makeup power cost.

Alstom, the developer of the chilled ammonia process described previously, performed cost of electricity comparisons for several CO_2 capture technologies in an

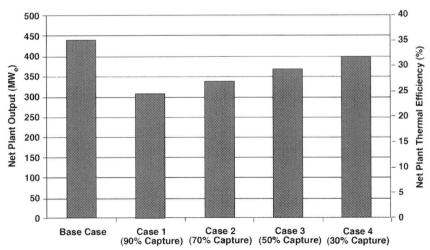

Figure 10.15 Plant performance impact of retrofitting a pulverized coal-fired plant at various levels of carbon capture.
Source: From Ramezan et al. (2007).

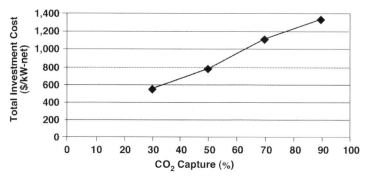

Figure 10.16 Effect of CO_2 capture rate on the total investment cost for retrofitting a pulverized coal-fired plant.
Source: From Ramezan et al. (2007).

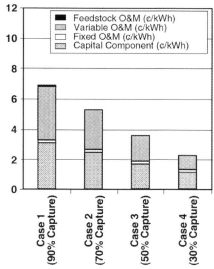

Figure 10.17 Incremental levelized cost of electricity for retrofitting a pulverized coal-fired plant.
Source: From Ramezan et al. (2007).

800 MWe pulverized coal–fired power plant [31], keeping in mind that economy of scale has a major influence on cost of electricity. Figure 10.18 shows the incremental reductions in the cost of electricity when comparing amine, future amine, and chilled ammonia.

The U.S. DOE performed an assessment on the prospects of retrofitting existing coal-fired power plants for CO_2 capture and sequestration using a generic model of retrofit costs as a function of basic plant characteristics, such as heat rate [32]. The cost for CO_2 retrofit included directs costs (capital and O&M), indirect costs (capacity and heat rate penalties), and a nominal cost for transportation, injection,

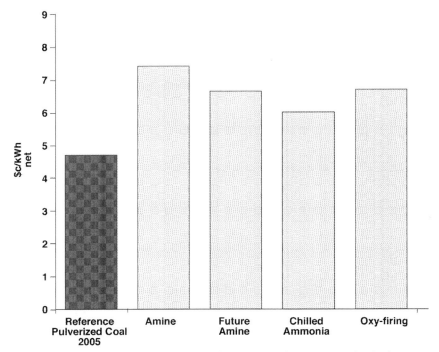

Figure 10.18 Cost of electricity comparison for various CO$_2$ capture technologies. *Source:* Modified from Otter (2007).

measurement, monitoring, and verification. The penetration of retrofits across the fleet was not observed to any significant extent until the carbon emission allowance prices exceeded $30/MTCO$_2$e (metric ton CO$_2$ equivalent). Retrofits did not occur through 2030 at $30/MTCO$_2$e, but about one-third of the starting coal fleet capacity was retrofitted at $45/MTCO$_2$e and about half at $60/MTCO$_2$e.

Oxy-Fuel Combustion Economics

As a retrofit option for existing coal-fired boilers, economic evaluations have shown that oxy-coal combustion retrofits for carbon dioxide capture are less expensive than other technologies such as amine scrubbing. This is illustrated in Figure 10.18, which shows the cost of electricity net in the Alstom study. Figure 10.19 shows the percent increase in COE from a DOE study [33] that compared current-state amine scrubbing with current-state oxy-fuel firing in a supercritical boiler and an ultra-supercritical boiler.

There is significant interest in reducing oxy-fuel combustion costs further. This is illustrated in Figure 10.19, where the last two bars show a reduction in cost as oxygen membranes are developed to reduce the cost of oxygen production and costorage of CO$_2$ and SO$_x$. Similarly, costorage is being explored where other pollutants are condensed in the CO$_2$ stream when the CO$_2$ is used for enhanced

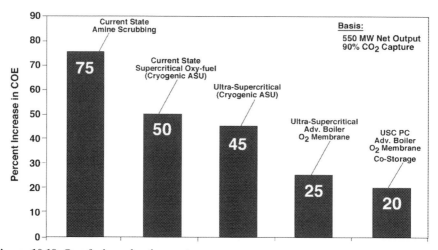

Figure 10.19 Oxy-fuel combustion costs.
Source: From Ciferno, Klara, and Wimer (2007).

oil recovery (EOR). Because CO_2 is in a liquid state for EOR, it provides a sink for other pollutants, and a small amount of SO_x in the CO_2 stream is acceptable because SO_x benefits the EOR process by improving oil miscibility [34]. The integrated emissions control approach has the potential to significantly reduce the costs related to pollution control and compensate for the increased investment for CO_2 capture. This technology is applicable to a new plant construction and might benefit a retrofit as well.

Economics of Biomass Cofiring

Biomass cofiring is technically feasible, but the costs are still a major barrier to increased coal and biomass cofiring. Cofiring economics depends on location, power plant type, and the availability of low-cost biomass fuels. Biomass is not economically competitive for electricity production in the current energy market except in niche applications where fuel costs are low or negative such as in integrated pulp and paper facilities [35]. Biomass is generally more expensive than coal; the delivered cost of biomass typically ranges from $1 to $4 per GJ, whereas coal costs around $1 per GJ. Delivered costs for dedicated energy crops can be significantly higher [21].

A typical cofiring installation includes modifications to the fuel handling and storage systems, possibly burner modifications, and sometimes boiler modifications or separate biomass feed injection into the boiler. Costs can increase significantly if the biomass needs to be dried, the size needs to be reduced, the pellets need to be produced, or the boiler requires a separate feeder. Retrofit costs can range from $100 to $300/kW of biomass generation in pulverized coal boilers. Cyclone boilers offer the lowest cost opportunities, as low as $50/kW [36].

Fuel supply is the most important cost factor. Costs for biomass depend on many factors such as climate, closeness to population centers, and the presence of industries that handle and dispose of wood. Low price, low shipping costs, and dependable supply are critical. Usually the cost of biomass must be equal to or less than the cost of coal, on heat basis, for cofiring to be economic. This, however, could change if carbon management legislation becomes enacted in the United States, for example.

Although coal and biomass cofiring is generally not cost competitive in the current energy market, cofiring offers benefits that could result in more widespread applications under certain policy scenarios. This is especially true as a near-term CO$_2$ mitigation strategy.

10.4.2   Costs of Transporting CO$_2$

The cost of CO$_2$ transport depends on the method of transport, the distance between the CO$_2$ source and the storage site (the single most important variable in the cost estimation), the volume transferred, the terrain (e.g., mountains, rivers, frozen ground), congested areas along the route, the existing regulations, the inlet pressure, and the availability of existing infrastructures [22, 37]. Costs have been estimated for both pipeline and marine transportation of CO$_2$, which is the most economical form of transport. Figure 10.20 shows the cost of pipeline transport for a distance of 155 miles (250 km) as a function of CO$_2$ mass flow rate [22]. The figure shows the ranges of costs for onshore and offshore pipelines.

In ship transport, the tanker volume and the type of loading and unloading systems are key factors in determining the overall transport cost [22]. Figure 10.21 compares pipeline and marine transportation costs, and it shows the breakeven

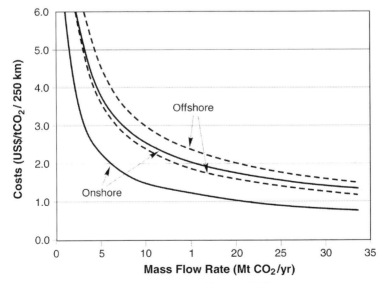

Figure 10.20 Transport costs for onshore and offshore pipelines.
Source: From Metz et al. (2005).

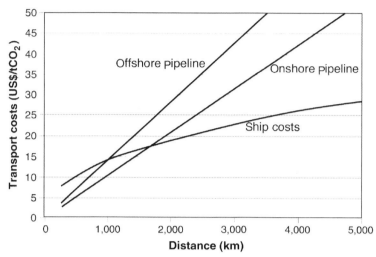

Figure 10.21 Costs versus distance for onshore pipelines, offshore pipelines, and ship transport.
Source: From Metz et al. (2005).

distance [22]. The marine option is typically less expensive than pipelines for distances greater than approximately 620 miles (about 1,000 km) and for amounts smaller than a few million tons of CO_2 per year.

10.4.3   Storage Costs

The determination of storage costs is very site dependent and is a function of the type of technology being considered. Ranges of storage are presented in this section for geologic, ocean, and mineralization storage.

Geologic Storage Costs

The technologies and equipment used for geologic storage are widely used in the oil and gas industries, so cost estimates for this option have a fairly high degree of confidence for storage capacity in the lower range of technical potential [22]. However, there is a wide range and variability of costs due to site-specific factors such as onshore versus offshore, reservoir depth, geologic characteristics of the storage formation, type of reservoir (aquifer, coal seam, oil or gas reservoir), and injection method used [22, 37].

Estimates for the cost for storage in saline formations and depleted oil and gas fields are between $0.5 and $8.0/metric ton of CO_2 injected [22]. In addition, monitoring the costs of $0.1 to $0.3/metric ton of CO_2 injected has been estimated. The lowest storage costs are for onshore, shallow, high-permeability reservoirs and/or storage sites where wells and infrastructure from existing oil and gas fields may be used [22].

Ocean Storage Costs

Although there is no experience with ocean storage, attempts have been made to estimate the cost of CO_2 storage projects that release CO_2 on the sea floor or in the deep ocean. Estimates range from $1 to $31/metric ton of CO_2 injected [22, 24]. Some specific estimates reported by the Intergovernmental Panel on Climate Change (IPCC) for ocean storage at depths deeper than 3,000 m ($\approx$2 miles) using a fixed pipeline are $6/metric ton of CO_2 injected and $31/metric ton of CO_2 injected for distances offshore of 100 km (62 miles) and 500 km (310 miles), respectively [22]. Similarly, the costs using a moving ship/platform are estimated to be $12 to $14/metric ton CO_2 injected and $13 to $16/metric ton CO_2 injected for distances offshore of 100 km (62 miles) and 500 km (310 miles), respectively.

Mineral Carbonation Costs

Although the costs of mineral, mineral preparation, and tailing disposal are well known and relatively inexpensive, the chemical reaction process is still too expensive for practical implementation. According to the IPCC, the best case studied to date is the wet carbonation of natural silicate olivine, where the estimated cost of this process is approximately $50 to $100/metric ton of CO_2 mineralized. To be more acceptable, costs below $30/metric ton are required. In addition, for each metric ton of CO_2 mineralized, 1.6 to 3.7 metric tons of silicates must be mined, and 2.6 to 4.7 metric tons of materials must be disposed for each metric ton of CO_2 stored as carbonates [22]. This will required a large operation with an environmental impact similar to a large-scale surface mining operation. Serpentine also often contains chrysotile, a natural form of asbestos; therefore, monitoring and mitigation measures used in the mining industry need to be implemented. It should be noted, however, that the products of mineral carbonation are chrysotile-free.

10.5 Permanence and Monitoring, Mitigation, and Verification

In addition to the major challenge of reducing carbon capture and storage costs, another important challenge is the long-term fate or permanence of the stored CO_2. To ensure that carbon sequestration represents an effective pathway for CO_2 management, permanence must be confirmed at a high level of accuracy. This is especially true for geologic and ocean storage; mineral carbonization is expected to ensure long-term storage.

Closely related to permanence is the issue of monitoring, mitigation, and verification (MMV). The success of CCS is dependent on the ability to measure the amount of CO_2 stored at a particular site, the ability to confirm that the stored CO_2 is not harming the host ecosystem, and the ability to effectively mitigate any impacts associated with a CO_2 leakage. Of the storage techniques discussed, ocean storage is the most problematic with this issue.

References

[1] J. Ciferno, DOE/NETL's Carbon Capture R&D Program for Existing Coal-Fired Power Plants, U.S. Department of Energy, National Energy Technology Laboratory, February 2009.

[2] J. Ciferno, DOE/NETL's Existing Plants Program CO_2 Capture R&D Overview, Presented at Annual NETL CO_2 Capture Technology for Existing Plants R&D Meeting, Pittsburgh, March 24–26 2009.

[3] DOE (U.S. Department of Energy), Energy Information Administration, Annual Energy Outlook 2008, Revised to Include the Impact of H.R. 6 "Energy Independence and Security Act of 2007" Enacted in December 2007, Washington, DC, 2008.

[4] DOE, Energy Information Administration, International Energy Outlook 2006, June 2006.

[5] Los Alamos, Clean Coal Compendium, The Products of Coal: Electricity and Carbon Dioxide, *www.lanl.gov/projects/cctc/climate/Coal_CO2.html*, last modified, December 17, 1999.

[6] A. Kather, S. Rafailidis, C. Hersforf, M. Klostermann, A. Maschmann, K. Mieske, et al., Research and development needs for clean coal deployment, IEA Clean Coal Centre, 2008.

[7] C. Henderson, Clean coal technologies, IEA Clean Coal Centre, 2003.

[8] J.D. Figueroa, T. Fout, S. Plasynski, H. McIlvried, R.D. Srivistava, Advances in CO_2 capture technology—The U.S. Department of Energy's Carbon Sequestration Program, International Journal of Greenhouse Gas Control (2008) 9–20.

[9] R.M. Davidson, Postcombustion carbon capture from coal fired plants – solvent scrubbing, IEA Clean Coal Centre, 2007.

[10] H.M. Krutka, S. Sjostrom, C.J. Bustard, M. Durham, K. Baldrey, R. Stewart, Summary of Postcombustion CO_2 Capture Technologies for Existing Coal-Fired Power Plants, Air & Waste Management Association Conference, June 24–26, 2008.

[11] X.C. Xu, C.S. Song, J.M. Andresen, B.G. Miller, A.W. Scaroni, Energy and Fuels 16 (2002) 1463.

[12] X.C. Xu, C.S. Song, J.M. Andresen, B.G. Miller, A.W. Scaroni, Microporous Mesoporous Materials, 62 (2003) 29.

[13] X.C. Xu, C.S. Song, B.G. Miller, A.W. Scaroni, Ind. Eng. Chem. Res. 44 (2005) 8113.

[14] X.C. Xu, C.S. Song, J.M. Andresen, B.G. Miller, A.W. Scaroni, Fuel Process. Technol. 86 (2005) 1457.

[15] X. Ma, X. Wang, C. Song, J. Am. Chem. Soc. 131 (2009) 5777.

[16] DOE, DOE to Provide $36 Million to Advance Carbon Dioxide Capture, *www.netl.doe.gov/publications/press*, July 31, 2008.

[17] D.A. Tillman, N.S. Harding, Fuels of Opportunity: Characteristics and Uses in Combustion Systems, Elsevier, 2005.

[18] B.G. Miller, D.A. Tillman (Eds.), Combustion Engineering Systems for Solid Fuel Systems, Elsevier, 2008.

[19] S. Patumsawad, Co-firing Biomass with Coal for Power Generation, in: Proc. of the Fourth Biomass-Asia Workshop "Biomass: Sources of Renewable Bioenergy and Biomaterial", November 20–22, 2007.

[20] M.K. Mann, P.L. Spath, A Life Cycle Assessment of Biomass Cofiring in a Coal-Fired Power Plant, Clean Prod. Processes 3 (2001) 81–91.

[21] R.P. Overend, A. Milbrandt, Tackling Climate Change in the U.S.: Potential Carbon Emissions Reductions from Biomass by 2030, in: C.F. Kutscher (Ed.), Tackling Climate

Change in the U.S.: Potential Carbon Emissions Reductions from Energy Efficiency and Renewable Energy by 2030, American Solar Energy Society, 2007.

[22] B. Metz, O. Davidson, H. Coninck, M. Loos, L. Meyer, Carbon Dioxide Capture and Storage, Intergovernmental Panel on Climate Change, Cambridge University Press, 2005.

[23] CCSReg Project, Carbon Capture and Sequestration: Framing the Issues for Regulation, Department of Engineering and Public Policy, Carnegie Mellon University, January 2009.

[24] F.P. Sioshansi (Ed.), Generating Electricity in a Carbon-Constrained World, Academic Press, 2010.

[25] DOE, National Energy Technology Laboratory, Carbon Sequestration, *www.netl.doe. gov/technologies/carbon_seq/core_rd/storage.html* (accessed November 11, 2009).

[26] DOE, National Energy Technology Laboratory, Carbon Sequestration Technology Roadmap and Program Plan, 2007.

[27] H. Sarv, Large-Scale CO$_2$ Transportation and Deep Ocean Sequestration, Phase I Final Report, Prepared for U.S. Department of Energy, National Energy Technology Laboratory, DE-AC26-98FT40412, March 1999.

[28] DOE, National Energy Technology Laboratory, Carbon Sequestration, *www.netl.doe. gov/coalpower/sequestration*.

[29] T.E. Fout, Carbon Capture R&D: DOE/NETL R&D Program, in: Proc. of the 7th Annual Conference on Carbon Capture and Sequestration, 2008.

[30] M. Ramezan, T.J. Skone, N. Nsakala, G.N. Liljedahl, L.E. Gearhart, R. Hertermann, et al., Carbon Dioxide Capture from Existing Coal-Fired Power Plants, DOE/NETL-401/110907, 2007.

[31] N. Otter, The Deployment of Clean Power Systems for Coal, in: Proc. of the Conference on Clean Coal Technology, May 15–18, 2007.

[32] R.A. Geisbrecht, Retrofitting Coal-Fired Power Plants for Carbon Dioxide Capture and Sequestration—Exploratory Testing of NEMS for Integrated Assessments, DOE/NETL-1309, 2008.

[33] J. Ciferno, J. Klara, J. Wimer, Outlook for Carbon Capture from Pulverized Coal and Integrated Gasification Combined Cycle Power Plants, in: Proc. of the Sixth Annual Conference on Carbon Capture and Sequestration, May 7–10, 2007.

[34] L. Zheng, Y. Tan, R. Pomalis, B. Clements, Integrated Emissions Control and its Economics for Advanced Power Generation Systems, in: Proc. of the 31st International Technical Conference on Coal Utilization & Fuel Systems, Coal Technology Association, May 21–25, 2006.

[35] A.L. Robinson, J.S. Rhodes, D.W. Keith, Assessment of Potential Carbon Dioxide Reductions Due to Biomass-Coal Cofiring in the United States, Environ. Sci. Technol. 37 (22) (2003) 5081–5088.

[36] E. Hughes, Biomass Cofiring in the U.S.—Status and Prospects, in: Proc. of the Third Chicago/Midwest Renewable Energy Workshop, June 19–20, 2003.

[37] J.S. Kessels, Bakker, A. Clemens, Clean Coal Technologies for a Carbon-Constrained World, IEA Clean Coal Centre, 2007.

11 U.S. and International Activities for Near-Zero Emissions during Electricity Generation

Clean coal technologies have become increasingly more important to many countries of the world because of concerns regarding future energy supplies, energy price volatility, and environmental effects. Many countries are working toward a long-term effort to achieve energy security. To do this, a balanced and diversified portfolio of energy resources is required, which includes fossil, nuclear, and renewable energy sources. Clean coal is recognized as a crucial element in an overall energy policy for many countries. Coal is a popular fuel choice because worldwide reserves sufficient to last two to three centuries are available, it is widely dispersed throughout the world, and it is among the most economic of energy resources. However, coal is among the most environmentally problematic of all energy resources, and much effort has gone into addressing this. Over the last two to three decades, the U.S. Department of Energy (DOE) has spearheaded the development of clean coal technologies, with many other countries working on it as well.

This chapter discusses U.S. and international programs that address technologies for reducing emissions to near-zero levels when utilizing coal in electricity generation plants. The chapter focuses on programs that have been implemented by the United States over the last 20 plus years to develop new technologies, which are also being exported and used by other countries. The various programs, under the umbrella of the DOE's Coal Power Program, address near- and long-range needs and include developing cost-effective environmental control technologies to comply with current and emerging regulations, and developing technologies for near-zero emissions power and clean fuels plants with carbon dioxide management capability, respectively. The DOE programs discussed in this chapter include the Clean Coal Technology Demonstration Program, Power Plant Improvement Initiative, Clean Coal Power Initiative, Vision 21, FutureGen, and carbon sequestration efforts. International clean coal and carbon sequestration programs are also discussed.

11.1 Introduction to U.S. Clean Coal Technology Programs

In the foreseeable future, the energy needed to sustain economic growth in the United States (as well as most countries of the world, for that matter) will continue to come largely from fossil fuels, with coal playing a leading role. In supplying this energy need,

Clean Coal Engineering Technology. DOI: 10.1016/B978-1-85617-710-8.00011-X
Copyright © 2011 by Elsevier Inc. All rights of reproduction in any form reserved.

however, the United States must address growing global and regional environmental concerns and energy prices. Maintaining low-cost electricity while demand grows and environmental pressures increase requires new technologies. These technologies must allow countries to use its indigenous resources wisely, cleanly, and efficiently.

The current fleet of coal-fired power plants in the United States is faced with increasingly stringent environmental regulations, with air emissions the primary focus. The Clean Air Act, particularly the 1990 amendments, addresses environmental performance of coal-based power systems, specifically targeting emissions of sulfur dioxide, nitrogen oxides, hazardous air pollutants (including mercury), and fine particulate matter. Carbon dioxide is also of concern because of its potential to contribute to global climate change.

As discussed in previous chapters, much progress has been made in reducing emissions from coal-fired power plants. However, with more stringent regulations anticipated in the near future, as well as the DOE's goal to develop near-zero emissions coal-fired power plants, major research and development activities are necessary to develop advanced coal power generation technologies, with full-scale demonstrations a key to realizing the benefits of these programs.

The Clean Coal Technology Program is providing a portfolio of technologies that will ensure that the U.S. recoverable coal reserves of 274 billion short tons can continue to supply the country's energy needs economically and in an environmentally sound manner. Under this program, cost-effective environmental control devices have been developed for existing power plants. In addition, a new generation of technologies that can produce electricity and other commodities, as well as provide the efficiencies and environmental performance responsive to global climate change concerns, have been created [1].

The PPII projects focus on technologies enabling existing coal-fired power plants to meet increasingly stringent environmental regulations at the lowest possible cost [2]. The CCPI Program is an innovative technology demonstration program that fosters more efficient clean coal technologies for use in existing and new power generation facilities [3]. Vision 21 is the DOE's initiative to effectively remove environmental concerns associated with the use of fossil fuels for producing electricity and transportation fuels through better technology with the design basis completed by 2015 and plant deployment by 2020 [4]. The DOE is particularly interested in coal-based energy plants. A specific integrated coproduction plant program is FutureGen, which is an integrated carbon dioxide capture and sequestration and hydrogen research initiative to design, build, and operate a nearly emission-free, coal-fired electric power and hydrogen production plant [5].

11.2 Clean Coal Technology Demonstration Program

More than 25 years ago, it was recognized that given the need to respond to environmental objectives, new technologies would be necessary if coal was to continue as a viable source of secure energy [6]. The Clean Coal Technology Demonstration Program (CCTDP) was initiated in 1985 with the objective to demonstrate a new generation of advanced coal utilization technologies. This investment in technology forms a solid foundation for addressing global and regional environmental concerns while providing low-cost energy that can compete in a deregulated electric power marketplace.

11.2.1 CCTDP Evolution

During the 1970s and the early 1980s, many of the government-sponsored technology demonstrations focused on synthetic fuels production technology. The Synthetic Fuels Corporation (SFC) was established under the Energy Security Act of 1980 to reduce U.S. vulnerability to disruptions of crude oil imports [1]. The SFC's purpose was accomplished by encouraging the private sector to build and operate synthetic fuel production facilities that used abundant domestic energy resources, specifically coal and oil shale. The strategy was for the SFC to be primarily a financier of pioneer commercial and near-commercial scale facilities [1]. By 1985, the market drivers for synthetic fuels dissolved as oil prices declined, world oil prices stabilized, and a short-term buffer was provided by the Strategic Petroleum Reserve. Congress responded to the decline of private-sector interest in the production of synthetic fuels and abolished the SFC in 1986.

The CCTDP—formerly known as the Clean Coal Technology (CCT) program and often referred to as such—was initiated in October 1984 through Public Law 98-473, Joint Resolution Making Continuing Appropriations for Fiscal Year 1985 and Other Purposes [1]. The United States moved from an energy policy based on synthetic fuels production to a more balanced policy. This policy established that the United States must have an adequate supply of energy maintained at a reasonable cost and consistent with environmental, health, and safety objectives. Energy stability, security, and strength were the foundations for this policy. Coal was recognized as an essential element in this energy policy for the foreseeable future because the following existed [1]:

- A well-understood coal resource base
- Available technology and a skilled labor base to safely and economically extract, transport, and use coal
- Existing multibillion-dollar infrastructure to gather, transport, and deliver coal to serve the domestic and international marketplace
- A secure and abundant energy resource within the country's borders that is relatively invulnerable to disruptions because the coal production is dispersed and flexible, the delivery network is vast, and the stockpiling capability is great
- The potential for export opportunities of U.S.-developed, coal-based technologies, since coal is the fuel of necessity in many lesser developed economies

Congress recognized that the continued viability of coal as an energy resource was dependent on the demonstration and commercial application of a new generation of advanced coal-based technologies with improved operational, economic, and environmental performance; consequently, the CCTDP was established [1]. The DOE issued the first solicitations (CCTDP-I) in 1986 for clean coal technology projects and selected a broad range of projects in four major product markets: environmental control devices, advanced electric power generation, coal processing for clean fuels, and industrial applications. In February 1988, the second solicitation (CCTDP-II) was issued and provided for the demonstration of technologies that were capable of achieving significant reductions in SO_2, NO_x, or both, from existing power plants that were to be more cost effective than current technologies and capable of commercial deployment in the 1990s. The emphasis of the solicitation was on SO_2 and NO_x, precursors of acid rain, since a major presidential initiative was launched to address acid rain. The DOE issued a third solicitation (CCTDP-III) in May 1989 with essentially the same objective as the second, but it also encouraged technologies that would produce clean fuels from run-of-mine coal.

The next two solicitations recognized emerging energy and environmental issues, such as global climate change and capping of SO_2 emissions, and focused on seeking highly efficient, economically competitive, and low-emissions technologies. Specifically, the fourth solicitation (CCTDP-IV), released in January 1991, had as its objective the demonstration of energy-efficient, economically competitive technologies capable of retrofitting, repowering, or replacing existing facilities while achieving significant reductions in SO_2 and NO_x emissions. The fifth solicitation (CCTDP-V) was released in July 1992 to provide for demonstration projects that significantly advanced the efficiency and environmental performance of technologies applicable to new or existing facilities. None of the projects selected in the fifth solicitation were completed, and all were withdrawn.

Projects selected over the course of the program included SO_2 control systems, NO_x control technologies, fluidized-bed combustion, gasification, advanced coal processing technologies to produce clean fuels, and coal utilization for industrial applications. These technologies have allowed for continued coal use in the United States, while reducing multiple emission levels by anywhere from 30 to 95 percent [7]. More than 20 of the technologies tested in the program have achieved commercial success. The final CCTDP project ended in 2006.

11.2.2 CCTDP Funding and Costs

The five CCTDP solicitations resulted in 33 successful demonstrations [7]. Fifty-seven projects were selected over the course of the program, with 24 projects withdrawn [8]. Although the projects were terminated for many reasons, often it was due to financial issues. The 33 successfully completed projects resulted in a combined commitment by the federal government and the private sector of approximately \$3.3 billion. The DOE's cost-share for these projects was about \$1.3 billion, or about 40 percent of the total [7]. The project participants (i.e., the nonfederal government participants) provided the remaining almost \$2 billion, or 60 percent of the total. Table 11.1 summarizes the costs, cost-sharing, and application category of the CCTDP projects.

11.2.3 CCTDP Projects

The CCTDP projects provided a portfolio of technologies that will enable coal to continue to provide low-cost, secure energy vital to the U.S. economy while satisfying energy and environmental goals. The projects were spread across the country in

Table 11.1 Costs and Cost-Sharing of Completed CCTDP Projects (dollars in thousands)

	Total Project Costs	%	Cost-Share Dollars		Cost-Share Percent	
			DOE	**Participant**	**DOE**	**Participant**
Subprogram						
CCT-I	844,363	26	239,640	604,723	28	72
CCT-II	318,577	10	139,195	179,382	44	56
CCT-III	1,138,741	35	483,665	655,076	42	58
CCT-IV	950,429	29	437,876	512,553	46	54
CCT-V	0	0	0	0	0	0
Total	3,252,110	100	1,300,376	1,951,734	40	60
Application Category						
Advanced electric power generators	1,978,492	61	812,912	1,165,580	41	59
Environmental control devices	620,110	19	252,832	367,278	41	59
Coal processing for clean fuels	431,810	13	192,029	239,781	44	56
Industrial applications	221,698	7	42,603	179,095	19	81
Total	3,252,110	100	1,300,376	1,951,734	40	60

Source: From DOE (2009).

16 states, as shown in Figure 11.1 [1]. Table 11.2 lists each project, nonfederal government participant, location of the project, solicitation under which the projects were awarded, and date the project was completed. The participants listed in Table 11.2 are the primary nonfederal government company, although each project had several supporting team members [1]. The projects are listed in Table 11.2 by four basic market sectors. A synopsis of the projects is provided in later sections in this chapter.

Environmental Control Devices

The initial thrust of the CCT Program addressed acid rain, and 18 projects have been completed that involved SO_2 and NO_x control for coal-fired boilers. The technologies demonstrated provide a suite of cost-effective control options for the full range of boiler types. The projects included seven NO_x emission control systems installed on more than 1,750 MW of utility generating capacity, five SO_2 emission control systems installed on almost 770 MW of capacity, and six combined SO_2/NO_x emission control systems installed on more than 665 MW of capacity [1].

SO_2 Control Technologies

The CCTDP successfully demonstrated two sorbent injection systems, one spray dryer system, and two advanced flue gas desulfurization (AFGD) systems. Sulfur dioxide reductions varying from 50 to 90+ percent were demonstrated.

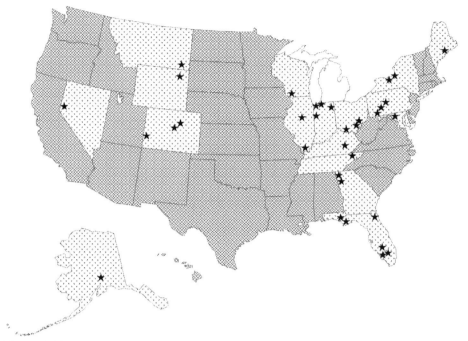

Figure 11.1 Location of the CCTDP projects.
Source: Modified from DOE (2002) and DOE (2010).

AirPol, Inc., demonstrated that FLS milfo, Inc.'s Gas Suspension Absorption system as an economic option for achieving Phase II 1990 CAAA SO_2 compliance in coal-fired boilers using high-sulfur coal [11]. The demonstration was performed using a vertical, single-nozzle reactor (i.e., spray dryer) with integrated sorbent particulate recycle in a 10 MW equivalent slipstream of flue gas from a Tennessee Valley Authority (West Paducah, Kentucky) 175 MW wall-fired boiler. Sulfur dioxide reductions of 60 to 90 percent were obtained firing 2.7 to 3.5 percent sulfur coal [1].

Bechtel Corporation demonstrated SO_2 removal capabilities of in-duct confined zone dispersion (CZD)/flue gas desulfurization (FGD) technology. This was done specifically to define the optimum process operating parameters and to determine CZD/FGD's operability, reliability, and cost-effectiveness during long-term testing and its impact on downstream operations and emissions [12]. The demonstration was performed using half of the flue gas from Pennsylvania Electric Company's Seward Station's 147 MW tangentially fired boiler. Sulfur dioxide reductions of 50 percent were achieved, firing 1.5 to 2.5 percent sulfur coal [1].

LIFAC-North America (a joint venture partnership between Tampella Power Corporation and ICF Kaiser Engineers, Inc.) demonstrated the LIFAC sorbent injection process, with furnace sorbent injection and sulfur capture occurring in a vertical

Table 11.2 CCTDP Projects

Project	Participant	Solicitation Round/ Completion Date
Environmental Control Devices *SO₂ Control Technologies* 10-MW Demonstration of Gas Suspension Absorption	AirPol, Inc.	CCT-III/completed 03/1994
Confined Zone Dispersion Flue Gas Desulfurization Demonstration	Bechtel Corporation	CCT-III/completed 06/1993
LIFAC Sorbent Injection Desulfurization Demonstration Project	LIFAC-North America	CCT-III/completed 06/1994
Advanced Flue Gas Desulfurization Demonstration Project	Pure Air on the Lake, L.P.	CCT-II/completed 06/1995
Demonstration of Innovative Applications of Technology for the CT-121 FGD Process	Southern Company Services, Inc.	CCT-II/completed 12/1994
NOₓ Control Technologies Demonstration of Advanced Combustion Techniques for a Wall-Fired Boiler	Southern Company Services, Inc.	CCT-II/completed 04/2003
Demonstration of Coal Reburning for Cyclone Boiler NOₓ Control	The Babcock & Wilcox Company	CCT-II/completed 12/1992
Full-Scale Demonstration of Low-NOₓ Cell Burner Retrofit	The Babcock & Wilcox Company	CCT-III/completed 04/1993
Evaluation of Gas Reburning and Low-NOₓ Burners on a Wall-Fired Boiler	Energy and Environmental Research Corporation	CCT-III/completed 01/1995
Micronized Coal Reburning Demonstration for NOₓ Control	New York State Electric & Gas Corporation	CCT-IV/completed 04/1999
Demonstration of Selective Catalytic Reduction Technology for the Control of NOₓ Emissions from High-Sulfur, Coal-Fired Boilers	Southern Company Services, Inc.	CCT-II/completed 07/1995
180-ME Demonstration of Advanced Tangentially Fired Combustion Techniques for the Reduction of NOₓ Emissions from Coal-Fired Boilers	Southern Company Services, Inc.	CCT-II/completed 12/1992

Continued

Table 11.2   CCTDP Projects—*Cont'd*

Project	Participant	Solicitation Round/ Completion Date
Combined SO_2/NO_x Control Technologies		
SNOX™ Flue Gas Cleaning Demonstration Project	ABB Environmental Systems	CCT-II/completed 12/1994
LIMB Demonstration Project Extension and Coolside Demonstration	The Babcock & Wilcox Company	CCT-I/completed 08/1991
SOx-NOx-Rox-Box™ Flue Gas Cleanup Demonstration Project	The Babcock & Wilcox Company	CCT-II/completed 05/1993
Enhancing the Use of Coals by Gas Reburning and Sorbent Injection	Energy and Environmental Research Corporation	CCT-I/completed 10/1994
Milliken Clean Coal Technology Demonstration Project	New York State Electric & Gas Corporation	CCT-IV/completed 06/1998
Integrated Dry NO_x/SO_2 Emissions Control System	Public Service of Colorado	CCT-III/completed 12/1996
Advanced Electric Power Generation *Fluidized-Bed Combustion*		
JEA Large-Scale CFB Combustion Demonstration Project	JEA	CCT-I/completed 06/2005
Tidd PFBC Demonstration Project	The Ohio Power Company	CCT-I/completed 03/1995
Nucla CFB Demonstration Project	Tri-State Generation and Transmission Association, Inc.	CCT-I/completed 01/1991
Integrated Gasification Combined Cycle		
Tampa Electric Integrated Gasification Combined Cycle Project	Tampa Electric Company	CCT-III/completed 10/2001
Piñon Pine IGCC Power Project	Sierra Pacific Power Company	CCT-IV/completed 01/2001
Wabash River Coal Gasification Repowering Project	Wabash River Coal Gasification Repowering Project Joint Venture	CCT-IV/completed 12/1999
Advanced Combustion/Heat Engines **Healy Clean Coal Project**	Alaska Industrial Development and Export Authority	CCT-III/completed 12/1999

Table 11.2 Cont'd

Project	Participant	Solicitation Round/ Completion Date
Coal Processing for Clean Fuels		
Commercial-Scale Demonstration of the Liquid Phase Methanol (LPMEOH™) Process	Air Products Liquid Phase Conversion Company, L.P.	CCT-III/completed 12/2002
Development of the Coal Quality Expert™	ABB Combustion Engineering, Inc. and CQ, Inc.	CCT-I/completed 12/1995
ENCOAL® Mild Coal Gasification Project	ENCOAL Corporation	CCT-III/completed 07/1997
Advanced Coal Conversion Process Development	Western SynCoal LLC	CCT-I/completed 01/2001
Industrial Applications		
Blast Furnace Granular-Coal Injection System Demonstration Project	Bethlehem Steel Corporation	CCT-III/completed 11/1998
Advanced Cyclone Combustor with Internal Sulfur, Nitrogen, and Ash Control	Coal Tech Corporation	CCT-I/completed 05/1990
Cement Kiln Flue Gas Recovery Scrubber	Passamaquoddy Tribe	CCT-II/completed 09/1993
Pulse Combustor Design Qualification Test	ThermoChem, Inc.	CCT-IV/completed 09/2001

Source: From DOE (2002), DOE (2003), and DOE (2006).

activation reactor [13]. The LIFAC process was shown to be easily retrofitted to power plants with space limitations and burning high-sulfur coals. The 60 MW demonstration was performed on Richmond (Indiana) Power & Light's Whitewater Valley Station and achieved 70 percent SO_2 removal firing 2.0 to 2.9 percent sulfur coal [1].

Pure Air on the Lake, L.P. (a subsidiary of Pure Air, which is a general partnership between Air Products and Chemicals, Inc. and Mitsubishi Heavy Industries America, Inc.) demonstrated Pure Air's AFGD process to reduce SO_2 emissions by 95 percent or more at approximately one-half the cost of conventional scrubbing technology, significantly reduce space requirements, and create no new waste streams [14]. A single SO_2 absorber was built for Baily (Indiana) Generating Station's Unit Nos. 7 and 8 (i.e., 528 MW) and achieved 95 percent SO_2 capture firing 2.3 to 4.7 percent sulfur coal [1].

Southern Company Services, Inc., demonstrated Chiyoda Corporation's Chiyoda Thoroughbred-121 AFGD process for combined particulate and SO_2 capture with high reliability [15]. Testing was performed using Georgia Power Company's Plant Yates, Unit No. 1 (100 MW) and achieved more than 90 percent SO_2 removal efficiency at SO_2 inlet concentrations of 1,000 to 3,000 ppm, with limestone utilization more than 97 percent, 97.7 to 99.3 percent particulate removal, more than 95 percent HCl and HF capture, 80 to 98 percent capture of most metals, more than 50 percent capture of mercury and cadmium, and over 70 percent capture of selenium [1].

NO_x Control Technologies

Under the CCTDP, seven NO_x control technologies were assessed encompassing low-NO_x burners (LNB), advanced overfire air (AOFA), reburning, selective catalytic reduction (SCR), selective noncatalytic reduction (SNCR), and combinations of them. NO_x reductions varying from 37 to 80 percent were demonstrated.

Southern Company Services, Inc. performed a demonstration using Foster Wheeler's LNB with AOFA and the Electric Power Research Institute's (EPRI's) Generic NO_x Control Intelligent System (GNOCIS) computer software to achieve 50 percent NO_x reduction with the LNB/AOFA system; to determine the contributions of AOFA and LNB to NO_x reduction and the parameters for optimal LNB/AOFA performance; and to assess the long-term effects of LNB, AOFA, combined LNB/AOFA, and the GNOCIS advanced digital controls on NO_x reduction, boiler performance, and auxiliary components [9, 16]. The demonstration was performed on Georgia Power's Plant Hammond, Unit No. 4, which is a 500 MW wall-fired boiler, and achieved 68 percent NO_x reduction with fly ash loss-on-ignition (LOI) increasing from a baseline of 7 percent to 8 to 10 percent [1].

The Babcock & Wilcox Company (B&W) demonstrated the technical and economic feasibility of their coal reburning system to achieve greater than 50 percent NO_x reduction with no serious impact on cyclone combustor operation, boiler performance, or other emission streams [17]. The demonstration was performed on Wisconsin Power and Light Company's 100 MW Nelson Dewey Station, Unit No. 2 and achieved 52 to 62 percent NO_x reduction with 30 percent heat input from the coal [1].

B&W also demonstrated the cost-effective reduction of NO_x from a large, baseloaded coal-fired utility boiler with their low-NO_x cell-burner (LNCB®) system to achieve at least 50 percent NO_x reduction without degradation of boiler performance at less cost than that of conventional low-NO_x burners [18]. The demonstration was performed on Dayton (Ohio) Power and Light Company's 605 MW J. M. Stuart Plant, Unit No. 4 and achieved 48 to 58 percent NO_x reduction, experienced average CO emissions of 28 to 55 ppm, increased fly ash production without affecting the electrostatic precipitator (ESP) performance, and increased unburned carbon (UBC) losses by about 28 percent [1].

The Energy and Environmental Research Corporation (EERC), currently GE Energy and Environmental Research, performed a demonstration to attain up to a 70 percent decrease in NO_x emissions from an existing wall-fired utility boiler firing low-sulfur coal using both natural gas reburning (GR) and LNBs and to assess the impact of GR-LNB technology on boiler performance [19]. The demonstration was performed on an 172 MW wall-fired boiler (i.e., Public Service Company of Colorado's Cherokee Station, Unit No. 3) and achieved 37 to 65 percent NO_x reduction, with 13 to 18 percent of the heat input from the natural gas, reduced SO_2 and particulate loadings by the percentage heat input by natural gas reburning, and it resulted in acceptable carbon-in-ash and CO levels with GR-LNB operation [1].

New York State Electric & Gas Corporation (NYSEG) demonstrated micronized coal reburning with the objective of achieving at least 50 percent NO_x reduction on a cyclone burner and 25 to 35 percent NO_x reduction on a tangentially-fired boiler [20]. Demonstrations were performed on NYSEG's Milliken Station (Lansing, New York) Unit No. 1, which is a 148 MW tangentially-fired boiler, and Eastman Kodak Company's Kodak Park (Rochester, New York) Power Plant Unit No. 1, which is 60 MW cyclone boiler. Nitrogen oxide reductions of 59 and 28 percent were achieved in the cyclone- and tangentially fired units, respectively. The micronized coal consisted of 17 and 14 percent of heat input, respectively. LOI was maintained at less than 5 percent at Milliken Station, but it increased from baseline levels of 10 to 15 percent to 40 to 50 percent at Kodak Park [1].

Southern Company Services, Inc. evaluated the performance of eight SCR catalysts with different shapes and chemical compositions when applied to operating conditions found in U.S. pulverized coal-fired utility boilers firing U.S. high-sulfur coal under various operating conditions, while achieving as much as 80 percent NO_x removal [21]. In this demonstration project, the SCR facility consisted of three 2.5 MW equivalent SCR reactors supplied by separate flue gas streams and six 0.20 MW equivalent reactors, for a total of 8.7 MW equivalent using flue gas from Gulf Power Company's Plant Crist (Pensacola, Florida), Unit No. 4. The reactors were sized to produce data that will allow the SCR process to be scaled up to commercial size. Nitrogen oxide reductions of more than 80 percent were achieved at an ammonia slip well under the 5 ppm level deemed acceptable for commercial operation [1].

Southern Company Services, Inc. also demonstrated short- and long-term NO_x reduction capabilities of ABB Combustion Engineering Inc.'s (now Alstom Power Inc.) low-NO_x concentric firing system (LNCFS™) at various combinations of OFA and coal nozzle positioning [22]. The demonstration was performed on Gulf Power Company's 180 MW tangentially fired Plant Lansing Smith (Lynn Haven, Florida), Unit No. 2. Reductions in NO_x of 37 to 45 percent were achieved [1].

Combined SO_2/NO_x Control Technologies

Six combined SO_2/NO_x control technologies were assessed under the CCT Program. These technologies used various combinations of technologies and demonstrated NO_x and SO_2 reductions of 40 to 94 and 50 to 95 percent, respectively.

ABB Environmental Systems demonstrated Haldor Topsoe's SNOX catalytic advanced flue gas cleanup system at an electric power plant using U.S. high-sulfur coals with the objective to remove 95 percent of the SO_2 and more than 90 percent of the NO_x from the flue gas and produce a salable by-product of concentrated sulfuric acid [23]. In the SNOX process, particulate is removed using a high-efficiency baghouse, NO_x is reduced in a catalytic reactor using ammonia, and SO_2 is oxidized to SO_3 in a second catalytic reactor and subsequently hydrolyzed to concentrated sulfuric acid. Testing was performed in a 35 MW equivalent slip-stream from Ohio Edison's Niles (Ohio) Station, Unit No. 2 (108 MW) and achieved SO_2 reductions in excess of 95 percent and NO_x reductions averaging 94 percent, produced a sulfuric acid that exceeded federal specifications for a Class I acid, eliminated CO and hydrocarbon emissions due to the presence of the SO_2 catalyst, and exhibited high-capture efficiency of most air toxics (except for mercury) in the high-efficiency baghouse [1].

B&W demonstrated that their limestone injection multistage burner (LIMB) process can achieve up to 50 percent NO_x and SO_2 reductions and that Consolidated Coal Company's Coolside duct injection of lime sorbents can achieve removal of up to 70 percent SO_2 [24]. The testing was performed in Ohio Edison's 105 MW Edgewater Station (Lorain, Ohio), Unit No. 4. The LIMB process reduces SO_2 by injecting dry sorbent into the boiler above the burners—in this case, B&W's DRB-XCL® low-NO_x burners—and SO_2 removal efficiencies varying from 45 to 60 percent with lime-based products and 22 to 40 percent with limestone were achieved, while nitrogen oxide reductions of 40 to 50 percent were obtained. The Coolside process, which is a humidified duct injection process, achieved 70 percent SO_2 reduction [1].

B&W also demonstrated their SOx-NOx-Rox Box™ (SNRB™) process with the objective of achieving greater than 70 percent SO_2 removal and 90 percent or higher reduction in NO_x emissions while maintaining particulate emissions below 0.03 lb/million (MM) Btu [25]. The SNRB process combines the removal of SO_2, NO_x, and particulates in one unit: a high-temperature baghouse. Sulfur dioxide is removed using sorbent injection, NO_x is reduced by injecting ammonia in the presence of an SCR catalyst inside the bags, and particulate is removed using high-temperature fiber filter bags. The testing was performed in a 5 MW equivalent slipstream from Ohio Edison Company's 156 MW R. E. Burger Plant (Dilles Bottom, Ohio), Unit No. 5, and SO_2 and NO_x reductions of 80 to 90 and 90 percent, respectively, were achieved. In addition, air toxic removal efficiency was comparable to that of the ESP at the plant, except that HCl and HF emissions were reduced by 95 and 84 percent, respectively [1].

The Energy and Environmental Research Corporation performed a demonstration in which natural gas reburning was combined with in-furnace sorbent injection with the objective of reducing NO_x by 60 percent and SO_2 by at least 50 percent in two different boiler configurations—tangentially and cyclone-fired units—while burning high-sulfur midwestern coal [26]. Testing was performed on Illinois Power Company's 71 MW Hennepin Plant, Unit No. 1 (tangentially fired boiler) and City Water, Light and Power's (Sprinfield, Illinois) 40 MW Lakeside Station, Unit No. 7 (cyclone-fired boiler). Nitrogen oxide reductions averaged 67 and 66 percent, respectively, for the tangentially and cyclone-fired boilers, while SO_2 reductions averaged 53 and 58 percent, respectively [1].

The NYSEG performed a demonstration using Saarberg-Hölter-Umwelttechnik, GmbH's (S-H-U) formic acid-enhanced, wet limestone scrubber technology, ABB Combustion Engineering Inc.'s LNCFS™ process, Stebbins Engineering and Manufacturing Company's split-module absorber, ABB Air Preheater's heat-pipe air preheater, and NYSEG's plant emissions optimization advisor (PEOA) with the objective of achieving high-sulfur capture efficiency and NO_x and particulate control at minimum power requirements, zero waste water discharge, and the production of by-products instead of wastes from the scrubber [27]. The flue gas from NYSEG's Milliken Station (Lansing, New York), Unit Nos. 1 and 2 (300 MW) was used in the project, and sulfur dioxide removals of 98 and 95 percent were demonstrated with and without formic acid, respectively, and 39 percent NO_x reduction was achieved [1].

The Public Service Company of Colorado demonstrated the integration of five technologies—B&W's DRB-XCL low-NO_x burners with OFA, in-duct sorbent injection, flue gas humidification, and furnace (urea) injection—with the objective of achieving 70 percent reduction in NO_x and SO_2 emissions and, more specifically, to assess the integration of a downfired low-NO_x burner with in-furnace urea injection and dry sorbent in-duct injection with humidification for SO_2 removal [28]. Testing was performed on Public Service Company of Colorado's Arapahoe Station (Denver, Colorado), 100 MW Unit No. 4 and demonstrated 70 percent SO_2 removal and 62 to 80 percent NO_x reduction [1].

Advanced Electric Power Generation Technology

The CCTDP provided a range of advanced electric power generation options for both repowering and new power generation in response to the need for load growth as well as environmental concerns. The emphasis of this Program category included technologies that could effectively repower aging power plants faced with the need to both control emissions and respond to growing power demands. Repowering is an important option because existing power generation sites have significant value and warrant investment because the infrastructure is in place and siting new plants represents a major undertaking.

These advanced systems offer greater than 20 percent reductions in greenhouse gas emissions; SO_2, NO_x, and particulate emissions far below New Source Performance Standards (NSPS); and salable solid and liquid by-products [1]. Over 1,800 MW of capacity were represented by 11 projects, including five fluidized-bed combustion systems (of which two terminated in June 2003 after completing designs), four integrated gasification combined cycle systems (three completed and one withdrawn), and two advanced combustion/heat engine systems (one completed and one withdrawn). The withdrawn projects are mentioned because some were in an operational stage when terminated. Although they were not listed in Table 11.2, some are summarized in the following sections because of their overall importance in clean coal technology development, since some information was obtained from them that can be applied to future systems.

The advanced electric power-generation technology projects selected under the CCT Program are characterized by high thermal efficiency, very low pollutant emissions, reduced CO_2 emissions, few solid waste problems, and enhanced economics. Five generic advanced electric power-generation technologies are demonstrated in the CCT Program: fluidized-bed combustion, integrated gasification combined cycle, integrated gasification fuel cell, coal-fired diesel, and slagging combustion.

Fluidized-Bed Combustion

Lakeland Electric, in the city of Lakeland, Florida, was selected by the DOE for two CCTDP projects in 1989 and 1993, but these projects were terminated in 2003 due to economic issues [9, 29]. The first project, a pressurized circulating fluidized-bed (PCFB) project, was to demonstrate Foster Wheeler Corporation's PCFB technology coupled with Siemens Westinghouse's ceramic candle type hot-gas cleanup system and power-generation technologies, which were to represent a cost-effective, high-efficiency, low-emissions means of adding generating capacity at greenfield sites or in repowering applications [1]. The second project, to be performed on the same boiler, was to demonstrate topped PCFB technology in a fully commercial power generation setting, thereby advancing the technology for future plants that will operate at higher gas turbine inlet temperatures and will be expected to achieve cycle efficiencies in excess of 45 percent.

JEA, formerly the Jacksonville (Florida) Electric Authority demonstrated atmospheric circulating fluidized-bed (ACFB) combustion at a scale larger than previously operated. The objective of the project was to demonstrate ACFB combustion at 297.5 MW, which represented a scaleup from previously constructed facilities; verify expectations of the technology's economic, environmental, and technical performance; provide potential users with the data necessary for evaluating a large-scale ACFB combustion as a commercial alternative; accomplish greater than 90 percent SO_2 removal; and reduce NO_x emissions below New Source Performance Standards (NSPS) [1, 10]. Over a two-year operational period (2003 and 2004), the unit operated at an average heat rate of 9,516 Btu/kWh (35.9 percent efficiency). It achieved greater than 90 percent full load boiler efficiencies on three fuels (Pittsburgh No. 8 bituminous coal; 50/50 blend of petroleum coke and Pittsburgh No. 8 coal; 20/80 blend of petroleum coke and Pittsburgh No. 8 coal) and 88 percent firing Illinois No. 6 bituminous coal. It achieved a typical 85 percent SO_2 removal efficiency in the boiler and 12.1 percent SO_2 removal efficiency in the polishing scrubber, and emissions were far below the target value of 0.15 lb SO_2/ MM Btu. The unit met its NO_x emissions target of 0.09 lb/MM Btu [10].

The Ohio Power Company performed a pressurized fluidized-bed (bubbling) combustion (PFBC) demonstration to verify expectations of PFBC economic, environmental, and technical performance in a combined-cycle repowering application at utility scale and to accomplish greater than 90 percent SO_2 removal and achieve an NO_x emission level of 0.3 lb/MM Btu at full load [30]. The demonstration was performed at Ohio Power Company's 70 MW Tidd Plant (Brilliant, Ohio), Unit No. 1 and was the first large-scale operational demonstration of PFBC in the United States. Sulfur dioxide removal efficiency of 90 to 95 percent was achieved at full load with calcium-to-sulfur (Ca/S) ratios of 1:1.4 and 1:5, respectively [1]. NO_x

emissions were 0.15 to 0.33 lb/MM Btu, CO emissions were less than 0.01 lb/MM Btu, and particulate emissions were less than 0.02 lb/MM Btu. Operationally, the PFBC boiler demonstrated commercial readiness.

Tri-State Generation and Transmission Association, Inc. demonstrated the feasibility of ACFB technology at the utility scale, evaluating the economic, environmental, and operational performance at that scale [31]. Three small, coal-fired stoker boilers at the Nucla Station (Nucla, Colorado) were replaced with a new 110 MW atmospheric CFB boiler. Environmentally, SO_2 capture efficiencies of 70 and 95 percent were achieved at Ca/S ratios of 1.5 and 4.0, respectively; NO_x emissions averaged 0.18 lb/MM Btu; CO emissions ranged from 70 to 140 ppm; particulate emissions ranged from 0.0072 to 0.0125 lb/MM Btu (or 99.9 percent removal efficiency); and solid waste was essentially benign and showed potential as an agricultural solid amendment, soil/roadbed stabilizer, or landfill cap [1].

Integrated Gasification Combined Cycle

As discussed in detail in Chapter 7 and summarized here, the integrated gasification combined cycle (IGCC) process has four basic steps:

1. Fuel gas is generated from a gasifier.
2. The fuel gas is either passed directly to a hot-gas cleanup system to remove particulates, sulfur, and nitrogen compounds, or the gas is first cooled to produce steam and then cleaned conventionally.
3. The clean fuel gas is combusted in a gas turbine generator to produce electricity.
4. The residual heat in the hot exhaust from the gas turbine generator is recovered in a heat recovery steam generator, and the steam is used to produce additional electricity in a steam turbine generator.

IGCC systems are among the cleanest and most efficient of the emerging clean coal technologies. Sulfur, nitrogen compounds, and particulate matter are removed before the fuel is combusted—that is, before combustion air is added, resulting in a much lower volume of gas to be treated in a postcombustion scrubber. With hot-gas cleanup, IGGC systems have the potential for efficiencies of more than 50 percent. Figure 11.2 shows Tampa Electric's Polk Power Station, which is an IGCC system [32].

Energy conversion in fuel cells is more efficient than traditional energy conversion devices and can be as high as 60 percent. Thus, there is interest in combining fuel cell technology with coal-fired gasifiers. A typical fuel cell system using coal as a fuel includes a coal gasifier with a gas cleanup system, a fuel cell to use the coal gas to generate electricity (direct current) and heat, an inverter to convert direct current to alternating current, and a heat recovery system that can be used to produce additional electric power in a bottoming steam cycle [1].

Fuel cells do not rely on combustion; instead, an electrochemical reaction generates electricity. Electrochemical reactions release the chemical energy that bonds atoms together—in this case, the atoms of hydrogen and oxygen [33]. The fuel cell is extremely clean and highly efficient. In a clean coal technology application, the fuel cell is fueled either by hydrogen extracted from the coal gas or

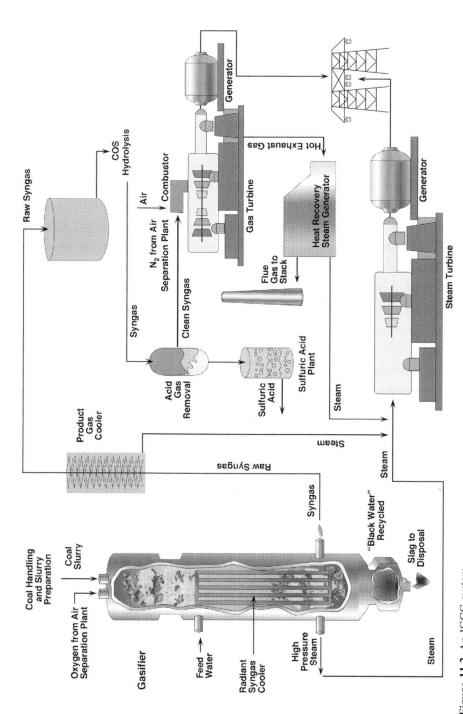

Figure 11.2 An IGCC system.
Source: From DOE (2000).

a mixture of synthesis gas (low-Btu gas consisting of CO and H_2). In a coal gasification/fuel cell application, coal gas is supplied to the anode, and air and CO_2 are supplied to the cathode to produce electricity and heat. The principal waste product from the fuel cell is water.

Fuel cells are often categorized by the material used to separate the electrodes, called the electrolyte. The most mature fuel cell concept is the phosphoric acid fuel cell [33]. Other concepts include the molten carbonate fuel cell (MCFC), which uses a hot mixture of lithium and potassium carbonate as the electrolyte, and the solid oxide fuel cell, which uses a hard ceramic material instead of a liquid electrolyte. The MCFC is integrated with one of the Clean Coal Technology IGCC projects described later.

The MCFC evolved from work in the 1960s aimed at producing a fuel cell which would operate directly on coal [34]. While direct operation on coal seems less likely today, operation on coal-derived fuel gases is both technically and economically viable. The MCFC, shown in Figure 11.3, uses a molten carbonate salt mixture as its electrolyte. The composition of the electrolyte varies, but it usually consists of lithium carbonate and potassium carbonate. At the operating temperature of about 1,200°F, the salt mixture is liquid and a good ionic conductor. The MCFC reactions that occur are [34]:

$$\text{Anode reactions: } H_2 + CO_3^{2-} \rightarrow H_2O + CO_2 + 2e^- \tag{11.1}$$

$$CO + CO_3^{2-} \rightarrow 2CO_2 + 2e^- \tag{11.2}$$

$$\text{Cathode reaction: } O_2 + 2CO_2 + 4e^- \rightarrow 2CO_3^{2-} \tag{11.3}$$

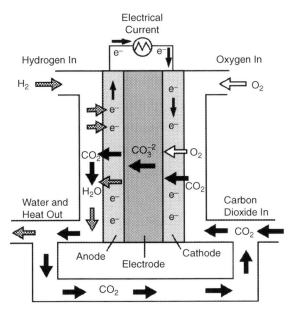

Figure 11.3 A molten carbonate fuel cell.
Source: From DOE (2003).

The anode process involves a reaction between hydrogen and carbon monoxide and the carbonate ions from the electrolyte, which produces water and carbon dioxide and releases electrons to the anode [34]. The cathode process combines oxygen and carbon dioxide from the oxidant stream with electrons from the cathode to produce carbonate ions that enter the electrolyte. The use of carbon dioxide in the oxidant stream requires a system for collecting carbon dioxide from the anode exhaust and mixing it with the cathode feed stream.

Of the four IGCC projects selected, the one that was terminated was to incorporate an MCFC with a coal gasifier. Consequently, this technology was not fully demonstrated but is one of much interest.

Tampa Electric Company demonstrated an advanced IGCC system using Texaco's (now Chevron-Texaco) pressurized, oxygen-blown entrained-flow gasifier technology [36]. The objective was to demonstrate IGCC technology in a greenfield commercial electric utility application at the 250 MW size using an entrained-flow, oxygen-blown gasifier with full heat recovery, conventional coal-gas cleanup, and an advanced gas turbine with nitrogen injection for power augmentation and NO_x control [1]. The IGCC system shown in Figure 11.2 is that of the Polk system [32]. The demonstration was performed at Tampa Electric Company's Polk Power Station (Mulberry, Florida) and achieved greater than 98 percent sulfur capture, while NO_x emissions were reduced by more than 90 percent compared with a conventional pulverized coal-fired power plant, particulate matter was well below the regulatory limits set for the Polk plant site, and carbon burnout exceeded 95 percent [32]. The plant is currently in commercial operation.

Sierra Pacific Power Company tested the integrated gasification combined cycle using the KRW air-blown pressurized fluidized-bed coal gasification system [37]. The objective was to demonstrate air-blown pressurized fluidized-bed IGCC technology incorporating hot-gas cleanup; evaluate a low-Btu gas combustion turbine; and assess long-term reliability, availability, maintainability, and environmental performance at a scale sufficient to determine commercial potential. The emissions targets were to remove more than 95 percent of the sulfur in the coal and emit less than 70 percent NO_x and 20 percent less CO than in a comparable conventional coal-fired plant [1]. The 107 MW demonstration (shown in the block diagram in Figure 11.4), which was performed at Sierra Pacific Power Company's Tracy Station (Reno, Nevada), experienced many operational difficulties, and steady-state operation was not reached in the course of the testing; therefore, environmental performance could not be evaluated. The project did succeed in identifying and working through a number of problems, made possible only through a full-scale demonstration, and positioned the technology for commercialization. In addition, the testing proved the ability of the KRW gasifier to produce coal-derived synthesis gas of design quality [1].

The Wabash River Coal Gasification Repowering Project Joint Venture (a joint venture of Dynegy, Inc.—formerly Destec Energy, Inc.—and PSI Energy, Inc.) demonstrated IGCC using Global Energy's two-stage pressurized, oxygen-blown, entrained-flow gasification system (i.e., E-Gas Technology^{TM}) [39]. A schematic diagram of the system is shown in Figure 11.5. The objective was to demonstrate utility repowering with the E-Gas Technology, including advancements in the technology relevant to the use

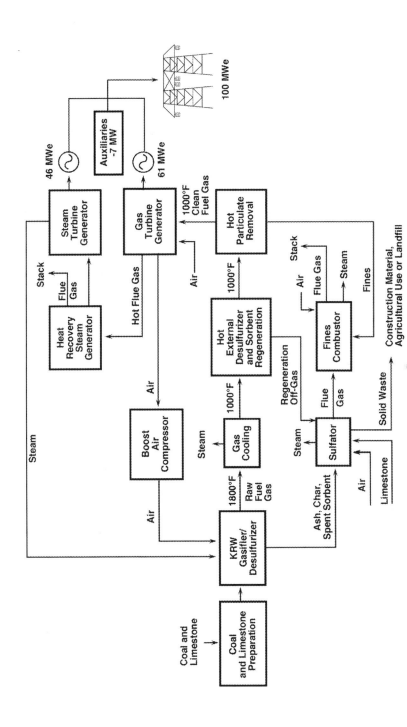

Figure 11.4 The Piñon Pine IGCC system. *Source:* DOE (1996).

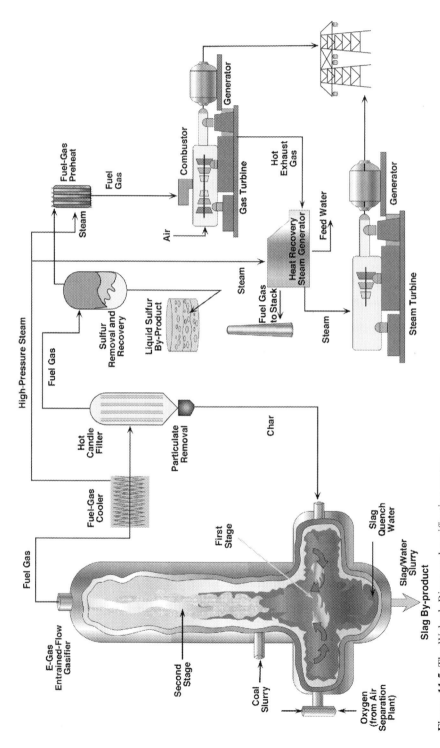

Figure 11.5 The Wabash River coal gasification system.
Source: From DOE (2000).

of high-sulfur bituminous coal, and to assess long-term reliability, availability, and maintainability of the system in a fully-commercial scale unit [1]. The 296 MW demonstration was successfully performed at PSI Energy's Wabash River Generating Station (Terre Haute, Indiana) and achieved sulfur capture efficiency greater than 99 percent. The sulfur-based pollutants were converted into 99.99 percent pure sulfur, NO_x emissions were 0.15 lb/MM Btu (which meets the 2003 target emission limits for ozone nonattainment areas), particulate emissions were below detectable limits, CO emissions averaged 0.05 lb/MM Btu, and coal ash was converted to a low-carbon vitreous slag that is valued as an aggregate in construction or as grit for abrasives and roofing metals [1]. The plant is currently in commercial operation.

Kentucky Pioneer Energy, LLC was awarded a CCTDP project to demonstrate and assess the reliability, availability, and maintainability of a utility-scale IGCC system that used a high-sulfur bituminous coal and refuse-derived fuel blend in oxy-gen-blown, fixed-bed, BGL slagging gasifiers and the operability of an MCFC fueled by coal gas [1]. The IGCC system, shown in Figure 11.6, was to be located at East Kentucky Power Cooperative's Smith site (Trapp, Kentucky), and its capacity was 580 MW IGCC. The MCFC portion of the project, which is a slipstream of fuel gas fed to the gas turbine (and rated at 2.0 MW), was to be moved to the Wabash River site [9] but did not proceed because of issues with a natural gas purchase agreement needed to support MCFC comparative testing on natural gas and synthesis gas [10]. The IGCC system that was to be demonstrated in this project was suitable for both repowering applications and new power plants; however, the Kentucky Public Service Commission withdrew approval of an electric power purchasing agreement, and the project was ended due to lack of progress [10].

Advanced Combustion/Heat Engines

One project was completed that demonstrated advanced combustion/heat engine technology. Alaska Industrial Development and Export Authority demonstrated TRW's clean coal combustion system integrated with B&W's spray dryer absorber (SDA) with sorbent recycle [41]. The demonstration was performed adjacent to Healy Unit No. 1. The objective was to demonstrate an innovative new power plant design featuring integration of an advanced combustor coupled with both high- and low-temperature emissions control processes. Emissions were controlled using TRW's advanced entrained/slagging combustors through staged fuel and air injection for NO_x control and limestone injection for SO_2 control. Additional SO_2 control was accomplished using B&W's activated recycle SDA. Carbon burnout goals of greater than 99 percent were achieved and emissions were successfully controlled; NO_x emissions averaged 0.245 lb/MM Btu, SO_2 removal efficiencies in excess of 90 percent were achieved with typical emissions of 0.038 lb/MM Btu, particulate matter emissions were 0.0047 lb/MM Btu, and CO emissions were less than 130 ppm at 3.0 percent O_2 [1].

Coal Processing for Clean Fuels Technology

The CCTDP also addressed approaches to converting raw run-of-mine coals to high-energy density, low-sulfur products. Four projects were completed in the coal processing for clean fuels category, which represents a diversified portfolio. These

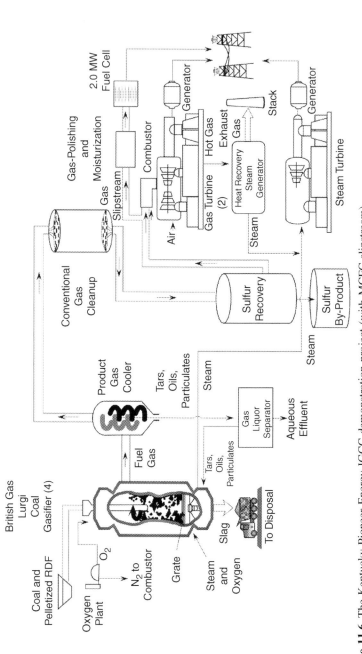

Figure 11.6 The Kentucky Pioneer Energy IGCC demonstration project (with MCFC slipstream). *Source:* From DOE (2002).

included two projects that produced high-energy density solid fuels (and discussed in Chapter 5), one of which also produced a liquid product equivalent to No. 6 fuel oil; one project demonstrated a new methanol production process; and one project complemented the process demonstrations by providing an expert computer model that enables a utility to assess the environmental, operational, and cost impact of utilizing coals not previously burned at a facility, including upgraded coal and coal blends [1].

The ENCOAL Corporation demonstrated SGI International's Liquids-From-Coal (LFC®) process at Triton Coal Company's Buckskin Mine (located near Gillette, Wyoming) [42]. The project objective was to demonstrate the integrated operation of a number of novel processing steps to produce two higher-heating value fuel forms with lower-sulfur contents from mild gasification of low-sulfur subbituminous coal, and to provide sufficient products for potential end users to conduct burn tests. The process, low-temperature carbonization (which is discussed in Chapter 5), produces a process-derived fuel (PDF®) and coal-derived liquid (CDL®). The LFC process consistently produced 250 short tons/day of PDF and 250 barrels/day of CDL from 500 short tons of run-of-mine coal per day. The PDF contains 0.26 percent sulfur with a heat content of 11,100 Btu/lb (compared with 0.45 percent sulfur and 8,300 Btu/lb for the feed coal) [1]. The CDL contains 0.6 percent sulfur and has a heating value of 140,000 Btu/gallon (compared with 0.8 percent sulfur and 150,000 Btu/gallon for No. 6 fuel oil) [1].

Western SynCoal LLC (formerly Rosebud SynCoal Partnership, a subsidiary of Montana Power Company's Energy Supply Division) demonstrated Western SynCoal LLC's advanced coal conversion process (ACCP) of upgrading low-rank subbituminous coal and lignite [43]. The process, described in Section 5.2.5, was performed to demonstrate the ACCP to produce a stable coal product having a moisture content as low as 1 percent, sulfur content as low as 0.3 percent, and heating value up to 12,000 Btu/lb [1]. The ACCP project processed more than 2.8 million short tons of raw coal at Colstrip, Montana to produce nearly 1.9 million short tons of SynCoal® products that were shipped to utility and industrial users. Lower emissions of SO_2 and NO_x were reported in addition to increased power plant output due to the higher grade of fuel burned.

Air Products Liquid Phase Conversion Company, L.P. (a limited partnership between Air Products and Chemicals, Inc., the general partner, and Eastman Chemical Company) demonstrated Air Products and Chemicals, Inc.'s liquid phase methanol process [44]. The objective was to demonstrate, on a commercial scale, the production of methanol from coal-derived synthesis gas using the LPMEOH process; to determine the suitability of methanol produced during this demonstration for use as a chemical feedstock or as a low SO_2 and NO_x emitting alternative fuel in stationary and transportation applications; and to demonstrate, if practical, the production of dimethyl ether (DME) as a mixed coproduct with methanol [1]. The LPMEOH process, illustrated in Figure 11.7, was successfully operated for 69 months, with the demonstration ending in December 2002 [45]. Over the entire operating period, the demonstration facility (located at Kingsport, Tennessee) operated at an onstream availability of 97.5 percent and produced nearly 104 million

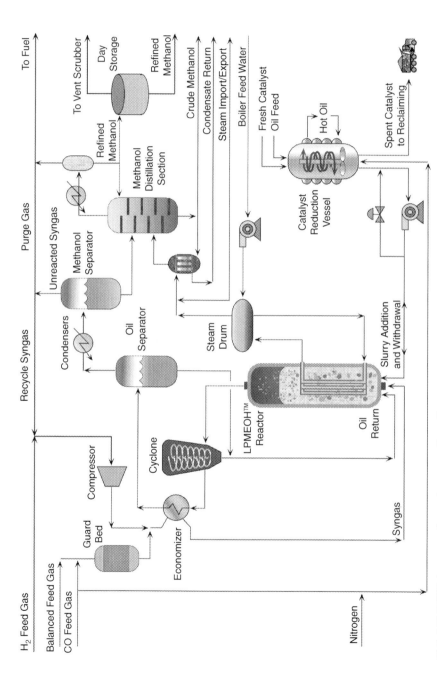

Figure 11.7 LPMEOH demonstration unit process flow diagram.
Source: From DOE (1999).

gallons of methanol, all of which was accepted by Eastman Chemical Company for use in downstream chemical processes. The facility is currently being operated in a commercial mode by Eastman Chemical Company [45]. The process was developed to enhance IGCC power generation by producing a clean-burning, storable liquid fuel from clean coal-derived gas. Methanol contains no sulfur and has exceptionally low NO_x characteristics when burned.

The final project in this category was the development of CQ Inc.'s (Homer City, Pennsylvania) EPRI Coal Quality Expert™ (CQE™) computer software [47]. The objective of the project was to provide the utility industry with a PC software program it could use to confidently and inexpensively evaluate the potential for coal cleaning, blending, and switching options to reduce emissions while producing the lowest-cost electricity [1]. Specifically, the project was supposed to: (1) enhance the existing Coal Quality Information Systems (CQIS™) database and Coal Quality Impact Model (CQIM™) to allow assessment of the effects of coal cleaning on specific boiler costs and performance; and (2) develop and validate CQE, a model that allows accurate and detailed prediction of coal quality impacts on total power plant operating cost and performance. The model that was developed evaluates the impacts of coal quality, capital improvements, operational changes, and environmental compliance alternatives on power plant emissions, performance, and productions costs [1].

Industrial Applications Technology

Projects were also undertaken to address pollution problems associated with using coal in the industrial sector. Five projects encompass substitution of coal for 40 percent of the coke in iron making, integration of a direct iron-making process with the production of electricity, reduction of cement kiln emissions and solid waste generation, demonstration of an industrial-scale slagging combustor, and demonstration of a pulse combustor system. Although electricity can be produced at the industrial scale, these projects are not discussed in this chapter because they have limited or no applicability in power generation and, except for the Blast Furnace Granular-Coal Injection System Demonstration Project [48] and Passamaquoddy Technology Recovery Scrubber™ [49], have not been considered commercial successes in that no domestic or international sales have been made of the demonstrated technologies, nor are they in continued operation at the demonstration site [50].

11.2.4 CCTDP Accomplishments

Over the past 20 to 25 years, the Clean Coal Technology Demonstration Program has successfully demonstrated technologies that do the following [51]:

- Increase efficiency and reduce emissions from coal-fired power plants and industrial facilities
- Expand the number of options, such as fluidized-bed boilers and gasifiers, available for the clean use of coal
- Produce coal-based fuels that burn cleaner and help reduce emissions

Many of the technologies that have been demonstrated under this program are now in commercial use. Table 11.3 lists the 33 projects discussed in the previous section, and those projects considered commercial successes to date are noted. Commercial success is considered if domestic or international sales are made or if the technology continues to operate commercially at the demonstration site. Others have identified

Table 11.3 Clean Coal Technology Program Commercial Successes to Date

Project	Participant	Location	Commercial Status
Gas Suspension Absorption	AirPol, Inc.	West Paducah, KY	Domestic sales International sales
Confined Zone Dispersion	Bechtel Corporation	Seward, PA	—[a]
LIFAC Sorbent Injection	LIFAC-North America	Richmond, IN	Domestic sales International sales Continued operation
Advanced Flue Gas Desulfurization	Pure Air	Chesterton, IN	Continued operation
CT-121 Flue Gas Scrubber	Southern Company Services	Newnan, GA	International sales Continued operation
NO_x Control—Wall-Fired Boiler	Southern Company Services	Coosa, GA	Domestic sales International sales Continued operation
Coal Reburning	B&W Company	Cassville, WI	Continued operation
Low-NO_x Cell Burner	B&W Company	Aberdeen, OH	Domestic sales Continued operation
Gas Reburning/Low-NO_x Burners	EERC	Denver, CO	Domestic sales International sales Continued operation
Micronized Coal Reburning	NYSEG	Lansing, NY	Continued operation
Selective Catalytic Reduction	Southern Company Services	Pensacola, FL	Domestic sales International sales
NO_x Control—T-Fired Boiler	Southern Company Services	Lynn Haven, FL	Domestic sales International sales Continued operation
SNOX™ Flue Gas Cleaning	ABB	Niles, OH	—
LIMB SO_2/NO_x Control	B&W Company	Lorain, OH	Domestic sales International sales

Table 11.3 *Cont'd*

Project	Participant	Location	Commercial Status
SNRB Process	B&W Company	Dilles Bottom, OH	—
Gas Reburning/ Sorbent Injection	EERC	Hennepin and Springfield, IL	Continued operation
Milliken Clean Coal	NYSEG	Lansing, NY	Domestic sales
Dry NO_x/SO_x Control System	Public Service of Colorado	Denver, CO	Domestic sales Continued operation
JEA FBC	JEA	Jacksonville, FL	—
Tidd PFBC	Ohio Power Company	Brilliant, OH	International sales
Nucla CFB	Tri-State	Nucla, CO	Domestic sales International sales
Tampa Electric IGCC	Tampa Electric	Mulberry, FL	Domestic sales International sales Continued operation
Piñon Pine Power	Sierra Pacific	Reno, NV	—
Wabash River Repowering	Wabash River Coal Gasification J.V.	West Terre Haute, IN	Continued operation
Healy Clean Coal	Alaska Ind. Dev. & Export Authority	Healy, AL	—
LPMEOH Process	Air Products	Kingsport, TN	Domestic sales Continued operation
Coal Quality Expert™	ABB and CQ, Inc.	Multiple sites	Domestic sales International sales
ENCOAL® Mild Gasification	ENCOAL Corporation	Gilette, WY	Domestic & international sales pending
Advanced Coal Conversion Process	Western SynCoal LLC	Colstrip, MT	Extended continued operation
Blast Furnace Coal Injection	Bethlehem Steel Corporation	Burns Harbor, IN	Domestic sales Continued operation
Advanced Cyclone Combustor	Coal Tech Corporation	Williamsport, PA	—
Cement Kiln Scrubber	Passamaquoddy Tribe	Thomaston, ME	Continued operation
Pulse Combustor	ThermoChem, Inc.	Baltimore, MD	—

[a]Nothing reported

success based on patents and awards granted to Clean Coal Technology Program projects [52]. Other economical benefits, including jobs created, are discussed in more detail later in this chapter.

Commercial sales (as of 2000), domestic and international, resulting from the CCTDP projects include approximately 130 gasifiers, 160 fluidized-bed units, 2,900 NO_x reduction units, and 200 SO_2 removal units [1, 52]. In addition, approximately 30 utilities have the model for coal processing for clean fuels. Detailed information about sales since 2000 is not readily available, but it is apparent from the data (presented later in this chapter) that sales of these technologies have increased significantly.

Prior to the CCTDP, scrubbers capable of high SO_2 removal were costly to build and difficult to maintain, placed a significant parasitic energy load on the plant output, and produced a sludge waste requiring disposal [53]. The demonstration projects conducted under the CCTDP have cut operating and capital costs in half, provided SO_2 removal efficiencies of 95 to 98 percent, produced valuable by-products, mitigated plant efficiency losses, and captured multiple air pollutants. If CCTDP-developed technologies were applied to all U.S. coal-fired boilers at an average efficiency of 90 percent, total SO_2 emissions could be further reduced by approximately 10 million short tons/year [53]. Currently about one-fourth of the total U.S. coal-fired capacity has flue gas desulfurization (FGD) units installed. The United States has about 260 units with a total capacity of 85,000 MW, which is the largest number of FGD installations in the world.

Prior to the CCTDP, NO_x control technology proven in U.S. utility service was essentially nonexistent. The CCT Program has met the regulatory challenge by developing and incorporating emerging NO_x control technologies into a portfolio of cost-effective compliance options for the full range of boiler types being used commercially [53]. Products of the CCT Program for NO_x control include the following:

- Low-NO_x burners, overfire air, and reburning systems that modify the combustion process to limit NO_x formation
- Postcombustion control options using SCR and SNCR
- Artificial intelligence–based control systems that effectively handle numerous dynamic parameters to optimize operational and environmental performance of boilers

As a result, more than three-fourths of U.S. coal-fired power plants have installed low-NO_x burners. Reburning and artificial intelligence systems have made significant market penetration as well. All sites that developed these NO_x control technologies have retained them for commercial use. In addition, several commercial installations of SCR and, to some extent SNCR, have been installed, with many planned for installation in the near future.

The CCTDP has provided the foundation for powering the twenty-first century through successful demonstration of FBC and IGCC projects on a commercial scale. These technologies are inherently clean, producing negligible emissions of SO_2, NO_x, and particulate matter. The IGCC demonstration projects have achieved excellent environmental performance, with emissions as low as 0.02 lb SO_2/MM Btu and 0.08 lb NO_x/MM Btu. In addition, the higher thermal efficiency processes result in significant reductions in CO_2 emissions.

11.3   Power Plant Improvement Initiative

The Clean Coal Technology Demonstration Program was a programmatic success and served as a model for other cooperative government/industry programs aimed at introducing new technologies into the commercial marketplace. Two follow-on programs have been developed that build on the successes of the CCTDP: the Power Plant Improvement Initiative (PPII) and the Clean Coal Power Initiative (CCPI). The PPII, established by the Department of the Interior and Related Agencies Appropriations for Fiscal Year 2001 (Public Law 106-291), was a cost-shared program patterned after the CCTDP that was directed toward improved reliability and environmental performance of the nation's existing coal-burning power plants [54]. Authorized by the U.S. Congress in 2001, eight projects were selected for negotiation; however, three projects withdrew during the negotiation phase prior to contract award, one project withdrew after award but prior to completion, and four projects were successfully completed [7]. Figure 11.8 is a map showing the location of the PPII projects [55].

The total funding commitment for the PPII projects was more than $71 million, $30 million of which DOE provided, as listed in Table 11.4 [7]. The industrial participants funded 57 percent of the total project costs. Note that Table 11.4 summarizes the four completed projects and the one project that was discontinued prior to completion in order to show all DOE funds expended.

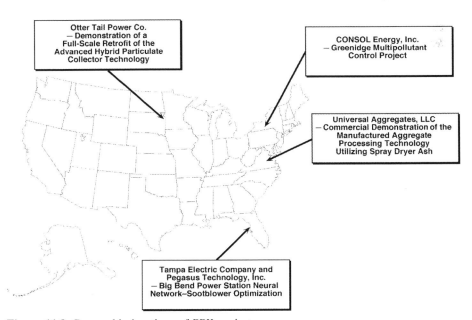

Figure 11.8 Geographic locations of PPII projects.
Source: From DOE (2010).

Table 11.4 PPII Project Costs (dollars)

	Total Project Cost	DOE Share	DOE Obligated	DOE Cost
Achieving NSPS emissions standards through integration of low-NO$_x$ burners with an optimization plan for boiler combustion (project discontinued)	3,005,169	1,387,530	1,387,530	1,387,530
Big Bend Power Station Neural Network-Sootblower Optimization (project completed)	2,381,614	905,013	905,013	905,013
Commercial demonstration of the manufactured aggregate processing technology utilizing spray dryer ash (project completed)	19,581,734	7,224,000	7,224,000	7,224,000
Demonstration of a full-scale retrofit of the advanced hybrid particulate collector technology (project completed)	13,353,288	6,490,585	6,490,585	6,490,585
Greenidge multi-pollutant control project (project completed)	32,742,976	14,341,423	14,341,423	14,341,423
Total PPII	71,064,781	30,348,551	30,348,551	30,348,551

Source: From DOE (2009).

11.3.1 PPII Projects

The PPII projects focused on technologies enabling coal-fired power plants to meet increasingly stringent environmental regulations at the lowest possible cost. With many plants threatened with shutdowns because of environmental concerns, more effective and lower-cost emission controls can keep generators operating while improving the quality of the nation's air and water [2]. Brief descriptions of the four projects that were completed and their objectives are provided.

Tampa Electric Company and Pegasus Technology, Inc. demonstrated control of boiler fouling on Big Bend Power Station's (Apollo Beach, Florida) 445 MW$_e$ unit using a neural-network soot-blowing system in conjunction with advanced controls and instruments [56]. Ash and slag deposition can compromise plant efficiency by impeding heat transfer to the working fluid, leading to higher fuel consumption and higher emissions. The process optimization was targeted to reduce total NO$_x$ generation by 30 percent or more, to improve heat rate by 2 percent, and to reduce particulate matter emissions by 5 percent. A schematic diagram of the project is shown in Figure 11.9 [57].

Compared to competing technologies, this system is estimated to be extremely cost-effective technology that can be readily adapted to virtually any pulverized coal-fired boiler. The project demonstrated: up to 8.5 percent NO$_x$ reduction under a variety of coal and operating conditions; opacity improvement of 1.0 to 1.5 percent during soot blowing events; and unit efficiency improvements of 10 Btu/kWh at high load and 50 Btu/kWh at low load. Based on the demonstration project, it has been shown that the neural network system demonstrated optimization of the sootblowing system and boiler combustion characteristics by stabilizing the unit through sootblowing operations; maintaining or improving heat rate within the required unit operational limits; optimizing the operation of the soot-blowing hardware throughout the full load range of the boiler; potentially reducing NO$_x$ emissions and opacity; reducing auxiliary power consumption, minimize over

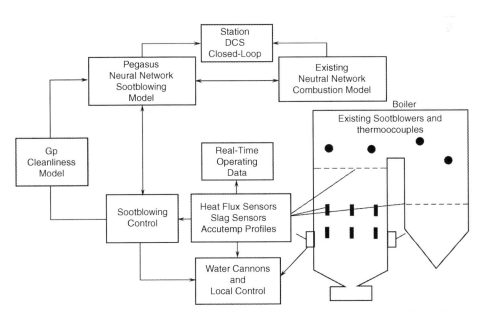

Figure 11.9 The Big Bend Power Station neural network–sootblower optimization project. *Source:* From DOE (2009).

sootblowing and associated tube wear; and remaining within safety and combustion constraints at all times [56].

The objective of the Universal Aggregates LLC project (a joint venture between CONSOL Energy, Inc., and SynAggs, Inc.) was to design, construct, and operate an aggregate manufacturing plant that converted 115,000 short tons/year of spray dryer by-products into 150,000 short tons/year of lightweight masonry blocks, lightweight concrete, or asphalt paving material [58]. Only about 30 percent of the 28 million short tons of flue gas desulfurization residue in the United States is recycled, with the remainder being landfilled. This process proposes to reduce plant disposal costs and the environmental drawbacks of landfilling by producing a salable by-product. The construction aggregate market in the United States is estimated to be about 2 billion short tons annually [59]. The demonstration was located by the 250 MW_e Birchwood Power Facility in King George County, Virginia. A schematic diagram of the system is shown in Figure 11.10 [59]. The manufactured aggregate plant successfully demonstrated the continuous and fully integrated process operation, including mixing, extrusion, curing, crushing, and screening. Lightweight aggregates with proper-size gradation and bulk density were produced from a spray dryer absorber for use in the production of concrete masonry units.

Otter Tail Power Company, with Montana Dakota Utilities, NorthWestern Public Service, W.L. Gore & Associates, Inc., and the University of North Dakota Energy

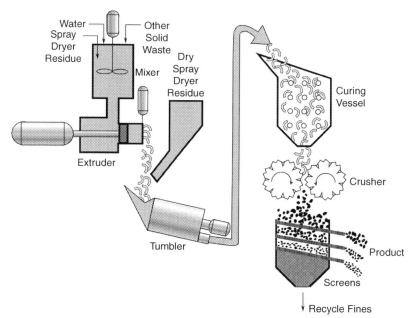

Figure 11.10 The Universal Aggregates, LLC aggregate processing project. *Source:* From DOE (2009).

and Environmental Research Center, demonstrated a hybrid technology with the potential to increase the particulate matter capture in coal plants up to 99.99 percent by integrating fabric filtration and electrostatic precipitation [60]. Performance targets were 99.99 percent capture of 0.1 to 50 μm particles while exhibiting low pressure drop, overall reliability of the technology, and long-term bag life. The advanced hybrid particulate collector (AHPC), discussed in detail and illustrated schematically in Chapter 9, combines the best features of an electrostatic precipitator (ESP) and a baghouse in the same housing, providing major synergism between the two methods to overcome the problem of excessive fine particulate emissions that escape collection in an ESP and the reentrainment of dust in a baghouse. The demonstration was performed on Big Stone Power Plant's (Big Stone City, South Dakota) 450 MW$_e$ cyclone-fired boiler and is a scaleup from a 2.5 MW$_e$ slipstream test program that was performed at the plant. The following project objectives were met [60]:

- Demonstrate that the APHC technology can be retrofitted into an existing ESP at the full-scale level was met.
- Demonstrate the ability of the AHPC to provide more than 99.99 percent particulate collection efficiency for particle sizes greater than 0.1 μm was met, although the tests were performed over a short time period. There was some concern that an objective of 99.99 percent collection efficiency after three to five years of operation would be problematic due to the failure of the filter bags.
- Demonstrate the reliability of the AHPC as defined by acceptable maintenance requirements that are the same or less than standard ESPs or baghouses were met.

The following project objectives were not met [60]:

- Demonstrate the ability of a retrofitted AHPC to meet performance specifications without derating the plant because of opacity was not met, since the plant had to derate as the bags began failing.
- Demonstrate the ability of the AHPC to achieve low-pressure drop at an air-to-cloth ratio of 12 ft/min was not met. Although the system was operated at 10.5 to 12.0 ft/min, the tubesheet differential pressure ranged from 8 to 10 inches H$_2$O, which is high and resulted in significant limitations in the plant output.
- Demonstrate the long-term operability of the AHPC was not met since the pressure drop/air-to-cloth ratio objective was not met.
- Demonstrate the economic viability of the AHPC was not because the objectives of air-to-cloth ratio and long-term operability were not met, along with the high cost of replacement bags that occurred.

The objective of the CONSOL Energy, Inc., project, with its participants AES Greenidge, LLC and Babcock Power Environmental Inc., was to demonstrate cost-effective multipollutant control for relatively small power plants using selective noncatalytic reduction (SNCR)/in-duct selective catalytic reduction (SCR) in combination with low-NO$_x$ burners and a circulating fluidized-bed dry scrubber (CFBDS) system with recycled baghouse ash and activated carbon injection to control NO$_x$ emissions to 0.10 lb/MM Btu at full load; reduce SO$_2$ emissions by 95 percent, mercury by 90 percent, and acid gases (SO$_3$, HCl, and HF) by 95 percent; and to evaluate the impact of biomass cofiring up to 10 percent heat input on the performance of

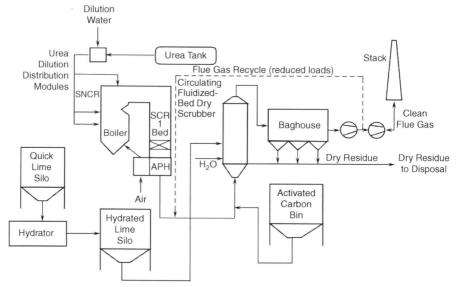

Figure 11.11 The Greenidge Multi-Pollutant Control Project.
Source: From DOE (2009).

the SNCR/SCR hybrid and CFBDS system [61]. The demonstration was performed on AES Greenidge LLC's 107 MW$_e$ Unit 4 located in Dresden, New York. A schematic of the system is shown in Figure 11.11 [61].

Test results demonstrated that the combination of technologies met all of the emissions reduction goals of the project with 96 percent SO$_2$ removal, 98 percent mercury removal with no activated carbon injection, 95 percent SO$_3$ removal, 97 percent HCl removal, 98 percent reduction in particulate matter relative to emissions prior to installation of the technology, and high-load NO$_x$ emissions averaged 0.14 lb/MM Btu during long-term operation, which was a 52 percent reduction in NO$_x$ emissions [62]. The testing was performed with a high-sulfur (>2 percent) eastern U.S. bituminous coal, except for some tests in which waste wood was cofired at less than 5 percent of the total heat input. No discernable effects on the performance of the multipollutant control system were observed; however, the duration of the cofiring was too limited to permit a thorough evaluation of its impact on the system.

11.3.2 Benefits of the PPII

The PPII, a precursor to CCPI (discussed in the next section), was designed to establish commercial-scale demonstrations of coal technologies to ensure energy supply reliability [63]. The PPII is poised to make near-term contributions to air quality improvements and focuses on technology that can be commercialized over the next few years.

11.4   Clean Coal Power Initiative

The second follow-on program to the CCTDP is the Clean Coal Power Initiative (CCPI). CCPI, initiated in 2002, is an innovative technology demonstration program that fosters more efficient clean coal technologies for use in existing and new power-generation facilities in the United States [3]. CCPI is to advance a broad spectrum of promising technologies that target today's most pressing environmental, economic, and energy security challenges. Candidate technologies are demonstrated at significant scale to ensure proof-of-operation prior to widespread commercialization. Technologies emerging from this program will help to meet new environmental objectives for the United States, detailed in the Clear Skies Initiative, Global Climate Change Initiative, and FutureGen, and to advance pollution control and coal utilization both in the United States and abroad. Early demonstrations emphasize technologies that are applicable to existing power plants and include construction of new plants. Later demonstrations will include systems comprising advanced turbines, membranes, fuel cells, gasification technologies, and hydrogen production [3].

The first solicitation (CCPI-1), which was issued in 2001, was open to any technology advancement related to coal-based power generation that results in efficiency, environmental, and economic improvement compared to currently available state-of-the-art alternatives. In February 2004, the second CCPI solicitation (CCP-2) was issued for proposals to demonstrate advances in coal gasification systems, technologies to permit improved management of carbon emissions, and advancements that reduce mercury and other power plant emissions. The third solicitation (CCPI-3A) was issued in 2008 and focused on the capture and sequestration, or beneficial reuse, of CO_2 emissions from coal-based power generation. On June 9, 2009, DOE issued an amendment that provided for a second application under the third round (CCPI-3B), with selected projects announced on December 4, 2009.

11.4.1   Program Importance

The strength and security of the U.S. economy is closely linked to the availability, reliability, and cost of electric power. Economic growth is linked to reliable and affordable electric power. Electricity requirements for the United States are steadily increasing, and coal will play a significant role in satisfying the United States' energy needs. CCPI will help meet these energy electricity demands by demonstrating new-generation technologies [3]. CCPI will also enable effective use of existing facilities and prepare for their retirement by demonstrating technological improvements in efficiency, advanced low-cost, high-performance emissions control technologies, and reliability at new and existing plants.

CCPI is closely aligned with research, development, and demonstration activities being performed under DOE's Coal and Power Systems core research and development programs that are working toward ultra-clean fossil fuel-based energy systems in the twenty-first century [6, 63]. CCPI technologies will address existing and new regulatory requirements, and it will complement the goals in the FutureGen Project, which is an initiative to create the world's first coal-based, zero emission electricity and hydrogen plant.

Table 11.5 New Plant Performance Targets That Represent the Best Integrated
Technology Capability

	Reference Plant[a]	2010	2020 (Vision 21)
Air emissions	98% SO_2 removal 0.15 lb NO_x/MM Btu 0.01 lb PM/MM Btu mercury (Hg)	99% SO_2 removal 0.05 lb NO_x/MM Btu[b] 0.005 lb PM/MM Btu[c] 90% Hg removal[d]	>99% SO_2 removal <0.01 lb NO_x/MM Btu 0.002 lb PM/MM Btu 95% Hg removal
By-product utilization	30%	50%	Near 100%
Plant efficiency (HHV)	40%	45–55%	50–60%
Availability[e]	>80%	>85%	≥90%
Plant capital cost ($/kW)[f]	900–1,300	900–1,000	800–900
Cost of electricity (¢/kWh)[g]	3.5	3.0-3.2	<3.0

[a]Plant can be built using current state-of-the-art technology; plant meets NSPS; NO_x levels below 0.15 lb/MM Btu can be achieved with a combination of advanced combustion and SCR technologies; some mercury (Hg) reduction achieved as cobenefit with existing environmental control technologies; by-product utilization represents an average for existing plant locations as actual plant utilization ranges from essentially zero to near 100%; no carbon capture and sequestration; reflects current cooling tower technology or use.
[b]For NO_x, reduce cost for achieving <0.10 and 0.05 lb/MM Btu to three-fourths of that of SCR by 2005 and 2010, respectively.
[c]Achieve particulate matter (PM) targets for existing plant in 2010 and 99.99% capture of 0.10–10 μm particles.
[d]Achieve 50 to 70% Hg reduction at less than three-fourths cost of activated carbon injection by 2005.
[e]Percent of time capable of generating power
[f]Range reflects projection for different plant technologies that will achieve environmental performance and energy cost targets.
[g]Bus-bar cost of electricity
Source: From Eastman (2003) and DOE (2004).

CCPI will help the United States to achieve improved power plant performance and near-zero emissions, and it is an integral part of achieving new plant performance targets identified in DOE's road map for existing and future energy plants [5, 63, 64]. Existing plant road map performance objectives include reduced cost for NO_x and high-efficiency mercury control and achieving particulate matter targets in 2010 of 99.99 percent capture of 0.1 to 10 μm particulates [63]. The long-term road map goals are aimed at achieving near-zero emissions power and clean fuels plants with CO_2 management capability. The long-term new plant performance targets are presented in Table 11.5 [63, 64].

11.4.2 Round 1 CCPI Projects

CCPI early demonstrations emphasize advanced technologies that are applicable to existing power plants but will also include construction of new, advanced, clean coal power plants. Eight projects were selected for funding, but negotiations ceased

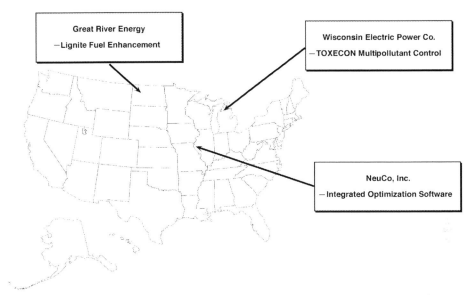

Figure 11.12 Geographical locations of the Clean Coal Power Initiative round 1 projects. *Source:* Modified from DOE (2009).

on one project, three were withdrawn, one was discontinued, two are active, and one has been completed. Figure 11.12 shows the locations of the three active or completed projects, and Table 11.6 summarizes the project cost and financial status of the CCP1-1 projects [7]. Note that two of the discontinued projects are included in Table 11.6 because federal funds were expended before the projects terminated. Brief descriptions of the technologies being demonstrated in the active/completed projects follow.

NeuCo, Inc. (Boston, Massachusetts) demonstrated the application of individual on-line optimization products at Dynergy Midwest Generation's Baldwin Energy Complex (Randolph County, Illinois) for combustion, sootblowing, SCR operations, overall unit thermal performance, and plantwide economic optimization; the integration of individual optimizers through NeuCo's ProcessLink® platform; and reduction of Baldwin Energy Complex NO_x emissions by 5 percent, increased efficiency by 1.5 percent, and improved reliability and availability to increase net annual electrical power production by 1.5 percent [7]. Five optimization systems (shown in Figure 11.13) were incorporated, with two 585 MW_e cyclone-fired boilers with SCR and a 595 MW_e tangentially fired boiler with low-NO_x burners. The project was successfully completed and demonstrated that advanced optimization technologies can play an important role in improving the environmental footprint of coal-based power generation while achieving other important operating objectives. Specific conclusions are [65]:

- The 5 percent target for NO_x reduction was exceeded with average NO_x reductions that were between 12 and 14 percent observed.
- The optimization systems delivered an average heat rate improvement of between 0.67 and 0.70 percent, which fell short of the 1.5 percent heat rate improvement target largely

Table 11.6 CCPI-1 Project Costs (in dollars)

	Total Project Cost	**DOE Share**	**DOE Obligated**	**DOE Cost**
Advanced multi-product coal utilization by-product processing plant (project discontinued)	1,245,305	621,407	621,407	617,366
Demonstration of integrated optimization software at the Baldwin energy complex (project completed)	19,904,733	8,592,630	8,592,630	8,592,630
Increasing power plant efficiency – lignite fuel enhancement (project active)	31,512,215	13,518,737	13,518,737	13,306,011
TOXECON retrofit for mercury and multi-pollutant control on three 90 MW$_e$ coal-fired boilers (project active)	52,978,113	24,859,578	24,859,578	21,736,758
Western Greenbrier co-production demonstration project (project discontinued)	16,256,940	8,128,470	8,128,470	7,861,662
Total CCPI-1	121,087,308	55,720,822	55,720,822	52,114,427

Source: Modified from DOE (2009).

because cyclone availability and NO$_x$ emissions were prioritized over heat rate in the event they needed to be traded off with one another.
- The target of increasing available MWh by 1.5 percent was met.
- Reductions in greenhouse gases, mercury, and particulates were observed, with the optimization leverage observed in the heat rate and NO$_x$ reductions.
- Commensurate improvements in costs, reliability, and availability resulted from the preceding benefits.

The objective of Great River Energy's project is to demonstrate a process for reducing the moisture content of lignite by 25 percent (from 40 to 30 percent moisture in the lignite) and, thereby, increase the value of high-moisture fuels in electrical generation plants by increasing the net generating capacity of units that burn high-moisture coal, increase the energy supply of units that burn high-moisture coal, increase the cost-effectiveness of the country's electrical generation industry, improve the environment by reducing emissions from coal-fired power plants, and

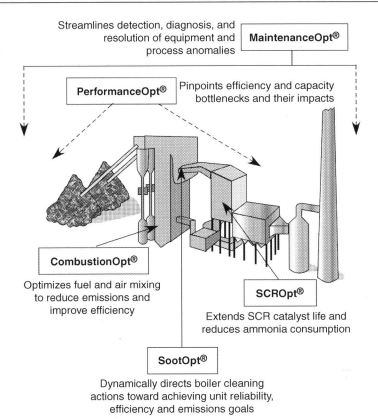

Figure 11.13 Optimization systems installed on the Baldwin Energy Complex. *Source:* From DOE (2009).

increase the value of the country's lignite reserves [7, 63]. The demonstration will be performed at Great River Energy's Coal Creek Station (i.e., two 546 MW$_e$ pulverized coal-fired units located at Underwood, North Dakota) and will use waste heat in two fluidized bed driers to reduce the lignite moisture content.

The project is being performed in two phases. In the first phase, which has been completed, a prototype dryer was constructed and successfully tested. In the second phase (shown schematically in Figure 11.14), full-scale dryers will be built to provide sufficient dryer capacity for fully fueling the 546 MW$_e$. The design of the integrated full-scale dryer was completed in December 2007. Fabrication and on-site assembly were completed in May 2008. Major dryer internals, such as water coils, air sparger, fire protection system, and explosion protection system were completed in March 2009. Installation of electrical cables, controls, and instrumentation; modifications to the coal handling system; and start of testing were completed in December 2009 [7, 66]. Commercial operation of the integrated dryer system at full load started in December 2009 and the demonstration is scheduled to be completed in 2010.

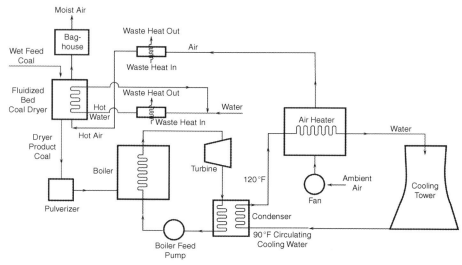

Figure 11.14 The GRE demonstration components.
Source: From DOE (2009).

The objective of Wisconsin Electric Power Company's (We Energies') project is to achieve 90 percent mercury removal through injection of activated carbon; increase particulate matter collection efficiency (specifically $PM_{2.5}$); reduce already low SO_2 and NO_x emissions at the plant by an additional 70 percent and 30 percent, respectively; recover 90 percent of the mercury captured in the sorbent; achieve 100 percent fly ash utilization; advance the reliability of mercury continuous emissions monitors; and successfully integrate the entire system [7, 67]. The project is demonstrating the TOXECON™ sorbent injection process for multipollutant control of a combined flue gas stream from three units totaling 270 MW$_e$. TOXECON, an Electric Power Research Institute-patented process (discussed in Chapter 9 and shown schematically in Figure 11.15), injects activated carbon and sodium-based sorbents into a pulsed-jet baghouse installed downstream of the air preheater to operate at relatively cool temperatures conducive to mercury and other pollutant absorption [7].

The project demonstrates long-term reliability by continuously operating the powdered activated carbon system. Over a two-year period, more than 90 percent mercury removal has been accomplished; ash handling and dust control issues have been resolved, and optimum bag cleaning cycles have been identified. Results from injection testing using a sodium-based sorbent (hydrated sodium bicarbonate) indicate 70 percent SO_2 removal, no effect on NO_x emission, and a decrease in mercury capture at normal activated carbon injection rates. The project is continuing to investigate cost improvements while maintaining greater than 90 percent mercury removal as well as improvements for control of particulate matter, NO_x, and SO_2 emissions.

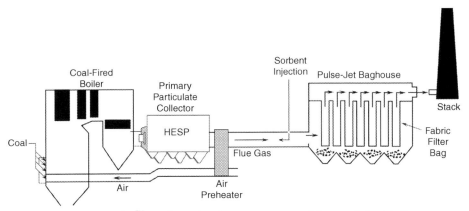

Figure 11.15 TOXECON™ configuration for mercury control in a coal-fired boiler. *Source:* From DOE (2009) and Eastman (2003).

11.4.3 Round 2 CCPI Projects

Four projects were selected for Round 2 funding. One project was withdrawn and three are active. Figure 11.16 shows the locations of the three active projects, and Table 11.7 summarizes the project cost and financial status of the round 2 CCPI (CCP1-2) active projects [7]. Brief descriptions of the technologies being demonstrated in the projects are provided following.

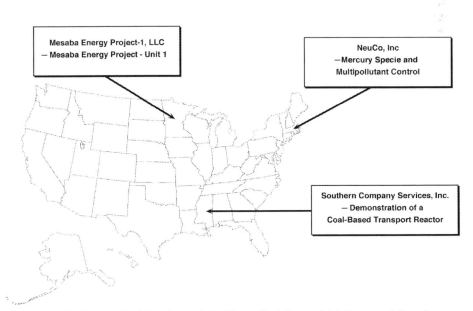

Figure 11.16 Geographical locations of the Clean Coal Power Initiative round 2 projects. *Source:* Modified from DOE (2009).

Table 11.7 CCPI-2 Project Costs (in dollars) for the Active Projects

	Total Project Cost	**DOE Share**	**DOE Obligated**	**DOE Cost**
Demonstration of a coal-based transport reactor	1,625,082,040	293,750,000	293,750,000	23,547,938
Mercury specie and multi-pollutant control	15,560,811	6,079,479	6,079,479	6,079,479
Mesaba energy project—Unit 1	2,155,680,783	36,000,000	22,245,505	17,776,616
Total CCPI-2	3,796,323,634	335,829,480	322,874,984	47,404,033

Source: From DOE (2009).

Southern Company Services, Inc. will be assessing the operational, environmental, and economic performance of an air-blown transport gasifier-based IGCC system with two transport gasifiers, two F-class combustion turbines, and one steam turbine [7, 68]. The transport gasifier is based on a simple, robust, and efficient technology similar in design to a fluidized catalytic cracking (FCC) unit that has been proved over 50 years in the petroleum refining industry. The transport gasifier operates at considerably higher circulation rates, velocities, and riser densities than does a conventional circulating fluidized-bed, resulting in higher throughput, better mixing, and higher mass transfer and heat transfer rates. This process technology makes possible the cost-effective production of syngas from low-rank, high-moisture, and high-ash coals. This project will demonstrate an advanced synags cleanup system that includes sulfur removal and recovery; high-temperature, high-pressure particulate filtration; ammonia recovery; and mercury control. The transport reactor has a fuel-flexible design projected to have higher efficiency and lower capital and operating costs compared to oxygen-blown entrained-flow gasifiers. The project is in the design phase (for a 285 MW_e net unit), and the plant, shown schematically in Figure 11.17, will be located in Kemper County, Mississippi using Mississippi lignite.

NeuCo, Inc. will demonstrate that state-of-the-art sensors and neural network-based optimization and controls can measure mercury species (elemental and oxidized mercury); control mercury emissions with existing flue gas desulfurization FGD) and ESP systems; and reduce pollutant emissions in general without major capital expenditures [7, 69]. The project will demonstrate nonintrusive advanced sensors and neural network-based optimization and control technologies for enhanced mercury and multipollutant control on a 890 MW_e tangentially fired boiler at the NRG Texas Limestone Plant in Jewett, Texas. The plant is equipped with a cold-side ESP and wet limestone FGD system and burns a blend of Texas lignite and Powder River Basin subbituminous coal, which are known to emit relatively high levels of elemental mercury emissions. NeuCo, Inc. will use sensors to evaluate the mercury species at key locations (shown generically in Figure 11.18), develop optimization software that results in the best plant conditions to promote mercury

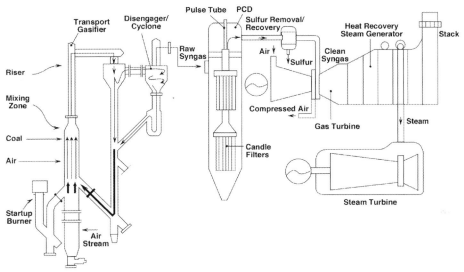

Figure 11.17 Tranport gasifier demonstration project.
Source: From DOE (2009).

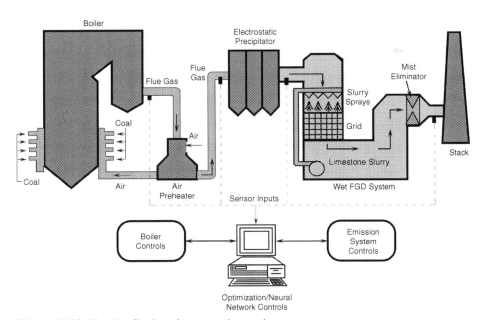

Figure 11.18 The NeuCo, Inc. demonstration project.

oxidation and minimize emissions in general, and use neural networks to effect the optimization conditions. The project entered the operational demonstration phase in early 2009 and is planned to continue through 2010.

The objectives of the Mesaba Energy Project-1, LLC (Excelsior Energy, Inc.) project is to demonstrate the ConocoPhillips E-Gas technology at twice the

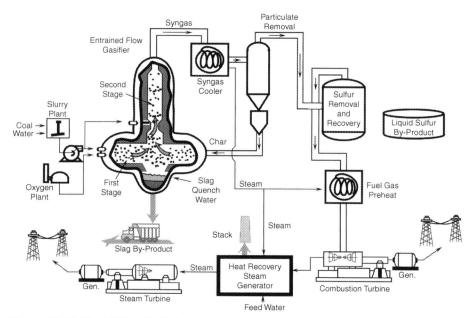

Figure 11.19 The MEP-1, LLC demonstration project.

generating capacity of the Wabash River Coal Gasification Repowering Project performed under the CCTDP; to achieve 90 percent or higher operational availability; and to demonstrate CO_2 emissions at 15 to 20 percent lower than the U.S. average in 2006 and emission levels of criteria pollutants and mercury equal to or below those of the lowest emission rates for utility-scale, coal-based generation [7, 70]. The project, which is shown schematically in Figure 11.19, will demonstrate the next-generation ConocoPhillips E-Gas technology in up to a 606 MW_e (net) IGCC application to be located in either Itasca or St. Louis County, Minnesota. The project will incorporate cost and performance improvements from more than a decade of experience with the predecessor design including gasifier scaleup, increased system pressure, increased slurry percentage to the second-stage gasifier, and enhanced by-product and containment removal systems. The project is near the end of the design phase and the beginning of the construction phase. Operation is planned for 2013 and 2014.

11.4.4   Round 3 CCPI (CCPI-3A and CCPI-3B) Projects

Five projects that advance carbon capture and storage were selected for Round 3 CCPI funding: two from the CCPI-3A solicitation and three from the CCPI-3B solicitation. The CCPI-3A projects were announced on July 1, 2009, and the CCPI-3B projects were announced on December 4, 2009. The geographical locations of the projects are shown in Figure 11.20. Summaries of the selected projects are provided.

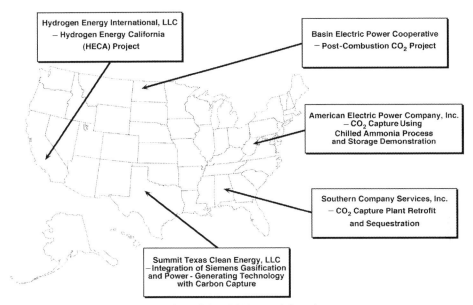

Figure 11.20 Geographical location of the CCPI round 3 projects.

Basin Electric Power Cooperative was awarded a postcombustion CO_2 capture project [71]. Basin Electric Power Cooperative will partner with Powerspan and Burns & McDonnell to demonstrate the removal of CO_2 from the flue gas of a lignite-fired boiler by adding CO_2 capture and sequestration to Basin Electric's existing Antelope Valley Station, located near Beulah, North Dakota. Powerspan's ECO2® ammonia-based technology will be used to capture CO_2 on a 120 MW_e equivalent flue gas stream from the 450 MW_e Antelope Valley Station Unit 1. The result will be 90 percent removal of CO_2 from the treated flue gas, yielding 3,000 short tons per day (1 million short tons per year) of pipeline quality CO_2. The ammonia-based SO_2 scrubbing system will also produce a liquid stream of ammonium sulfate that will be processed into a fertilizer by-product. DOE is providing $100 million toward the project.

Hydrogen Energy International LLC, a joint venture owned by BP Alternative Energy and Rio Tinto, was awarded a project to demonstrate advanced IGCC with full carbon capture [71]. Hydrogen Energy International LLC will design, construct, and operate an IGCC power plant that will take blends of coal and petroleum coke, combined with nonpotable water, and convert them into H_2 and CO_2. The CO_2 will be separated from the H_2 using the methanol-based Rectisol process. The H_2 gas will be used to fuel a power station, and the CO_2 will be transported by pipeline to nearby oil reservoirs, where it will be injected for storage and used for enhanced oil recovery. The project will be located in Kern County, California, and will capture more than 2 million short tons per year of CO_2. DOE is providing $308 million toward the project.

American Electric Power Company, Inc. was awarded a CO_2 capture and storage demonstration in which they will design, construct, and operate a chilled ammonia process that is expected to effectively capture at least 90 percent of the CO_2 (1.5 million metric tons per year) in a 235 MW_e flue gas stream at the existing 1,300 MW_e Appalachian Power Company (APCo) Mountaineer Power Plant near New Haven, West Virginia [71]. The captured CO_2 will be treated, compressed, and then transported by pipeline to proposed injection sites located near the capture facility. During the operation phase, AEP plans to permanently store the entire amount of captured CO_2 in two separate saline formations located approximately 1.5 miles below the earth's surface. The project team includes AEP, APCo, Schlumberger Carbon Services, Battelle Memorial Institute, CONSOL Energy, Alstom, and an advisory team of geologic experts. DOE's share of the project is $334 million.

Southern Company Services, Inc. has been awarded a carbon capture and sequestration demonstration in which they will retrofit a CO_2 capture plant on a 160 MW_e flue gas stream at an existing coal-fired power plant, Alabama Power's Plant Barry [71]. The captured CO_2 will be compressed and transported through a pipeline, and up to 1 million metric tons per year of CO_2 will be sequestered in deep saline formations. Southern Company Services will also explore and utilize potential opportunities for beneficial use of the CO_2 for enhanced oil recovery. The project team also includes Mitsubishi Heavy Industries America, Schlumberger Carbon Services, Advanced Resources International, the Geological Survey of Alabama, Electric Power Research Institute, Stanford University, the University of Alabama, AJW Group, and the University of Alabama at Birmingham. DOE's share of the project is $295 million.

Summit Texas Clean Energy, LLC, has been awarded a project to integrate the Siemens gasification and power-generating technology with carbon capture technologies to effectively capture 90 percent of the CO_2 (2.7 million metric tons per year) at a 400 MW_e plant to be constructed new Midland-Odessa, Texas [71]. The captured CO_2 will be treated, compressed, and then transported by pipeline to oilfields in the Permian Basin of West Texas for use in enhanced oil recovery operations. The Bureau of Economic Geology at the University of Texas will design and ensure compliance with a state-of-the-art CO_2 sequestration monitoring, verification, and accounting program. DOE's share of the project is $350 million.

11.4.5 CCPI Benefits

CCPI bridges the gap following the CCTDP and PPII and the implementation of Vision 21 systems, ensuring the ongoing development of advanced systems for power production emerging from DOE's core fossil-fuel research programs. Successful completion of the CCPI Program will introduce technologies to the U.S. marketplace that can achieve compliance with emerging air regulations and National Energy Policy (NEP) priorities. CCPI provides an important platform to implement the NEP recommendation to increase investment in clean coal technology [63]. The program will also help to ensure that upcoming regulations are science and engineering based and exploit emerging technologies developed under CCPI. CCPI will mitigate costs and reduce the technical and environmental risks associated with advanced technology development and will serve as a proving ground in the United States to speed technologies to market both in the United States and abroad, thereby assuring the realization of early environmental benefits [63].

The CCPI Program benefits are expected to be substantial when compared to the investment costs. Unless advanced technologies achieve widespread commercial use, which must occur through demonstrations, the projected benefits will not be achieved. The following benefits are expected from the program [63]:

- Reduced fuel costs due to higher plant efficiencies
- Lower capital costs for construction of new plants and repowered facilities
- Lower capital and operating costs for existing plants
- Reduced costs of environmental compliance
- Avoidance of environmental costs (e.g., health, infrastructure, agriculture)
- Enhanced industrial competitiveness, leading to increased domestic and international sales
- Additional jobs

11.5   Benefits of DOE's Clean Coal Technology Programs

Previous sections presented benefits from individual DOE Clean Coal Technology Programs (i.e., CCTDP, PPII, and CCPI). Several studies have been performed over the years to estimate economic and environmental benefits of the programs as a whole, including the following:

- National Research Council of the National Academy of Science (NRC/NAS) in 2001
- U.S. DOE National Energy Technology Laboratory (NETL) in 2002
- Electric Power Research Institute (EPRI) in 2002
- DOE, EPRI, and Coal Utilization Research Council in 2004
- NRC/NAS in 2005
- NETL in 2006
- Management Information Services, Inc. in 2009

DOE has reported success stories (as discussed in previous sections), and some generalized benefits are shown in Table 11.8, which is from their 2006 study of the Clean Coal Technology Program [72]. Management Information Services, Inc. has performed a recent, in-depth study that encompasses CCTDP, PPII, CCPI, and clean coal–related research and development (which totals about $100 to $300 million annually through in-house research and Small Business Innovative Research (SBIR) and Small Business Technology Transfer (STTR) programs), in which total costs for government and industry are estimated and benefits are quantified for the following [73]:

- Reduced capital costs of advanced technologies in new plants
- Reduced capital and operating costs at existing plants to remain compliant with environmental regulations
- Reduced fuel costs due to higher efficiencies
- Avoidance of environmental costs
- The value of clean coal technology export sales
- More jobs created

Table 11.8 Clean Coal Technology Benefits

Technology	Impact
Low-NO$_x$ burners	Now on 75 percent of existing U.S. coal power plants; 10–50% of the cost of older systems; 25 million short ton reduction in U.S. NO$_x$ emissions through 2005; $25 billion national benefit
Selective catalytic reduction	Achieves NO$_x$ reduction of 80–90% or more; technology today costs 50% of what it did in the 1980s; is deployed on about 30% of U.S. coal plants
Flue Gas Desulfurization (FGD)	Systems now cost one-third of what they did in the 1970s; more than 400 commercial units deployed; 7 million short ton reduction in SO$_2$ (beyond what would have occurred without DOE R&D) through 2005; overall $50 billion savings from lower FGD costs and environmental improvements
Fluidized-bed combustion	170 units deployed in the U.S.; 400 units deployed worldwide; highly commercialized, with more than $6 billion in domestic sales and nearly $3 billion in overseas sales; inherently low NO$_x$ emitting technology capable of using coal waste fuels not previously usable, providing an economic/ environmental benefit of $2 billion through 2020
IGCC	In early stage, but 7.5 GW$_e$ projected to be operating in U.S. by 2020; estimated economic/environmental benefits of more than $12 billion by 2020

Source: From DOE (2006).

The data presented in the remainder of this section are a summary of Management Information Services, Inc.'s study, where they have gone into great detail in their evaluation [73]. They have shown that the return on investment to the U.S. government from the Clean Coal Technology program is favorable and growing rapidly, as illustrated in Figure 11.21. The program has also been successful in creating jobs, which is projected to continue, as shown in Figure 11.22. These are predominately manufacturing-orientated, well-paying jobs.

The major findings of Management Information Services, Inc. are summarized in Table 11.9 and Figures 11.23 through 11.25 [73]; the table and figures show the following:

- Benefits over the 20-year period 2000–2020 total $111 billion (in 2008 dollars).
- The benefits in individual categories range from $15 billion in fuel cost savings to $39 billion for capital and technology cost savings in new and existing plants.
- Total jobs created exceed 1.2 million, with an annual average of about 60,000 jobs created.
- Cumulative benefits exceed cumulative DOE costs after 2005 and cumulative total costs after 2008.
- By the end of the forecast period, cumulative benefits exceed cumulative DOE costs and cumulative industry costs by more than 13 to 1.

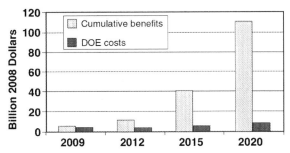

Figure 11.21 Benefits and costs of the Clean Coal Technology Program.
Source: From Management Information Services, Inc. (2009).

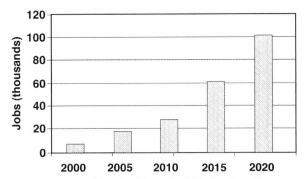

Figure 11.22 Annual jobs created by the Clean Coal Technology Program.
Source: From Management Information Services, Inc. (2009).

Table 11.9 Summary of Clean Coal Technology
Benefits, 2000–2020

Benefits Category	Benefits (in billions of 2008 dollars)
Capital and technology cost savings in new and existing plants	$39
Fuel cost savings	$15
Avoided environmental costs	$25
Clean coal technology exports	$32
Total monetary benefits	$111
Total cumulative jobs created (thousands)	1,200
Average annual jobs created (thousands)	60

Source: From Management Information Services, Inc. (2009).

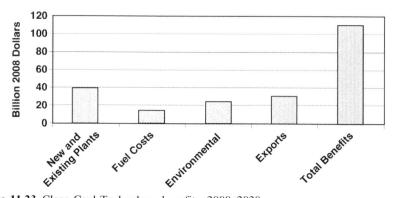

Figure 11.23 Clean Coal Technology benefits, 2000–2020.
Source: From Management Information Services, Inc. (2009).

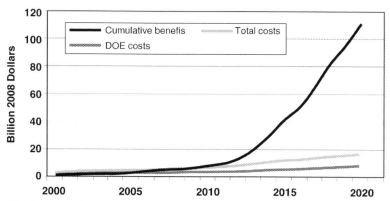

Figure 11.24 Cumulative Clean Coal Technology costs and benefits, 2000–2020.
Source: From Management Information Services, Inc. (2009).

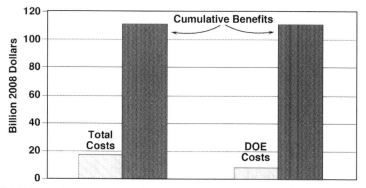

Figure 11.25 Clean Coal Technology benefit-cost ratios, 2000–2020.
Source: From Management Information Services, Inc. (2009).

11.6   Vision 21

The U.S. DOE is providing the foundation needed to build a future generation of fossil energy-based power systems capable of meeting the energy and environmental demands of the twenty-first century. This initiative, called Vision 21, is the DOE's approach for developing the technology needed for ultra-clean twenty-first-century, energy plants. The overall goal of Vision 21 is to develop the core modules for a fleet of fuel-flexible, multiproduct energy plants that boost power efficiencies to 60+ percent, emit virtually no pollutants, and with carbon sequestration release minimal or zero carbon emissions [4, 64].

Achieving this goal will require an intensive, long-range (i.e., 15 to 20 years) research and development effort that emphasizes innovation and commercialization of revolutionary technologies. First-generation systems emerging from the Clean Coal Technology Development, PPII, and CCPI programs provide, or will provide, the basis for Vision 21, including (1) the knowledge base from which to launch commercial systems, which will experience increasingly improved cost and performance over time through design refinement; and (2) platforms on which to test new components, which will result in improvements in cost and performance.

Vision 21 is based on three premises: the United States will rely on fossil fuels for a major share of its energy needs well into the twenty-first century; that a diverse mix of energy resources including coal, gas, oil, biomass, and other renewables and nuclear must be used for strategic and security reasons, rather than using a limited subset of these resources; and that research and development directed at resolving energy and environmental issues can find an affordable way to make energy conversion systems meet ever-stricter environmental standards [64]. Vision 21 plants will effectively remove environmental constraints as an issue in the use of fossil fuels.

Emissions of traditional pollutants, including smog- and acid-raining species, will be near zero, and the greenhouse gases, carbon dioxide, will be reduced 40 to 50 percent by efficiency improvements and reduced to zero if combined with sequestration. In addition, Vision 21 plants will address water use, by-product utilization, sustainability (i.e., no future legacies), timely deployment of new technology, and affordable, competitive systems with other energy options [4, 64, 74]. The Vision 21 energy plant performance targets are listed in Table 11.10 [4, 64, 74], and the technology concept is shown in Figure 11.26 [4, 74].

11.6.1   Vision 21 Technologies

Vision 21 energy plants will utilize a modular design philosophy and will comprise technology modules selected and configured to produce the desired products from the feedstocks, which would include fossil fuels combined with opportunity feedstocks, such as biomass, when appropriate. The configuration of the complete plant, feedstocks, products, environmental controls, and plant size will be site

Table 11.10 Vision 21 Energy Plant Performance Targets

Efficiency—electricity generation[a]	Coal based systems 60% (HHV); natural gas-based systems 75% (LHV) with no credit for cogenerated steam.[b]
Efficiency—combined heat and power	Overall thermal efficiency above 85% (HHV); also meets efficiency goals for electricity.[b]
Efficiency—fuels plant only	Fuel utilization efficiency of 75% (LHV) when producing coal-derived fuels such as H_2 or liquid transportation fuels alone from coal.[b]
Environmental	Near-zero emissions of sulfur (<0.01 lb/MM Btu), nitrogen oxides (<0.01 lb/MM Btu), particulate matter (<0.005 lb/MM Btu), trace elements (<1 lb Hg/trillion Btu), and organic compounds (less than one-half of emission rates for organic compounds listed in the Utility HAPS Report (EPA, 1998)); 40–50% reduction in CO_2 emissions by efficiency improvement; 100% reduction with sequestration.
Costs	Cost of electricity 10% lower than conventional systems; Vision 21 plant products cost-competitive with market clearing prices.
Timing	Major spinoffs such as improved gasifiers, advanced combustors, high-temperature filters and heat exchangers, and gas separation membranes begin by 2006; designs for most Vision 21 subsystems and modules available by 2012; Vision 21 modules available for commercial plant designs available by 2015.

[a]HHV—based on fuel higher heating value; LHV—based on fuel lower heating value.
[b]The efficiency goal for a plant cofeeding coal and natural gas will be calculated on a pro rata basis. Likewise, the efficiency goal for a plant producing both electricity and fuels will be calculated on a pro rata basis.
Source: From DOE (2009), DOE (2004), and DOE (1998).

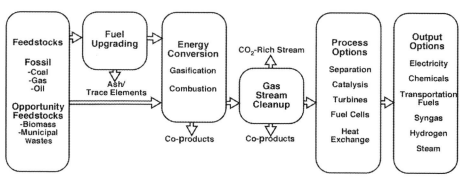

Figure 11.26 Vision 21 technology modules.
Source: From DOE (2009) and DOE (1998).

specific and determined by prevailing market and economic conditions. The technology modules will be based on key technologies, including the following:

- Combustion and high-temperature heat exchange
- Gasification
- Gas purification
- Gas separation
- Turbines
- Fuel cells
- Synthesis gas conversion
- Environmental control
- Materials
- Controls and sensors
- Computational modeling and virtual simulation
- Systems analysis and systems integration

The approach is to develop a suite of technology modules that can be interconnected in different configurations to produce selected products. Vision 21 builds upon a portfolio of technologies already being developed, including low-polluting combustion, gasification, high-efficiency furnaces and heat exchangers, advanced gas turbines, fuels cells, and fuel synthesis, and adds other critical technologies and system integration techniques. When coupled with CO_2 capture and storage or utilization, Vision 21 systems would release no net CO_2 emissions and have no adverse environmental impacts [4].

11.6.2   Vision 21 Benefits

Many benefits will be realized by Vision 21 successes. Environmental barriers to fossil fuel use will be removed. This includes advances in control of smog- and acid rain–forming pollutants, and particulate and hazardous air pollutants, capture and sequestration of carbon dioxide, and minimization and utilization of solid wastes. Affordable energy costs will be maintained by using a wide range of low-cost fossil fuel options. Useful coproducts, including transportation fuels, will be produced in the energy plant, which will reduce reliance on imported oil, stabilize oil prices, and improve the U.S. balance of trade. The United States will continue its leadership role in clean energy technology by promoting export of fossil energy and environmental technology, equipment, and sales.

11.7   FutureGen

On February 27, 2003, the federal government announced plans to build a prototype of the fossil fuel power plant of the future: FutureGen. FutureGen is a cost-shared venture with the private sector and international partners that will combine electricity and hydrogen production with the virtual total elimination of harmful emissions, including greenhouse gases through sequestration [74–79]. FutureGen will improve on the already low emissions of IGCC power plants in terms of particulate, SO_2, NO_x, and mercury, and it will geologically sequester CO_2.

Pollutants such as SO_2 and NO_x would be cleaned from the coal gases and converted to usable by-products such as fertilizers and soil enhancers [74]. Mercury pollutants would be removed, and CO_2 would be captured and sequestered in deep underground geologic formations. Candidate reservoirs could include depleted oil and gas reservoirs, unmineable coal seams, deep saline aquifers, and basalt formations [78]. The reservoirs will be intensively monitored to verify the permanence of the CO_2 storage.

The prototype plant would be sized to generate approximately 275 MW_e of electricity—equivalent to an average midsize coal-fired power plant [79]. The plant would be a stepping stone toward a future coal-fueled power plant that not only would be emissions free but would also operate at unprecedented fuel efficiencies. Figure 11.27 is a schematic of the conceptual design of the system, although a final design has not been completed yet.

The goals of the project include the following [78]:

- Design, construct, and operate a nominal 275 MW_e (net equivalent output) prototype plant that produces electricity and hydrogen with near-zero emissions. The size of the plant is dictated by the need for producing commercially relevant data, including the requirement for producing 1 million metric tons per year of CO_2 to adequately validate the integrated operation of the gasification plant and the receiving geologic formation.
- Sequester at least 90 percent of the CO_2 emissions from the plant with the future potential to capture and sequester nearly 100 percent.
- Prove the effectiveness, safety, and permanence of CO_2 sequestration.
- Establish standardized technologies and protocols for CO_2 measuring, monitoring, and verification.
- Validate the engineering, economic, and environmental viability of advanced coal-based, near-zero emission technologies by 2020.

The project is being led by the FutureGen Industrial Alliance, Inc., a nonprofit industrial consortium representing the coal and power industries, with the project results being shared among all participants, and industry as a whole. Table 11.11 lists the Alliance members (as of January 2010). The Alliance has prepared a conceptual design, site qualification, and environmental analysis. Mattoon, Illinois, was selected as the site in December 2007. The program was delayed shortly thereafter when DOE announced in early 2008 that it was revising its FutureGen approach.

In mid-2009, after discussions with President Obama's administration, the project was reinstated as originally planned. On July 14, 2009, the DOE issued a National Environmental Policy Act (NEPA) Record of Decision to move forward. The Alliance and the U.S. DOE signed a Cooperative Agreement on September 1, 2009, allowing the continued development of the FutureGen power plant in Matoon, Illinois. The $17.3 million shared-cost Cooperative Agreement covers preliminary design activities through the end of 2009. Under the Agreement, the Alliance will work with DOE and other partners to continue electric grid interconnected studies, work on securing environmental permits, define alliance operational activities, update plant design and project cost estimates, and make additional subsurface characterizations. Following these activities, the Alliance and DOE will make a decision on taking the project forward to the final design and construction in early

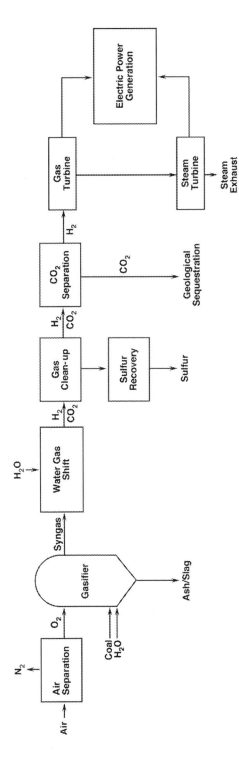

Figure 11.27 Conceptual design of the FutureGen system.
Source: Modified from Sarkus (2008) and FutureGen Alliance, Inc.

Table 11.11 FutureGen Industrial Alliance, Inc., Members

	Headquarters
Current Member (as of January 2010) Companies	
Alpha Natural Resources, Inc. (formerly Foundation Coal Corporation)	Linthicum Heights, MD
Anglo American Services (UK) Limited	London, UK
BHP Bilton Energy Coal Inc.	Melbourne, AUS
China Huaneng Group	Beijing, CN
CONSOL Energy Inc.	Pittsburgh, PA
E.ON U.S. LLC	Louisville, KY
Peabody Energy Corp.	St. Louis, MO
Rio Tinto Energy America Services	Gillette, WY
Xstrata Coal Pty Limited	Sydney, AUS
Former Member Companies	
American Electric Power Service Corp.	Columbus, OH
Luminant	Dallas, TX
PPL Energy Services Group, LLC	Allentown, PA
Southern Company Services, Inc.	Atlanta, GA

2010. If the project proceeds, plans are for construction to begin in 2012, construction to be completed and the demonstration to begin in 2015, and demonstration project operations to be completed in 2020 (although the plant would continue to operate upon completion of the demonstration).

11.8   DOE Carbon Sequestration Programs

DOE's Carbon Sequestration Program was launched in 1997 as a small-scale research effort to determine the technical viability of CCS [80]. The Carbon Sequestration Program has grown into a multifaceted research, development, and deployment initiative, with the objective to provide the means by which fossil fuels can continue to be used in a carbon-constrained world. The Carbon Sequestration Program is helping to develop technologies to capture, separate, and store CO_2 in order to reduce greenhouse gas emissions without adversely affecting energy use or hindering economic growth [81]. DOE envisions having a portfolio of capture, storage, and mitigation technologies that are safe, cost-effective, and available for commercial deployment in 2020. The U.S. DOE's primary research and development (R&D) objectives are lowering the cost and energy penalty associated with CO_2 capture from large point sources and improving the understanding of factors affecting CO_2 storage permanence, capacity, and safety in geologic formations and terrestrial ecosystems [81]. Once these objectives are met, new and existing power plants worldwide will have the potential to be retrofitted with CO_2 capture technologies.

The Carbon Sequestration Program encompasses three main elements: Core R&D, Infrastructure, and Global Collaborations. These are depicted in Figure 11.28

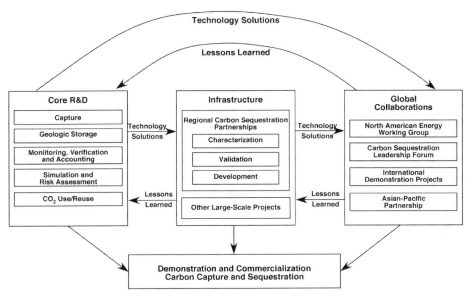

Figure 11.28 U.S. DOE Carbon Sequestration Program.
Source: Modified from DOE (2007) and DOE (2010).

and culminate in the demonstration and commercialization of CCS including such programs as FutureGen [80, 81].

Core R&D involves laboratory- and pilot-scale research aimed at developing new technologies and new systems for greenhouse gas mitigation. Core R&D integrates basic research and computational sciences to study advanced materials and energy systems and has five focus areas: CO_2 capture; geologic storage; monitoring, verification, and accounting (MVA); simulation and risk assessment; and CO_2 use/reuse. (CO_2 capture and storage are discussed in Chapter 10.) MVA efforts focus on the development and deployment of technologies that can provide an accurate accounting of stored CO_2, and high level of confidence that the CO_2 will remain permanently sequestered. Risk assessment research focuses on indentifying and quantifying potential risks to humans and the environment associated with CO_2 sequestration and ensuring that these risks remain low. Core R&D is an active effort with new projects being added on a continuous basis. Figure 11.29 shows the locations (based on the primary contractors) of many carbon sequestration projects as of mid-2009 [81]. While this list is continually being added to, Figure 11.29 illustrates the types of projects (by major category) and the range of their implementation.

The infrastructure element includes large-scale projects and the Regional Carbon Sequestration Partnerships (RCSPs), which is a collaboration between government, industry, universities, and international organizations funded by DOE and tasked with developing guidelines for the most suitable technologies, regulations, and infrastructure needs for CCS in different regions of the United States and Canada [81, 82]. The energy sectors of both countries are closely related. Geographical differences in fossil fuel use and CO_2 storage potential across the United States and Canada require

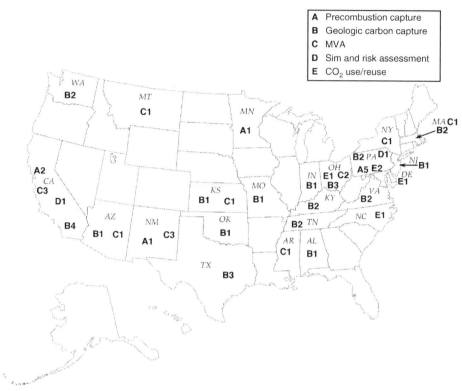

Figure 11.29 Carbon sequestration projects by locations of primary contractors. *Source:* From DOE (2010). (Note that the number after the letter denotes number of projects; that is, A2 refers to two precombustion capture projects, C3 refers to three MVA projects, etc.)

regional approaches to sequester CO_2. Seven RCSPs form a network that includes more than 350 state agencies, universities, and private companies, covering 42 states and four Canadian Provinces. Table 11.12 lists the Regional Carbon Sequestration Partnerships and their members [82]. The RCSPs' efforts consist of three phases: the Characterization Phase (2003–2005), the Validation Phase (2005–2009), and the Development Phase (2008–2018). The Characterization Phase focused on developing the necessary framework to validate and potentially deploy CCS technologies. The Validation Phase focused on validating the most promising regional opportunities to deploy CCS technologies using two different CO_2 storage approaches: geologic and terrestrial carbon sequestrations. In the Development Phase, the injection of 1 million short tons or more of CO_2 by each RCSP into regionally significant geologic formations of different depositional environments will be demonstrated.

DOE participates in international collaborations with technology solutions developed in the Core R&D and Infrastructure elements being tested internationally, thereby helping to advance CCS overall at a lower cost and quicker timeframe. DOE participates in the Carbon Sequestration Leadership Forum (CSLF), an international

Table 11.12 Regional Carbon Sequestration Partnerships and Members

Regional Carbon Sequestration Partnership	Lead Organization	Member States/Provinces
Big Sky Carbon Sequestration Partnership (BSCSP)	Montana State University	Montana, Idaho, South Dakota, Wyoming, Eastern Oregon, and Washington, and adjacent areas in British Columbia and Alberta
Midwest Geological Sequestration Partnership (MGSC)	Illinois State Geological Survey	Illinois, Western Indiana, and Western Kentucky
Midwest Regional Carbon Sequestration Partnership (MRCSP)	Battelle Memorial Institute	Eastern Indiana, Eastern Kentucky, Maryland, Michigan, New York, Ohio, Pennsylvania, and West Virginia
Plains CO_2 Reduction (PCOR) Partnership	University of North Dakota, Energy and Environmental Research Center	Eastern Montana, Eastern Wyoming, Nebraska, Eastern South Dakota, North Dakota, Minnesota, Wisconsin, Iowa, Missouri, Alberta, Saskatchewan, Manitoba, and Northeastern British Columbia
Southeast Regional Carbon Sequestration Partnership (SECARB)	Southern States Energy Board	East Texas, Arkansas, Louisiana, Mississippi, Alabama, Tennessee, Florida, Georgia, South Carolina, North Carolina, and Virginia
Southwest Regional Partnership (SWP)	New Mexico Institute of Mining and Technology	Western Texas, Oklahoma, Kansas, Colorado, Utah, and Eastern Arizona
West Coast Regional Carbon Sequestration Partnership (WESTCARB)	California Energy Commission	Alaska, Western Arizona, Western British Columbia, California, Hawaii, Nevada, Western Oregon, and Western Washington

Source: From DOE (2008).

climate change initiative that is focused on the development of improved, cost-effective technologies for the separation, capture, transport, and long-term storage of CO_2. The purpose of the CSLF is to make these technologies available internationally and to identify and address wider issues relating to carbon capture and storage, such as regulatory and policy options [82]. U.S. technological advances and expertise in greenhouse gas mitigation are being shared in many initiatives, including the Australian Otway Basin project, the European Union–funded CO_2 SINK project in Germany, the Algerian In Salah industrial-scale CO_2 storage project, and IEA Weyburn-Midale CO_2 Monitoring and Storage project in Canada [81, 83].

11.9   International Carbon Sequestration Programs

In addition to the DOE's carbon sequestration program/efforts, worldwide CCS programs/projects are on the increase [83–86]. An effort to document these is underway via an online database set up by DOE [83]. As of November 2009, the database contained 192 proposed and active CCS projects worldwide located in 20 countries across five continents. The projects include 38 capture, 46 storage, and 108 for capture and storage. It is anticipated that the number of projects will increase substantially over the next several years making this type of database important for those following the progress of the CCS projects. Most of the projects in the database are still in the planning and development state or have just been recently proposed, while eight are actively capturing and injecting CO_2 [83]:

- In Salah Gas Storage Project—Algeria
- CRUST Project—K12-B Test—The Netherlands
- Sleipner Project—Norway
- Snøhvt Field LNG and CO_2 Storage Project—Norway
- Zama Field—Canada
- SECARB Cranfield—United States
- Weyburn-Midale—Canada
- Mountaineer CCS Project—United States

Several programs and projects are ongoing in the European Union [84]. Funding for these is coming from various sources, including government, industry, and the United Nations through the UN Framework Convention on Climate Change. Detailed information on these numerous and various types of projects can be found elsewhere, since the list is rapidly changing [83, 84, 86].

In the previous section, the Carbon Sequestration Leadership Forum (CSLF) was introduced. The CLSF is a voluntary climate initiative of developed and developing nations that account for 77 percent of all human-made CO_2 emissions [85, 86]. The CSLF was formed in 2003 and directs intellectual, technical, and financial resources from all parts of the world to support the long-term goal of the UN Framework Convention on Climate Change, which is the stabilization of atmospheric CO_2 concentration in this century [85]. Members are dedicated to collaboration and information sharing in developing, proving safe, demonstrating, and fostering the worldwide deployment of multiple technologies for the capture and long-term geologic storage of CO_2 at low costs; and to establishing a foundation of legislative, regulatory, administrative, and institutional practices that will ensure safe, verifiable storage for as long as a millennia.

11.10   International Clean Coal Technology and Carbon Sequestration Activities

Most countries that use coal have had coal R&D activities for many years [87, 88]. Much of this research was directed toward fundamental studies of coal behavior and pollutant formation, emissions characterization and reduction, technology development, and improved/optimized system performance. With increased interest and concern regarding future energy supplies and climate change, clean coal technology (CCT) started in some countries in the 1980s (e.g., in the United States) with a number of roadmaps emerging in the early 2000s [88, 89]. CCT is now an important component of many countries' energy policies and future. Economic instruments such as emissions trading and carbon taxes have CCT as an integral component of their schemes [90]. This section presents some of the major activities worldwide (with the exception of the United States, which was discussed earlier) in the development and deployment of clean coal technologies. In some cases, roadmaps/programs will be presented; however, because these have changed so extensively over the last three to five years (in name, objectives, and schedule), more emphasis is given on specific major demonstrations or technologies that are being focused on. This section is not all-inclusive but rather focuses on major activities in key coal utilizing countries.

11.10.1 Canada

The two primary CCT programs in Canada are the Canadian Clean Power Coalition (CCPC) activities and the Natural Resources Canada's CanmetENERGY clean energy program. The CCPC, established in 2000, is an association of Canadian Coal and coal-fired electricity producers whose aim is to secure a future for coal-fired electricity generation by proactively addressing environmental issues with governments and stakeholders [88, 91]. Members (as of 2009) include, EPCOR, Sherrit International, Nova Scotia Power, SaskPower, TransAlta, along with a North Dakota utility, Basin Electric Power Cooperative, and the California-based Electric Power Research Institute (EPRI) [91]. The CCPC established a goal to develop projects to demonstrate technology at a commercial utility scale for retrofit to existing plants or for use in new coal-fired power plants that would allow all emissions, including CO_2, to be controlled to meet all foreseeable new regulatory requirements. The emissions target was to allow a coal-fired plant to be as clean as a modern natural gas–fired gas turbine plant. The goal was to accomplish this while maintaining overall efficiency at or above current levels, maintaining costs competitive with other generation technologies, and enabling the CO_2 to be captured [92].

A study was undertaken (i.e., Phase I) that was comprised of conceptual engineering and feasibility studies that were undertaken from mid-2001 to early 2004. The objective of the conceptual engineering and feasibility studies was to determine the most appropriate technologies for demonstration. Implementation plans, preliminary designs and cost estimates were developed for those technologies, recognizing the geographical variability of coal: western lignite and subbituminous coals, and eastern bituminous coals. This led to Phase II feasibility studies where the objective was to investigate supercritical pulverized coal plants with CO_2 capture using amine scrubbing or oxy-fuel combustion processes, and gasification technology optimization. In addition, upgrading of the coal by blending with petroleum coke was evaluated for the polygeneration of power, hydrogen, and CO_2 [93]. This phase was initiated in 2004 and completed in 2007. One outcome of the study was the identification of key areas where future work is recommended. The study also confirmed that for carbon storage technology to make significant advances, demonstration plants are needed to show that the technology can work and to further optimize the processes based on actual plant operation data. CCPC's plans are to construct a clean coal demonstration plant, which they have estimated will cost more than $1 billion over a ten-year period to develop and construct.

CanmetENERGY is conducting two major programs: clean coal and carbon capture and storage [94]. The vision in their Clean Coal Technology Roadmap is of a Canadian industry that is a leader in adapting and integrating technology and knowledge for the effective utilization of coal and other low-value carbon fuels as an energy source for the production of electricity, hydrogen, heat, and chemical feedstock with zero or minimal environmental impacts on land, air, and water [94]. This Roadmap is a snapshot of key information for policy and decision makers to understand regarding the strategic advantage of Canada's large endowment of coal, the challenges and expectations that clean coal will face, the various CCT options available, and projections of what clean coal might look like by 2035. The Roadmap concludes with a set of primary objectives that must be achieved and the strategic steps necessary to achieve these objectives. Industry and government commitment is essential for this vision of clean coal to become a reality in Canada.

Similarly, CanmetENERGY is exploring CCS to lessen the environmental impact of fossil fuel combustion technologies, which currently comprise a substantial portion of the Canadian energy supply. CCS involves capturing CO_2 emissions from large point sources and storing them underground in suitable geological formations [94]. The removal of CO_2 from flue gas streams for storage is achieved using CO_2 capture processes. The most promising capture processes, such as precombustion, postcombustion, and oxy-fuel combustion, produce a highly concentrated CO_2 stream that is, after compression, ready for transport and storage. CanmetENERGY is actively pursuing the research and development of near-zero emissions oxy-fuel technologies for pulverized coal-fired power plants, oxy-fuel circulating fluidized-bed combustion, high-pressure oxy-fuel combustion, feasibility studies of oxy-fuel combustion, gasification for CCS, and computation fluid dynamics (CFD) modeling of oxy-fuel combustion. Other examples include CO_2 capture and compression technologies, CO_2 postcombustion capture processes, advanced power cycles, and advanced fuel cell research and development. Storage options such as enhanced oil recovery (EOR) and/or enhanced coal-bed methane (ECBM) recovery provide short-term opportunities for storing the captured CO_2, while longer-term options, such as storage in saline aquifers, are researched and assessed. CanmetENERGY is also involved in funding and collaborative research in the following areas of CO_2 storage: CO_2 injection; monitoring, measurement, and verification; CO_2 storage sites and opportunities; storage integrity; and capacity estimation [94].

11.10.2 Australia

The Australian coal industry has made the development and demonstration of low-emissions coal technologies the central focus of its response to the global challenge of climate change [95]. In 2003, the industry established the COAL21 initiative, bringing together the coal and electric power industries, unions, federal and state governments, and research organizations. The COAL21 National Action Plan adopted in 2004 provided a blueprint for demonstrating the technologies that would allow Australians to continue to use affordable coal-fired electricity but with significantly lower greenhouse gas emissions.

In 2006, the Australian Coal Association member companies established the COAL21 Fund, a voluntary company levy based on production volume. This levy is separate from and additional to existing levies such as the Australian Coal Association Research Program levy. The levy is a $1 billion plus commitment by the Australian coal industry (see Table 11.13 for a list of the COAL21 Fund participating companies as of 2009). The aim of the COAL21 Fund is to demonstrate the technical and economic viability of low-emissions coal technology, leading to commercial deployment by 2015.

To date, the COAL21 Fund has made major commitments to the following active and in-development CCS research projects [95]:

- Up to $300 million for a Queensland Integrated Gasification Combined Cycle (IGCC) project, including $26 million for a feasibility study for the ZeroGen Project, which is a two-phase project (Phase I, demonstration-scale 120 MW$_e$ (gross) IGCC power plant with

Table 11.13 COAL21 Fund Participating Companies

Anglo Coal Australia Pty Ltd
Bloomfield Collieries Pty Ltd
BM Alliance Coal Operations Pty Ltd
Caledon Coal Pty Ltd
Centennial Coal Company Ltd
Donaldson Coal Pty Ltd
Enhance Place Pty Ltd
Ensham Resources Pty Ltd
Felix Resources Pty Ltd
Foxleigh Mining Pty Ltd
Gloucester Coal Ltd
Hunter Valley Energy Coal
Idemitsu Australia Resources Pty Ltd
Illawarra Coal Holdings Pty Ltd
Jellinbah Resources Pty Ltd
Macarthur Coal Limited
New Hope Corporation
Peabody Pacific Pty Ltd
Rio Tinto Coal Australia Pty Ltd
TEC Coal Pty Ltd
Wesfarmers Resources Ltd
Whitehaven Coal Mining Pty Ltd
Xstrata Coal Pty Ltd

Source: From COAL21 Fund (2010).

75 percent CO_2 capture to begin operations in 2012; Phase II, large-scale 450 MW_e (gross) IGCC power plant with CCS for deployment in 2017).

- $68 million to the Callide Oxy-Fuel Project in Queensland.
- $50 million for a postcombustion capture (PCC) project in New South Wales, starting with the Munmorah PCC Project.
- $20 million for Queensland geologic sequestration initiatives.

11.10.3 Germany

Germany is a strong supporter of international treaties, such as the United Nations Framework Convention on Climate Change [88]. Energy security is also an important policy objective, and coal is an important source of energy for Germany. Consequently, Germany is very active in research and development activities that address emissions reduction and efficiency improvements. This has lead to one of the world's first coal-fired power plants equipped with CCS technology starting up in September 2009 [96]. The oxy-fuel fired power plant, the 30 MW_t (thermal) Scharze Pumpe Power Station, was built and is operated by Vattenfall, a Swedish power company. The $101 million ($70 million euros) pilot plant will produce power along with 10 tons of highly concentrated CO_2 an hour. The CO_2 will be loaded onto tankers and taken to a nearby gas field for sequestration. Construction of the pilot plant started in May 2006. During 2007, Vattenfall announced the first demonstration plant (250–350 MW_e), which will use results from the pilot plant in order to improve the CCS process and decrease the cost. The construction of the 30 MW_t pilot plant is an important milestone for the Vattenfall project. It is the necessary scaleup link between initial engineering and successful operation of the future demonstration plant. The pilot plant testing program will run for three years. Thereafter, the pilot plant will be available for other tests. The plant is planned to be in operation for at least ten years.

Lignite and hard coal will be combusted in a mixture of oxygen and recirculated CO_2. The flue gas will then be treated and sulfur oxides, particles, and other contaminants removed. Finally, the water will be condensed and the concentrated CO_2 compressed into liquid.

The purpose of the pilot plant is to validate engineering work, to learn and better understand the dynamics of oxy-fuel combustion, and to demonstrate the capture technology. Eventually, the pilot plant will be used for backup testing for the demonstration plant testing program, once the latter has started.

11.10.4 Japan

Japan has a history of government support for long-term R&D, including the full range of coal utilization technologies [88]. This is important to Japan because coal provides a significant proportion of Japan's energy needs (e.g., utility and industrial such as the iron and steel industry) and is expected to continue to do so in the future. In addition, Japan recognizes that coal will be a dominant energy source in the entire Asia-Pacific region and must be used economically, efficiently, and in an environmentally responsible manner. As a result, Japan has become a leader in technology development for coal utilization and emissions reduction.

The New Energy and Industrial Technology Development Organization (NEDO) was established by the Japanese government in 1980 after the second oil embargo to develop new oil-alternative energy technologies. Its purpose was to coordinate the R&D strengths of the public and private sectors in Japan [88, 97]. In 1988, NEDO's activities were expanded to include industrial technology research and development, and in 1990, environmental technology research and development. Activities to promote new energy and energy conservation technology were subsequently added in 1993. Following its reorganization as an incorporated administrative agency in October 2003, NEDO is now also responsible for R&D project planning and formation, project management and postproject technology evaluation functions. Activities include the development and promotion of new energy and energy conservation technologies and promotion of internal cooperation involving joint R&D and information exchange (such as the joint Australian-Japanese Callide Oxy-Fuel Project). Clean coal technologies are important within their overall program. Along with developing specific technologies for gasification, liquefaction, or emissions control, NEDO has established programs such as the Strategic Technical Platform for Clean Coal Technology (STEP CCT), the Clean Coal Technology Promotion Program, and the Innovative Zero-Emission Coal Gasification Power Generation Project [98].

Another agency active in the clean coal technology arena is the Japan Coal Energy Center (JCOAL). JCOAL was established in 1990, and their mission is to promote coal activities from mining to utilization, ensuring a stable energy supply, sustainable economic growth, and the reduction of environmental impacts including CO_2 emissions [99]. JCOAL was integrated with another agency, CCUJ (Center for Coal Utilization, Japan), in 2005. JCOAL consists of 108 organizations.

11.10.5 China

China is becoming an increasingly important factor in clean coal technology deployment. China is the largest producer of coal in the world and its carbon emissions are estimated to be 10 percent of the world's total. This will only increase, since China is rapidly bringing new power plants on line to improve its standard of living. Consequently, many Western countries are working with China to deploy clean coal technologies. Technology transfer is key in allowing emerging nations to adjust to climate change. In addition, China has agreed to set national targets to limit its carbon outputs and introduce a range of environmental policies, and it is believed the China-European Union (EU) Joint Declaration on Climate Change will only help this process along.

The European Commission will provide China with $85.85 million for a clean coal project that will have almost no emissions. Both parties said they would work toward the China-EU near-zero emissions coal (NZEC) project [100]. The project is important if China is to meet ongoing emissions reductions targets.

The United States and China have announced the establishment of the U.S.-China Clean Energy Research Center [101]. The Center will facilitate joint research and the development of clean energy technologies by teams of scientists and engineers from the United States and China, as well as serve as a clearinghouse to help researchers in each country. The Center will be supported by public and private funding of at least $150 million over five years, split evenly between the two countries. Initial research priorities include clean coal technology including carbon capture and storage.

Work is underway on China's first clean coal-based power plant [102]. The US$1 billion project, called GreenGen, will be the country's first commercial-scale plant to use carbon capture and storage. It is being funded by a group of investors consisting of Peabody Energy, five of China's largest power companies, two domestic coal companies and government entity State Development and Investment Corp, which operates the country's state-owned assets. All components for the plant are being manufactured domestically, with the exception of a gas turbine power unit. GreenGen is a 650 megawatt project, a commercial scale near-zero emissions power project that is under construction near the northern city of Tianjin. It is China's first IGCC power project and will serve as a carbon management research center. GreenGen is being developed in phases and ultimately will capture and store CO_2.

The first phase of the project is expected to be completed in 2011 at a cost of about US$290 million. It will be capable of generating 250 MW_e of electricity and gasifying 2,000 metric tons of coal daily. The plant's capacity will then be expanded to 650 MW_e by 2016, when it will be able to gasify 3,500 metric tons of coal per day.

The European Commission is also planning to invest up to US$70 million to help build a facility in China to test CCS technology [101]. Currently, a test CCS facility is being developed in Inner Mongolia—an area rich in coal mines. The Mongolia project is being led by Shenhua Group, China's largest coal producer and a stakeholder in GreenGen. It will inject an estimated 100,000 metric tons of carbon emissions underground by the end of this year and ultimately aims to capture 3 million metric tons annually.

11.10.6 Others

In 2009, the European Union, which consists of 27 countries, announced that they will be investing $50 billion euros into research and development of solar and carbon capture at coal plants [103]. Europe relies on 40 percent of its power supply from coal burning power plants, and the EU hopes that with the investment in carbon capture, they can catch up with the United States and Canada, who are already investing billions of dollars in funding and other tax credits for carbon capture projects.

Similarly, the United Kingdom has made a commitment to the use of clean coal technologies and has set aside funding to establish four demonstration plants in the United Kingdom in the near future [104]. The United Kingdom has not constructed a new coal-fired power plant since 1981, although 40 percent of its electricity is generated by coal. The four new plants would test different technologies—namely, precombustion, and postcombustion CCS technologies. The United Kingdom has also set out a framework stating that any new-build coal fire power stations in the country should demonstrate 25 percent CCS capability on its output, and when the technology is proven and accepted, all coal-fired power stations in operation in 2025 should be CCS compliant.

South Africa, through its South African Centre for CCS, is preparing a comprehensive CCS storage atlas for South Africa, which would show in detail potential storage sites for sequestrated carbon dioxide [104]. The South African storage atlas project was started in 2008, and the completed atlas was expected by mid-2010. Following the completion of the storage atlas, the Centre would aim at commissioning a test injection site by 2016 and hoped to have a demonstration plant up and running by 2020, with the target of 2025 for commercial operation of a fully integrated CCS plant. CCS is viewed as a vital technology if global greenhouse gas emission reduction targets were to be met, particularly in South Africa, where some 90 percent of the country's electricity is generated from coal-fired power.

References

[1] DOE (U.S. Department of Energy), Clean Coal Technology Demonstration Program: Program Update 2001 Including Power Plant Improvement Initiative Projects, Office of Fossil Energy, July 2002.

[2] DOE, Power Plant Improvement Initiative (PPII), *www.netl.doe.gov/coalpower/ccpi/ppii_projects.html*, last modified, January 5, 2004.

[3] DOE, Clean Coal Technology Demonstrations Program Facts: Clean Coal Power Initiative (CCPI), Office of Fossil Energy, July 2003.

[4] DOE, Vision 21—the Ultimate Power Plant Concept, *www.fossil.energy.gov/programs/powersystems/vision21/*, last modified, April 17, 2009.

[5] DOE, Clean Coal Technology Roadmap "CURC/EPRI/DOE Consensus Roadmap," *www.netl.doe.gov/coalpower/ccpi/main.html*, January 6, 2004a.

[6] DOE, Coal & Power Systems Strategic & Multi-Year Program Plans, Office of Fossil Energy, February 2001.

[7] DOE, Clean Coal Technology Programs: Program Update 2009, Includes Power Plant Improvement Initiative (PPII) and Clean Coal Power Initiatives (CCPI) Projects, As of June 2009, Office of Fossil Energy, October 2009.

[8] DOE, Clean Coal Demonstrations, *www.netl.doe.gov/technologies/coalpower/cctc/cctdp/index.html* (accessed 2010).

[9] DOE, Clean Coal Today, Office of Fossil Energy, DOE/FE-0215P-54, Issue No. 54 Summer 2003.

[10] DOE, Clean Coal Technology Programs: Program Update 2006, Includes Clean Coal Technology Demonstration Program (CCCTDP), Power Plant Improvement Initiative (PPII) and Clean Coal Power Initiatives (CCPI) Projects, As of June 2006, Office of Fossil Energy, September 2006.

[11] F.E. Hsu, 10 MW Demonstration of Gas Suspension Absorption Final Project Performance and Economics Report, DE-F22-90PC90542, June 1995.

[12] Bechtel (Bechtel Corporation), Confined Zone Dispersion Project Final Technical Report, DE-FC22-90PC90546, June 1994.

[13] LIFAC (LIFAC North America), LIFAC Demonstration at Richmond Power and Light Whitewater Valley Unit No. 2 Project Performance and Economics Final Report, DE-FC22-9090548, April 1988.

[14] Pure Air (Pure Air on the Lake, L.P.), Advanced Flue Gas Desulfurization (AFGD) Demonstration Project Final Technical Report, DE-FC22-90PC89660, April 1996.

[15] SCS (Southern Company Services, Inc.), Demonstration of Innovative Applications of Technology for Cost Reductions to the CT-121 FGD Process Final Report, DE-FC22-90PC89650, January 1997.

[16] SCS, Demonstration of Advanced Combustion NO_x Control Techniques for a Wall-Fired Boiler, Project Performance Summary, DE-FC22-90PC89651, January 2001.

[17] B&W (The Babcock & Wilcox Company), Demonstration of Coal Reburning for Cyclone Boiler NO_x Control Final Project Report, DE-FC22-90PC89659, February 1994.

[18] B&W, Full-Scale Demonstration of Low-NO_x Cell™ Burner Retrofit Final Report, DE-FC22-90PC90545, July 1994.

[19] EERC (Energy and Environmental Research Corporation), Evaluation of Gas Reburning and Low-NO_x Burners on a Wall-Fired Boiler Final Report, DE-FC91-PC90547, July 1998.

[20] CONSOL (CONSOL, Inc.), Milliken Clean Coal Technology Demonstration Project Final Report, DE-FC22-93PC92642, October 1999.

[21] SCS, Demonstration of Selective Catalytic Reduction (SCR) Technology for the Control of Nitrogen Oxide (NO_x) Emissions from High-Sulfur Coal-Fired Boilers Final Report, DE-FC22-90PC89652, October 1996.

[22] ETE (Energy Technologies Enterprises Corp.), 180 MW Demonstration of Advanced Tangentially Fired Combustion Techniques for the Reduction of Nitrogen Oxide (NO_x) Emissions from Coal-Fired Boilers, DE-FC22-90PC89653, December 1993.

[23] ABB (Asea Brown Boveri Environmental Systems), SNOX Demonstration Project, Project Performance and Economics Final Report, DE-FC22-90PC89655, July 1996.

[24] T.R. Goots, M.J. DePero, P.S. Nolan, LIMB Demonstration Project Extension and Coolside Demonstration, DE-FC22-87PC79798, November 1992.

[25] B&W, SOx-NOx-Rox Box™ Flue Gas Cleanup Demonstration Project Performance Summary, DE-FC22-90PC89656, June 1999.

[26] EERC, Enhancing the Use of Coals by Gas Reburning—Sorbent Injection Final Report, DE-FC22-87PC79796, February 1997.

[27] NYSEG (New York State Electric & Gas Corporation), Milliken Clean Coal Technology Demonstration Project, Project Performance and Economics, Final Report, DE-FC22-93PC92642, April 1999.

[28] T. Hunt, T.J. Hanley, Integrated Dry NO_x/SO_2 Emissions Control System Final Report, DE-FC22-91PC90550, November 1997.

[29] N. Raskin, Foster Wheeler Power Group, Inc., personal communication, January 8, 2004.

[30] Ohio Power Company, Tidd PFBC Demonstration Project Final Report, DE-FC21-87MC24132, August 1995.

[31] Colorado-Ute Electric Association, Inc., NUCLA Circulating Atmospheric Fluidized Bed Demonstration Project Final Report, DE-FC21-89MC25137, October 1991.

[32] DOE, Clean Coal Technology, Tampa Electric Integrated Gasification Combined-Cycle Project An Update, Office of Fossil Energy, July 2000.

[33] DOE, Clean Coal Technology, The New Coal Era, Office of Fossil Energy, June 1990.

[34] DOD (U.S. Department of Defense), Fuel Cell Information Guide—Molten Carbonate Fuel Cells, United States Corps of Engineers, Engineer Research and Development Center, *www.dodfuelcell.com/molten.html.* last updated January 6, 2004.

[35] DOE, Energy Efficiency and Renewable Energy—Hydrogen, Fuel Cells, & Infrastructure Technologies Program, *www.eere.energy.gov/hydrogenandfuelcells/fuelcells/types.html#mcfc*, last updated, January 27, 2003.

[36] M.J. Hornick, J.E. McDaniel, Tampa Electric Polk Power Station Integrated Gasification Combined Cycle Project, DE-FC21-91MC27363, August 2002.

[37] P. Cargill, G. DeJonghe, T. Howsley, B. Lawson, L. Leighton, M. Woodward, Piñon Pine IGCC Project Final Technical Report, DE-FC21-92MC29309, January 2001.

[38] DOE, Clean Coal Technology, The Piñon Pine Power Project, Technical Report Number 6 Office of Fossil Energy, December 1996.

[39] Wabash (Wabash River Energy Ltd.), Wasbash River Coal Gasification Repowering Project Final Technical Report, DE-FC21-91MC29310, August 2000.

[40] DOE, Clean Coal Technology, The Wabash River Coal Gasification Repowering Project An Update, Office of Fossil Energy, September 2000.

[41] Alaska Industrial Development and Export Authority, Healy Clean Coal Project, Project Performance and Economics Final Report, DE-FC22-91PC90544, April 2001.

[42] ENCOAL, ENCOAL Mild Gasification Project, DE-FC21-90MC27339, September 1997.

[43] DOE, Clean Coal Technology Compendium CCT Program, *www.lanl.gov/projects/cctc/factsheets/rsbud/adcconvdemo.html*, last modified, December 2, 2002.

[44] E.C. Heydorn, B.W. Diamond, R.D. Lilly, Commercial-Scale Demonstration of the Liquid Phase Methanol (LPMEOH™) Process Final Report, DE-FC22-92PC90543, June 2003.

[45] DOE, Clean Coal Today, Office of Fossil Energy, DOE/FE-0215P-53, Issue No. 53 Spring 2003.

[46] DOE, Clean Coal Technology, Commercial-Scale Demonstration of the Liquid Phase Methanol (LPMEOH™) Process, Technical Report Number 11, Office of Fossil Energy, April 1999.

[47] CQ Inc., Development of a Coal Quality Expert, DE-FC22-90PC89663, June 20, 1998.

[48] Bethlehem Steel Corporation, Blast Furnace Granular Coal Injection System, DE-FC21-MC27362, October 1999.

[49] Passamaquoddy Technology, L.P., Passamaquoddy Technology Recovery Scrubber™ Final Report, DE-AC22-90PC89657, December 20, 1989.

[50] NMA (National Mining Association), Clean Coal Technology: Current Progress, Future Promise, *www.nma.org*.

[51] DOE, Institutional Plan FY 2003-2007, Office of Fossil Energy, National Energy Technology Laboratory, October 2002.

[52] Energetics, Inc., Clean Coal Technology Commercial Successes, Draft Report for United States Department of Energy, Office of Fossil Energy, January 2000.

[53] DOE, Clean Coal Technology: Environmental Benefits of Clean Coal Technology, Technical Report Number 18, Office of Fossil Energy, April 2001.

[54] DOE, Clean Coal Technology: The JEA Large-Scale CFB Combustion Demonstration Project, Technical Report Number 22, Office of Fossil Energy, March 2003.

[55] DOE, Clean Coal Demonstrations, *www.netl.doe.gov/technologies/coalpower/cctc/PPII/index.html* (accessed 2010).

[56] Tampa Electric Company and Pegasus Technologies, Inc., Tampa Electric Company Big Bend Unit #2 Neural Network Based Intelligent Sootblowing System Project Performance and Review, DOE Award No. DE-FC26-02NT41425, April 2005.

[57] DOE, Big Bend Power Station Neural Network-Sootblower Optimization, Fact Sheet, March 2009.

[58] M. Wu, P. Yuran, Commercial Demonstration of the Manufactured Aggregate Processing Technology Utilizing Spray Dryer Ash, Final Technical Report, DOE Award No. DE-FC26-02NT41421, November 12, 2007.

[59] DOE, Commercial Demonstration of the Manufactured Aggregate Processing Technology Utilizing Spray Dryer Ash, Fact Sheet, March 2009.

[60] T. Hrdlicak, W. Swanson, Demonstration of a Full-Scale Retrofit of the Advanced Hybrid Particulate Collector Technology, DOE Award No. DE-FC26-02NT41420, August 14, 2006.

[61] DOE, Greenidge Multi-Pollutant Control Project, Fact Sheet, March 2009.

[62] D.P. Connell, Greenidge Multi-Pollutant Control Project, Final Report of Work Performed May 19, 2006–October 18, 2008, DOE Cooperative Agreement No. DE-FC26-06NT41426, April 2009.

[63] M.L. Eastman, Clean Coal Power Initiative, presented at the Clean Coal Power Conference, www.netl.doe.gov/coalpower/ccpi/program_info.html, November 18, 2003.

[64] DOE, Clean Coal Technology Roadmap "CURC/EPRI/DOE Consensus Roadmap" Background Information, www.netl.doe.gov/coalpower/ccpi/main.html, January 6, 2004b.

[65] R. James, J. McDermott, S. Patnik, S. Piché, Demonstration of Integrated Optimization Software at the Baldwin Energy Complex, Final Report, DOE Cooperative Agreement, DE-FC26-04NT41768, September 5, 2008.

[66] DOE, Increasing Power Plant Efficiency—Lignite Fuel Enhancement, Fact Sheet, March 2009.

[67] DOE, TOXCON Retrofit for Mercury and Multi-Pollutant Control on three 90-MW Coal-Fired Boilers, Fact Sheet, March 2009.

[68] DOE, Demonstration of a Coal-Based Transport Gasifier, Fact Sheet, April 2009.

[69] DOE, Mercury Specie and Multi-Pollutant Control, Fact Sheet, March 2009.

[70] DOE, Mesaba Energy Project—Unit 1, Fact Sheet, March 2009.

[71] DOE, Fossil Energy Techline, www.fossil.energy.gov/news (accessed 2010).

[72] DOE, Clean Coal Technology from Research to Reality, 2006.

[73] Management Information Services, Inc., Benefits of Investments in Clean Coal Technology, October 25, 2009.

[74] DOE, Vision 21: Clean Energy for the 21st Century, Office of Fossil Energy, November 1998.

[75] DOE, A Prospectus for Participation by Foreign Governments in FutureGen, Office of Fossil Energy, June 20, 2003.

[76] DOE, Coal and Power Systems, Future Clean Coal Projects, www.netl.doe.gov/technologies/coalpower/futuregen/index.html last updated, June 12, 2009.

[77] T.A. Sarkus, Fossil Energy, Clean Coal Technology, and FutureGen, Coal Age 113 (7) 2008 56–59.

[78] DOE, FutureGen—A Sequestration and Hydrogen Research Initiative Fact Sheet, Office of Fossil Energy, February 2003.

[79] FutureGen Alliance, Inc., http://futuregenalliance.org (accessed 2010).

[80] DOE, Carbon Sequestration Technology Roadmap and Program Plan 2007, Office of Fossil Energy, April 2007.

[81] DOE, NETL Carbon Sequestration Program, www.netl.doe.gov/technologies/carbon_seq (accessed 4.01.10).

[82] DOE, 2008 Carbon Sequestration Atlas, second ed., Office of Fossil Energy, November 2008.

[83] DOE, Worldwide Carbon Capture and Storage Projects on the Increase, http://netl.doe.gov/publications, November 13, 2009.

[84] IEA (International Energy Agency), CO_2 Capture and Storage R, D&D Projects Database, *www.co2captureandstorage.info/research_programmes.php* (accessed 4.01.10).

[85] DOE, Carbon Sequestration Leadership Forum, *http://fossil.energy.gov/programs/sequestration* (accessed 4.01.10).

[86] Carbon Sequestration Leadership Forum, CSLF Projects, *www.cslforum.org/projects/index/html* (accessed 4.01.10).

[87] S.J. Mills, Coal in an enlarged European Union, IEA Clean Coal Centre, 2004.

[88] C. Henderson, Clean coal technologies roadmaps, IEA Clean Coal Centre, 2003.

[89] J. Kessels, S. Bakker, A. Clemens, Clean coal technologies for a carbon-constrained world, IEA Clean Coal Centre, 2007.

[90] J. Kessels, Economic instruments and clean coal technologies, IEA Clean Coal Centre, 2008.

[91] Clean Coal Power Coalition, *www.canadiancleanpowercoalition.com* (accessed 15.01.10).

[92] Canadian Clean Power Coalition (CCPC), Report on the Phase I Feasibility Studies, May 2004.

[93] CCPC Technical Committee, CCPC Phase II Summary, September 2008.

[94] CanmetENERGY, Clean Fossil Fuels, *http://canmetenergy-canmetenergie.nrcan.gc.ca/eng/clean_fossils* (accessed 15.01.10).

[95] COAL21 Fund, *www.newgencoal.com* (accessed 15.01.10).

[96] Vattenfall's Project on CCS, *www.vattenfall.com*, December 9, 2009.

[97] NEDO, *www.nedo.go.jp/* (accessed 15.01.10).

[98] Clean Coal Technologies in Japan, Technological Innovation in the Coal Industry, 2006.

[99] JCOAL Coal Technology Development, *www.jcoal.or.jp/index-en.html* (accessed 15.01.10).

[100] China Signs Up for Clean Coal Project with EU, *www.power-technology.com/news*, December 1, 2009.

[101] U.S.–China Clean Energy Research Center, *www.energy.gov/news*, November 17, 2009.

[102] China's first clean coal plant underway, *www.busniessgreen.com*, June 29, 2009.

[103] Today EU invests in solar and clean coal carbon capture, *http://examiner.com*, October 7, 2009.

[104] UK's Miliband sees clean coal forming part of future energy mix, *www.engineeringnews.co.za*, August 6, 2009.

12 Coal and Energy Security

America's economic engine is fueled primarily by fossil fuels, a trend that is expected to continue for several decades. Coal, oil, and natural gas supplied 84 percent of the nation's total energy, 69 percent of its electricity, and 97 percent of its transportation fuels in 2008 [1]. The contribution of fossil fuels to the U.S. energy supply is expected to increase; the DOE forecasts that fossil fuels will be supplying 90 percent of the nation's energy by 2020 because of projected growth in natural gas consumption. Of the 99.2 quadrillion Btu of energy consumed in the United States in 2008, 23 percent (22.5 quadrillion Btu) was attributed to coal, while natural gas, petroleum, nuclear power, and renewable energy consumption was 24, 37, 7, and 9 percent, respectively [1]. The important role of fossil fuels in general, and coal in particular, in the energy picture in the United States is further illustrated in Figure 12.1, which shows coal's contributions to the various end-use sectors: transportation, residential and commercial, industrial, and electric power generation.

America's economic strength was established largely due to abundant and inexpensive energy. Consumers benefit from some of the lowest electricity rates of any free-market economy, and coal has been a major contributor to the surge in electrification of the country. However, increasing electricity rates and natural gas prices are affecting most Americans. The volatility in natural gas prices, as well as its availability, are becoming more troublesome with frequent episodes of low supply and high prices. Rising crude oil prices are reflected in increased gasoline and heating oil prices, creating an economic strain on consumers. Conflicts in the Middle East affect the world oil markets and prices and result in economic and emotional strains on U.S. citizens. Threats to the energy security of the United States include concerns of terrorist activities targeting nuclear power facilities and oil and gas infrastructure, ever increasing dependency on oil and natural gas imports, aging infrastructures such as power transmission lines, insufficient infrastructures such as natural gas pipelines, and low reserve capacity.

Many of these issues can be addressed through the use of coal. This chapter discusses coal's role in providing U.S. energy security. Specifically, it discusses the impact of volatile energy prices and fuel availability on the economy, the stability of coal prices and its economic impact, the use of natural gas rather than coal in power generation, coal's potential to reduce U.S. dependence on imported oil, and the use of a coal for generating hydrogen. Coal's role in international energy security and sustainable development is also discussed.

Clean Coal Engineering Technology. DOI: 10.1016/B978-1-85617-710-8.00012-1
Copyright © 2011 by Elsevier Inc. All rights of reproduction in any form reserved.

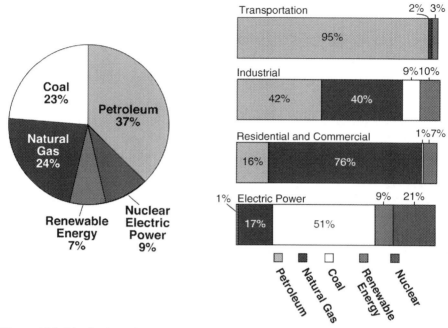

Figure 12.1 Distribution of energy consumption in the United States in 2008.
Source: From EIA (2009).

12.1   Overview of U.S. Energy Security Issues

Energy security is a complex issue and, in the case of the United States, is assured when the nation can deliver energy economically, reliably, environmentally soundly and safely, and in quantities sufficient to support the growing economy and defense needs [2]. This will require policies that support expansion of the energy supply and delivery infrastructure (with sufficient storage and generating reserves), diversity of fuels, and redundancy of infrastructure to meet the demands of economic growth. The United States will need increased contributions toward energy security from all available sources over the long term, including conservation, traditional sources of energy, renewable resources, and new energy sources, such as hydrogen. In order to enhance energy security, the United States must do the following [2]:

- Encourage conservation and energy efficiency
- Maintain diverse energy supplies while enhancing domestic production and delivery
- Maximize economic efficiency
- Accelerate research and development to create and deploy advanced energy technologies
- Develop and implement effective contingency and emergency plans
- Develop policies based on sound science and realistic economic, national security, and environmental needs to make decisions that are timely, consistent, and coordinated with energy security, economic, and environmental objectives

U.S. energy security is closely linked to global energy markets and trends, North American energy resources, energy production, transportation, storage systems, and changing patterns of consumption [2]. One energy source of particular concern is petroleum. Approximately 62 percent of the crude oil consumed in the United States (as in 2008) is imported (i.e., 11 million barrels per day (bbl/day) imported; 6.7 million bbl/day produced domestically) [1]. World liquid demand, of which conventional crude is the majority, is projected to increase by nearly 22 million barrels per day by 2030. More than 80 percent of this increased demand will be in non–Organization for Economic Cooperation and Development (OECD) Asia (see Appendix A for a complete listing of the countries) and the Middle East countries, and 80 percent of the total demand will be in the transportation sector. Approximately 40 percent of U.S. oil will be supplied by OPEC (Organization of Petroleum Exporting Countries), and 68 percent of that will come from the Middle East [3]. These trends have enormous implications for U.S. energy security as global competition and potential disruptions to the United States will impact the availability and cost to the consumers.

Energy patterns in the United States are continually shifting, which have energy security implications [1, 4]. Total U.S. energy consumption is projected to increase from 99 quadrillion Btu in 2008 to almost 114 quadrillion Btu in 2030. Petroleum is expected to remain the dominant fuel in U.S. markets, maintaining about 37 percent market share with domestic production remaining constant and the United States relying heavily on imported oil. Petroleum demand is expected to increase from 21 million bbl/day in 2008 to 22 million bbl/day in 2030. The transportation sector, where 95 percent of the energy is supplied by oil, is expected to grow by only 1 million bbl/day, increasing from about 13 million bbl/day in 2008 to almost 14 million bbl/day in 2030. Natural gas consumption is projected to increase from 23 trillion cubic feet in 2008 to 24 trillion cubic feet in 2030. Renewable energy consumption is projected to increase from 6 quadrillion Btu in 2007 to 6.7 quadrillion Btu in 2030. Coal consumption is projected to increase from almost 1.1 billion short tons in 2007 to almost 1.3 billion short tons in 2030, with about 90 percent of the total coal demand used for electricity generation. Total U.S. electricity demand is projected to increase from 4,190 billion kilowatt-hours (kWh) in 2007 to 4,398 kWh in 2030.

The energy security implications of imported oil and the use of natural gas in the power-generation industry are discussed later in this chapter. The availability and price of natural gas, especially if used in large quantities in the power industry, and the use of natural gas as a feedstock in the chemical/fertilizer industry, for residential heating, and in the transportation sector are also discussed, as is the application of reforming natural gas to produce hydrogen.

The U.S. energy infrastructure is the keystone of the American way of life. It is the foundation for all other critical infrastructures, including information and telecommunications, postal and shipping, public health, and agriculture. Ensuring the safety, security, and reliability of the energy infrastructure is vital to homeland security and economic prosperity.

12.2 The Future of Energy in the United States

The U.S. government, aware of its growing demand for energy, is trying to make rational decisions to avoid any energy shortfalls. These shortfalls result in high prices at the gasoline pumps, high home heating bills due to high natural gas and fuel oil prices, rolling blackouts during periods of cold weather when natural gas-fired power plants cannot obtain sufficient quantities of gas, and blackouts due to aging electricity distribution infrastructure, to name a few.

The National Research Council (NRC) assessed future energy needs in the United States in a recently released report [5]. The report addressed a potential portfolio of energy-supply and end-use technologies—their states of development, costs, implementation barriers, and impacts—both at present and projected over the next two to three decades. The report's aim is to inform policymakers about technology options for transforming energy production, distribution, and use to increase sustainability, support long-term economic prosperity, promote energy security, and reduce adverse environmental impacts. It made the following eight key findings [5]:

1. With a sustained national commitment, the United States could obtain substantial energy-efficiency improvements, new sources of energy, and reductions in greenhouse gas (GHG) emissions through the accelerated deployment of existing and emerging energy-supply and end use technologies.

2. The deployment of existing energy-efficiency technologies is the nearest-term and lowest-cost option for moderating U.S. demand for energy, especially over the next decade. The potential energy savings available from the accelerated deployment of existing energy-efficiency technologies in the buildings, industry, and transportation sectors could more than offset the Energy Information Administration's projected increases in energy consumption through 2030.

3. The United States has many promising options for obtaining new supplies of electricity and changing its supply mix during the next two to three decades, especially if carbon capture and storage and evolutionary nuclear technologies can be deployed at required scales. However, the deployment of these new supply technologies is very likely to result in higher consumer prices for electricity.

4. Expansion and modernization of U.S. electrical transmission and distribution systems (i.e., the power grid) are urgently needed. Expansion and modernizations would enhance reliability and security, accommodate changes in load growth and electricity demand, and enable the deployment of energy efficiency and supply technologies, especially intermittent wind and solar energy.

5. Petroleum will continue to be an indispensable transportation fuel during the time periods considered in the NRC report. Maintaining current rates of domestic petroleum production (about 6.7 million barrels per day in 2008) will be challenging. There are limited options for replacing petroleum or reducing petroleum use by 2020, but there are more substantial longer-term options that could begin to make significant contributions in the 2030 to 2035 timeframe.

6. Substantial reductions in greenhouse gas emissions from the electricity sector are achievable over the next two to three decades through a portfolio approach involving the widespread deployment of energy efficiency technologies; renewable energy; coal, natural gas, and biomass with carbon capture and storage; and nuclear technologies. Achieving

substantial greenhouse gas reductions in the transportation sector over the next two to three decades will also require a portfolio approach involving the widespread deployment of energy efficiency technologies, alternative liquid fuels with low CO_2 emissions, and light-duty vehicle electrification technologies.

7. To enable accelerated deployments of new energy technologies starting around 2020 and to ensure that innovative ideas continue to be explored, the public and private sectors will need to perform extensive research, development, and demonstration over the next decade.

8. A number of current barriers are likely to delay or even prevent the accelerated deployment of the energy-supply and end-use technologies described in the report. Policy and regulatory actions, as well as other incentives, will be required to overcome these barriers.

12.3   Energy and the Economy

Energy prices have a significant impact on the U.S. economy, as evidenced by the oil embargoes and the recent rising energy—natural gas and oil—prices. Prior to the oil embargo of 1973–1974, total energy expenditures comprised 8 percent of the U.S. GDP, the share of petroleum expenditures was slightly less than 5 percent, and natural gas expenditures accounted for 1 percent [6]. The price shocks of the 1970s and early 1980s resulted in these rising dramatically to 14, 8, and 2 percent, respectively, by 1981. For the next two decades, the shares have decreased consistently to approximately preembargo levels. However, in the late 1990s, these shares began to increase again due to higher natural gas and petroleum prices.

High energy prices result in increased inflation. Viewed from a long-term perspective, inflation, measured by the rate of change in the consumer price index (CPI), tracks movements in the world oil price [6]. Oil and other energy prices constitute a portion of the actual CPI but also impact other (downstream) commodity prices that will have a lagged effect on the CPI inflation. Since the 1970s, observable and dramatic changes in GDP growth have occurred as the world oil price has undergone dramatic changes as well [6]. The price shocks of 1973–1974, the late 1970s to early 1980s, and the early 1990s were all followed by recessions. More recently (2003 and 2008/2009), higher energy prices contributed to a downturn in the U.S. economy.

The strength and security of the U.S. economy is closely linked to the availability, reliability, and cost of electric power. Since 1970, real GDP in the United States and electricity generation have been clearly linked, as illustrated in Figure 12.2 [7]. Since economic growth is linked to reliable and affordable electric power, continued use of domestic coal resources will play a significant role in satisfying the energy needs of the United States. This is likely to continue through the middle of the twenty-first century and beyond [7].

The lowest-cost electrical generation plants are coal fired. Figure 12.3 shows that most states with low-cost electricity receive a large amount of their generation from coal, with the noted exception of the three Pacific Northwest states—Washington, Oregon, and Idaho—which have ample hydroelectric resources [8]. Conversely,

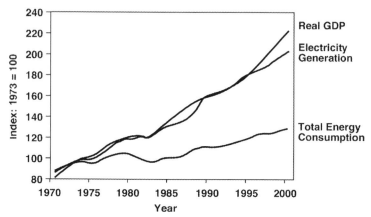

Figure 12.2 Economic growth and energy security are linked to electricity consumption. *Source:* From Eastman (2003).

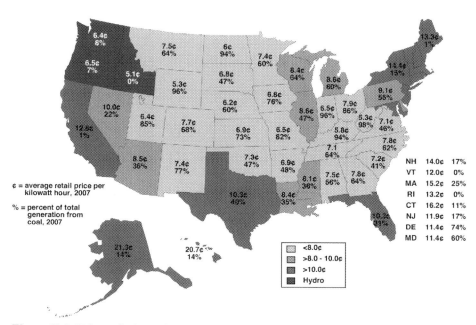

Figure 12.3 U.S. retail electricity costs for 2007. Values shown for each state are the average retail price (¢/kWh) and the percent of total generation from coal for 2007. *Source:* From EIA (2008).

the highest-cost electricity states are those that generate less than 30 percent of their electricity from coal. Kentucky, West Virginia, and Wyoming), which have the lowest electricity costs—below 6¢/kWh—all generate more than 94 percent of their electricity from coal. The exception is Idaho, which generates all of its electricity from hydroelectric resources.

The abundance of coal helps keep prices low and stable, both in long-term contracts and the spot market. States that rely on coal for most of their generation are insulated from wholesale power price spikes that have followed the volatility of the natural gas market. This is illustrated in Figure 12.4, which shows average fossil fuel receipts at electric generating plants for coal (averages of all coal prices), natural gas, and fuel oil for the period from 1990 to 2008 [9], and Figure 12.5, which shows oil, coal (PRB and Appalachian Basin), and natural gas daily spot prices

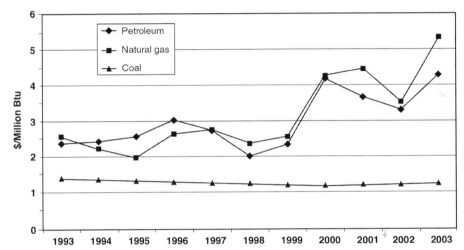

Figure 12.4 Delivered fuel prices for the period January 1993 through May 2003. *Source:* From Roberts (2003).

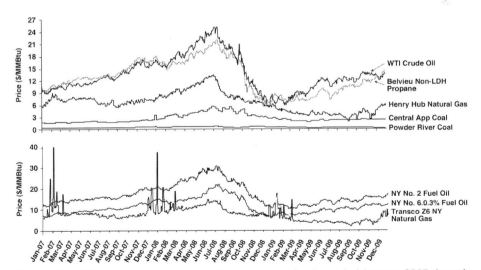

Figure 12.5 Oil, coal, and natural gas daily spot prices for the period January 2007 through December 2009. *Source:* From Federal Energy Regulatory Commission (2010).

for 2007 through 2009 [10]. These figures illustrate the price stability of coal and, even during periods where coal prices increased (e.g., February 2008 through May 2009), the increases were small compared to oil and gas price increases.

12.4 Natural Gas Use in Power Generation

Natural gas is a premium fossil fuel that is easily transported, can be used in many applications, is often the least expensive option from a capital investment viewpoint, and burns with low levels of emissions. Of the fossil fuels, natural gas releases the lowest quantity of carbon dioxide per million Btu of energy consumed. For these reasons, natural gas has become a popular choice among the residential, commercial, industrial, and electricity generation sectors, and it is also becoming increasingly popular in compressed natural gas vehicles.

With this popularity, however, have come serious supply and demand issues. In 2008, 23.8 quadrillion Btu of natural gas was consumed in the United States [1]. Of this, approximately 34, 29, 34, and 3 percent were consumed by the industrial, electric power, residential and commercial, and transportation sectors, respectively (see Figure 12.1). Natural gas consumption is projected to increase to almost 25 quadrillion Btu in 2035, with electric power generation responsible for much of the increase [11].

Electric power generation from natural gas is projected to increase from about 900 billion kilowatthours (kWh) in 2008 to about 1,100 billion kWh in 2035 after decreasing to about 750 billion kWh around 2012. Using natural gas instead of coal in new power plants is an attractive option since the plants have lower construction costs, shorter construction times, and reduced environmental impacts, specifically about half that of CO_2 emissions (per kW produced) compared to coal (e.g., natural gas, bituminous coal, subbituminous coal, and lignite emit approximately 117, 205, 213, and 215 lb CO_2 per million Btu burned, respectively). This is a concern, since supply is currently not keeping pace with demand as natural gas imports are increasing, price volatility is being experienced, and shortages during periods of extreme weather are occurring.

The greater demand on natural gas could increase U.S. reliance on liquefied natural gas imports and cause a significant rise in domestic prices, although sustained significant increases in production of natural gas from shales could limit that reliance. A recent major gas find in the Marcellous shale in the eastern United States has resulted in increased production; however, the extent to which the natural gas will be removed is uncertain due to environmental concerns with the drilling technique used (i.e., fracturing the shale with high-pressure water) [12].

Natural gas prices are increasing as demand exceeds supply, with electric power generation one of the causes of the current market strain. The natural gas crisis impacts all sectors that utilize gas; however, the crisis becomes especially newsworthy when residential and industrial users are severely impacted. In the fairly recent past (beginning in the winter of 2000–2001), natural gas price spikes and their impact on the residential and industrial sectors have received much attention, including

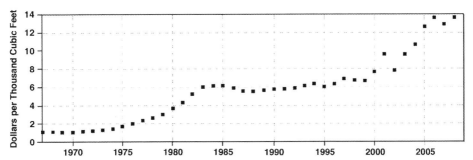

Figure 12.6 Average yearly prices for natural gas delivered to residential customers. *Source:* From EIA (2010).

interest from politicians. Natural gas prices have been rising for the past ten years, starting in winter 1999–2000, and often with spikes during the last few winters.

Figure 12.6 shows the average yearly prices for natural gas delivered to residential customers from 1980 through 2008 [13]. The price trend is similar to that of natural gas delivered to power plants in that they too are rising. The U.S. Energy Information Administration predicts that natural gas prices will fluctuate less and be more predictable in the future, given the significant increase in gas reserves [14, 15]. However, it was expected in the 1990s that natural gas prices would remain low into the future [16], but, as can be seen from Figures 12.4 and 12.6, this did not occur. It is not clear if the new shale gas reserves that have been found will be able to reduce and stabilize the natural gas prices.

Currently, natural gas combined cycle plants generate about 630,400,000 MWh (compared to 2,016,500,000 MWh for coal); however, the average capacity factor was 42 percent in 2007, which is primarily a result of fluctuating and high gas prices [16]. If these units doubled their capacity factor, this would lead to a substantial increase in the use of natural gas for power production. The potential for displacing coal by making greater use of existing natural gas–fired power plants depends on several factors [16]:

• The amount and excess natural gas–fired generating capacity available
• The current operating patterns of coal and gas plants, and the amount of flexibility power system operators have for changing those patterns
• Whether the transmission grid can deliver power from existing gas power plants to loads currently served by coal plants
• Whether there is a sufficient supply of natural gas, as well as pipeline and transmission storage capacity, to deliver large amounts of additional fuel to gas-fired power plants

A balanced approach to new power generation is needed. Natural gas is a premium fuel, and its use in base-loaded power generation needs to be questioned. Coal, with its abundance and stable prices, should be a fuel of choice for new power-generation plants. Coal technologies are becoming more environmentally sound and, as discussed in previous chapters, will be making even greater strides in efficiency and pollutant reduction in the future. Coal technologies are cost competitive with natural gas

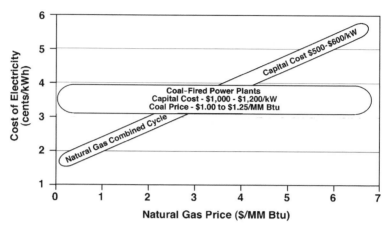

Figure 12.7 Comparison of coal technologies with natural gas combined cycle power plant for various natural gas prices.
Source: From DOE (2001) [17].

(Figure 12.7) and need to be seriously considered in lieu of natural gas-fired power plants. A sound energy policy with the goal of energy security mandates this.

12.5 Coal's Potential to Reduce U.S. Dependence on Imported Crude Oil

Inexpensive crude oil contributes to U.S. economic prosperity. However, increasing reliance on imported crude oil makes the United States vulnerable to oil supply disruptions and threatens the nation's economic and energy security. As evidenced by the 1973 Arab oil embargo and the 1979 Iranian Revolution, abrupt and prolonged loss of crude oil from the Persian Gulf region drastically affects the U.S. economy, increases unemployment, and boosts inflation [18]. Even shorter periods of spiking prices, such as that experienced during the 1990 Gulf War, resulted in a downturn in the U.S. economy. One of the major factors in the current (2009–2010) economic recession (mini-depression) is the massive military spending in Iraq, which is a direct result of maintaining oil security in the Persian Gulf region. In 1979, President Carter called the heavy reliance on foreign oil "a clear and present danger to our national security" [18].

Since the 1970s, there has been a lot of discussion about energy security and national security by the U.S. Congress and the White House; however, little has truly been done to address this, as the United States is more dependent on foreign oil today than it was in the 1970s. In 1973 the United States imported about 34 percent of its crude oil [19]. Currently, the United States imports approximately 62 percent of the oil it uses [1]. The United States' current dependency on imported crude oil is illustrated in Figure 12.8, which shows the petroleum flow in the United States in 2008 [1].

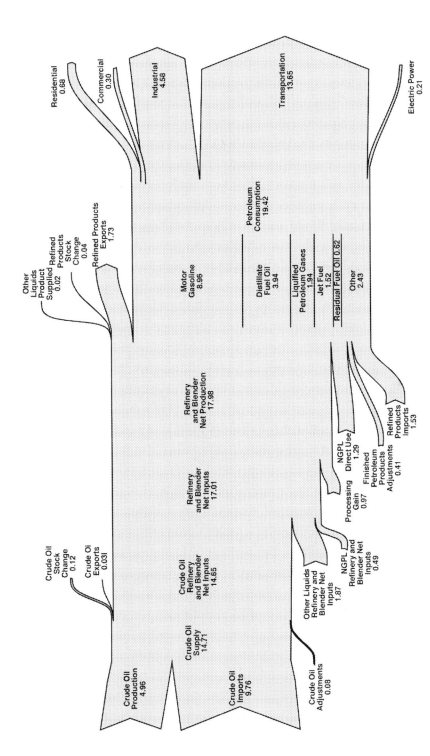

Figure 12.8 Petroleum flow in the United States in 2008.
Source: From EIA (2008).

The solution to achieving energy security in the United States does not involve isolationism or switching to importing crude oil only from non–Persian Gulf countries, even if either of these options could be realized. Taylor [20] points out that being energy independent with respect to oil consumption does not insulate oneself from a global crisis. The oil price spike of 1979 affected Great Britain, where all of the oil Great Britain consumed came from the North Sea, as much as it affected Japan, a country that imports all of its oil. The reason for this is the market for crude oil is global, not regional.

The United States is increasingly dependent on foreign crude oil and is very vulnerable to an oil supply disruption. Presently, five major producers provide almost 40 percent of the petroleum the United States uses: Canada (2.5 million barrels per day), Saudi Arabia (1.5 million barrels per day), Mexico (1.3 million barrels per day), Venezuela (1.2 million barrels per day), and Nigeria (1.0 million barrels per day). These countries account for 68 percent of the petroleum that is imported into the United States. Therefore, reducing the quantity of imported crude oil consumed in the United States is of utmost importance and is a crucial step to attaining energy and economic security.

While the price of crude oil may be influenced by global events, reducing the total amount of imported crude oil needs to be pursued. Approximately 71 percent of the petroleum consumed in the United States is in the transportation sector, with another 23 percent consumed in the industrial sector.

Reducing the transportation sector's reliance on oil is clearly a key to improving energy security. Options to lessen the dependency on imported crude oil include improving vehicle fuel efficiency and diversifying the feedstocks to produce transportation fuels, transportation fuel additives, and liquid fuels/feedstocks to the industrial sector. Feedstock diversification can be achieved by using biomass and coal. Utilizing biomass for producing biofuels and additives is important and needs to be pursued; however, coal can provide the greatest and quickest impact in reducing dependence on imported crude oil due to the vast coal resources and proven technological capability to produce liquid fuels from coal.

Technologies are available to convert coal to liquid products (as discussed in previous chapters), with activities underway to further improve these processes. Gasification followed by Fischer-Tropsch synthesis is the leading processing candidate for producing the liquid fuels. The importance of coal to produce a variety of products, including transportation fuels, is evidenced by the direction of DOE's research and development programs addressing future plants that produce power, fuels, and chemicals.

12.6   Coal's Role in Future U.S. Electric Power Generation

The North American electricity market, which consists of the United States, Canada, and Mexico, is the largest in the world, accounting for almost 30 percent of global electricity production [21]. The United States is the largest consumer of electricity and is projected to remain in that position through 2030, although by then, its share of world electricity generation will decline as other nations experience higher growth in electricity demand. Consequently, fuel diversity for power generation, with coal as a major contributor, is necessary for energy security, which is recognized by both industry and lawmakers.

Tom Ridge, a previous governor of Pennsylvania, was a supporter of energy security and supported the development of new generating capacity in Pennsylvania with an emphasis on fuel diversity [22]. The first major coal plant to be built in Pennsylvania in 20 years was dedicated during his tenure and began operation in the spring of 2004. A 561 MW fluidized-bed power plant burning bituminous coal wastes was constructed at the Seward Station and replaced an aging 30 MW wall-fired pulverized coal-fired boiler. Pennsylvania lawmakers are not the only ones who realize that new capacity fueled by a stable coal supply is essential and a much better alternative to making virtually all new capacity dependent on natural gas–fired units considering natural gas's inherent price volatility. Regulators, lawmakers, and various industries in Wisconsin, Illinois, Iowa, Kentucky, North Dakota, Wyoming, South Carolina, Louisiana, and elsewhere are aggressively promoting diversification with coal [22].

In 2007, coal accounted for 360 gigawatts (GW) of installed capacity [4, 21] and about 50 percent of the net U.S. power-generating sources, which has been steady through 2009. For the foreseeable future, coal-fired plants will continue to dominate the electricity generation sector. In the period up to 2030, U.S. electricity demand is expected to increase at an average rate of about 1.0 percent per year or 26 percent over the period 2007 through 2030 [4, 21]. This translates into almost 240 GW of additional capacity, of which 91 GW is anticipated to come from coal.

The new coal-fired capacity is being installed on a variety of sites. Much of the new capacity is being built on brownfield sites (i.e., sites where industrial activities have been performed) because of existing permitting coupled with the presence of infrastructure such as rail or barge, water, and transmissions access [22]. Some new capacity is being installed as replacements, such as the Seward plant, while the rest is being installed at greenfield sites (i.e., new industrial sites). The plants being installed are using advanced combustion and emissions technology. The Seward plant, constructed by Reliant Energy, is utilizing fluidized-bed technology with ultra-low emissions. Prairie State Energy Campus (Washington County, Illinois) is installing two 800 MW supercritical units, and these boilers will be among the cleanest coal plants east of the Mississippi River [23].

These are but a couple of the projects currently underway, and more are anticipated as DOE's Clean Coal Power Initiative and FutureGen proceed. However, the number of coal-fired power plants that will be constructed will probably be less than half of the number that has been reported [24, 25]. Project cancellations and delays have been attributed to regulatory uncertainty (regarding climate change) or strained project economics due to escalating costs in the industry. These delays are illustrated in Figure 12.9, which shows past capacity announcements versus actual construction [25]. These data are from DOE NETL's tracking of new coal-fired power plants, which is done on a yearly basis. Figure 12.9 shows that actual capacity installed is significantly less than proposed capacity. For example, DOE's

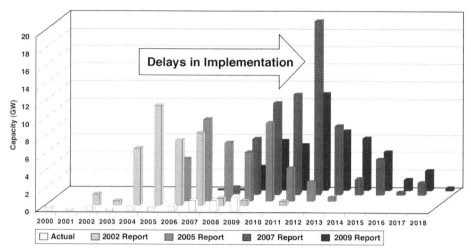

Figure 12.9 Past capacity announcements versus actual construction.
Source: From DOE (2009).

2002 report listed 11,455 MW of proposed capacity for the year 2005, when actually only 328 MW were constructed. Despite the setbacks, a respectable amount of new coal-fired generation remains under construction in the United States. As of late 2009, 32 new coal-fired units accounting for 15,919 MW of capacity are under construction, as shown in Figure 12.10 [24].

Decisions regarding the depth, scope, and speed of greenhouse gas emission reductions will have the greatest impact on the industry over the next several decades [26]. These decisions were expected to be finalized sometime in 2011. Until then, there will be continued delays on new coal-fired power construction. In addition, should a cap-and-trade system be implemented to address climate change, North American generators will likely expect severely altered coal-fired power plant operating profiles [27]. It is being predicted that coal-fired units will be used in a manner other than what was originally intended: for intermediate load rather than baseload generation. This change in utilization will likely result in more starts and less run time per year. The problem with this type of operation is that most North American coal units are not designed for intermediate-load duty and, in fact, have never operated in this manner [27].

An issue with the delays in new plant construction is that it translates into an aging boiler industry. The older power plants are those that tend to operate with lower efficiencies and higher emissions.

Figures 12.11 and 12.12 show the number of boilers and the average boiler size as a function of boiler age, respectively [28]. Figure 12.13 shows the age (in increments of five years) and size distribution (in increments of 150 MW) of plants [28]. Over 80 percent of the U.S. coal plants are 20 to 50 years old [29]. Emission rates (lb pollutant/million Btu fired) for NO_x and SO_2 are higher for the older plants,

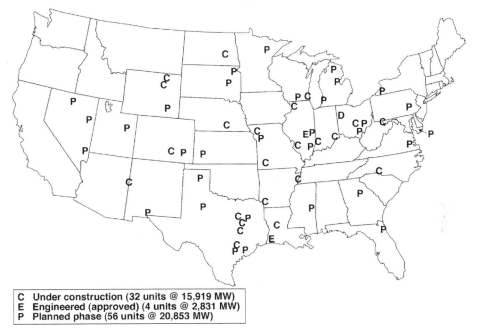

C Under construction (32 units @ 15,919 MW)
E Engineered (approved) (4 units @ 2,831 MW)
P Planned phase (56 units @ 20,853 MW)

Figure 12.10 New coal-fired power generation construction.
Source: Modified from Burt (2009).

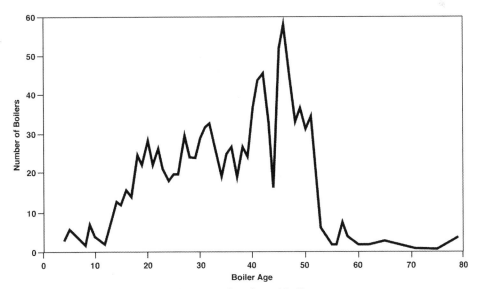

Figure 12.11 Number of U.S. boilers as a function of boiler age.
Source: From DOE (2002).

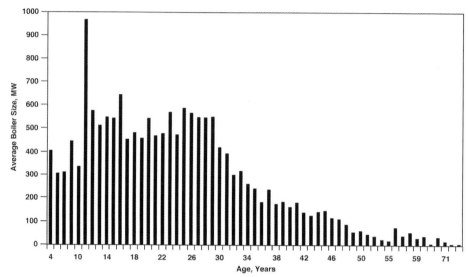

Figure 12.12 Average boiler size (MW) as a function of boiler age.
Source: From DOE (2002).

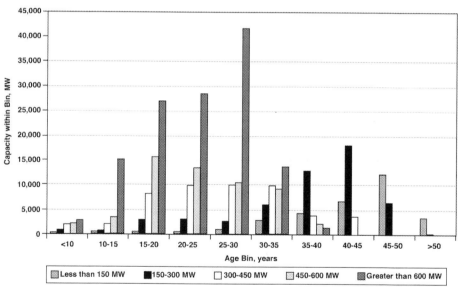

Figure 12.13 Boiler capacity (MW) as a function of boiler age.
Source: From DOE (2002).

as shown in Figure 12.14 and 12.15, respectively [29]. Similarly, overall power plant efficiencies are lower for the older plants, as shown in Figure 12.16 [29]. As discussed in previous chapters, increasing overall efficiencies is one method

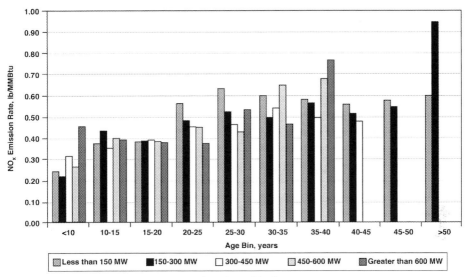

Figure 12.14 NO_x emission rates as a function of boiler age.
Source: From DOE (2002).

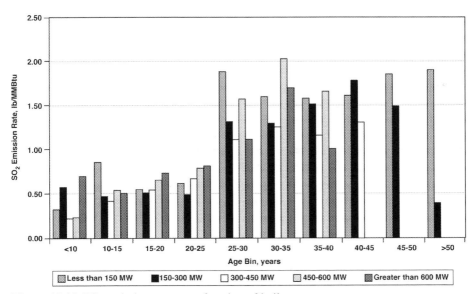

Figure 12.15 SO_2 emission rates as a function of boiler age.
Source: From DOE (2002).

for reducing CO_2 emissions (i.e., less coal burned per MW produced, resulting in less CO_2 being generated). However, as demonstrated in Figure 12.16, the older units exhibit the lowest efficiencies. In a recent study, the DOE estimates that a reduction in U.S. greenhouse gas emissions on the order of 150 million metric tons of CO_2 per year could be realized by driving fleet performance levels to that of the

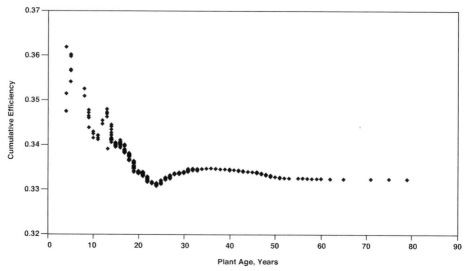

Figure 12.16 Boiler efficiency as a function of power plant age.
Source: From DOE (2002).

highest performers [30]. In 2007, 10 percent of the coal-fired generating capacity in the United States operated at an average efficiency of 37.9 percent, significantly higher than the industry average of 32.5 percent. Table 12.1 lists the coal-fired power plants ranked by efficiency from the DOE's study [30]. Efficiency improvements can be realized through operational changes; system modifications; and replacing aging units with newer, higher-efficiency units.

Table 12.1 U.S. Coal-Fired Power Plants Ranked by Efficiency

Decile	Number of Units	Net Nameplate Capacity (GW)	2007 Total Generation (billion kWh)	2007 Generation-Weighted Efficiency (% based on HHV)
1	181	30	177	26.5
2	108	30	180	30.0
3	90	30	189	31.0
4	73	30	189	31.7
5	84	30	194	32.4
6	75	30	181	33.2
7	79	29	182	34.0
8	70	30	186	34.9
9	57	29	184	35.9
10	46	30	192	37.9
Overall	863	298	1,854	32.5

Source: Modified from Stallard and DiPeitro (2009).

12.7   Production of Hydrogen from Coal

Previous chapters provided a discussion of hydrogen production from coal gasification, along with the DOE's leading activities in this area; therefore detailed coverage of hydrogen production from coal is not presented here. Future energy systems will produce several products: electricity, chemicals, and other fuels such as hydrogen. Hydrogen is a clean-burning fuel, since the product of combustion is water, and much research is currently underway in the United States exploring hydrogen production, storage, and utilization, with a focus on fuel cells. Any discussion of U.S. energy security involving hydrogen utilization must include coal as the feedstock.

The DOE has a Hydrogen from Coal Program with the goals of developing technologies that use domestic coal in a clean manner to produce hydrogen to help ensure energy security and reduce emissions of greenhouse gases [31]. The program's mission is to facilitate the transition to a hydrogen economy and the use of the abundant coal resources in the United States to produce, store, deliver, and utilize affordable hydrogen in an environmentally acceptable manner. These are the program's goals:

- Complete the development of hydrogen and hydrogen-natural gas mixture engine modifications and operations by the end of 2010.
- By the end of 2014, make available an alternative hydrogen production pathway, including a product reforming system, for decentralized production of hydrogen from high hydrogen content hydrocarbon liquids and/or substitute or natural gas (SNG) that can be delivered through the existing fuel distribution infrastructure.
- By 2015, make available processes to enhance coal facility profitability by producing a variety of high-value, coal-derived chemicals and/or carbon materials that can be incorporated into the central or alternative pathway hydrogen production systems.
- By the end of 2016, prove the feasibility of a 60 percent efficient, near-zero emissions, coal-fueled hydrogen and power coproduction facility that reduces the cost of hydrogen by 25 percent compared to current coal-based technology.

The world economy currently consumes about 42 million tons of hydrogen per year, with about 60 percent of this used as feedstock for ammonia production and subsequent use in fertilizer [32]. The United States uses more than 9 million tons of hydrogen, 7.5 million tons of which are consumed at the place of manufacture, with the rest used in a variety of commercial applications. Currently, nearly all hydrogen production is based on fossil fuels. Worldwide, 48 percent of hydrogen is produced from natural gas, 30 percent from oil (and mostly consumed in refineries), 18 percent from coal, and the remaining 4 percent via water electrolysis [32]. Most of the hydrogen produced in the United States is done by steam reforming or as a by-product of petroleum refining and chemicals production [33]. Steam reforming uses thermal energy to separate the hydrogen from the carbon components in methane (and sometimes methanol) and involves the reaction of the fuel with steam on catalytic surfaces. The first step of the reaction (when using natural gas) decomposes the gas into hydrogen and carbon monoxide, which is then followed by the shift reaction that occurs when the carbon monoxide mixes with steam to produce carbon dioxide and hydrogen via the reactions

$$CH_4 + H_2O \rightarrow 2H_2 + CO_2 \tag{12.1}$$

$$CO + H_2O \rightarrow CO_2 + H_2 \tag{12.2}$$

Currently, natural gas provides a more affordable feedstock for producing hydrogen [33]. However, continued use of natural gas to produce hydrogen, especially as the demand for hydrogen increases, does not make any sense. Natural gas is a premium fuel for the residential and industrial sectors, and it may have an increased role in the transportation sector in compressed natural gas mass-transit systems. Depending on natural gas for hydrogen production has many of the same negative consequences to the economy as does using natural gas to generate power.

Hydrogen has many uses, including a feedstock to fuel cells, a fuel for power generation, and as a transportation fuel. However, much work needs to be done before most of these applications will become commercial. A major hurdle to overcome is the necessity to reduce the cost of production. In short, the reality of a full hydrogen economy is many years in the future. During the development of these hydrogen technologies, the DOE is aggressively pursuing "clean" technologies for hydrogen production using coal. This option provides the most energy and economic security to the United States because of coal's abundance and price.

12.8 Coal's Role in International Energy Security and Sustainable Development

Coal will play a major role in providing energy security and sustainable development throughout the world. This will occur because of coal's abundant reserves, low and stable prices, and increased electrification in many regions of the world.

12.8.1 International Demand for Electricity

As previously discussed, coal is dispersed throughout the world (found in about 70 countries), it does not exhibit volatile price swings as oil and natural gas do, and it has abundant reserves (see Table 12.2) [3, 34]. These factors make coal the fuel of choice for most countries of the world, especially for power generation.

Table 12.2 2008 World Fossil Fuel Reserves

Fossil Fuel	Quantity of Reserves (gigatons)	Reserve-to-Production Ratio; R/P (years of resource availability)
Coal	444	147
Oil	186	42
Natural gas	175	60

Source: From BP (2009).

Electricity demand is increasing in all major regions, and large investments are planned or being made in both developed and developing countries [3, 21]. The largest generating capacity additions will be in Asia, specifically China and India, which will require increasing the amounts of electricity to continue their rapid development. Much of this increased capacity will be provided by coal-fired power plants. The North American market for electricity is the largest in the world and will remain so for some time [21]. In the United States, coal provides about 50 percent of the electricity generation—a trend that is expected to continue for many decades. Despite uncertainties over carbon management legislation, much of the increased demand for coal-fired generation in the United States is expected to be coal fired. In Europe, growth in electricity demand is expected to be modest (see Figure 12.17). In some countries, coal will remain crucial in the supply of electricity [21].

World electricity generation is projected to increase by 77 percent during the period of 2006 to 2030 [3]. World net electricity generation is projected to increase by an average of 2.4 percent per year from 2006 to 2030. Electricity is projected to supply an increasing share of the world's total energy demand and is the fastest-growing form of end-use energy worldwide in the midterm. Since 1990, growth in net electricity generation has outpaced the growth in total energy consumption (2.9 percent per year and 1.9 percent per year, respectively), and the growth in demand for electricity continues to outpace growth in total energy use throughout the projection [3]. World net electricity generation is expected to increase by 77 percent from 18 trillion kilowatt-hours in 2006 to 23.2 trillion kilowatt-hours in 2015 and 31.8 trillion kilowatt-hours in 2030. In general, growth in the OECD countries, where electricity markets are well established and consuming patterns are mature, is slower than in the non-OECD countries, where a large amount of demand goes unmet at present (see Figure 12.17).

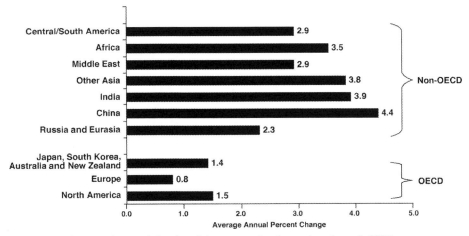

Figure 12.17 Annual growth in electricity generation by region through 2030.
Source: From EIA (2009).

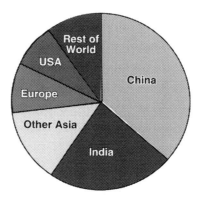

Figure 12.18 Coal-fired generation that is planned or under construction by region.
Source: From The National Coal Council (2008).

It is estimated that nearly 32 percent of the population in the developing non-OECD countries (excluding non-OECD Europe and Eurasia) did not have access to electricity in 2005—a total of about 1.6 billion people [3]. Regionally, sub-Saharan Africa fares the worst: more than 75 percent of the population remains without access to power. High projected economic growth rates support strong increases in demand for electricity among the developing regions of the world through the end of the projection period. The non-OECD nations consumed 45 percent of the world's total electricity supply in 2006, and their share of world consumption is poised to increase over the projection period. In 2030, non-OECD nations are projected to account for 58 percent of world electricity use, while the OECD share declines to 42 percent. More than 660 gigawatts of new coal-based generation is planned or under construction, as illustrated in Figure 12.18 by region, to address this growing need for electricity [35].

12.8.2   Advanced Coal Technology Application Support for Sustained Coal Utilization

As discussed in Chapter 11, major clean coal and carbon management efforts are underway worldwide to preserve the option for utilizing coal in the future and to ensure that coal usage is performed in an environmentally acceptable manner. Because of this tremendous interest in advanced coal technologies, the Electric Power Research Institute (EPRI) started a Global Coal Initiative (GCI) in 2000, which has transitioned into the creation of the CoalFleet for Tomorrow program (CoalFleet) in 2004 [36–39]. The initiative is addressing the technical, economic, and institutional challenges to making clean coal power systems with higher efficiency, near zero emissions, CO_2-capture-amenable coal plants a prudent investment option for both the short term and long term.

Working with advanced coal power project owners and developers, power industry equipment and service suppliers, and independent world-class experts, the program develops evaluation tools and technologies to help guide the design of innovative coal plant systems that manage cost and risk. EPRI works with the DOE, the Coal Utilization Research Council (CURC), and numerous international organizations to include technology and information from public and private sources in coordinating advanced coal research, development, and demonstrations. The program focuses on developing a portfolio of advanced coal technologies, including IGCC, ultra-supercritical pulverized coal, circulating fluidized-bed combustion (CFBC), and oxygen combustion for pulverized coal and CFBC units. The following specific industry needs and issues must be addressed [39]:

- Tools to efficiently accommodate future CO_2 emission reduction technologies in plants that may initially start service without such controls
- Technical means to counteract sharply the higher construction costs and high natural gas prices
- Energy security concerns
- Large projected increases in demand for clean, affordable coal-powered generation worldwide

This initiative builds on and supplements DOE initiatives and worldwide coal combustion advances that are aimed at maintaining the strategic value of coal as a power generating fuel worldwide. The long-term viability of coal as a worldwide generating fuel depends on finding ways to further reduce or even eliminate coal's environmental impact, including CO_2 emissions.

12.8.3 Energy Security

Energy security has been defined by the United Nations as "the continuous availability of energy in varied forms, in sufficient quantities, and at affordable prices" [40]. In terms of energy security, a key issue is resource availability, which is the actual physical amount of the resource available around the world (i.e., long-term security). Another aspect of energy security is the need for system reliability, such as the continuous supply of energy, particularly electricity, to meet consumer demand at any given time (i.e., short-term security). There are many drivers governing the secure supply of energy [41]:

- Diversification of generation capacity—to allow prices to remain reasonably stable
- Prices—energy must be affordable
- Levels of investment required—a significant investment is needed to meet the forecast growth in energy demand and the availability of that investment can be problematic in developing countries
- Ease of transport—energy must be readily available
- Concentration of suppliers—the reliance on imported fuels from a limited number of suppliers increases the risk of adverse market influence
- Availability of infrastructure expertise—countries must have access to different energy sources to achieve a diverse energy mix
- Interconnection of energy systems—the interconnection of energy systems, particularly electricity, must be considered
- Fuel substitution—diversification in the uses of fuels may also be important for energy security
- Political threats—the energy supply system can be vulnerable to disruptions caused by political interests and terrorist attacks

Coal has an important role to play in meeting the demand for a secure energy source [40]. It has many attributes that specifically address the drivers just listed. Coal reserves are large and will be available for the foreseeable future without raising geopolitical or safety issues. Indigenous coal resources enable economic development and can be transformed to guard against import dependence and price shocks. Coal is readily available from a wide variety of sources in a well-supplied worldwide market. It is an affordable source of energy. Coal is easily transportable, and it does not need high-pressure pipelines or dedicated supply routes—routes that must be protected at enormous expense. Coal can be easily stored at power stations, and stocks can be drawn on in emergencies. Coal-based power generation is well established and reliable, and it is not dependent on the weather. Studies have shown that substituting primary energy sources such as oil and natural gas with coal will increase energy security of supply for most world regions [42].

12.8.4 Sustainable Development

Several of the clean coal and environmental technologies discussed in Chapter 11, along with the CoalFleet program, complement the United Nations initiative to address coal and sustainable development [43]. Even though *coal* and *sustainable development* may be contradictory terms to some, this was not the message that came from the World Summit on Sustainable Development held in Johannesburg, South Africa, in August/September 2002, where it was agreed that coal is not only compatible with sustainable development but is essential [43]. Sustainable development was defined as "development that meets the needs of the present generation without undermining the capacity of future generations to meet their needs."

At the Summit, the World Bank noted that improved energy services can enhance indoor air quality and reduce health hazards (by switching from traditional biomass and fossil fuels for cooking and heating), increase income (by reducing the time in collecting traditional biomass and implementing small-scale manufacturing and service activities), bring environmental benefits (by stopping deforestation and its subsequent soil degradation), and provide educational opportunities (by providing lighting or other services of direct educational relevance). At the Summit, it was estimated that 1.6 billion people in developing countries currently have no access to electricity, while 2.4 billion rely on primitive biomass for cooking and heating, and 30 years from now 1.4 billion people will still be without electricity and 2.6 billion people will still be relying on traditional biomass. It was these challenges that the Summit addressed, and a Plan of Implementation was developed. One aspect of the plan asks governments to diversify energy supply by developing advanced, cleaner, more efficient, affordable, and cost-effective energy technologies, including fossil fuel technologies, and to transfer these technologies to developing countries.

At the Summit, it was clear that coal has been, and must continue to be, a major contributor of sustainable development, since coal can be found in most nations of the world. On a global scale, coal accounts for almost 25 percent of the world's primary energy consumption and it fuels 40 percent of the world electricity generation. Almost 5 billion short tons/year of coal are consumed and the international coal trade consists of almost 600 million short tons/year, about 150 years' worth of reserves remain at the current rates of consumption. In addition, coal is a "conflict-free" energy source, and there is a diversity of reserves and production [34, 44]. Coal must be used efficiently and cleanly, and it was recognized that coal technologies continue to improve and effective options are now available for countries of all levels of economic development with respect to emissions reduction.

Coal continues to be the foundation for economic and social development of the world's largest economies in both the developed and developing world. Coal is a major component in securing an improved quality of life for billions of people worldwide who gain access to it through the energy services they need for daily life, industrial development, and social advancement. A sustainable energy policy can be achieved by encouraging electrification, establishing sound environmental regulations, safeguarding energy supplies through diversification, and continuing to support advanced coal technologies.

12.9   Concluding Statements

U.S. consumers are feeling the impact of volatile foreign oil supplies with high prices at the gas pumps and through increased costs for goods and services that use oil and related fuels in their production processes. Volatile natural gas prices are also negatively impacting consumers. These issues are challenging U.S. policy makers to implement an energy policy that ensures greater independence from foreign energy interests and thereby greater energy security. Low-cost and abundant coal is a domestic resource of great strategic value.

Coal is the most abundant energy resource in the United States, and as a fuel source it generates as much electricity as all other energy sources combined. The United States has more coal reserves than all of the oil reserves in the world. The energy value in Montana's coal reserves, for instance, are greater than the oil reserves in Saudi Arabia, Kuwait, Iran, and Iraq combined [45].

Coal's strategic value as a fuel source is due to its role in increasing economic growth and development and in providing energy security at local, national, and international levels [46]. The use of coal provides jobs, supports infrastructure through taxes, supports economic growth and electrification, encourages productivity through electric technologies, provides a reliable energy source, is resistant to energy shocks, and stabilizes power prices. Coal-fired power plants provide system security through infrastructure reliability, which prevents sudden disruptions because there are many diverse sources of coal, which are less vulnerable to supply disruption; coal-fired plants have better storage capability near the power generation site because they are less vulnerable to transportation disruption; coal has a more certain availability during peak demand; coal-fired power plants are less vulnerable to outages, since they are a mature, reliable technology; and coal-fired power plants are less vulnerable to terrorism compared to nuclear power plants, natural gas pipelines, or liquefied natural gas facilities [46].

Energy security in the United States cannot be overemphasized, and its importance is clearly evident in DOE's Coal Power Program. One of the DOE's strategic goals is to protect national and economic security by promoting a diverse supply of reliable, affordable, and environmentally sound energy [47]. This goal is accomplished by developing technologies that foster a diverse supply of affordable and environmentally sound energy, improving energy efficiency, providing for reliable delivery of energy, exploring advanced technologies that make fundamental changes in energy options, and guarding against energy emergencies. Benefits of the Coal Power Program include reducing dependence on imported oil, which can be achieved by coproduction of power and environmentally attractive fuels such as Fischer-Tropsch liquids and hydrogen. Additional benefits include maintaining diversity of energy resource options to avoid overreliance on natural gas for power generation, encouraging economical use of natural gas in other sectors, and reducing energy price volatility and supply uncertainty. The Coal Power Program also retains domestic manufacturing capabilities and U.S. energy technology leadership to enhance economic growth and security.

The U.S. energy security/sustainability depends on sufficient energy supplies to support United States and global economic growth, and coal is a major contributor to this security. Coal-fired electricity generation will enable and stimulate economic growth and social welfare. Diversifying energy production, through the use coal in power generation and production of chemicals and hydrogen provides the United States with energy security. Economic growth and energy security will enable cost-effective environmental controls and continued energy affordability.

References

[1] EIA (Energy Information Administration), Annual Energy Review 2008, U.S. Department of Energy, June 2009.
[2] USEA (U.S. Energy Association), National Energy Security Post 9/11, U.S. Energy Association, June 2002.
[3] EIA, International Energy Outlook 2009, U.S. Department of Energy, May 2009.
[4] EIA, Annual Energy Outlook 2009, U.S. Department of Energy, March 2009.
[5] The National Research Council, America's Energy Future: Technology and Transformation, The National Academies Press, 2009.
[6] EIA, Energy Price Impacts on the U.S. Economy, *www.eia.doe.gov/oiaf/economy/energy_price.html*, April 2001.
[7] M.L. Eastman, Clean Coal Power Initiative, presented at the Clean Coal Power Conference, *www.netl.doe/gov/coalpower/ccpi/program_info.html*, November 18, 2003.
[8] EIA, 2007 State Electricity Profiles, *www.eia.doe.gov/fuelelectric.html*, March 2008.
[9] A. Roberts, Four Common Sense Reasons to Burn Coal, and One More, Coal Age 108 (10) (2003) 37–38.
[10] Federal Energy Regulatory Commission, Oil, Coal and Natural Gas Daily Spot Prices, *www.ferc.gov/market-oversight/*, January 8, 2010.
[11] EIA, Annual Energy Outlook 2010 Early Release Overview, U.S. Department of Energy, *www.eia.doe.gov/oiaf/aeo/index.html*, December 2009.

[12] T. Engelder, G.G. Lash, Marcellus Shale Play's Vast Resource Potential Creating a Stir in Appalachia, The American Oil & Gas Reporter, May 2008.

[13] EIA, Natural Gas Navigator, *http://tonto.eia.doe.gov/dnav/ng/ng_sum_top.asp* (accessed 2010).

[14] EIA, Short-Term Energy Outlook, *www.eia.doe.gov/*, November 2009.

[15] K. Maize, R. Peltier, The U.S. Gas Rebound, Power, *www.powermag.com*, January 2010.

[16] S.M. Kaplan, Displacing Coal with Generation from Existing Natural Gas-Fired Power Plants, Congressional Research Service, January 19, 2010.

[17] DOE (U.S. Department of Energy), Coal Technologies are Cost Competitive, 2001.

[18] C. Coon, Strengthening National Security through Energy Security, *www.heritge.org/Research/EnergyandEnvironment/WM94.cfm*, April 2002.

[19] NREL (National Renewable Energy Laboratory), Energy Security, *www.nrel.gov/clean_energy/security.html* (accessed 18.1.04).

[20] J. Taylor, Don't Worry About Energy Security, *www.cato.org/cgi-bin/scripts/printtech.cgi/dailys/10-18-01.html*, October 18, 2001.

[21] S.J. Mills, Meeting the Demand for New Coal-Fired Power Plants, IEA Clean Coal Centre, 2008.

[22] Power Engineering, Homeland Security: U.S. Brownfield, Power Engineering 106 (6) (2002) 28–34.

[23] Prairie State Energy Campus, Power Plant, 2010; *www.prairiestateenergycampus.com*.

[24] B. Burt, Coal-Fired Generators Worried about Getting Burned, Power, *www.powermag.com*, October 2009.

[25] DOE, Tracking New Coal-Fired Power Plants, *www.netl.doe.gov/technologies/coalpower/index.html*, October 8, 2009.

[26] D. Eppinger, R. Smith, A New Foundation for Future Growth, *www.powermag.com*, January 2010.

[27] W.E. Platt, R.B. Jones, The Impact of Carbon Trading on Performance: What Europe's Experience Can Teach North American Generators, *www.powermag.com*, January 2010.

[28] DOE, Power Plant Aging Characterization, NETL Coal Power Database Analysis, *www.netl.doe.gov/energy-analyses/technology.html*, March 2002.

[29] CCSReg Project, Carbon Capture and Sequestration: Framing the Issues for Regulation, Department of Engineering and Public Policy, Carnegie Mellon University, January 2009.

[30] S. Stallard, P. DiPeitro, An Opportunity to Improve Coal-Fired Generation Efficiency, Power Engineering, *www.power-eng.com*, November 2009.

[31] DOE, Hydrogen from Coal Program: Research, Development, and Demonstration Plan for the Period 2009 through 2016, September 2009.

[32] National Hydrogen Association, Hydrogen General Facts, *http://www.hydrogenassocation.org* (accessed 25.2.10).

[33] DOE, Hydrogen Production Research, *www.fossil.energy.gov/index.html* (accessed January 3, 2010).

[34] British Petroleum, BP Statistical Review of World Energy, June 2009.

[35] The National Coal Council, The Urgency of Sustainable Coal, May 2008.

[36] T. Armor, Coal-Fired Power Plants, Increasingly Lean and Green, Power Engineering 105 (9) (2001) 40–43.

[37] EPRI (Electric Power Research Institute), Bush Energy Policy Resonates with Global Coal Initiative, EPRI Journal 26 (1) (2001) 21–23.

[38] EPRI, CoalFleet Advanced Coal Technology Application Support, *www.epri.com*, March 2007.

[39] EPRI, 66 CoalFleet for Tomorrow—Future Coal Genration Options Program Overview, *www.epri.com*, 2009 Portfolio.

[40] World Coal Institute, Coal: Secure Energy, 2005.

[41] World Coal Institute, Coal & Energy Security, *www.worldcoal.org*, 2008.

[42] J.S. Kessels, Bakker, B. Wetzelaer, Energy Security and the Role of Coal, International Energy Association Clean Coal Centre, 2008.

[43] World Coal Institute, Coal and Sustainable Development: The World Summit on Sustainable Development and its Implications, prepared for the United Nations Economic and Social Council Committee on Sustainable Energy, *www.wci-coal.com*, October 16, 2002.

[44] World Coal Institute, Coal: Delivering Sustainable Development, 2007.

[45] V. Svec, Again, a Crisis with a Solution: Energy in America & Coal-Based Generation, American Coal Council, 2003, pp. 15–19.

[46] W.A. Bruno, Coal's Role in International Energy Security and Sustainable Development, 2003 Conference on Unburned Carbon on Utility Fly Ash, October 28, 2003.

[47] DOE, Clean Coal Technology Roadmap "CURC/EPRI/DOE Consensus Roadmap" Background Information, *www.netl.doe.gov/coalpower/ccpi/main.html*, January 6, 2004.

Appendix A
Regional Definitions

The four country groupings used in this book are shown in Figure A.1 and defined as follows:

A.1 Organization for Economic Cooperation and Development (OECD)

This includes 18 percent of the 2008 world population:

- North America—Canada, Mexico, and United States
- OECD Europe—Austria, Belgium, Czech Republic, Denmark, Finland, France, Germany, Greece, Hungary, Iceland, Italy, Luxembourg, the Netherlands, Norway, Poland, Portugal, Slovakia, Spain, Sweden, Switzerland, Turkey, and the United Kingdom
- OECD Asia—Australia, Japan, New Zealand, South Korea

A.2 Non-Organization for Economic Cooperation and Development (Non-OECD)

This includes 82 percent of the 2008 world population:

- Non-OECD Europe and Eurasia (5 percent of the 2008 world population)—Albania, Armenia, Azerbaijan, Belarus, Bosnia and Herzegovina, Bulgaria, Croatia, Estonia, Georgia, Kazakhstan, Kyrgyzstan, Latvia, Lithuania, Macedonia, Malta, Moldova, Montenegro, Romania, Russia, Serbia, Slovenia, Tajikistan, Turkmenistan, Ukraine, and Uzbekistan
- Non-OECD Asia (53 percent of the 2008 world population)—Afghanistan, Bangladesh, Bhutan, Brunei, Cambodia (Kampuchea), China, Fiji, French Polynesia, Guam, Hong Kong, India, Indonesia, Kiribati, Laos, Malaysia, Macau, Maldives, Mongolia, Myanmar (Burma), Nauru, Nepal, New Caledonia, Niue, North Korea, Pakistan, Papua New Guinea, Philippines, Samoa, Singapore, Solomon Islands, Sri Lanka, Taiwan, Thailand, Tonga, Vanuatu, and Vietnam
- Middle East (3 percent of the 2008 world population)—Bahrain, Cyprus, Iran, Iraq, Israel, Jordan, Kuwait, Lebanon, Oman, Qatar, Saudi Arabia, Syria, the United Arab Emirates, and Yemen
- Africa (14 percent of the 2008 world population)—Algeria, Angola, Benin, Botswana, Burkina, Burundi, Cameroon, Cape Verde, Central African Republic, Chad, Comoros, Congo (Brazzaville), Congo (Kinshasa), Côte d'Ivoire, Djibouti, Egypt, Equatorial

Clean Coal Engineering Technology. DOI: 10.1016/B978-1-85617-710-8.00018-2
Copyright © 2011 by Elsevier Inc. All rights of reproduction in any form reserved.

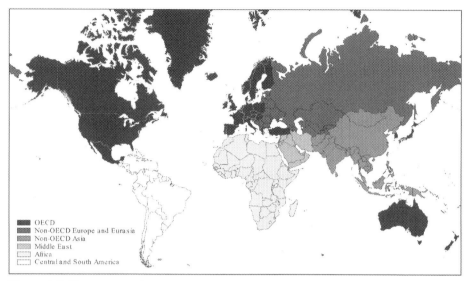

Figure A.1 Country groupings.

Guinea, Eritrea, Ethiopia, Faso, Gabon, Gambia, Ghana, Guinea, Guinea-Bissau, Kenya, Lesotho, Liberia, Libya, Madagascar, Malawi, Mali, Mauritania, Mauritius, Morocco, Mozambique, Namibia, Niger, Nigeria, Réunion, Rwanda, Sao Tome and Principe, Senegal, Seychelles, Sierra Leone, Somalia, South Africa, St. Helena, Sudan, Swaziland, Tanzania, Togo, Tunisia, Uganda, Western Sahara, Zambia, and Zimbabwe

- Central and South America (7 percent of the 2008 world population)—Antarctica, Antigua and Barbuda, Argentina, Aruba, Bahama Islands, Barbados Islands, Belize, Bolivia, Brazil, British Virgin Islands, Cayman Islands, Chile, Colombia, Costa Rica, Cuba, Dominica, Dominican Republic, Ecuador, El Salvador, Falkland Islands, French Guiana, Grenada, Guadeloupe, Guatemala, Guyana, Haiti, Honduras, Jamaica, Martinique, Montserrat, Netherlands Antilles, Nicaragua, Panama Republic, Paraguay, Peru, Puerto Rico, St. Kitts-Nevis, St. Lucia, St. Vincent/Grenadines, Suriname, Trinidad and Tobago, Turks and Caicos Islands, Uruguay, U.S. Virgin Islands, and Venezuela

A.3   European Union (EU)

- Austria, Belgium, Bulgaria, Cyprus, Czech Republic, Denmark, Estonia, Finland, France, Germany, Greece, Hungary, Ireland, Italy, Latvia, Lithuania, Luxembourg, Malta, the Netherlands, Poland, Portugal, Romania, Slovakia, Slovenia, Spain, Sweden, and the United Kingdom

A.4   Organization of Petroleum Exporting Countries (OPEC)

- Algeria, Angola, Ecuador, Indonesia, Iran, Iraq, Kuwait, Libya, Nigeria, Qatar, Saudi Arabia, the United Arab Emirates, and Venezuela

Appendix B
Commercial Gasification Facilities Worldwide

Clean Coal Engineering Technology. DOI: 10.1016/B978-1-85617-710-8.00019-4
Copyright © 2011 by Elsevier Inc. All rights of reproduction in any form reserved.

Table B.1 Worldwide Gasification Units in Operation

Plant Owner	Country	Gasifier Technology	Total Gasifiers	Syngas Capacity (Nm^3/d)	MWth Output	Feedstock	Products
Sasol Chemical Industries Ltd./ Sasol Ltd.	South Africa	Sasol Lurgi	17	7,100,100	970.6	Coal	FT liquids
Beijing No. 4 Chemical	China	GE Energy	1	320,000	43.7	Petroleum	Oxcochemicals
China National Petrochemical Corp./ Sinopec	China	GE Energy	1	210,000	28.7	Petroleum	Oxcochemicals
CNPC Ningxia Dayuan Refining & Chemical Ind. Co. Ltd.	China	GE Energy	3	2,500,000	341.8	Petroleum	Gases
China National Petrochemical Corp./ Sinopec	China	GE Energy	3	2,100,000	286.6	Petroleum	Ammonia
Dalian Chemical Industrial Corp.	China	GE Energy	2	2,096,500	286.6	Petroleum	Ammonia
Lu Nan Chemical Industry Co./ CNTIC	China	GE Energy	2	525,000	71.8	Coal	Ammonia
Shanghai Coking & Chemical (Shanghai Pacific)	China	GE Energy	3	1,530,000	209.2	Coal	Methanol, town gas, acetic acid
Weihe Fertilizer Co.	China	GE Energy	3	2,040,000	278.9	Coal	Ammonia
Zhnhai Refining & Chemical Co.	China	GE Energy	3	2,100,000	287.1	Petroleum	Ammonia
Ube Ammonia Industry Co. Ltd.	Japan	GE Energy	4	2,150,000	293.9	Coal	Ammonia
Shell MDS (Malaysia) Sdn. Bhd.	Malayasia	Shell	6	7,552,000	1,032.4	Gas	Mid-distillates
Linde AG	Singapore	GE Energy	2	1,610,000	220.1	Petroleum	H_2, CO
Air Products & Chemicals, Inc.	United States	GE Energy	2	1,850,000	252.7	Gas	H_2, CO

Company	Country	Technology				Feedstock	Products
Dakota Gasification Co.	United States	Sasol Lurgi	14	1,390,000,000	1,900.3	Coal	SNG, CO_2
Eastman Chemical Co.	United States	GE Energy	2	1,600,000	218.7	Coal	Acetic anhydride, methanol
Frontier Oil & Refining Co. (Texaco Inc.)	United States	GE Energy	1	80,559	11	Petcoke	Electricity, HP steam
Valero Energy Corp.	United States	GE Energy	2	3,800,000	519.5	Petcoke	Electricity, steam
Motiva Enterprises LLC	United States	GE Energy	2	1,880,000	257	Petroleum	H_2
Global Energy, Inc.	United States	E-Gas	2	4,320,000	590.6	Petcoke	Electricity
Coffeyvile Resources Refining & Marketing, LLC	United States	GE Energy	2	2,141,200	292.7	Petcoke	Ammonia, UAN
T&P Syngas (Texaco/ Praxair)	United States	GE Energy	1	1,920,000	278	Gas	H_2, CO
Tampa Electric Co.	United States	GE Energy	1	3,330,000	451.1	Coal	Electricity
Dow (former Union Carbide Corp.)	United States	GE Energy	1	432,000	59.1	Gas	Oxochemicals
Air Liquide (Rhone-Poulenc)	France	GE Energy	1	278,000	38	Gas	H_2, CO
BASF AG	Germany	GE Energy	4	980,208	134	Petroleum	H_2
Henan	China	Sasol Lurgi	4	2,282,494	312	Coal	Ammonia
Air Liquide (Dow Stade GmbH)	Germany	GE Energy	1	264,000	36.1	Gas	CO

Continued

Table B.1 Worldwide Gasification Units in Operation—Cont'd

Plant Owner	Country	Gasifier Technology	Total Gasifiers	Syngas Capacity (Nm³/d)	MWth Output	Feedstock	Products
Mitteldeutsche Erdöl-Raffinerie GmbH	Germany	Shell	6	7,200,000	984.3	Petroleum	H₂, methanol, electricity
Rheinbraun	Germany	GE Energy	3	2,231,500	305.1	Coal	Methanol
SAR GmbH	Germany	GE Energy	1	1,200,000	164	Petroleum	Oxochemicals, H₂
api Energia S.p.A.	Italy	GE Energy	2	3,845,000	525.6	Petroleum	Electricity, steam
ISAB Energy	Italy	GE Energy	2	8,80,000	1,203	Petroleum	Electricity
SARLUX srt	Italy	GE Energy	3	10,400,000	1,271.2	Petroleum	Electricity, H₂, steam
Nuon Power Buggenum	Netherlands	Shell	1	3,408,000	465.9	Coal	Electricity
Shell Nederland Raffinaderij BV	Netherlands	Shell	3	4,662,200	637.3	Petroleum	Electricity, H₂, steam
Elcogas SA	Spain	PRENFLO	1	4,300,00	587.8	Coal	Electricity
GE Plastics España	Spain	GE Energy	1	160,800	22	Gas	CO
BP Chemicals, Ltd.	United Kingdom	GE Energy	1	910,900	124.5	Gas	Acetyls
Mitsubishi Petrochemicals	Japan	Shell	2	400,000	54.7	Petroleum	Syngas
Kemira Chemicals Oy	Finland	Shell	1	300,000	41	Petroleum	Syngas
Lucky Goldstar Chemical Ltd.	South Korea	Shell	1	700,000	68.4	Petroleum	Ammonia
Chemopetrol a.s.	Czech Republic	Shell	6	3,600,000	492.1	Petroleum	Methanol, ammonia

DEA Mineraloel AG	Germany	Shell	2	1,300,000	177.7	Petroleum	Methanol
Falconbridge Dominicania	Dominican Republic	Shell	12	1,440,000	96.9	Petroleum	Reducing gas
Veba Oil Refining & Petrochemicals GmbH	Germany	Shell	4	4,300,000	587.8	Petroleum	Ammonia, methanol
Fertilizer Corp. of India Ltd.	India	Shell	3	2,100,00	287.1	Petroleum	Ammonia
Sekundärrohstoff-Verwertungszentrum Schwarze Pumpe GmbH	Germany	GSP	1	1,200,000	164	Petroleum	Electricity, methanol
Oxochimie S.A.	France	GE	1	590,976	80.8	Gas	Oxochemicals
Nippon Petroleum Refining Co.	Japan	GE	2	5,800,000	792.9	Petroleum	Electricity
Nanjing Chemical Industry Co.	China	GE	2	2,200,000	300.8	Petroleum	Ammonia
BOC Gases	Australia	GE	2	805,000	110	Gas	H_2
Nitrogen Works of Societé el Nasr d' Engrois	Egypt	Koppers-Totzek	3	778,000	106.4	Gas	Ammonia
Sasol (Pty) Ltd.	South Africa	Sasol Lurgi Dry Ash	40	39,600,000	7048	Coal	Gas, chemicals
Sasol (Pty) Ltd.	South Africa	Sasol Lurgi Dry Ash	40	39,600,000	7048	Coal	Gas, chemicals
China National Technology Import Co. (CNTIC)	China	Sasol Lurgi Dry Ash	4	2,282,492	312	Coal	Ammonia
IBIL Energy Systems Ltd. (IES)	India	GTI U-GAS	1	804,000	109.1	Coal	Electricity, steam
Shanghai Pacific Chemical (Group) Co., Ltd.	China	GTI U-GAS	8	3,000,000	410.1	Coal	Fuel gas, town gas

Continued

Table B.1 Worldwide Gasification Units in Operation—Cont'd

Plant Owner	Country	Gasifier Technology	Total Gasifiers	Syngas Capacity (Nm³/d)	MWth Output	Feedstock	Products
Chinese Petroleum Corp.	Taiwan	GE	2	2,143,000	293	Petroleum	H₂, CO, methanol syngas
Gujarat Narmada Valley Fertilizers Co. Ltd.	India	GE	3	2,964,600	405.3	Petroleum	Ammonia, methanol
Ube Ammonia Industry Co. Ltd.	Japan	GE	1	200,000	27.3	Petcoke	CO
Jilin Chemical Industrial Corp.	China	GE	2	2,096,500	286.6	Petroleum	Ammonia
Huainan General Chemical Works	China	GE	3	1,400,000	191.4	Coal	Ammonia
PRAOIL	Italy	GE	2	1,150,000	157.2	Gas	H₂
Praxair (EniChem)	Italy	GE	2	700,000	95.7	Gas	CO
BASF AG	Germany	GE	1	1,000,000	136.7	Petroleum	Oxochemicals
Chemische Werke Hüls AG	Germany	GE	1	600,000	82	Petroleum	Oxochemicals
Mitsui	Japan	GE	2	600,000	82	Petroleum	CO
BASF AG	Germany	GE	4	2,500,000	341.8	Petroleum	Methanol
Celanese Chemical (Ruhrchemie)	Germany	GE	1	960,000	131.2	Petroleum	Oxochemicals
Millenium (Quantum)	United States	GE	2	4,800,000	656.2	Gas	Methanol, CO
Hoechst Celanese	United States	GE	2	500,000	68.4	Petroleum	Oxochemicals
Akzo Nobel/Berol-Kemi	Sweden	GE	1	200,000	27.3	Petroleum	Oxochemicals
Daicel	Japan	GE	2	550,000	75.2	Petroleum	Methanol
Ultrafertil S.A.	Brazil	Shell	3	3,300,000	451.1	Petroleum	Ammonia

Qilu Petrochemical Ind.	China	Shell	2	715,000	97.7	Petroleum	Methanol, oxochemicals
Quimigal Adubos	Portugal	Shell	2	2,400,000	328.1	Petroleum	Ammonia
Fushun Detergent Co.	China	Shell	1	60,000	8.2	Petroleum	Oxochemicals
Inner Mongolia Fertilizer Co.	China	Shell	2	2,100,000	287.1	Petroleum	Ammonia
Lanzhou Chemical Industrial Co.	China	Shell	2	2,100,000	287.1	Gas	Ammonia
Jujiang Petrochemical Co.	China	Shell	2	2,100,000	287.1	Petroleum	Ammonia
Lucky Goldstar Chemical Ltd.	South Korea	Shell	1	384,000	52.5	Petroleum	Oxochemicals
Sekundärrohstoff-Verwertungszentrum Schwarze Pumpe GmbH	Germany	Sasol Lurgi Dry Ash	7	3,000,000	410.1	Biomass/Waste	Electricity, methanol
Rüdersdorfer Zement GmbH	Germany	Sasol Lurgi CFB	1	732,000	100	Biomass/Waste	Fuel gas
EPZ	Netherlands	Sasol Lurgi CFB	1	614,300	84	Biomass/Waste	Electricity
Fabrika Azotnih Jendinjenja	Former Yugoslavia	LP Winkler	1	120,000	16.4	Coal	Ammonia
Sekundärrohstoff-Verwertungszentrum Schwarze Pumpe GmbH	Germany	BGL (Allied Syngas)	1	1,138,000	155.6	Biomass/Waste	Electricity, methanol
ExxonMobil	United States	GE	2	2,540,000	347.2	Petroleum	Syngas
Esso Singapore Pty. Ltd.	Singapore	GE	2	2,660,000	363.6	Petroleum	Electricity, H$_2$, steam
Sokolovska Uhelna, A.S.	Czech Republic	Sasol Lurgi Dry Ash	26	4,700,000	636.4	Coal	Electricity, steam

Continued

Table B.1 Worldwide Gasification Units in Operation—Cont'd

Plant Owner	Country	Gasifier Technology	Total Gasifiers	Syngas Capacity (Nm³/d)	MWth Output	Feedstock	Products
Sydkraft AB	Sweden	FW PCFBG	1	105,000	14.4	Biomass/ Waste	Electricity, District heat
Corenso United Oy Ltd.	Finland	FW ACFBG	1	234,133	32	Biomass/ Waste	Syngas
Sekundärrohstoff- Verwertungszentrum Schwarze Pumpe GmbH	Germany	Sasol Lurgi MPG	5	1,440,000	196.9	Petroleum	Electricity, methanol
Hoechst Celanese	United States	Shell	3	2,100,000	287.1	Gas	Oxochemicals
Hydro Agri Brunsbüttel GmbH	Germany	Shell	4	4,700,000	642.5	Petroleum	Ammonia
Exxon Chemical Co.	United States	Shell	3	570,000	77.9	Petroleum	Oxochemicals
National Fertilizer Ltd.	India	Shell	3	2,100,000	287.1	Petroleum	Ammonia
Neyveli Lignite Corp. Ltd.	India	Shell	2	800,000	109.4	Petroleum	Syngas
Air Products (ICI)	United Kingdom	GE	3	600,000	82	Gas	Oxochemicals
Sunoco	United States	GE	1	400,000	54.7	Gas	Oxochemicals
Dow (former Union Carbide)	United States	GE	2	830,000	113.5	Gas	Oxochemicals

MSK-Radna	Former Yugoslavia	GE	1	1,540,000	210.5	Gas	Methanol
Texas Eastman	United States	GE	1	350,000	47.8	Gas	Oxochemicals
BP Samsung	South Korea	GE	1	600,000	82	Petroleum	CO
Lahden Lämpövoima Oy	Finland	FW ACFBG	1	351,118	48	Biomass/Waste	Electricity, District heat
Oy W. Schauman Ab Mills	Finland	FW ACFBG	1	204,819	28	Biomass/Waste	Syngas
Norrsundet Bruks Ab	Sweden	FW ACFBG	1	146,299	20	Biomass/Waste	Syngas
Portucel	Portugal	FW ACFBG	1	87,800	15	Biomass/Waste	Syngas
DEA Mineraloel AG	Germany	GE	1	1,500,000	205.1	Petroleum	Methanol
Air Liquide America Corp.	United States	GE	2	1,558,000	213	Gas	Syngas, steam
Unspecified Owner	Germany	ThermoSelect	3	250,000	34.2	Biomass/Waste	Electricity
Shuanghuan Chemical	China	Shell	1	1,320,000	197	Coal	—[a]
Sinopec-Shell	China	Shell	1	3,410,000	509	Coal	—
Sinopec	China	Shell	—	—	273.4	Coal	—
Liuzhou Chemical Industry Corp. Ltd.	China	Shell	1	1,720,000	256	Coal	Ammonia
China 1 (Expansion)	China	GE	1	2,045,000	279.6	Coal	Methanol

Continued

Table B.1 Worldwide Gasification Units in Operation—Cont'd

Plant Owner	Country	Gasifier Technology	Total Gasifiers	Syngas Capacity (Nm³/d)	MWth Output	Feedstock	Products
China 2	China	GE	1	1,275,000	174.3	Coal	Methanol
BP/Formosa	Taiwan	GE	1	1,080,000	147.6	Petroleum	Methanol
China 5	China	GE	3	2,080,000	284.3	Coal	Methanol
Sinopec Jinling Plant	China	GE	1	2,100,000	287.1	Coal	Ammonia
China 4	China	GE	1	2,100,000	287.1	Coal	Ammonia, urea
China 3	China	GE	3	2,100,000	287.1	Coal	Ammonia, H_2
Haolianghe Ammonia Plant	China	GE	1	2,200,000	300	Coal	Ammonia
Shanghai Chemical & Coking (Shanghai Pacific)	China	GE	1	765,000	104.6	Coal	Methanol, town gas, acetic acid
Zhong Yuan Dahua Group Ltd.	China	Sasol Lurgi Dry Ash	—	2,282,492	312	Coal	—
Sekundärrohstoff-Verwertungszentrum Schwarze Pumpe GmbH	Germany	BGL (Allied Syngas)	—	1,120,000	164	Biomass/Waste	—
Shaanxi Shenmu Chemical Company	China	GE	—	1,925,000	263	Coal	—

[a]Not reported

Appendix C
Coal-Fired Emission Factors

Clean Coal Engineering Technology. DOI: 10.1016/B978-1-85617-710-8.00020-0
Copyright © 2011 by Elsevier Inc. All rights of reproduction in any form reserved.

Table C.1 Emission Factors for SO_x, NO_x, and CO from Bituminous and Subbituminous Coal Combustion[a]

Firing Configuration	SO_x[b] Emission Factor (lb/ton)	Emission Factor Rating	NO_x[c] Emission Factor (lb/ton)	Emission Factor Rating	CO[d,e] Emission Factor (lb/ton)	Emission Factor Rating
PC, dry bottom, PC, dry bottom, wall-fired[f], bituminous pre-NSPS[g]	38S	A	22	A	0.5	A
PC, dry bottom, wall-fired[f], bituminous pre-NSPS[g] with low-NO_x burner	38S	A	11	A	0.5	A
PC, dry bottom, wall-fired[f], bituminous NSPS[g]	38S	A	12	A	0.5	A
PC, dry bottom, wall-fired[f], subbituminous pre-NSPS[g]	35S	A	12	C	0.5	A
PC, dry bottom, wall-fired[f], subbituminous NSPS[g]	35S	A	7.4	A	0.5	A
PC, dry bottom, cell burner fired, bituminous	38S	A	31	A	0.5	A
PC, dry bottom, cell burner fired, subbituminous	35S	A	14	E	0.5	A
PC, dry bottom, tangentially fired, bituminous, pre-NSPS[g]	38S	A	15	A	0.5	A
PC, dry bottom, tangentially fired, bituminous, pre-NSPS[g] with low-NO_x burner	38S	A	9.7	A	0.5	A
PC, dry bottom, tangentially fired, bituminous, NSPS[g]	38S		10	A	0.5	A
PC, dry bottom, tangentially fired, subbituminous, pre-NSPS[g]	35S	A	8.4	A	0.5	A
PC, dry bottom, tangentially fired, subbituminous, pre-NSPS[g]	35S	A	7.2	A	0.5	A
PC, wet bottom, wall-fired[f], bituminous, pre-NSPS[g]	38S	A	31	D	0.5	A
PC, wet bottom, tangentially fired, bituminous, NSPS[g]	38S	A	14	E	0.5	A
PC, wet bottom, wall-fired, subbituminous	35S	A	24	E	0.5	A
Cyclone furnace, bituminous	38S	A	33	A	0.5	A
Cyclone furnace, subbituminous	35S	A	17	C	0.5	A

Spreader stoker, bituminous	38S	B	11	B	5	A
Spreader stoker, subbituminous	35S	B	8.8	B	5	A
Overfeed stoker[h]	38S(35S)	B	7.5	A	6	B
Underfeed stoker	31S	B	9.5	A	11	B
Hand-fed units	31S	D	9.1	E	275	E
FBC, circulating bed	—[i]	E	5.0	D	18	E
FBC, bubbling bed	—[i]	E	15.2	D	18	D

[a]*Source*: AP-42. Factors represents uncontrolled emissions unless otherwise specified and should be applied to coal feed, as fired. Tons are short tons.

[b]Expressed as SO_2, including SO_2, SO_3, and gaseous sulfates. Factors in parentheses should be used to estimate gaseous SO_x emissions for subbituminous coal. In all cases, S is weight % sulfur content of coal as fired. Emission factor would be calculated by multiplying the weight percent sulfur in the coal by the numerical value preceding S. For example, if fuel is 1.2% sulfur, then $S = 1.2$. On average for bituminous coal, 95% of fuel sulfur is emitted as SO_2, and only about 0.7% of fuel sulfur is emitted as SO_3 and gaseous sulfate. An equally small percent of fuel sulfur is emitted as particulate sulfate. Small quantities of sulfur are also retained in bottom ash. With subbituminous coal, about 10% more fuel sulfur is retained in the bottom ash and particulate because of the more alkaline nature of the coal ash. Conversion to gaseous sulfate appears about the same as for bituminous coal.

[c]Expressed as NO_2. Generally, 95 vol.% or more of NO_x present in combustion exhaust will be in the form of NO, the rest NO_2. To express factors as NO, multiply factors by 0.66. All factors represent emissions at baseline operation (i.e., 60 to 110% load and no NO_x control measures).

[d]Nominal values achievable under normal operating conditions. Values 1 or 2 orders of magnitude higher can occur when combustion is not complete.

[e]Emission factors for CO_2 emissions from coal combustion should be calculated using lb CO_2/ton coal $= 72.6C$, where C is the weight % carbon content of the coal. For example, if carbon content is 85%, then C equals 85.

[f]Wall-fired includes front and rear wall-fired units, as well as opposed wall-fired units.

[g]Pre-NSPS boilers are not subject to any NSPS. NSPS boilers are subject to Subpart D or Subpart Da. Subpart D boilers are boilers constructed after August 17, 1971, and with a heat input rate greater than 250 million Btu per hour (MMBtu/h). Subpart Da boilers are boilers constructed after September 18, 1978, and with a heat input rate greater than 250 MMBtu/h.

[h]Includes traveling grate, vibrating grate, and chain grate stokers.

[i]SO_2 emission factors for fluidized bed combustion are a function of fuel sulfur content and calcium-to-sulfur ratio. For both bubbling bed and circulating bed design, use: lb SO_2/ton coal $= 39.6(S)(Ca/S)^{1.9}$. In this equation, S is the weight percent sulfur in the fuel and Ca/S is the molar calcium-to-sulfur ratio in the bed. This equation may be used when the Ca/S is between 1.5 and 7. When no calcium-based sorbents are used and the bed material is inert with respect to sulfur capture the emission factor for underfeed stokers should be used to estimate the SO_2 emissions. In this case, the emission factor ratings are E for both bubbling and circulating units.

Table C.2 Emission Factors for CH_4, TNMOC, and N_2O from Bituminous and Subbituminous Coal Combustion[a]

Firing Configuration	CH_4[b] Emission Factor (lb/ton)	CH_4[b] Emission Factor Rating	TNMOC[b,c] Emission Factor (lb/ton)	TNMOC[b,c] Emission Factor Rating	N_2O Emission Factor (lb/ton)	N_2O Emission Factor Rating
PC-fired, dry bottom, wall fired	0.04	B	0.06	B	0.03	B
PC-fired, dry bottom, tangentially fired	0.04	B	0.06	B	0.08	B
PC-fired, wet bottom	0.05	B	0.04	B	0.08	E
Cyclone furnace	0.01	B	0.11	B	0.09[c]	E
Spreader stoker	0.06	B	0.05	B	0.04	E
Spreader stoker, with multiple cyclones, and reinjection	0.06	B	0.05	B	0.04	E
Spreader stoker, with multiple cyclones, no reinjection	0.06	B	0.05	B	0.04	E
Overfeed stoker	0.06	B	0.05	B	0.04	E
Overfeed stoker, with multiple cyclones	0.06	B	0.05	B	0.04	E
Underfeed stoker	0.8	B	1.3	B	0.04	E
Underfeed stoker, with multiple cyclones	0.8	B	1.3	B	0.04	E
Hand-fed units	5	E	10	E	0.04	E
FBC, bubbling bed	0.06	E	0.05	E	3.5	B
FBC, circulating bed	0.06	E	0.05	E	3.5	B

[a]Source: AP-42. Tons are short tons. Factors represent uncontrolled emissions unless otherwise specified and should be applied to coal feed, as fired.
[b]Nominal values achievable under normal operating conditions; values 1 or 2 orders of magnitude higher can occur when combustion is not complete.
[c]TNMOC are expressed as C_2 to C_{16} alkane equivalents. Because of limited data, the effects of firing configuration on TNMOC emission factors could not be distinguished. As a result, all data were averaged collectively to develop a single average emission factor for pulverized coal units, cyclones, spreaders, and overfeed stokers.

Table C.3 Uncontrolled Emission Factors for PM and PM$_{10}$ from Bituminous and Subbituminous Coal Combustion[a]

Firing Configuration	Filterable PM[b]		Filterable PM$_{10}$	
	Emission Factor (lb/ton)	Emission Factor Rating	Emission Factor (lb/ton)	Emission Factor Rating
PC-fired, dry bottom, wall-fired	10A	A	2.3A	E
PC-fired, dry bottom, tangentially fired	10A	B	2.3A[c]	E
PC-fired, wet bottom	7A[d]	D	2.6A	E
Cyclone furnace	2A[d]	E	0.26A	E
Spreader stoker	66[e]	B	13.2	E
Spreader stoker, with multiple cyclones, and reinjection	17	B	12.4	E
Spreader stoker, with multiple cyclones, no reinjection	12	A	7.8	E
Overfeed stoker[f]	16[g]	C	6.0	E
Overfeed stoker, with multiple cyclones[f]	9	C	5.0	E
Underfeed stoker	15[h]	D	6.2	E
Underfeed stoker, with multiple cyclones	11	D	6.2[h]	E
Hand-fed units	15	E	6.2[i]	E
FBC, bubbling bed	—[j]	E	—[j]	E
FBC, circulating bed	—[j]	E	—[j]	E

[a]*Source:* AP-42. Factors represents uncontrolled emissions unless otherwise specified and should be applied to coal feed, as fired.Tons are short tons.

[b]Based on EPA Method 5 (front half catch). Where particulate is expressed in terms of coal ash content, the A factor is determined by multiplying weight % ash content of coal (as fired) by the numerical value preceding the A. For example, if coal with 8% ash is fired in a PC-fired, dry bottom unit, the PM emission factor would be 10 × 8, or 80 lb/ton.

[c]No data found; emission factor for PC-fired dry bottom boilers used.

[d]Uncontrolled particulate emissions, when no fly ash reinjection is employed. When control device is installed, and collected fly ash is reinjected to boiler, particulate from boiler reaching control equipment can increase up to a factor of 2.

[e]Accounts for fly ash settling in an economizer, air heater, or breaching upstream of control device or stack. (Particulate directly at boiler outlet typically will be twice this level.) Factor should be applied even when fly ash is reinjected to boiler form air heater or economizer dust hoppers.

[f]Includes traveling grate, vibrating grate, and chain grate stokers.

[g]Accounts for fly ash settling in breaching or stack base. Particulate loadings directly at boiler outlet typically can be 50% higher.

[h]Accounts for fly ash settling in breaching downstream of boiler outlet.

[i]No data found; emission factor for underfeed stoker used.

[j]No data found; use emission factor for spreader stoker with multiple cyclones and reinjection.

Table C.4 Condensable Particulate Matter Emission Factors for Bituminous and Subbituminous Coal Combustion[a]

Firing Configuration[b]	Controls[c]	CPM-TOT[d,e]		CPM-IOR[d,e]		CPM-ORG[d,e]	
		Emission Factor (lb/MM Btu)	Emission Factor Rating	Emission Factor (lb/MM Btu)	Emission Factor Rating	Emission Factor (lb/MM Btu)	Emission Factor Rating
All pulverized coal-fired boilers	All PM controls (without FGD controls)	0.1S–0.03[f]	B	80% of CPM-TOT emission factor[e]	E	20% of CMP-TOT emission factor[e]	E
All pulverized coal-fired boilers	All PM controls combined with FGD controls	0.02	E	ND		ND	E
Spreader stoker, traveling grate overfeed stoker, underfeed stoker	All PM controls, or uncontrolled	0.04	C	80% of CPM-TOT emission factor[e]	E	20% CPM-TOT emission factor[e]	E

[a]Source: AP-42. All condensable PM is assumed to be less than 1.0 μm in diameter.
[b]No data are available for cyclone boilers or for atmospheric fluidized bed combustion (AFBC) boilers. For cyclone boilers, use the factors provided for pulverized coal-fired boilers and applicable control devices. For AFBC boilers, use the factors provided for pulverized coal-fired boilers with PM and FGD controls.
[c]FGD = flue gas desulfurization.
[d]CPM-TOT, total condensable particulate matter; CPM-IOR, inorganic condensable particulate matter; CPM-ORG, organic condensable particulate matter; ND, no data.
[e]Factors should be multiplied by fuel rate on a heat input basis (MM Btu), as fired. To convert to lb/ton of bituminous coal, multiply by 26 MM Btu/ton. To convert to lb/ton of subbituminous coal, multiply by 20 MM Btu/ton.
[f]S = coal sulfur percent by weight, as fired. For example, if the sulfur percent is 1.04, then S = 1.04. If the coal sulfur percent is 0.4 or less, use a default emission factor of 0.01 lb/MM Btu rather than the emission equation.

Table C.5 Emission Factors for Trace Elements, POM, and HCOH From Uncontrolled Bituminous and Subbituminous Coal Combustion[a] (Emission Factor Rating: E)

Firing Configuration	Emission Factor, lb/10^{12} Btu									
	As	Be	Cd	Cr	Pb[b]	Mn	Hg	Ni	POM	HCOH
Pulverized coal, configuration unknown	ND	ND	ND	1,922	ND	ND	ND	ND	ND	112[c]
Pulverized coal, wet bottom	538	81	44–70	1,020–1,570	507	808–2,980	16	840–1,290	ND	ND
Pulverized coal, dry bottom	684	81	44.4	1,250–1,570	507	228–2,980	16	1,030–1,290	2.08	ND
Pulverized coal, dry bottom, tangential	ND	ND	ND	ND	ND	ND	ND	ND	2.4	ND
Cyclone furnace	115	<81	28	212–1,502	507	228–1,300	16	174–1,290	ND	ND
Stoker, configuration unknown	ND	73	ND	19–300	ND	2,170	16	775–1,290	ND	ND
Spreader stoker	264–542	ND	21–43	942–1,570	507	ND	ND	ND	ND	221[d]
Overfeed stoker, traveling grate	542–1,030	ND	43–82	ND	507	ND	ND	ND	ND	140[c]

[a]*Source*: AP-42. The emission factors in this table represent the ranges of factors reported in the literature. If only 1 data point was found, it is still reported in this table. To convert from lb/10^{12} Btu to pg/J, multiply by 0.43. ND = no data.
[b]Lead emission factors were taken directly from an EPA background document for support of the National Ambient Air Quality Standards.
[c]Based on 2 units; 133×10^6 Btu/h and 155×10^6 Btu/h.
[d]Based on 1 unit; 59×10^6 Btu/h.

Table C.6 Cumulative Particle Size Distribution and Size-Specific Emission Factors for Dry Bottom Boilers Burning Pulverized Bituminous and Subbituminous Coal[a]

Particle Size[b] (μm)	Cumulative Mass % ≤ Stated Size						Cumulative Emission Factor[c] (lb/ton)				
	Uncontrolled	Controlled					Uncontrolled[d]	Controlled[c]			
		Multiple Cyclones	Scrubber	ESP	Baghouse			Multiple Cyclones[e]	Scrubber[f]	ESP[g]	Baghouse[f]
15	32	54	81	79	97		3.2A	1.08A	0.48A	0.064A	0.02A
10	23	29	71	67	92		2.3A	0.58A	0.42A	0.054A	0.02A
6	17	14	62	50	77		1.7A	0.28A	0.38A	0.024A	0.02A
2.5	6	3	51	29	53		0.6A	0.06A	0.3A	0.024A	0.01A
1.25	2	1	35	17	31		0.2A	0.02A	0.22A	0.01A	0.006A
1.00	2	1	31	14	25		0.2A	0.02A	0.18A	0.01A	0.006A
0.625	1	1	20	12	14		0.10A	0.02A	0.12A	0.01A	0.002A
Total	**100**	**100**	**100**	**100**	**100**		**10A**	**2A**	**0.6A**	**0.08A**	**0.02A**

[a]*Source:* AP-42. Tons are short tons. To convert from lb/ton to kg/Mg, multiply by 0.5. Emission factors are lb of pollutant per ton of coal combusted, as fired. ESP = electrostatic precipitator.

[b]Expressed as aerodynamic equivalent diameter.

[c]A = coal ash weight percent as fired. For example, if coal ash weight is 8.2%, then A = 8.2.

[d]Emission factor rating = C.

[e]Estimated control efficiency for multiple cyclones is 80%; for scrubber, 94%; for ESP, 99.2%; and for baghouse, 99.8%.

[f]Emission factor rating = E.

[g]Emission factor rating = D.

Table C.7 Cumulative Particle Size Distribution and Size-Specific Emission Factors for Wet Bottom Boilers Burning Pulverized Bituminous Coal (Emission Factor Rating: E)[a]

Particle Size[b] (μm)	Cumulative Mass % ≤ Stated Size			Cumulative Emission Factor[c] (lb/ton)		
		Controlled			Controlled[d]	
	Uncontrolled	Multiple Cyclones	ESP	Uncontrolled	Multiple Cyclones	ESP
15	40	99	83	2.8A	1.38A	0.046A
10	37	93	75	2.6A	1.3A	0.042A
6	33	84	63	2.32A	1.18A	0.036A
2.5	21	61	40	1.48A	0.86A	0.022A
1.25	6	31	17	0.42A	0.44A	0.01A
1.00	4	19	8	0.28A	0.26A	0.004A
0.625	2	—[e]	—[e]	0.14A	—[e]	—[e]
Total	**100**	**100**	**100**	**7.0A**	**1.4A**	**0.056A**

[a]*Source:* AP-42. Tons are short tons. To convert from lb/ton to kg/Mg, multiply by 0.5. Emission factors are lb of pollutant per ton of coal combusted as fired. ESP = Electrostatic precipitator.
[b]Expressed as aerodynamic equivalent diameter.
[c]A = coal ash weight %, as fired. For example, if coal ash weight is 2.4%, then A = 2.4.
[d]Estimated control efficiency for multiple cyclones is 94%, and for ESPs, 99.2%.
[e]Insufficient data.

Table C.8 Cumulative Size Distribution and Size-Specific Emission Factors for Cyclone Furnaces Burning Bituminous Coal (Emission Factor Rating: E)[a]

Particle Size[b] (μm)	Cumulative Mass % ≤ Stated Size			Cumulative Emission Factor[c] (lb/ton)		
		Controlled			Controlled[d]	
	Uncontrolled	Multiple Cyclones	ESP	Uncontrolled	Multiple Cyclones	ESP
15	33	95	90	0.66A	0.114A	0.013A
10	13	94	68	0.26A	0.112A	0.011A
6	8	93	56	0.16A	0.112A	0.009A
2.5	5.5	92	36	0.11A[c]	0.11A	0.006A
1.25	5	85	22	0.10A[e]	0.10A	0.04A
1.00	5	82	17	0.10A[e]	0.10A	0.004A
0.625	0	e	—[e]	0	—[f]	—[f]
Total	**100**	**100**	**100**	**2A**	**0.12A**	**0.016A**

[a]*Source:* AP-42. Tons are short tons. To convert from lb/ton to kg/Mg, multiply by 0.5. Emission factors are lb of pollutant per ton of coal combusted as fired.
[b]Expressed as aerodynamic equivalent diameter.
[c]A = coal ash weight %, as fired. For example, if coal ash weight is 2.4%, then A = 2.4.
[d]Estimated control efficiency for multiple cyclones is 94%, and for ESPs, 99.2%.
[e]These values are estimates based on data from controlled source.
[f]Insufficient data.

Table C.9 Cumulative Particle Size Distribution and Size-Specific Emission Factors for Overfeed Stokers Burning Bituminous Coal[a]

| Particle Size[b] (μm) | Cumulative Mass % ≤ Stated Size | | Cumulative Emission Factor (lb/ton) | | | |
| | Uncontrolled | Multiple Cyclones Controlled | Uncontrolled | | Multiple Cyclones Controlled[c] | |
			Emission Factor	Emission Factor Rating	Emission Factor	Emission Factor Rating
15	49	60	7.8	C	5.4	E
10	37	55	6.0	C	5.0	E
6	24	49	3.8	C	4.4	E
2.5	14	43	2.2	C	3.8	E
1.25	13	39	2.0	C	3.6	E
1.00	12	39	2.0	C	3.6	E
0.625	d	16	d		1.4	E
Total	**100**	**100**	**16.0**	**C**	**9.0**	**E**

[a]*Source:* AP-42. Tons are short tons. To convert from lb/ton to kg/Mg, multiply by 0.5. Emission factors are lb of pollutant per ton of coal combusted as fired. ESP = Electrostatic precipitator.

[b]Expressed as aerodynamic equivalent diameter.

[c]A = coal ash weight %, as fired. For example, if coal ash weight is 2.4%, then A = 2.4.

[d]Estimated control efficiency for multiple cyclones is 80%.

Table C.10 Cumulative Particle Size Distribution and Size-Specific Emission Factors for Underfeed Stokers Burning Bituminous Coal (Emission Factor Rating: C)[a]

Particle Size [b] (μm)	Cumulative Mass % ≤ Stated Size	Uncontrolled Cumulative Emission Factor[c] (lb/ton)
15	50	7.6
10	41	6.2
6	32	4.8
2.5	25	3.8
1.25	22	3.4
1.00	21	3.2
0.625	18	2.7
Total	**100**	**15.0**

[a]*Source:* AP-42. Tons are short tons. To convert from lb/ton to kg/Mg, multiply by 0.5. Emission factors are lb of pollutant per ton of coal combusted, as fired.
[b]Expressed as aerodynamic equivalent diameter.
[c]May also be used for uncontrolled hand-fired units.

Table C.11 Cumulative Particle Size Distribution and Size-Specific Emission Factors for Spreader Stokers Burning Bituminous Coal[a]

| Particle Size[b] (μm) | Cumulative Mass % ≤ Stated Size | | | | | Cumulative Emission Factor (lb/ton) | | | | |
| | Uncontrolled | Controlled | | | | Uncontrolled[e] | Controlled | | | |
		Multiple Cyclones[c]	Multiple Cyclones[d]	ESP	Baghouse		Multiple Cyclones[e,f]	Multiple Cyclones[d,e]	ESP	Baghouse[e,g]
15	28	86	74	97	72	18.5	14.6	8.8	0.46	0.086
10	20	73	65	90	60	13.2	12	7.8	0.44	0.072
6	14	51	52	82	46	9.2	8.6	6.2	0.40	0.056
2.5	7	8	27	61	26	4.6	1.4	3.2	0.30	0.032
1.25	5	2	16	46	18	3.3	0.4	2.0	0.22	0.022
1.00	5	2	14	41	15	3.3	0.4	1.6	0.20	0.018
0.625	4	1	9	C[h]	7	2.6	0.2	1.0	C[h]	0.006
Total	**100**	**100**	**100**	**100**	**100**	**66.0**	**17.0**	**12.0**	**0.48**	**0.12**

[a]*Source*: AP-42. Tons are short tons. To convert from lb/ton to kg/Mg, multiply by 0.5. Emissions are lb of pollutant per ton of coal combusted, as fired.
[b]Expressed as aerodynamic equivalent diameter.
[c]With flyash reinjection.
[d]Without flyash reinjection.
[e]Emission factor rating = C.
[f]Emission factor rating = E.
[g]Estimated control efficiency for ESP is 99.22%; and for baghouse, 99.8%.
[h]Insufficient data.

Table C.12 Emission Factors for SO_x, NO_x, CO, and CO_2 from Uncontrolled Lignite Combustion (Emission Factor Rating: C (except as noted))[a]

Firing Configuration	SO_x Emission Factor[b] (lb/ton)	NO_x Emission Factor (lb/ton)	CO Emission Factor (lb/ton)	CO_2 Emission Factor[e] (lb/ton)	TNMOC[g,h,i] Emission Factor (lb/ton)
Pulverized coal, dry bottom, tangential	30S	7.1[f]	ND	72.6C	0.04
Pulverized coal, dry bottom, wall fired[c], Pre-NSPS[d]	30S	13	0.25	72.6C	0,04
Pulverized coal, dry bottom, wall fired[c], NSPS[d]	30S	6.3	0.25	72.6C	0.04
Cyclone	30S	15	ND	72.6C	0.07
Spreader stoker	30S	5.8	ND	72.6C	0.03
Traveling grate overfeed stoker	30S	ND	ND	72.6C	0.03
Atmospheric fluidized bed combustor	10S[i]	3.6	0.15	72.6C	0.03

[a]*Source:* AP-42. Tons are short tons. To convert from lb/ton to kg/Mg, multiply by 0.5. To convert from lb/ton to lb/MM Btu, multiply by 0.0625. ND = no data.

[b]S = weight % sulfur content of lignite, wet basis. For example, if the sulfur content equals 3.4%, then S = 3.4. For high sodium ash (Na_2O > 8%), use 22S. For low sodium ash (Na_2O < 2%), use 34S. If ash sodium content is unknown, use 30S.

[c]Wall-fired includes front and rear wall-fired units, as well as opposed wall-fired units.

[d]Pre-NSPS boilers are not subject to an NSPS. NSPS boilers are subject to Subpart D or Subpart Da. Subpart D boilers are boilers constructed after August 17, 1971 and with a heat input greater than 250 million Btu per hour (MM Btu/h). Subpart Da boilers are boilers constructed after September 18, 1978 and with a heat input rate greater than 250 MM Btu/h.

[e]Emission Factor Rating: B. C = weight % carbon of lignite, as-fired basis. For example, if carbon content equals 63%, then C = 63. If the %C value is not known, a default CO_2 emission value of 4,600 lb/ton may be used.

[f]Emission Factor Rating = A.

[g]TNMOC: Total nonmethane organic compounds. Emission factors were derived from bituminous coal data in the absence of lignite data assuming emissions are proportional to coal heating value. TNMOC are expressed as C_2 to C_{16} alkane equivalents. Because of limited data, the effects of firing configuration on TNMOC emission factors could not be distinguished. As a result, all data were averaged collectively to develop a single average emission factor for pulverized coal, cyclones, spreaders, and overfeed stokers.

[h]Nominal values achievable under normal operating conditions; values 1 or 2 orders of magnitude higher can occur when combustion is not complete.

[i]Using limestone bed material.

Table C.13 Emission Factors for NO_x and CO from Lignite Combustion with NO_x Controls[a]

		NO_x		CO	
Firing Configuration	Control Device	Emission Factor (lb/ton)	Emission Factor Rating	Emission Factor (lb/ton)	Emission Factor Rating
Subpart D boilers:[b] Pulverized coal, tangential-fired	Overfire air	6.8	C	ND	NA
Pulverized coal, wall-fired	Overfire air and low NO_x burners	4.6	C	0.48	D
Subpart Da boilers:[b] Pulverized coal, tangential-fired	Overfire air	6.0	C	0.1	D

[a]Source: AP-42. Tons are short tons. To convert from lb/ton to kg/Mg, multiply by 0.5. To convert from lb/ton to lb/MM Btu, multiply by 0.0625. ND = no data. NA = not applicable.
[b]Subpart D boilers are boilers constructed after August 17, 1971, and with a heat input rate greater than 250 million Btu per hour (MM Btu/h). Subpart Da boilers are boilers constructed after September 18, 1978, and with a heat input rate greater than 250 MM Btu/h.

Table C.14 Emission Factors for POM from Controlled Lignite Combustion (Emission Factor Rating: E)[a]

Firing Configuration	Control Device	Emission Factor (lb/10^{12} Btu)
Pulverized coal	High efficiency cold-side ESP	2.3
Pulverized dry bottom	Multi-cyclones ESP	1.8–18[b] 2.6[b]
Cyclone furnace	ESP	0.11[c]–1.6[b]
Spreader stoker	Multicyclones	15[c]

[a]Source: AP-42. To convert from lb/10^{12} Btu to pg/J, multiply by 0.43.
[b]Primarily trimethyl propenyl napthalene.
[c]Primarily biphenyl.

Table C.15 Emission Factors for Filterable PM and N_2O from Uncontrolled Lignite Combustion (Emission Factor Rating: E (except as noted))[a]

Firing Configuration	Filterable PM Emission Factor[b] (lb/ton)	N_2O Emission Factor[c] (lb/ton)
Pulverized coal, dry bottom, tangential	6.5A	ND
Pulverized coal, dry bottom, wall fired	5.1A	ND
Cyclone	6.7A[c]	ND
Spreader stoker	8.0A	ND
Other stoker	3.4A	ND
Atmospheric fluidized-bed combustor	ND	2.5

[a]*Source:* AP-42. Tons are short tons. To convert from lb/ton to kg/Mg, multiply by 0.5. To convert from lb/ton to lb/MM Btu, multiply by 0.0625. ND = no data.
[b]A = weight % ash content of lignite, wet basis. For example, if the ash content is 5%, then A = 5.
[c]Emission factor rating: C.

Table C.16 Emission Factors for Filterable PM Emissions from Controlled Lignite Combustion (Emission Factor Rating: C (except as noted))[a]

Firing Configuration	Control Device	Filterable PM Emission Factor (lb/ton)
Subpart D boilers[b]	Baghouse	0.08A
	Wet Scrubber	0.05A
Subpart Da boilers[b]	Wet Scrubber	0.01A
Atmospheric fluidized bed combustor[c]	ESP	0.07A

[a]Source: AP-42. Tons are short tons. A = weight % ash content of lignite, wet basis. For example, if lignite is 2.3% ash, then A = 2.3. To convert from lb/ton to kg/Mg, multiply by 0.5. To convert from lb/ton to lb/MM Btu, multiply by 0.0625.
[b]Subpart D boilers are boilers constructed before August 17, 1971, and with a heat input rate greater than 250 million Btu per hour (MM Btu/h). Subpart Da boilers are boilers constructed after September 18, 1978, and with a heat input rate greater than 250 MM Btu/h.
[c]Emission factor rating: D.

Table C.17 Condensable Particulate Matter Emission Factors for Lignite Combustion[a]

Firing Configuration[b]	Controls[c]	CPM-TOT[d,e]		CPM-IOR[d,e]		CPM-ORG[d,e]	
		lb/MM Btu	Rating	lb/MM Btu	Rating	lb/MM Btu	Rating
All pulverized coal-fired boilers	All PM controls (without FGD controls)	0.1S–0.03[f]	C	80% of CPM-TOT emission factor[e]	E	20% of CPM-TOT emission factor[e]	E
All pulverized coal-fired boilers	All PM controls combined with an FGD control	0.02[f]	E	ND		ND	
Traveling grate overfeed stoker, spreader stoker	All PM controls, or uncontrolled	0.04	D	80% of CPM-TOT emission factor	E	20% of CPM-TOT emission factor	E

[a]*Source*: AP-42. All condensable PM is assumed to be less than 1.0 micron in diameter.

[b]No data are available for cyclone boilers. For cyclone boilers, use the factors provided for pulverized coal-fired boilers and applicable controls.

[c]FGD = flue gas desulfurization.

[d]CPM-TOT = total condensable particulate matter.

CPM-IOR = inorganic condensable particulate matter.

CPM-ORG = organic condensable particulate matter.

ND = no data.

[e]Factors should be multiplied by fuel rate on a heat input basis (MM Btu), as fired. To convert to lb/short ton of lignite, multiply by 16 MM Btu/short ton.

[f]S = coal sulfur percent by weight, as fired. For example, if the sulfur percent is 1.04, then S = 1.04. If the coal sulfur percent is 0.4 or less, use a default emission factor of 0.01 lb/MM Btu rather than the emission equation.

Table C.18 Emission Factors for Trace Elements From Uncontrolled Lignite Combustion (Emission Factor Rating: E)[a]

Firing Configuration	Emission Factor (lb/10^{12}Btu)						
	As	Be	Cd	Cr	Mn	Hg	Ni
Pulverized, wet bottom	2,730	131	49–77	1,220–1,880	4,410–16,250	21	154–1,160
Pulverized, dry bottom	1,390	131	49	1,500–1,880	16,200	21	928–1,160
Cyclone furnace	235–632	131	31	253–1,880	3,760	21	157–1,160
Stoker configuration unknown	ND	118	ND	ND	11,800	21	ND
Spreader stoker	538–1,100	ND	23–47	1,130–1,880	ND	ND	696–1,160
Traveling grate (overfed) stoker	1,100–2,100	ND	47–90	ND	ND	ND	ND

[a]*Source*: AP-42. To convert from lb/10^{12} Btu to pg/J, multiply by 0.43. ND = no data.

Table C.19 Cumulative Particle Size Distribution and Size-Specific
Emission Factors for Boilers Firing Pulverized Lignite
(Emission Factor Rating: E)[a]

Particle Size[b] (μm)	Cumulative Mass % $\leq$ Stated Size		Cumulative Emission[c] (lb/ton)	
	Uncontrolled	Multiple Cyclone Controlled	Uncontrolled	Multiple Cyclone Controlled[d]
15	51	77	3.4A	1.0A
10	5	67	2.3A	0.88A
6	26	57	1.7A	0.75A
2.5	10	27	0.66A	0.36A
1.25	7	16	0.47A	0.21A
1.00	6	14	0.40A	0.19A
0.625	3	8	0.19A	0.11A
Total			**6.6A**	**1.3A**

[a]*Source:* AP-42. Tons are short tons. Based on tangentially fired units. For wall-fired units multiply emission factors in the table by 0.79.
[b]Expressed as aerodynamic equivalent diameter.
[c]A = weight % ash content of lignite, wet basis. For example, if lignite is 3.4% ash, then A = 3.4. To convert from lb/ton to kg/Mg, multiply by 0.5. To convert from lb/ton to lb/MM Btu, multiply by 0.0625.
[d]Estimated control efficiency for multiple cyclone is 80%, averaged over all particle sizes.

Table C.20 Cumulative Particle Size Distribution and Size-Specific
Emission Factors for Lignite-Fired Spreader Stokers
(Emission Factor Rating: E)[a]

Particle Size[b] (μm)	Cumulative Mass % $\leq$ Stated Size		Cumulative Emission[c] (lb/ton)	
	Uncontrolled	Multiple Cyclone Controlled	Uncontrolled	Multiple Cyclone Controlled[d]
15	28	55	2.2A	0.88A
10	20	41	1.6A	0.66A
6	14	31	1.1A	0.50A
2.5	7	26	0.56A	0.42A
1.25	5	23	0.40A	0.37A
1.00	5	22	0.40A	0.35A
0.625	4	—[e]	0.33A	—[e]
Total			**8.0A**	**1.6A**

[a]*Source:* AP-42. Tons are short tons.
[b]Expressed as aerodynamic equivalent diameter.
[c]A = weight % ash content of lignite, wet basis. For example, if lignite is 5% ash, then A = 5. To convert from lb/ton to kg/Mg, multiply by 0.5. To convert from lb/ton to lb/MM Btu, multiply by 0.0625.
[d]Estimated control efficiency for multiple cyclone is 80%.
[e]Insufficient data.

Table C.21 Default CO_2 Emission Factors for U.S. Coals (Emission Factor Rating: C)[a]

Coal Type	Average %C[b]	Conversion Factor[c]	Emission Factor[d] lb/ton Coal
Subbituminous	66.3	72.6	4,810
High-volatile bituminous	75.9	72.6	5,510
Medium-volatile bituminous	83.2	72.6	6,040
Low-volatile bituminous	86.1	72.6	6,250

[a]*Source:* AP-42. Tons are short tons. This table should be used only when an ultimate analysis is not available. If the ultimate analysis is available, CO_2 emissions should be calculated by multiplying the % carbon (%C) by 72.6. This resultant factor would receive a quality rating of B.
[b]Based on average carbon contents for each coal type (dry basis) based on extensive sampling of U.S. coals.
[c]Based on the following equation:

$$\frac{44 \text{ton } CO_2}{12 \text{ton } C} \times 0.99 \times 2000 \frac{\text{lb } CO_2}{\text{ton } CO_2} \times \frac{1}{100\%} = 72.6 \frac{\text{lb } CO_2}{\text{ton}\%C}$$

where 44 = molecular weight of CO_2; 12 = molecular weight of carbon; and 0.99 = fraction of fuel oxidized during combustion.

[d]To convert from lb/ton to kg/Mg, multiply by 0.5.

Table C.22 Emission Factors for Various Organic Compounds
from Controlled Coal Combustion[a]

Pollutant[b]	Emission Factor (lb/ton)[c]	Emission Factor Rating
Acetaldehyde	5.7E-04	C
Acetophenone	1.5E-05	D
Acrolein	2.9E-04	D
Benzene	1.3E-03	A
Benzyl chloride	7.0E-04	D
Bis(2-ethylhexyl)phthalate (DEHP)	7.3E-05	D
Bromoform	3.9E-05	E
Carbon disulfide	1.3E-04	D
2-Chloroacetophenone	7.0E-06	E
Chlorobenzene	2.2E-05	D
Chloroform	5.9E-05	D
Cumene	5.3E-06	E
Cyanide	2.5E-03	D
2,4-Dinitrotoluene	2.8E-07	D
Dimethyl sulfate	4.8E-05	E
Ethyl benzene	9.4E-05	D
Ethyl chloride	4.2E-05	D
Ethylene dichloride	4.0E-05	E
Ethylene dibromide	1.2E-06	E
Formaldehyde	2.4E-04	A
Hexane	6.7E-05	D
Isophorone	5.8E-04	D
Methyl bromide	1.6E-04	D
Methyl chloride	5.3E-04	D
Methyl ethyl ketone	3.9E-04	D

Continued

Table C.22 Emission Factors for Various Organic Compounds
from Controlled Coal Combustion[a]—Cont'd

Pollutant[b]	Emission Factor (lb/ton)[c]	Emission Factor Rating
Methyl hydrazine	1.7E-04	E
Methyl methacrylate	2.0E-05	E
Methyl tert-butyl ether	3.5E-05	E
Methylene chloride	2.9E-04	D
Phenol	1.6E-05	D
Propionaldehyde	3.8E-04	D
Tetrachloroethylene	4.3E-05	D
Toluene	2.4E-04	A
1,1,1-Trichloroethane	2.0E-05	E
Styrene	2.5E-05	D
Xylenes	3.7E-05	C
Vinyl acetate	7.6E-06	E

[a]*Source:* AP-42. Tons are short tons. Factors were developed from emissions data from 10 sites firing bituminous coal, eight sights firing subbituminous coal, and from one site firing lignite. The emission factors are applicable to boilers using both wet limestone scrubbers or spray dryers and an electrostatic precipitator (ESP) or fabric filter (FF). In addition, the factors apply to boilers utilizing only an ESP or FF.
[b]Pollutants sampled for but not detected in any sampling run include: carbon tetrachloride, 2 sites; 1,3-dichloropropyene, 2 sites; N-nitrosodimethylamine, 2 sites; ethylidene dichloride, 2 sites; hexachlorobutadiene, 2 sites; hexachloroethane, 1 site; propylene dichloride, 2 sites; 1,1,2,2-tetrachloroethane, 2 sites; 1,1,2-trichloroethane, 2 sites; vinyl chloride, 2 sites; and hexachlorobenzene, 2 sites.
[c]Emission factor should be applied to coal feed, as fired.

Table C.23 Emission Factors for Polynuclear Aromatic Hydrocarbons
from Controlled Coal Combustion[a]

Pollutant	Emission Factor[b] (lb/ton)	Emission Factor Rating
Biphenyl	1.7E-06	D
Acenaphthene	5.1E-07	B
Acenaphthylene	2.5E-07	B
Anthracene	2.1E-07	B
Benzo(a)anthracene	8.0E-08	B
Benzo(a)pyrene	3.8E-08	D
Benzo(b,j,k)fluoranthene	1.1E-07	B
Benzo(g,h,i)perylene	2.7E-08	D
Chrysene	1.0E-07	C
Fluoranthene	7.1E-07	B
Fluorene	9.1E-07	B
Indeno(1,2,3-cd)pyrene	6.1E-08	C
Naphthalene	1.3E-05	C
Phenanthrene	2.7E-06	B
Pyrene	3.3E-07	B
5-Methyl chrysene	2.2E-08	D

[a]*Source:* AP-42. Tons are short tons. Factors were developed from emissions data from six sites firing bituminous coal, four sights firing subbituminous coal, and from one site firing lignite. Factors apply to boilers utilizing both wet limestone scrubbers or spray dryers and an electrostatic precipitator (ESP) or fabric filter (FF). The factors apply to boilers utilizing only an ESP or FF.
[b]Emission factor should be applied to coal feed, as fired. To convert from lb/ton to lb/MM Btu, multiply by 0.0625. To convert from lb/ton to kg/Mg, multiply by 0.5. Emissions are lb of pollutant per ton of coal combusted.

Table C.24 Emission Factors for Hydrogen Chloride and Hydrogen Fluoride from Coal Combustion (Emission Factor Rating: B)[a]

Firing Configuration	HCl Emission Factor (lb/ton)	HF Emission Factor (lb/ton)
PC-fired	1.2	0.15
PC-fired, tangential	1.2	0.15
Cyclone furnace	1.2	0.15
Traveling grate (overfeed stoker)	1.2	0.15
Spreader stoker	1.2	0.15
FBC, Circulating bed	1.2	0.15

[a]*Source:* AP-42. Tons are short tons. The emission factors were developed from bituminous coal, subbituminous coal, and lignite emissions data. To convert from lb/ton to kg/Mg, multiply by 0.5. To convert from lb/ton to lb/MM Btu, multiply by 0.0625. The factors apply to both controlled and uncontrolled sources.

Table C.25 Emission Factors for Trace Metals from Controlled Coal Combustion[a]

Pollutant	Emission Factor[b] (lb/ton)	Emission Factor Rating
Antimony	1.8E-05	A
Arsenic	4.1E-04	A
Beryllium	2.1E-05	A
Cadmium	5.1E-05	A
Chromium	2.6E-04	A
Chromium (VI)	7.9E-05	D
Cobalt	1.0E-04	A
Lead	4.2E-04	A
Magnesium	1.1E-02	A
Manganese	4.9E-04	A
Mercury	8.3E-05	A
Nickel	2.8E-04	A
Selenium	1.3E-03	A

[a]*Source:* AP-42. Tons are short tons. The emission factors were developed from emissions data at eleven facilities firing bituminous coal, fifteen facilities firing subbituminous coal, and from two facilities firing lignite. The factors apply to boilers utilizing either venturi scrubbers, spray dryer absorbers, or wet limestone scrubbers with an electrostatic precipitator (ESP) or fabric filter (FF). In addition, the factors apply to boilers using only an ESP, FF, or venturi scrubber. Firing configurations include pulverized coal-fired, dry bottom boilers; pulverized coal, dry bottom, tangentially fired boilers; cyclone boilers; and atmospheric fluidized bed combustors, circulating bed.
[b]Emission factor should be applied to coal feed, as fired. To convert from lb/ton to kg/Mg, multiply by 0.5.

Table C.26 Emission Factor Equations for Trace Elements from
Coal Combustion[a] (Emission Factor Equation Rating: A)[b]

Pollutant	Emission Equation (lb/10^{12}Btu)[c]
Antimony	$0.92\,[(C/A)PM]^{0.63}$
Arsenic	$3.1\,[(C/A)PM]^{0.85}$
Beryllium	$1.2\,[(C/A)PM]^{1.1}$
Cadmium	$3.3\,[(C/A)PM]^{0.5}$
Chromium	$3.7\,[(C/A)PM]^{0.58}$
Cobalt	$1.7\,[(C/A)PM]^{0.69}$
Lead	$3.4\,[(C/A)PM]^{0.80}$
Manganese	$3.8\,[(C/A)PM]^{0.60}$
Nickel	$4.4\,[(C/A)PM]^{0.48}$

[a]*Source:* AP-42. The equations were developed from emissions data from
bituminous coal combustion, subbituminous coal combustion, and from lignite
combustion. The equations may be used to generate factors for both controlled and
uncontrolled boilers. The emission factor equations are applicable to all typical
firing configurations for electric generation (utility), industrial, and commercial/
industrial boilers for bituminous coal, subbituminous coal, and lignite.
[b]AP-42 criteria for rating emission factors were used to rate the equations.
[c]The factors produced by the equations should be applied to heat input. To convert
from lb/10^{12} Btu to kg/joules, multiply by 4.31×10^{-16}. C = concentration of metal
in the coal, parts per million by weight (ppmwt); A = weight fraction of ash in the
coal. For example, 10% ash is 0.1 ash fraction; PM = site-specific emission factor
for total particulate matter, lb/10^6 Btu.

Table C.27 Emission Factors for SO_x and NO_x Compounds from Uncontrolled
Anthracite Combustors[a]

Source Category	SO_x		NO_x	
	Emission Factor (lb/ton)	Emission Factor Rating	Emission Factor (lb/ton)	Emission Factor Rating
Stoker-fired boilers	$39S$[b]	B	9.0	C
FBC boilers[c]	2.9	E	1.8	E
Pulverized coal boilers	$39S$[b]	B	18	B

[a]*Source:* AP-42. Tons are short tons. Units are lb of pollutant/ton of coal burned. To convert from lb/ton to kg/Mg,
multiply by 0.5.
[b]S = weight percent sulfur. For example, if the sulfur content is 3.4%, then S = 3.4.
[c]FBC boilers burning culm fuel; all other sources burning anthracite.

Table C.28 Emission Factors for CO and Carbon Dioxide (CO_2)
from Uncontrolled Anthracite Combustors[a]

Source Category	CO		CO_2	
	Emission Factor (lb/ton)	Emission Factor Rating	Emission Factor (lb/ton)	Emission Factor Rating
Stoker-fired boilers	0.6	B	5,680	C
FBC boilers[b]	0.6	E	ND	NA

ND = no data. NA = not applicable.
[a]*Source:* AP-42. Tons are short tons. Units are lb of pollutant/ton of coal burned. To convert from lb/ton to kg/Mg, multiply by 0.5.
[b]FBC boilers burning culm fuel; all other sources burning anthracite.

Table C.29 Emission Factors for Speciated Organic Compounds
from Anthracite Combustors (Emission Factor Rating: E)[a]

Pollutant	Stoker-Fired Boilers Emission Factor (lb/ton)
Acenaphthene	ND
Acenaphthylene	ND
Anthrene	ND
Anthracene	ND
Benzo(a)anthracene	ND
Benzo(a)pyrene	ND
Benzo(e)pyrene	ND
Benzo(g,h,i)perylene	ND
Benzo(k)fluoranthrene	ND
Biphenyl	2.5 E-02
Chrysene	ND
Coronene	ND
Fluoranthrene	ND
Fluorene	ND
Indeno(123-cd)perylene	ND
Naphthalene	1.3 E-01
Perylene	ND
Phenanthrene	6.8 E-03
Pyrene	ND

[a]*Source:* AP-42. Tons are short tons. Units are lb of pollutant/ton of coal burned. To convert from lb/ton to kg/Mg, multiply by 0.5. ND = no data.

Table C.30 Emission Factors for TOC and Methane (CH$_4$) from Anthracite Combustors (Emission Factor Rating: E)[a]

Source Category	TOC Emission Factor (lb/ton)	CH$_4$ Emission Factor (lb/ton)
Stoker-fired boilers	0.30	ND

[a]*Source:* AP-42. Tons are short tons. Units are lb of pollutant/ton of coal burned. To convert from lb/ton to kg/Mg, multiply by 0.5. ND = no data.

Table C.31 Emission Factors for Speciated Metals from Anthracite Combustion in Stoker-Fired Boilers (Emission Factor Rating: E)[a]

Pollutant	Emission Factor Range (lb/ton)	Average Emission Factor (lb/ton)
Arsenic	BDL–2.4 E-04	1.9 E-04
Antimony	BDL	BDL
Beryllium	3.0 E-05–5.4 E-04	3.1 E-04
Cadmium	4.5 E-05–1.1 E-04	7.1 E-05
Chromium	5.9 E-03–4.9 E-02	2.8 E-02
Manganese	9.8 E-04–5.3 E-03	3.6 E-03
Mercury	8.7 E-05–1.7 E-04	1.3 E-04
Nickel	7.8 E-03–3.5 E-02	2.6 E-02
Selenium	4.7 E-04–2.1 E-03	1.3 E-03

[a]*Source:* AP-42. Tons are short tons. Units are lb of pollutant/ton of coal burned. To convert from lb/ton to kg/Mg, multiply by 0.5. BDL = below detection limit.

Table C.32 Emission Factors for PM and Lead (Pb) from Uncontrolled Anthracite Combustors[a]

| Source Category | Filterable PM | | Condensable PM | | Pb | |
	Emission Factor (lb/ton)	Emission Factor Rating	Emission Factor (lb/ton)	Emission Factor Rating	Emission Factor (lb/ton)	Emission Factor Rating
Stoker-fired boilers	0.8A[b]	C	0.08A[b]	C	8.9E-03	E
Hand-fired units	10	B	ND	NA	ND	NA

[a]*Source*: AP-42. Tons are short tons. Units are lb of pollutant/ton of coal burned. To convert from lb/ton to kg/Mg, multiply by 0.5. ND = no data. NA = not applicable.
[b]A = ash content of fuel, weight %. For example, if the ash content is 5%, then A = 5.

649

Table C.33 Cumulative Particle Size Distribution and Size-Specific Emission Factors for Dry Bottom Boilers Burning Pulverized Anthracite (Emission Factor Rating: D)[a]

Particle Size[b] (μm)	Cumulative Mass % ≤ Stated Size			Cumulative Emission Factor as Fired[c] (lb/ton)		
		Controlled[d]			Controlled[d]	
	Uncontrolled	Multiple Cyclone	Baghouse	Uncontrolled	Multiple Cyclone	Baghouse
15	32	63	79	3.2A[e]	1.26A	0.016A
10	23	55	67	2.3A	1.10A	0.013A
6	17	46	51	1.7A	0.92A	0.010A
2.5	6	24	32	0.6A	0.48A	0.006A
1.25	2	13	21	0.2A	0.26A	0.004A
1.00	2	10	18	0.2A	0.20A	0.004A
0.625	1	7	[f]	0.1A	0.14A	[f]
Total	**100**	**100**	**100**	**10A**	**2A**	**0.02A**

[a]*Source:* AP-42. Tons are short tons.
[b]Expressed as aerodynamic equivalent diameter.
[c]Units are lb of pollutant/ton of coal burned. To convert from lb/ton to kg/Mg, multiply by 0.5.
[d]Estimated control efficiency for multiple cyclone is 80%; for baghouse, 99.8%.
[e]A = coal ash weight %, as fired. For example, if ash content is 5%, then A = 5.
[f]Insufficient data.

Appendix D
Original List of Hazardous
Air Pollutants

CAS Number	Chemical Name
75070	Acetaldehyde
60355	Acetamide
75058	Acetonitrile
98862	Acetophenone
53963	2-Acetylaminofluorene
107028	Acrolein
79061	Acrylamide
79107	Acrylic acid
107131	Acrylonitrile
107051	Allyl chloride
92671	4-Aminobiphenyl
62533	Aniline
90040	o-Anisidine
1332214	Asbestos
71432	Benzene (including benzene from gasoline)
92875	Benzidine
98077	Benzotrichloride
100447	Benzyl chloride
92524	Biphenyl
117817	Bis(2-ethylhexyl)phthalate (DEHP)
542881	Bis(chloromethyl)ether
75252	Bromoform
106990	1,3-Butadiene
156627	Calcium cyanamide
105602	Caprolactam[a]
133062	Captan
63252	Carbaryl
75150	Carbon disulfide
56235	Carbon tetrachloride
463581	Carbonyl sulfide
120809	Catechol

Continued

Clean Coal Engineering Technology. DOI: 10.1016/B978-1-85617-710-8.00024-8
Copyright © 2011 by Elsevier Inc. All rights of reproduction in any form reserved.

CAS Number	Chemical Name
133904	Chloramben
57749	Chlordane
7782505	Chlorine
79118	Chloroacetic acid
532274	2-Chloroacetophenone
108907	Chlorobenzene
510156	Chlorobenzilate
67663	Chloroform
107302	Chloromethyl methyl ether
126998	Chloropene
1319773	Cresols/Cresylic acid (isomers and mixture)
95487	o-Cresol
108394	m-Cresol
106445	p-Cresol
98828	Cumene
94757	2,4-D, salts and esters
3547044	DDE
334883	Diazomethane
132649	Dibenzofurans
96128	1,2-Dibromo-3-chloropropane
84742	Dibutylphthalate
106467	1,4-Dichlorobenzene(p)
91941	3,3-Dichlorobenzidene
111444	Dichloroethyl ether (Bis(2-chloroethyl)ether)
542756	1,3-Dichloropropene
62737	Dichlorvos
111422	Diethanolamine
121697	N,N-Diethyl aniline (N,N-Dimethylaniline)
64675	Diethyl sulfate
119904	3,3-Dimethoxybenzidine
60117	Dimethyl aminoazobenzene
119937	3,3'-Dimethyl benzidine
79447	Dimethyl carbamoyl chloride
68122	Dimethyl formamide
57147	1,1-Dimethyl hydrazine
131113	Dimethyl phthalate
77781	Dimethyl sulfate
534521	4,6-Dinitro-o-cresol, and salts
51285	2,4-Dinitrophenol
121142	2,4-Dinitrotoluene
123911	1,4-Dioxane (1,4-Diethyleneoxide)
122667	1,2-Diphenylhydrazine
106898	Epichlorohydrin (l-Chloro-2,3-epoxypropane)
106887	1,2-Epoxybutane
140885	Ethyl acrylate

CAS Number	Chemical Name
100414	Ethyl benzene
51796	Ethyl carbamate (Urethane)
75003	Ethyl chloride (Chloroethane)
106934	Ethylene dibromide (Dibromoethane)
107062	Ethylene dichloride (1,2-Dichloroethane)
107211	Ethylene glycol
151564	Ethylene imine (Aziridine)
75218	Ethylene oxide
96457	Ethylene thiourea
75343	Ethylidene dichloride (1,1-Dichloroethane)
50000	Formaldehyde
76448	Heptachlor
118741	Hexachlorobenzene
87683	Hexachlorobutadiene
77474	Hexachlorocyclopentadiene
67721	Hexachloroethane
822060	Hexamethylene-1,6-diisocyanate
680319	Hexamethylphosphoramide
110543	Hexane
302012	Hydrazine
7647010	Hydrochloric acid
7664393	Hydrogen fluoride (Hydrofluoric acid)
7783064	Hydrogen sulfide[b]
123319	Hydroquinone
78591	Isophorone
58899	Lindane (all isomers)
108316	Maleic anhydride
67561	Methanol
72435	Methoxychlor
74839	Methyl bromide (Bromomethane)
74873	Methyl chloride (Chloromethane)
71556	Methyl chloroform (1,1,1-Trichloroethane)
78933	Methyl ethyl ketone (2-Butanone)
60344	Methyl hydrazine
74884	Methyl iodide (Iodomethane)
108101	Methyl isobutyl ketone (Hexone)
624839	Methyl isocyanate
80626	Methyl methacrylate
1634044	Methyl tert butyl ether
101144	4,4-Methylene bis(2-chloroaniline)
75092	Methylene chloride (Dichloromethane)
101688	Methylene diphenyl diisocyanate (MDI)
101779	4,4-Methylenedianiline
91203	Naphthalene
98953	Nitrobenzene

Continued

CAS Number	Chemical Name
92933	4-Nitrobiphenyl
100027	4-Nitrophenol
79469	2-Nitropropane
684935	N-Nitroso-N-methylurea
62759	N-Nitrosodimethylamine
59892	N-Nitrosomorpholine
56382	Parathion
82688	Pentachloronitrobenzene (Quintobenzene)
87865	Pentachlorophenol
108952	Phenol
106503	p-Phenylenediamine
75445	Phosgene
7803512	Phosphine
7723140	Phosphorus
85449	Phthalic anhydride
1336363	Polychlorinated biphenyls (Aroclors)
1120714	1,3-Propane sultone
57578	β-Propiolactone
123386	Propionaldehyde
114261	Propoxur (Baygon)
78875	Propylene dichloride (1,2-Dichloropropane)
75569	Propylene oxide
75558	1,2-Propylenimine (2-Methyl aziridine)
91225	Quinoline
106514	Quinone
100425	Styrene
96093	Styrene oxide
1746016	2,3,7,8-Tetrachlorodibenzo-p-dioxin
79345	1,1,2,2-Tetrachloroethane
127184	Tetrachloroethylene (Perchloroethylene)
7550450	Titanium tetrachloride
108883	Toluene
95807	2,4-Toluene diamine
584849	2,4-Toluene diisocyanate
95534	o-Toluidine
8001352	Toxaphene (chlorinated camphene)
120821	1,2,4-Trichlorobenzene
79005	1,1,2-Trichloroethane
79016	Trichloroethylene
95954	2,4,5-Trichlorophenol
88062	2,4,6-Trichlorophenol
121448	Triethylamine
1582098	Trifluralin
540841	2,2,4-Trimethylpentane
108054	Vinyl acetate

CAS Number	Chemical Name
593602	Vinyl bromide
75014	Vinyl chloride
75354	Vinylidene chloride (1,1,-Dichloroethylene)
1330207	Xylenes (isomers and mixture)
95476	o-Xylenes
108383	m-Xylenes
106423	p-Xylenes
0	Antimony Compounds
0	Arsenic Compounds (inorganic including arsine)
0	Beryllium Compounds
0	Cadmium Compounds
0	Chromium Compounds
0	Cobalt Compounds
0	Coke Oven Emissions
0	Cyanide Compounds[c]
0	Glycol ethers[d]
0	Lead Compounds
0	Manganese Compounds
0	Mercury Compounds
0	Fine Mineral Fibers[e]
0	Nickel Compounds
0	Polycylic Organic Matter[f]
0	Radionuclides (including radon)[g]
0	Selenium Compounds

Note: The following applies to all listings that contain the word "compounds" and for glycol ethers: Unless otherwise specified, these listings are defined as including any unique chemical substance that contains the named chemical (i.e., antimony, arsenic, etc.) as part of that chemical's infrastructure.

[a]Caprolactam was delisted on June 18, 1996.

[b]Hydrogen sulfide was inadvertently added to the Section 112(b) list of HAPs through a clerical error. A joint resolution was passed by Congress and approved by the president on December 4, 1991, removing hydrogen sulfide from Section 112(b). Hydrogen sulfide is included in Section 112(r) and is subject to accidental release provisions.

[c]X'CN where X = H' or any other group where a formal dissociation may occur—for example, KCN or $Ca(CN)_2$.

[d]Includes mono- and di-ethers of ethylene glycol. Diethylene glycol, and triethylene glycol $R-(OCH_2CH_2)n-OR'$ where

n = 1, 2, or 3

R = alkyl or aryl groups

R' = R, H, or groups which, when removed, yield glycol ethers with the structure: $R-(OCH_2CH)n-OH$. Polymers are excluded from the glycol category.

[e]Includes mineral fiber emissions from facilities manufacturing or processing glass, rock, or slag fibers (or other mineral derived fibers) of average diameter 1 micrometer or less.

[f]Includes organic compounds with more than one benzene ring and that have a boiling point greater than or equal to 100°C.

[g]A type of atom which spontaneously undergoes radioactive decay.

Appendix E
Initial 263 Units Identified in Phase I (SO₂) of the Acid Rain Program

Plant Name	Unit Number	State
Albright	3	West Virginia
Allen	1, 2, 3	Tennessee
Armstrong	1, 2	Pennsylvania
Asbury	1	Missouri
Ashtabula	7	Ohio
Avon Lake	11, 12	Ohio
B L England	1, 2	New Jersey
Bailly	7, 8	Indiana
Baldwin	1, 2, 3	Illinois
Big Bend	BB01, BB02, BB03	Florida
Bowen	1BLR, 2BLR, 3BLR, 4BLR	Georgia
Breed	1	Indiana
Brunner Island	1, 2, 3	Pennsylvania
Burlington	1	Iowa
C P Crane	1, 2	Maryland
Cardinal/Tidd	1, 2	Ohio
Cayuga	1, 2	Indiana
Chalk Point	1, 2	Maryland
Cheswick	1	Pennsylvania
Clifty Creek	1, 2, 3, 4, 5, 6	Indiana
Coffeen	1, 2	Illinois
Colbert	1, 2, 3, 4, 5	Alabama
Coleman	C1, C2, C3	Kentucky
Conemaugh	1, 2	Pennsylvania
Conesville	1, 2, 3, 4	Ohio
Cooper	1, 2	Kentucky
Crist	6, 7	Florida
Cumberland	1, 2	Tennessee
Des Moines	11	Iowa

Clean Coal Engineering Technology. DOI: 10.1016/B978-1-85617-710-8.00021-2
Copyright © 2011 by Elsevier Inc. All rights of reproduction in any form reserved.

Plant Name	Unit Number	State
Dunkirk	3, 4	New York
E C Gaston	1, 2, 3, 4, 5	Alabama
E W Brown	1, 2, 3	Kentucky
Elmer W Stout	50, 60, 70	Indiana
Eastlake	1, 2, 3, 4, 5	Ohio
Edgewater	4	Wisconsin
Edgewater	13	Ohio
Elmer Smith	1, 2	Kentucky
F B Culley	2, 3	Indiana
Fort Martin	1, 2	West Virginia
Frank E Ratts	1SG1, 2SG1	Indiana
Gallatin	1, 2, 3, 4	Tennessee
Gen J M Gavin	1, 2	Ohio
Genoa	1	Wisconsin
George Neal North	1	Indiana
Ghent	1	Kentucky
Gibson	1, 2, 3,4	Indiana
Grand Tower	9	Illinois
Green River	5	Kentucky
Greenidge	6	New York
H L Spurlock	1	Kentucky
H M P&L Station 2	H1, H2	Kentucky
H T Pritchard	6	Indiana
Hammond	1, 2, 3, 4	Georgia
Harrison	1, 2, 3	West Virginia
Hatfield's Ferry	3	Pennsylvania
Hennepin	2	Illinois
High Bridge	6	Minnesota
J H Campbell	1, 2	Michigan
Jack McDonough	MB1, MB2	Georgia
Jack Watson	4, 5	Mississippi
James River	5	Missouri
Johnsonville	1, 2, 3 4, 5, 6, 7, 8, 9, 10	Tennessee
Joppa Steam	1, 2, 3, 4, 5, 6	Illinois
Kammer	1, 2, 3	West Virginia
Kincaid	1, 2	Illinois
Kyger Creek	1, 2, 3, 4, 5	Ohio
Labadie	1, 2, 3, 4	Missouri
Martins Creek	1, 2	Pennsylvania
Meredosia	5	Illinois
Merrimack	1, 2	New Hampshire
Miami Fort	5-1, 5-2, 6, 7	Ohio
Michigan City	12	Indiana
Milliken	1, 2	New York
Milton L Kapp	2	Iowa

Continued

Plant Name	Unit Number	State
Mitchell	1, 2	West Virginia
Montrose	1, 2, 3	Missouri
Morgantown	1, 2	Maryland
Mt Storm	1, 2, 3	West Virginia
Muskingum River	1, 2, 3, 4, 5	Ohio
Nelson Dewey	1, 2	Wisconsin
New Madrid	1, 2	Missouri
Niles	1, 2	Ohio
North Oak Creek	1, 2, 3	Wisconsin
Northport	1, 2, 3	New York
Paradise	3	Kentucky
Petersburg	1, 2	Indiana
Picway	9	Ohio
Port Jefferson	3, 4	New York
Portland	1, 2	Pennsylvania
Prairie Creek	4	Iowa
Pulliam	8	Wisconsin
Quindaro	2	Kansas
R E Burger	5, 6, 7, 8	Ohio
R Gallagher	1, 2, 3, 4	Indiana
Riverside	9	Iowa
Shawnee	10	Kentucky
Shawville	1, 2, 3, 4	Pennsylvania
Sibley	3	Missouri
Sioux	1, 2	Missouri
South Oak Creek	5, 6, 7, 8	Wisconsin
Sunbury	3, 4	Pennsylvania
Tanners Creek	U4	Indiana
Thomas Hill	MB1, MB2	Missouri
Vermillion	2	Illinois
W H Sammis	5, 6, 7	Ohio
Wabash River	1, 2, 3, 4, 5, 6	Indiana
Walter C Beckjord	5, 6	Ohio
Wansley	1, 2	Georgia
Warrick	4	Indiana
Yates	Y1BR,Y2BR,Y3BR,Y4BR, Y5BR, Y6BR,Y7BR	Georgia

Index

Note: Page numbers followed by *f* indicate figures and *t* indicate tables.